The International Atlas of Mars Exploration

From Spirit to Curiosity
Volume 2: 2004 to 2014

Beginning with the landing of the Spirit and Opportunity rovers in 2004 and concluding with the end of the Curiosity primary mission in 2014, this second volume of *The International Atlas of Mars Exploration* continues the story of Mars exploration in spectacular detail. It is an essential reference source on Mars and its moons, combining scientific and historical data with detailed and unique illustrations to provide a thorough analysis of twenty-first-century Mars mission proposals, spacecraft operations, landing site selection and surface locations. Combining a wealth of data, facts and illustrations, most created for this volume, the atlas charts the history of modern Mars exploration in more detail than ever before. Like its predecessor, the atlas is accessible to space enthusiasts, but the bibliography and meticulous detail make it a particularly valuable resource for academic researchers and students working in planetary science and planetary mapping.

PHILIP J. STOOKE is a cartographer and imaging expert at the University of Western Ontario, whose interest in mapping the Moon and planets began during the Apollo missions. He has developed novel methods for mapping asteroids, and many of his asteroid maps are now accessible from NASA's Planetary Data System. He has studied spacecraft locations on the Moon and Mars, notably locating Viking 2 on Mars. He is the author of many papers and articles on planetary mapping, planetary geology and the history of cartography and planetary science. His book *The International Atlas of Lunar Exploration* was published by Cambridge University Press in 2008. This was followed by *The International Atlas of Mars Exploration: The First Five Decades*, published by Cambridge University Press in 2012, which was selected as an Outstanding Academic Title by the American Library Association in 2013.

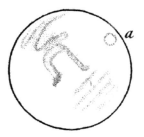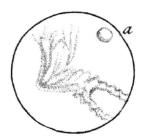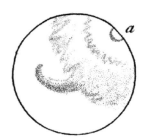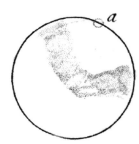

Figures 16–19 from Herschel (1784).

The analogy between Mars and the Earth is, perhaps, by far the greatest in the whole solar system. Their diurnal motion is nearly the same; the obliquity of their respective ecliptics, on which the seasons depend, not very different; of all the superior planets the distance from Mars to the Sun is by far the nearest alike to that of the Earth; nor will the length of the martial year appear very different from that which we enjoy, when compared to the surprising duration of the years of Jupiter, Saturn and the Georgium Sidus ... From other phaenomena it appears, however, that this planet is not without a considerable atmosphere; for, besides the permanent spots on its surface, I have often noticed occasional changes of partial bright belts ... and also once a darkish one, in a pretty high latitude ... And these alterations we can hardly ascribe to any other cause than the variable disposition of clouds and vapours floating in the atmosphere of that planet. ... [Mars] has a considerable but moderate atmosphere, so that its inhabitants probably enjoy a situation in many respects similar to ours.

> On the Remarkable Appearances at the Polar Regions of the Planet Mars, the Inclination of Its Axis, the Position of Its Poles, and Its Spheroidical Figure; With a Few Hints Relating to Its Real Diameter and Atmosphere. By William Herschel, Esq. F.R.S. *Philosophical Transactions of the Royal Society of London*, Vol. 74 (1784), pp. 233–273.

... if Mars be, indeed, untenanted by any forms of life, then these processes going on year after year, and century after century, represent an exertion of Nature's energies which appears absolutely without conceivable utility. If one cloud, out of a hundred of those which shed their waters upon Mars, supplies in any degree the wants of living creatures, then the purport of those clouds is not unintelligable; but if not a single race of beings peoples that distant world, then indeed we seem compelled to say that, in Mars at least, Nature's forces are wholly wasted. Such a conclusion, however, the true philosopher would not care needlessly to adopt.

> *Other Worlds than Ours*. Richard A. Proctor, B.A., F.R.A.S. New York: D. Appleton and Company, 1889.

The International Atlas of Mars Exploration

From Spirit to Curiosity

Volume 2: 2004 to 2014

PHILIP J. STOOKE

University of Western Ontario

CAMBRIDGE
UNIVERSITY PRESS

CAMBRIDGE
UNIVERSITY PRESS

University Printing House, Cambridge CB2 8BS, United Kingdom

Cambridge University Press is part of the University of Cambridge.

It furthers the University's mission by disseminating knowledge in the pursuit of education, learning and research at the highest international levels of excellence.

www.cambridge.org
Information on this title: www.cambridge.org/9781107030930

© Philip J. Stooke 2016

First published 2016

Printed in the United Kingdom by TJ International Ltd. Padstow Cornwall

A catalog record for this publication is available from the British Library

Library of Congress Cataloging in Publication data
Stooke, Philip.
The international atlas of mars exploration : From Spirit to Curiosity: Volume 2: 2004 to 2014/ Philip Stooke.
    p.   cm.
Includes bibliographical references and index.
ISBN 978-1-107-03093-0 (Hardback)
1. Mars (Planet) – Remote-sensing maps.   2. Mars (Planet) – Exploration.   3. Mars (Planet) – Maps.   4. Space flight to Mars – Maps.   5. Space flight to Mars – History.   I. Title.
G1000.5.M3A4S8 2015
912.99′23–dc23   2012007339

ISBN 978-1-107-12033-4   2-volume Hardback set
ISBN 978-0-521-76553-4   volume 1 Hardback
ISBN 978-1-107-03093-0   volume 2 Hardback

Contents

Missions and events – chronological list

Foreword

As I write these words, an SUV-sized rover called Curiosity roams the rusted deserts of Mars, making its way across the floor of the 154 km diameter crater Gale. With its bevy of tools, cameras and sensors, it is reaching back billions of years to document what was once a complex of lakes and streams. On the other side of the planet, the smaller rover Opportunity continues its exploration of a vast plain called Meridiani, more than 11 years after it landed there in early 2004 and discovered the telltale geologic signature of an ancient, salty sea. Meanwhile, Opportunity's twin, Spirit, its explorations long finished, sits motionless in the Columbia Hills, where its own discoveries revealed yet another place on Mars where water once shaped what is now a bone-dry world.

Spirit, Opportunity and Curiosity are just three of the robots that have explored Mars in the first decades of the twenty-first century. They will never come home to Earth, never receive the hero's welcome they certainly deserve. But, with this fascinating volume, planetary cartographer Phil Stooke has done the next best thing. He has woven their stories into an extraordinarily detailed and comprehensive chronicle of Mars exploration.

On these pages we relive the rovers' explorations as they made the first overland treks on another planet – sol by sol, drive by drive, rock by rock. Detailed descriptions are illustrated by Stooke's own meticulously constructed maps based on the rovers' own images and the incredibly detailed overhead views from the Mars Reconnaissance Orbiter. We track the rovers as they surmount difficulties, including Spirit's year-long climb to the summit of Husband Hill, and Opportunity's record-setting trek across the dark plains of Meridiani, where it was stuck for several weeks in a low sand dune. Through their electronic eyes we see the vast Martian plains, the distant hills. We see dust devils whirling across the windswept landscape. We see the small Martian moons Phobos and Deimos moving across the sky, even silhouetted against a distant Sun.

Stooke also details the explorations of the Martian arctic by the Phoenix lander, which touched down there in 2008, and flyby observations from the Dawn and Rosetta missions. For his chronicles of lander and rover missions, Stooke begins with the selection of their landing sites, a painstaking effort to extract the greatest possible scientific return from each mission. Altogether, this book provides a record of more than a decade of discoveries that have transformed our understanding of our alluring neighbor world. Phil Stooke has taken it upon himself to document, in rich detail, a history that otherwise would likely have gone unrecorded. For that, we owe him a great debt.

Andrew Chaikin
Author of *A Man on the Moon* and
A Passion for Mars

Preface and acknowledgments

Preface

Like my previous atlases of space exploration, this book is about places, in this case places on Mars: where they are, what they look like, what happened (or might have happened) there and why they were chosen. It is a historical atlas, describing and illustrating events primarily through the medium of maps. It is not a book about science, the geology of Mars, the people who work on spacecraft and missions, technology or politics. Those topics may be mentioned in passing here, but can be explored more fully in the books, journal articles, online presentations and websites cited throughout the atlas. This book fills a different niche not well represented anywhere else, and was in fact created specifically to fill that void. This is a reference work, not a novel, and if it is to be followed from beginning to end, the sequence of illustrations rather than the text might form a more satisfying narrative.

The first volume of *The International Atlas of Mars Exploration* described missions and events up to the Mars Express mission and its lander, Beagle 2. The Mars Express and Mars Odyssey missions were in progress when that book was compiled, but every other mission had already concluded. This volume includes two rover missions that are still active as the book nears completion, and for which route maps and day-to-day activities had to be brought as far up to date as possible prior to publication, compiled as the events were unfolding.

Work on this atlas began as the first volume was submitted to the publisher in January 2012, prior to the Curiosity landing and on about sol 2825 for Opportunity. The Opportunity material prior to that was historical, based on pre-existing sources. After that, and throughout the surface mission of Curiosity, maps were updated every time a rover moved or a feature name became known. The tables listing activities, necessarily very abbreviated for such long missions, were based on tabulations in the online Analyst's Notebooks in NASA's Planetary Data System, with additional details taken from the Science Operations Working Group (SOWG) documents from the same website. (In some tables, underscores are used to turn a proper name into a similar computer filename, as used in the technical documents.) Those sources are always many months behind the current activity in each mission, so coverage of both missions ends in the middle of 2014, on sol 3700 for Opportunity and sol 669, the end of the primary mission, for Curiosity. Planning for future missions and other Mars activities after the launch of Curiosity are not covered in this volume.

In the past, books like my *International Atlas of Lunar Exploration* and much of the first Mars volume were compiled from standard library resources, including journal articles, books and technical reports. Access to physical archives such as the collections at the Lunar and Planetary Institute provided more material, including unpublished committee minutes and obscure reports difficult to find elsewhere. Compiling history as the events are happening today is very different, and though those traditional sources are still useful, many others also become necessary.

The open publication of detailed abstracts (*e.g.* the Lunar and Planetary Science Conferences, LPSC) to anybody, not just attendees, is very useful. The open distribution of conference or workshop presentation files (*e.g.* meetings of the Mars Exploration Program Analysis Group, MEPAG, or the Mars Science Laboratory Landing Site Workshops) is invaluable. The past practice of a meeting distributing a proceedings volume to its attendees was no doubt useful to the attendees, but is of limited help to others looking for facts a decade or two later, unless the proceedings find their way into a library or appear in a journal. As mentioned above, the Analyst's Notebook is essential to understanding a mission. NASA has become very good at making information openly available, with the exception of proposals to the Discovery Program and other completed missions, which are considered proprietary, but some other agencies have a closed culture that severely limits historical study. This is beginning to change, especially in Europe, but other space agencies are very restricted, which will hamper space historians as space exploration becomes a global endeavour and perhaps more of a commercial activity.

Apart from meeting resources, the writer of "real-time" history must now monitor numerous online forums,

blogs, tweets and other resources to keep up to date. These often provide not just useful details, such as the name of a rock on Mars, but also links to publications or meetings which might otherwise be missed. Perhaps the best example of this is the work of Emily Lakdawalla of the Planetary Society, whose blog is invaluable. Other helpful sources of information come from people working on missions, who usually operate under constraints that prevent the release of sensitive details but can still say much that is useful.

Martian coordinates are discussed in the Preface to Volume 1. A change in the preferred coordinate system for Mars occurred gradually around the time of the Mars Exploration Rover (MER) mission landings, leading to occasional confusion in the literature. The most significant point in most cases is whether longitudes are measured to the east or west from the Prime Meridian. West longitudes were widely used before MER, east longitudes after. Confusion can be avoided by always specifying a direction with a longitude (*e.g.* 60° E or 30° W). One example of confusion caused by this change is seen in the literature relating to Mars Science Laboratory (MSL) landing site selection. At the first MSL Landing Site Workshop, James Dohm (University of Arizona) and his colleagues promoted a site called Northwest Slope Valleys (Table 43 in this atlas), which is southwest of Tharsis near 146.5° W, but in every written account of the site selection process and at each subsequent workshop the site was recorded at 146.5° E.

At risk of confusing matters further, I have chosen to use west longitudes in Volume 1 and east longitudes here. The advantage is that the atlas volumes correspond, more or less, with the contemporary literature. A disadvantage is that they conflict with each other. To help alleviate that potential problem, Table 58 gives landing and impact sites for all Mars missions in east longitudes for comparison with Table 80 in Volume 1. I do not indicate north in my maps, but those with grids are unambiguous and those without (such as Figures 99 and 176) are always shown with north at or near the top. Only the close-up images of brush or drill holes (including Figures 38 and 189) are shown in the camera's perspective rather than north-up.

Explorers like to assign names to features, and space explorers do so as well. Official names of planetary surface features are assigned and managed by the

Working Group for Planetary System Nomenclature of the International Astronomical Union. Since the days of Apollo, it has been customary for human or robotic exploration missions to use names for features and instrument targets, but these are mostly informal. Nevertheless, they find their way into the literature and news reports, and should be recorded. I have made a point of recording as many as I could in this atlas, but this is not always straightforward, as there are many inconsistencies between sources. Some features are given multiple names (*e.g.* The Dugout, Eastern Valley and Silica Valley near Home Plate at the Spirit research site), and all are recorded here so the varied sources can be reconciled. Some names are used more than once in the same landing region (*e.g.* Cape Upright on Murray Ridge at the Opportunity research site, used for two imaging targets on sols 3596 and 3689), or repeated between sites (Home Plate at both Spirit and Opportunity sites, Figures 30 and 51 in this atlas). Complete consistency with every source is not to be expected, but I have attempted to record everything I could so that inconsistencies can be recognized if not resolved.

In terrestrial applications the term "soil" has often been formally restricted to materials with a significant organic component (roots, decaying organic matter, microorganisms, etc.), but where the word "soil" is used in this atlas no organic or biological component is implied. Although Mars does not have Earth-like organic-rich soils, it also lacks a Moon-like impact-generated regolith, and no other convenient term is obvious (Certini and Ugolini, 2013). This was once a cause of conflict, but the term "soil" is now accepted for Martian surface materials and is often used here. The terms "dune," "drift" and "ripple" have distinct meanings in aeolian geomorphology. I have tried to use the word "dune" only where appropriate, but "drift" and "ripple" may be applied more loosely, or even interchangeably, based on sources which are themselves not always consistent.

As in my previous Mars atlas, I have adopted the Mars calendar of Clancy *et al.* (2000) to describe the timing of events on Mars. This can be useful when events on different missions need to be compared. To help tie mission sols, Earth dates and Mars dates together, I have specified all three at 100 sol intervals through the rover mission descriptions, and at 10 sol intervals for Phoenix. The dates of many events in Mars exploration history are listed in Tables 58–60. As in Volume 1, I make no attempt to distinguish between the Mars-wide sols (sol at the Prime Meridian) in which an orbiter event might be recorded and the local sols for a rover at a particular landing site. Sols for Spirit and Curiosity are both about half a sol out of phase with Opportunity sols, as they are on opposite sides of the planet.

Local site maps are often made from reprojected surface panoramas. The larger reprojections (*e.g.* inside Endurance crater or Duck Bay) are tied to HiRISE geometric control, but many smaller maps are not and will contain significant distortions caused by relief. Rover positions in these maps are always fixed relative to surrounding features in the images, not to calculated coordinates.

Acknowledgments

I would like to extend thanks to many people who have assisted me during work on this atlas. Matt Golombek and Bruce Banerdt (JPL) helped me with information on the Cerberus network mission site selection process. Leslie Tamppari (JPL) and Peter Smith (University of Arizona) provided information or sources for Phoenix landing site selection. Jeff Plescia (Johns Hopkins University Applied Physics Laboratory) helped with the Urey Mars Scout mission section, Robert Grimm (Southwest Research Institute) for the Naiades mission proposal, Dawn Turney and Scott Murchie (Johns Hopkins University Applied Physics Laboratory) for CRISM image coverage information, Michael Ravine (Malin Space Science Systems Inc.) for CTX image coverage data and Ari Espinoza (Lunar and Planetary Laboratory, University of Arizona) for HiRISE information. In Moscow, Irina P. Karachevtseva and colleagues, including Maxim V. Nyrtsov at the Moscow State University for Geodesy and Cartography (MIIGAiK), helped with Phobos and Deimos material and corrected a misunderstanding of mine concerning earlier work on mapping Phobos by Professor Lev M. Bugaevsky. Atlanta architect Chuck Clark allowed me to use two of his maps of Deimos to illustrate his innovative cartographic work.

In addition to those researchers, I must also thank the endlessly creative and productive members of the online forum at www.unmannedspaceflight.com, a constant source of support, knowledge and wisdom. Preliminary versions of many of my maps of Curiosity activities were first posted there, and some recent maps of Opportunity activities. Members of the forum frequently alert others to the appearance of a new report, publication or data source, which in itself is a great help in a rapidly changing field. I particularly appreciate the wonderful image processing and mosaic or panorama construction done by Damia Bouic, James Canvin, Jan van Driel, Michael Howard, Ed Truthan, James Sorensen, Iñaci Docio and others.

Rover route maps in this atlas were compiled using several pre-existing maps as a starting point, especially maps by Tim Parker and Fred Calef (JPL), Rongxing Li (Ohio State University) and Larry Crumpler (New Mexico Museum of Natural History and Science). These came primarily from mission websites, but also including material posted to www.unmannedspaceflight.com by members Eduardo Tesheiner, "Pando," Joe Knapp and Michael Howard. Rover positions marked with a date label (*e.g.* 443, meaning sol 443 of that mission) represent the position at the end of that sol. The rover would have begun the sol at the previous location along the route, often making observations there before departing. Every rover position was checked by comparing a circular projection of a surface panorama (similar to those in Figures 14 and 46, for example) with a HiRISE image. Routes between those positions are based on tracks visible in HiRISE images, or seen in rover images looking backwards along the traverse. The routes are not generalized and may be assumed to be precise in most areas. Comparison with older maps will reveal differences, for example for Opportunity at Endurance crater where mission maps using calculated traverses showed the rover on sol 128 suspended in the thin Martian air over the rim of the crater. The error was caused by not taking wheel slip into account, but it is corrected here by using HiRISE images of the fading tracks and comparisons with surface imaging.

At the University of Western Ontario, I have been welcomed and assisted frequently by the staff of the Map and Data Centre, and I have made extensive use of its Serge A. Sauer Map Collection. I have benefited from the support and encouragement of many members, both faculty and students, of the Centre for Planetary Science and Exploration. Mahdia Ibrahim helped prepare some panoramic images for site location work.

Lastly, I extend my thanks to Vince Higgs and colleagues at Cambridge University Press for their continuing support, and for making the production of this book so straightforward, for me at least.

Data sources

Most of the data used in the creation of the illustrations in this atlas come from NASA planetary missions, and I have processed raw data from NASA's Planetary Data System or elsewhere to create unique images, rather than relying on standard press release images. I use a map derived for Volume 1 of this atlas from an early chart by the US Air Force's Aeronautical Chart and Information Center in Figure 214. I make frequent use of the web-accessible global photomosaics produced by USGS (*e.g.* Figures 1 and 2) and Arizona State University (*e.g.* Figure 173). All of these raw materials are in the public domain, and are credited in captions.

A small number of images require specific credits. I especially thank Irina P. Karachevtseva of the Moscow State University for Geodesy and Cartography (known by its older Russian acronym MIIGAiK) for allowing me to use a map of Phobos on a projection designed by Professor Lev M. Bugaevsky (Figure 210A), and Maxim V. Nyrtsov, also of MIIGAiK, for one of his maps of Phobos (Figure 210B). Chuck Clark of Atlanta, GA, kindly allowed me to use his maps of Deimos in Figure 210C.

All other credits below, listed by mission in roughly chronological order, are for specific spacecraft instruments and their Principal Investigators. Without their dedicated work, none of this exploration would be possible.

Viking Orbiters: NASA/JPL and Michael Carr.

Mars Global Surveyor: Mars Orbiter Laser Altimeter (MOLA): NASA/Goddard Space Flight Center and David Smith.

2001 Mars Odyssey: Thermal Emission Imaging System: NASA/JPL/Arizona State University and Philip Christensen.

Mars Express: High Resolution Stereo Camera (HRSC): ESA and G. Neukum. OMEGA: Institut d'Astrophysique Spatiale, Orsay, France, and Jean-Pierre Bibring.

Mars Exploration Rovers: Pancam: NASA/JPL/ Cornell University/Arizona State University and James Bell. Navcam: NASA/JPL.

Mars Reconnaissance Orbiter: CRISM: NASA/JPL/ Johns Hopkins University and Scott Murchie. High Resolution Imaging Science Experiment: NASA/JPL/ University of Arizona and Alfred McEwen. MCS: NASA/JPL and Daniel McCleese. MARCI: Malin Space Science Systems, Inc. and Michael Malin.

Phoenix: Robotic Arm Camera (RAC): NASA/JPL/ University of Arizona and Robert Bonitz. Surface Stereo Imager (SSI): NASA/University of Arizona and Mark Lemmon.

Rosetta: Comet Infrared and Visible Analyser (CIVA): Institut d'Astrophysique Spatiale, Orsay, France, and Jean-Pierre Bibring. Optical, Spectroscopic, and Infrared Remote Imaging System (OSIRIS): Max Planck Institute for Solar System Research and Holger Sierks.

Mars Science Laboratory: Navcam: NASA/JPL. Mastcam: Malin Space Science Systems, Inc. and Michael Malin. Mars Hand Lens Imager (MAHLI): Malin Space Science Systems, Inc. and Ken Edgett. ChemCam Remote Micro-Imager (RMI): NASA/ JPL-Caltech/LANL/CNES/IRAP and Roger Wiens.

Dawn: Framing Camera: German Aerospace Centre (DLR) and Horst Uwe Keller.

1. Chronological sequence of missions and events

2003: MER landing site selection

The Mars Exploration Rover (MER) mission was intended to land two identical rovers at different sites showing evidence of past water activity. This description of the landing site selection process is based on the extensive public documentation at the Mars Landing Site website at NASA Ames Research Center (marsoweb.nas.nasa.gov/ landingsites/index.html) and a summary by Golombek *et al.* (2003). Maps and tables released at the time contain minor discrepancies, which are corrected here as noted.

MER landing site selection began in September 2000. Engineering considerations stipulated landings below -1.3 km elevation for the parachute system and between latitudes $15°$ S and $5°$ N (MER-A) and between latitudes $5°$ S and $15°$ N (MER-B) for solar power. Thermal inertia data further limited the available area by excluding very rocky or dusty regions. The entry procedure and accuracy defined the size and orientation of the landing ellipse. This varied from 80 by 30 km for MER-A at $15°$ S to 360 by 30 km for MER-B at $15°$ N. The orientation also varied with latitude. The two rover sites had to be at least $37°$ of longitude apart to minimize communication conflicts. A Landing Site Steering Committee headed by Matt Golombek (JPL) made an initial assessment of possible sites. Maps of the accessible area were overlaid with ellipses at every location free of hazards in Viking images. In this way 85 candidate ellipses were defined for MER-A and 100 for MER-B (Tables 1 and 2, Figures 1 and 2). Some pairs of sites have identical coordinates but different orientations.

Next, a shortlist of high-priority sites was assembled from abstracts submitted to the First Landing Site Workshop, augmented by sites for which team members requested new high-resolution images (Table 3). A region in Sinus Meridiani shown by Mars Global Surveyor (MGS) to be rich in hematite (Christensen *et al.*, 2001) was a clear favorite from the start, as it had been for Mars Surveyor 2001. Crater lakes were the next most popular targets. Soon after that shortlist was circulated on 13 December 2000, new analyses of the landing process increased the length of the landing ellipses. Some ellipses that would no longer fit safely between obstacles were eliminated on 21 December. These sites are labeled (*) in Table 3.

The shortlisted sites were illustrated for the workshop in maps which differed in several instances from the earlier site maps depicted in Figures 1 and 2. Three additional ellipses were drawn adjacent to existing ellipses at Gusev (EP55A), at Melas Chasma (VM53A) and at VM41A in the Valles Marineris outflow area. The first two of these are labeled as alternative ellipses in Figure 3, and all three are appended to Table 3. These new maps omitted TM24B and added TM13A/24B and VM47A. The EP52B site in Table 3 was illustrated as EP51B, and the Isidis site called IP85B/96B in Table 3 was now mapped as IP95B/96B, suggesting errors in the source from which Table 3 was compiled.

The First Landing Site Workshop was held on 24 and 25 January 2001. After the participants in the meeting prioritized the shortlisted sites, three lists were drawn up to indicate those sites with the highest priority for further study and data collection, those considered of medium priority, and those which could be eliminated (Table 4). Additional imaging of sites by MGS was helped by an unusual dynamical situation at this time in which the ground tracks "walked" in longitude very slowly from orbit to orbit, making mosaic production easier.

On 30 April 2001 the list of ellipses in Table 5a was distributed for consideration. The Ganges Chasma site was eliminated shortly after this list was circulated. Some sites which had been eliminated earlier were still on this list in case they might later become acceptable again. Another version of the landing ellipse list appeared on 16 July 2001 in a memorandum from Golombek and Timothy Parker to the MER Project and the Landing Site Steering Committee. Table 5b indicates the changes in priority or coordinates between the April list and the July memorandum. In particular, Ellipse TM21B moved from highest to medium priority.

Table 1. *Potential Landing Sites for MER-A*

Site	Location±0.2°	Site	Location±0.2°	Site	Location±0.2°
TM1A	4.6° N, 352.2° E	XT30A	8.8° S, 348.3° E	EP59A	10.4° S, 168.4° E
TM2A	4.0° N, 352.9° E	XT31A	10.8° S, 344.2° E	EP60A	13.1° S, 165.9° E
TM3A	3.7° N, 349.9° E	XT32A	10.7° S, 334.3° E*	EP61A	11.8° S, 164.5° E
TM4A	4.2° N, 346.9° E	XT33A	4.2° N, 333.7° E	EP62A	9.7° S, 161.8° E
TM5A	2.0° N, 349.0° E	XT34A	1.9° S, 337.3° E	EP63A	12.6° S, 163.5° E
TM6A	0.9° N, 353.4° E	VM35A	9.4° S, 325.5° E	EP64A	14.8° S, 162.5° E
TM7A	0.2° S, 351.1° E	VM36A	10.2° S, 323.4° E	EP65A	8.7° S, 157.4° E
TM8A	0.2° S, 349.6° E	VM37A	11.1° S, 322.1° E	EP66A	4.0° N, 153.5° E
TM9A	1.2° S, 354.4° E	VM38A	8.2° S, 319.5° E	EP67A	4.6° N, 150.4° E
TM10A	2.2° S, 353.4° E	VM39A	12.8° S, 320.0° E	EP68A	3.5° N, 150.8° E
TM11A	3.4° S, 353.1° E	VM40A	8.8° S, 317.1° E	EP69A	9.3° S, 150.5° E
TM12A	3.6° S, 357.1° E	VM41A	14.0° S, 318.0° E	EP70A	4.0° N, 146.5° E
TM13A	2.9° S, 349.5° E	VM42A	7.7° S, 309.3° E	EP71A	1.2° N, 148.0° E
TM14A	3.8° S, 349.9° E	VM43A	13.9° S, 302.5° E	EP72A	4.2° S, 148.6° E
TM15A	8.6° S, 353.3° E	VM44A	13.1° S, 297.5° E	EP73A	5.5° S, 148.0° E
TM16A	9.4° S, 353.4° E	VM45A	12.0° S, 292.3° E	EP74A	4.2° N, 143.4° E
TM17A	11.0° S, 352.8° E	VM46A	13.6° S, 289.4° E	EP75A	0.7° S, 142.8° E
XT18A	0.4° S, 335.5° E	VM47A	6.2° S, 289.9° E	EP76A	1.4° S, 142.5° E
XT19A	1.1° S, 334.8° E	VM48A	7.1° S, 287.5° E	EP77A	4.5° N, 139.5° E
XT20A	1.0° S, 332.3° E	VM49A	9.2° S, 286.5° E	EP78A	1.7° N, 139.0° E
XT21A	1.5° S, 329.3° E	VM50A	9.7° S, 286.2° E	EP79A	4.5° N, 135.6° E
XT22A	4.7° S, 343.3° E	VM51A	10.1° S, 287.0° E	EP80A	3.2° N, 135.7° E
XT23A	5.4° S, 342.3° E	VM52A	10.5° S, 287.5° E	EP81A	0.2° S, 135.3° E
XT24A	5.3° S, 347.2° E	VM53A	8.8° S, 282.3° E	EP82A	5.8° S, 137.6° E
XT25A	5.7° S, 347.6° E	EP54A	11.4° S, 177.1° E*	EP83A	3.5° N, 125.3° E
XT26A	7.7° S, 344.5° E	EP55A	14.2° S, 175.2° E	IP84A	4.5° N, 88.1° E
XT27A	7.7° S, 346.4° E	EP56A	14.6° S, 171.9° E	IP85A	4.6° N, 84.5° E*
XT28A	7.2° S, 348.6° E	EP57A	9.0° S, 168.6° E		
XT29A	9.3° S, 344.5° E	EP58A	9.9° S, 168.5° E		

Notes: Site designations begin with a two-letter location code: CP, Chryse Planitia; EP, Elysium Planitia; IP, Isidis Planitia; TM, Terra Meridiani; VM, Valles Marineris; XT, Xanthe Terra. The number is a unique site identifier, and the final A in each designation refers to MER-A.
* Entry corrects an error in the source.

By August 2001 improvements in navigation and trajectory planning, a combination of simultaneous spacecraft tracking by two Deep Space Network stations and a fifth trajectory correction 48 hours before landing, allowed the landing ellipses to shrink again, making nine additional sites from Tables 1 and 2 feasible in Ares Vallis, Crommelin crater, Margaritifer Valles, SE Melas Chasma, Sinus Meridiani, one more in Isidis, two in Elysium and one in the highlands. The nine ellipses were not identified on the website.

In October 2001 the Second Landing Site Workshop narrowed the list of sites still further, to four primary and two backup sites (Table 6). These were still regions of interest, not individual ellipses, and work continued to define the various ellipses within them. These sites were recommended to NASA and the MER project team at JPL. The site called Athabasca Vallis in Table 6 was previously named Elysium Outflow (Table 4), though the ellipse would be smaller and further north (Figure 3B). The name of the valley in Figure 3 (Athabasca Valles, the plural form) reflects a change after the site selection process concluded. There are numerous small inconsistencies between tables and maps throughout the site selection process

Table 2. *Potential Landing Sites for MER-B*

Site	Location±0.2°	Site	Location±0.2°	Site	Location±0.2°
wA1B	14.4° N, 353.0° E	XT34B	6.2° N, 319.4° E	EP68B	5.6° N, 137.1° E
wA2B	13° N, 349.3° E	CP35B	0.4° S, 326.2° E	EP69B	3.6° N, 136.6° E
wA3B	9.2° N, 354.0° E	CP36B	8.2° N, 327.0° E	EP70B	3.1° N, 135.0° E
wA4B	7.8° N, 349.2° E	CP37B	10.5° N, 314.4° E	EP71B	1.6° N, 139.4° E
wA5B	11.2° N, 342.4° E	CP38B	11.8° N, 314.6° E	EP72B	0.4° S, 135.1° E
wA6B	8.6° N, 341.4° E	CP39B	14.1° N, 313.0° E	EP73B	12.9° N, 134.4° E
wA7B	6.5° N, 344.8° E	EP40B	14.2° N, 158.7° E	EP74B	8.8° N, 134.1° E
wA8B	10.7° N, 336.3° E	EP41B	14.3° N, 156.0° E	EP75B	6.8° N, 132.4° E
TM9B	6.0° N, 357.3° E	EP42B	13.4° N, 156.9° E	EP76B	14.1° N, 129.9° E
TM10B	5.6° N, 351.7° E	EP43B	12.4° N, 157.0° E	EP77B	13.0° N, 127.5° E
TM11B	4.2° N, 352.2° E	EP44B	11.5° N, 155.9° E	EP78B	11.2° N, 123.9° E
TM12B	4.0° N, 346.5° E	EP45B	7.2° N, 159.0° E	EP79B	3.7° N, 124.5° E
TM13B	2.3° N, 348.0° E	EP46B	12.6° N, 151.3° E	EP80B	13.9° N, 115.9° E
TM14B	0.9° N, 353.3° E	EP47B	11.2° N, 152.6° E	EP81B	12.1° N, 114.8° E
TM15B	0.3° N, 350.3° E	EP48B	9.2° N, 153.2° E	IP82B	14.1° N, 96.4° E*
TM16B	0.6° S, 348.8° E	EP49B	7.4° N, 154.4° E	IP83B	13.5° N, 94.9° E
TM17B	0.5° S, 351.0° E	EP50B	6.6° N, 152.4° E	IP84B	12.8° N, 94.2° E
TM18B	1.7° S, 350.5° E*	EP51B	3.8° N, 150.6° E*	IP85B	11.6° N, 94.5° E
TM19B	1.2° S, 354.7° E	EP52B	5.3° N, 152.0° E	IP86B	11.0° N, 91.3° E
TM20B	2.3° S, 353.8° E	EP53B	11.0° N, 148.2° E	IP87B	9.7° N, 93.6° E
TM21B	2.5° S, 356.7° E	EP54B	8.2° N, 148.3° E	IP88B	8.3° N, 91.9° E
TM22B	3.2° S, 352.9° E	EP55B	5.6° N, 146.3° E	IP89B	13.7° N, 88.8° E
TM23B	3.4° S, 356.9° E	EP56B	4.4° N, 146.4° E	IP90B	14.0° N, 85.4° E
TM24B	2.8° S, 349.9° E	EP57B	4.2° S, 147.9° E	IP91B	11.2° N, 84.5° E
TM25B	3.6° S, 349.9° E	EP58B	10.7° N, 140.5° E	IP92B	8.9° N, 87.9° E
TM26B	4.6° S, 349.9° E	EP59B	7.2° N, 143.0° E	IP93B	8.2° N, 85.4° E
XT27B	1.9° S, 337.5° E	EP60B	6.5° N, 140.4° E	IP94B	6.9° N, 89.0° E
XT28B	1.2° S, 335.5° E	EP61B	3.7° N, 144.0° E	IP95B	5.6° N, 88.0° E
XT29B	1.0° S, 332.3° E	EP62B	3.6° N, 141.5° E	IP96B	4.6° N, 87.7° E
XT30B	1.2° S, 329.4° E	EP63B	0.7° S, 143.1° E	IP97B	5.7° N, 83.4° E
XT31B	4.0° N, 333.8° E	EP64B	1.5° S, 142.2° E	IP98B	4.7° N, 83.6° E
XT32B	4.5° N, 325.4° E	EP65B	14° N, 136.2° E	SM99B	8.1° N, 80.3° E
XT33B	3.2° N, 321.6° E	EP66B	11.5° N, 135.9° E	SM100B	7.1° N, 79.7° E
		EP67B	9.2° N, 138.8° E		

Notes: Site designations begin with a two-letter location code: CP, Chryse Planitia; EP, Elysium Planitia; IP, Isidis Planitia; SM, Syrtis Major; TM, Terra Meridiani; VM, Valles Marineris; wA, western Arabia Terra; XT, Xanthe Terra. The number is a unique site identifier. B in each designation refers to MER-B.

* Entry corrects an error in the source.

because of the frequent changes to ellipse locations, sizes and orientations.

By November 2001 the choice of ellipses had narrowed and some had been moved. In particular, the Gusev ellipse was moved to the west to avoid most of the rim of an old crater called Thira and nearby rough terrain (Table 7a, Figure 3A).

During 2002 continuing evaluation of radar data and analysis of landing safety and rover mobility relegated the Athabasca site to backup status and promoted Isidis to prime status. The effects of winds during the parachute descent caused concern, and the canyon sites (Melas and Eos) were dropped from further consideration for that reason despite considerable scientific interest in those

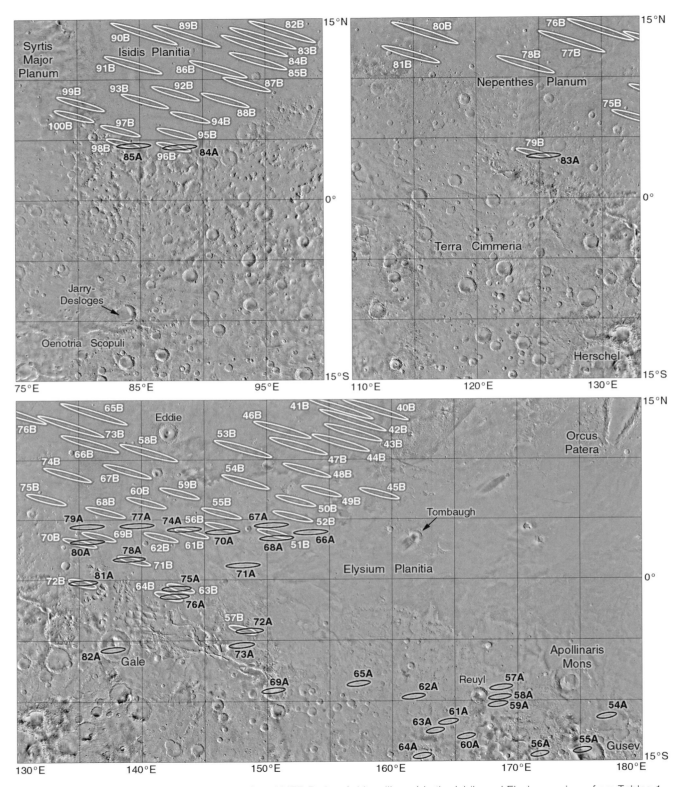

Figure 1. Potential MER-A sites (black ellipses) and MER-B sites (white ellipses) in the Isidis and Elysium regions, from Tables 1 and 2. For scale, the 5° grid squares are 300 km across at the equator.

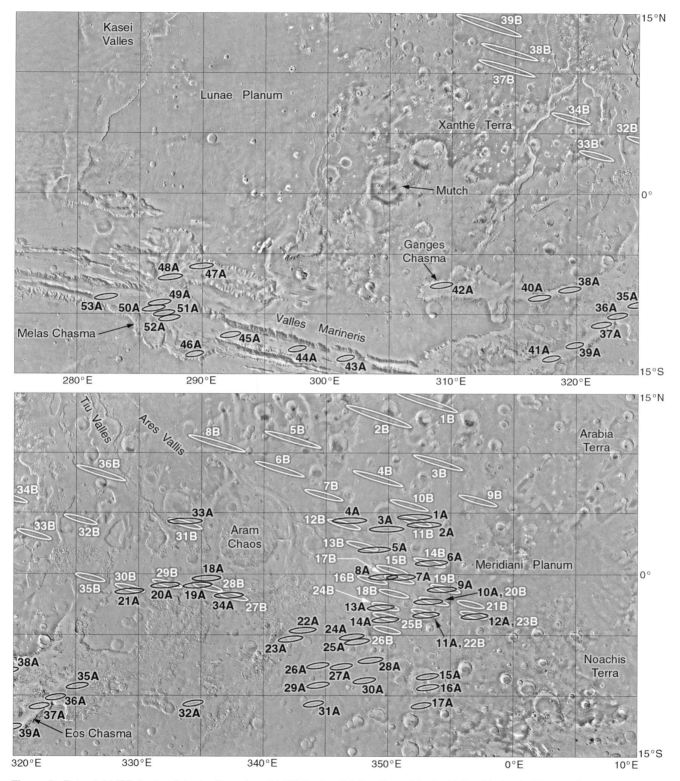

Figure 2. Potential MER-A sites (black ellipses) and MER-B sites (white ellipses) in the Valles Marineris, Xanthe, Chryse and Meridiani regions, from Tables 1 and 2. For scale, the 5° grid squares are 300 km across at the equator.

Table 3. *High-Priority MER Sites, December 2000*

Region or site type	Ellipse	Region or site type	Ellipse
Hematite	TM10A/20B	Valles Marineris	VM53A
	TM11A/22B(*)		VM48A(*)
	TM21B(*)		VM44A(*)
	TM12A/23B(*)	Valles Marineris outflow	VM42A
	TM9A/19B		VM41A
Gale	EP82A(*)		VM37A(*)
Gusev	EP55A	Elysium	EP52B
Meridiani crater	TM15A		EP74A
	TM16A		EP71A
Unnamed crater	EP69A(*)		EP62B
Boeddicker	EP64A(*)		EP77A
Durius Vallis outflow area	EP56A		EP61B
Apollinaris Chaos (new site)	11.1° S, 171.5° E(*)		EP49B
Cratered terrain	TM24B	Isidis	IP85A/96B
Chryse Planitia	CP35B(*)		IP98B

Notes: The following additional sites are illustrated in planning maps at the First Landing Site Workshop: Gusev, 13.6° S, 175.1° E; Valles Marineris, 8.9° S, 283.2° E; Valles Marineris outflow, 13.4° S, 318.5° E.

Some sites at Meridiani (Hematite) and Isidis are accessible to both MER-A and MER-B. Sites eliminated on 21 December are marked (*). Chryse (CP35B) was added after the shortlist was initially compiled, and then eliminated. See text for further discussion of these sites.

locations. Global climate modeling was used to locate two new ellipses, slightly modified from the original lists, as the site numbers in Tables 1, 2 and 7 suggest, in a low-wind region in Elysium as a backup despite the area's lower scientific interest (Table 7b, Figure 3E). A Third Landing Site Workshop in March 2002 examined each site, ranking them for safety and scientific value.

NASA announced the final targets for MER on 11 April 2003. The Hematite site in Meridiani and the apparent lake site in Gusev crater were the prime targets. An Elysium low-wind site and two ellipses in Isidis Planitia served as backups. The final ellipse details are shown in Figures 3, 4 and 43, and are listed in Table 7c.

10 June 2003: MER-A (Spirit)

The Mars Exploration Rover (MER) mission consisted of two identical rovers designed to investigate the geology, past environmental conditions and habitability of their landing sites, which had been chosen on the basis of evidence of past water activity in the surface materials of Mars. The mission's instruments were not designed to seek evidence of life. The rovers were given names chosen through a competition sponsored by NASA and the Danish toy manufacturer Lego, and won by Arizona school student Sofi Collis. The rovers were designed for a nominal lifetime of 90 days and a range of about 600 m, with the possibility of covering up to 100 m in a day if necessary, but they both greatly exceeded these goals while exploring complex and difficult landscapes. Spirit operated in Gusev crater for 2210 sols (3.3 Mars years or 6.2 Earth years) and drove 7.73 km. Asteroid 37452, a member of the Hilda family of asteroids in a 3 : 2 orbital resonance with Jupiter, was named "Spirit" in October 2004 to commemorate the rover.

Each 180 kg MER rover was 1.5 m high, 1.6 m long and 2.3 m wide, with solar panels and camera mast deployed. The width from wheel to wheel, or the width of visible tracks on the surface of Mars, was 1.2 m. The body was attached to six wheels on an articulated bogey system, and contained an insulated electronics box which was kept warm to protect its equipment. The top of the body was covered with solar panels, including folding sections giving a plan shape like a short arrowhead. Above the solar panels were the camera mast, the high- and low-gain antennae, a camera calibration target with a small sundial and a magnet array. The panoramic

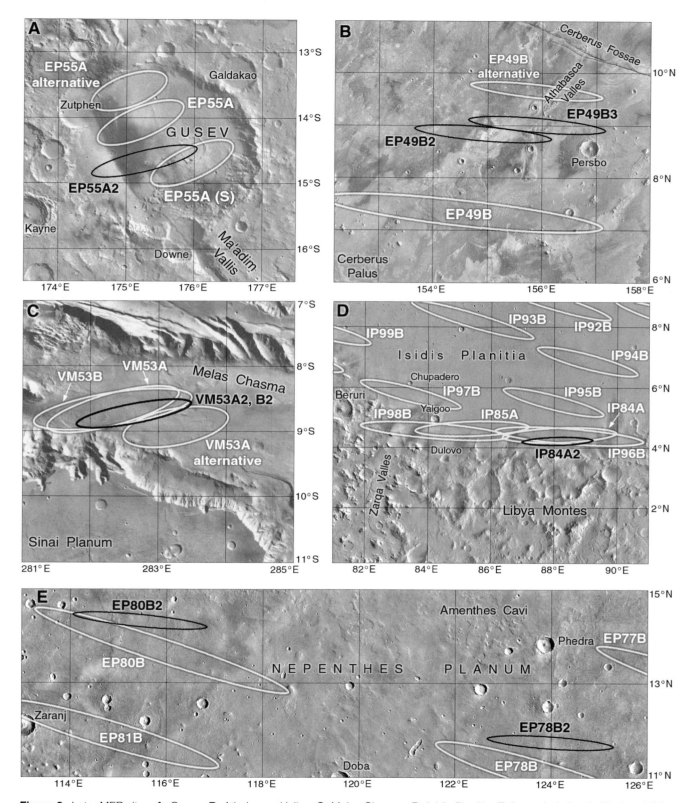

Figure 3. Later MER sites. **A:** Gusev. **B:** Athabasca Valles. **C:** Melas Chasma. **D:** Isidis Planitia. **E:** Low-wind sites in Elysium. White ellipses are older sites; black ellipses are those considered in the later stages of site selection and data collection. Two alternative sites in Gusev and Melas were shown on mission planning maps in 2001. Images are Viking MDIM2.1 mosaics except at Athabasca which incorporates Mars Odyssey THEMIS infrared data with inverted shading. For scale, 1° is approximately 60 km.

Table 4. *MER Sites From the First Landing Site Workshop, January 2001*

Highest-priority sites (12)		Medium-priority sites (19)	
Ellipse	Location	Ellipse	Location
Hematite:		Hematite:	
TM10A	2.2° S, 353.4° E	TM22B	3.2° S, 352.9° E
TM20B	2.3° S, 353.8° E	TM11A	3.4° S, 353.1° E
TM21B	2.5° S, 356.7° E	TM23B	3.4° S, 356.9° E
TM9A	1.2° S, 354.4° E	TM12A	3.6° S, 357.1° E
TM19B	1.2° S, 354.7° E		
Gale: EP82A	5.8° S, 137.6° E	Unnamed crater: EP69A	9.3° S, 150.5° E
Gusev south (new site)	15.5° S, 175.5° E	Boeddicker: EP64A	14.8° S, 162.5° E
		Durius Valles: EP56A	14.6° S, 171.9° E
Valles Marineris:		Meridiani crater:	
VM53A	8.8° S, 282.3° E	TM15A	8.6° S, 353.3° E
VM41A	14.0° S, 318.0° E	TM16A	9.4° S, 353.4° E
Elysium Outflow:		Gusev: EP55A	14.2° S, 175.2° E
EP49B	7.4° N, 154.4° E	Elysium: EP74A	4.2° N, 143.4° E
Isidis:		Valles Marineris:	
IP85A	4.5° N, 88.1° E	VM44A	13.1° S, 297.5° E
IP98B	4.7° N, 83.6° E	VM47A	6.2° S, 289.9° E
Sites to be eliminated (7)		VM48A	7.1° S, 287.5° E
Vallis Marineris: CP35B		VM37A	11.1° S, 322.1° E
Elysium Planitia: EP52B/EP68A, EP71A, EP62B,		VM42A	7.7° S, 309.3° E
EP77A, EP19B, EP61B		Apollinaris (new site)	9.5° S, 169.8° E
		Isidis:	
		IP84A	4.5° N, 88.1° E*
		IP96B	4.6° N, 87.7° E
		Cratered terrain:	
		TM13A	2.9° S, 349.5° E
		TM24B	2.8° S, 349.9° E

Note: * Because of a misprint in the source material for Table 1, the earlier site numbers IP84A and IP85A are sometimes switched in subsequent tables.

cameras (Pancams), navigation cameras (Navcams) and the Miniature Thermal Emission Spectrometer (Mini-TES) were mounted on the camera mast 140 cm above the ground.

An arm (Instrument Deployment Device, IDD) carrying a science instrument package referred to as Athena was mounted at the front of the body. Athena consisted of an Alpha Particle X-ray Spectrometer (APXS, an improved version of the Alpha Proton X-ray Spectrometer carried on Mars Pathfinder's Sojourner rover), a Mössbauer Spectrometer (MB), a Microscopic Imager (MI) and the Rock Abrasion Tool (RAT), which could brush dust off targets or grind several millimeters into a rock to cut through any outer weathered layer. Two pairs

of hazard-avoidance cameras (Hazcams) were mounted on the rover body, looking forwards and backwards. The rover was folded to fit into a tetrahedral lander similar to that used by Pathfinder, enclosed in an aeroshell for atmospheric entry, and was carried to Mars by a small cruise stage. The lander carried no power system or instruments other than a small descent camera, and it relied on the rover for power and control during entry and landing. It was abandoned as soon as the rover departed.

Figure 4A shows the regional setting of the Spirit landing site, including some nearby candidate landing ellipses. Other ellipses are shown in Figure 3A. The 800 km long Ma'adim Vallis enters Gusev crater from the south, suggesting that water might have flowed into

Table 5. *MER Ellipse Priorities, Mid 2001*

5a. MER ellipses, April 2001

Medium-priority sites			Highest-priority sites		
Region	Ellipse	Location	Region	Ellipse	Location
Hematite	TM22B	3.40° S, 352.8° E	Hematite	TM20B	1.99° S, 353.99° E
	TM23B	3.10° S, 356.9° E		TM21B	2.50° S, 356.6° E
	TM12A	3.60° S, 357.1° E		TM19B	1.20° S, 354.7° E
	TM11A	3.40° S, 353.1° E		TM10A	2.20° S, 353.7° E
Boedickker	EP64A	15.30° S, 162.56° E		TM9A	1.20° S, 354.4° E
Unnamed crater	EP69A	9.20° S, 150.4° E	Melas Chasma	VM53A	9.07° S, 283.57° E
Isidis	IP84A	4.50° N, 88.10° E		New B Site	9.07° S, 283.57° E
	IP96B	4.48° N, 88.40° E	Gale crater	EP82A	5.81° S, 137.77° E
Meridiani crater	TM15A	8.60° S, 352.9° E	Gusev crater	EP55A (S)	15.00° S, 175.13° E
	TM16A	9.36° S, 353.25° E	Eos Chasma	VM41A	13.34° S, 318.61° E
Meridiani highlands	TM13A	3.00° S, 350.00° E	Isidis	IP98B	4.64° N, 84.12° E
	TM24B	2.80° S, 349.90° E		IP85A	4.7° N, 85.32° E
Valles Marineris	VM37A	11.10° S, 321.95° E	**Note:** The coordinates for the Melas Chasma sites are identical but the		
	VM44A	13.10° S, 297.50° E	ellipse sizes and orientations differed		
Ganges Chasma	VM42A	7.66° S, 308.70° E			

5b. Changes to the MER ellipses, July 2001

Region	Ellipse	Location	Change
Hematite	TM21B	2.50° S, 356.70° E	Moved from high to medium priority
	TM10A	2.20° S, 353.40° E	Coordinates changed
Meridiani crater	TM16A	9.36° S, 353.24° E	Coordinates changed
Melas Chasma	VM53A	8.8° S, 282.2° E	Coordinates changed
	New B site	8.8° S, 282.2° E	Coordinates changed
Gusev crater	EP55A (S)	14.85° S, 175.84° E	Coordinates changed

Table 6. *MER Prime and Backup Sites, October 2001*

Site	Name	Notes
Prime	Hematite – Meridiani	Geochemical anomaly associated with water, detected from orbit
	Melas Chasma	Canyon floor site, possible lakebed
	Athabasca Vallis – Elysium	Recent flow of water from Cerberus Rupes
	Gusev crater	Episodically flooded crater
Backup	Isidis Planitia	Fluvial deposits originating from nearby highlands
	Eos Chasma	Canyon floor with chaotic terrain, paleolake outflow region

the crater in the distant past to form a lake (Cabrol *et al.*, 2003), or at least might have deposited sediments. The large volcanic shield Apollinaris Patera 200 km north of Gusev, and two smaller volcanoes (Apollinaris Tholus and Zephyria Tholus) southwest of the crater, might also have contributed material to this site (Figure 4A). Gusev was included in the Mars Landing Site Catalog (Greeley and Thomas, 1995) as site 138 (Figure 136 in Stooke, 2012) and had been considered previously as a landing site for MESUR, InterMarsnet and Mars Surveyor 2001.

Table 7. *Final MER Ellipse Selection, 2001 to 2003*

7a. MER ellipse details, November 2001

Site	Name	Ellipse	Location
Prime	Hematite – Meridiani	TM10A2, 119 by 17 km	2.07° S, 353.92° E
		TM20B2, 117 by 18 km	
	Melas Chasma	VM53A2, 103 by 18 km	8.88° S, 282.52° E
		B2, 105 by 20 km	
	Gusev crater	EP55A2, 96 by 19 km	14.82° S, 175.15° E
	Athabasca Vallis	EP49B2, 152 by 16 km	8.92° N, 154.79° E
	revised, April 2002	EP49B3, 152 by 16 km	9.08° N, 155.80° E
Backup	Isidis Planitia	IP84A2, 133 by 16 km	4.31° N, 88.03° E
		IP96B2, 136 by 16 km	
	Eos Chasma	VM41A2, 98 by 19 km	13.34° S, 318.61° E

7b. MER low-wind ellipses in Elysium, 2002

Site	Name	Ellipse	Location
Backup	Elysium low wind	EP80B2, 165 by 15 km	14.50° N, 115.37° E
		EP78B2, 155 by 16 km	11.91° N, 123.90° E

7c. Final MER target ellipses, April 2003

Site	Name	Ellipse	Location
Prime	Hematite – Meridiani	TM20B3, 81.5 by 11.5 km	1.98° S, 354.06° E
	Gusev crater	EP55A3, 81 by 12 km	14.59° S, 175.30° E
Backup	Elysium Planitia	EP78B2, 155 by 16 km	11.91° N, 123.9° E
	Isidis Planitia	IP84A2, 133 by 16 km	4.31° N, 88.03° E
	(orientations differ)	IP96B2, 136 by 16 km	4.31° N, 88.03° E

A group of flat-topped hills at the mouth of Ma'adim Vallis was interpreted as an eroded remnant of a delta, further evidence of a possible lake (Figure 4B). Wrinkle ridges north of the ellipse suggested lava plains by analogy with lunar mare ridges. Two low-albedo wind streaks crossed the crater floor from northwest to southeast, indicating areas where winds and dust devils had removed bright dust from the surface. The streaks change appearance frequently while retaining the same general form and location. Figure 3A shows the albedo markings during the Viking mission (Viking MDIM2.1 mosaic) and Figure 4B shows their appearance at the time of Spirit's landing, using Mars Orbiter Camera (MOC) image R13-04079, which was obtained just after the landing. Albedo markings changed significantly in the two decades between images.

The 96 by 19 km landing ellipse was centered at 14.82° S, 175.15° E and oriented at 76° azimuth from north. It crossed central Gusev crater, between several sites considered previously (Figure 3, Table 1). Figure 4C depicts the whole MER-A ellipse on a background of MOC images (Malin Space Science Systems, 2004a). It extends 81 km from the interior of Thira crater across the plains north of the 2.5 km diameter crater Castril to a point southwest of the 6 km wide crater Cravitz. The broad dark wind streaks break up into numerous small patches and streaks at this resolution, each formed by an individual dust devil and many of them associated with topographic obstacles. The nominal target was the center of the ellipse. Spirit landed in the eastern wind streak region, an area of plains with scattered hills and small craters.

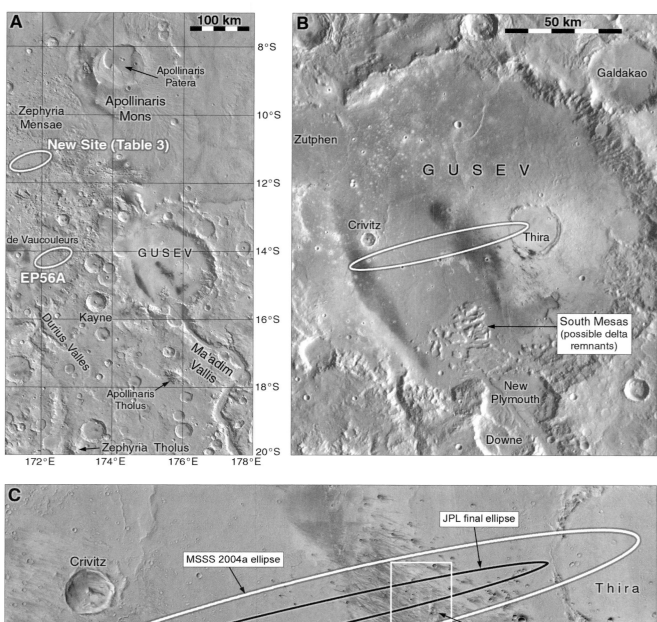

Figure 4. Spirit landing site in Gusev crater. **A:** Context map with two nearby candidate ellipses from Table 3 (Viking MDIM2.1 mosaic plus MOC image R13-04079). **B:** Gusev crater with the possible delta remnants at the mouth of Ma'adim Vallis (Mars Odyssey THEMIS infrared mosaic with inverted shading). **C:** Spirit landing ellipses on a mosaic of MGS MOC images (NASA/JPL/ Malin Space Science Systems). The final ellipse is from Knocke *et al.* (2004).

Figure 5 locates the landing site in images of increasing resolution, and Figure 6 shows the spacecraft components on the surface. Soon after landing, the hills in the vicinity were named informally to commemorate the astronauts killed in two NASA space accidents, Apollo 1 (27 January 1967) and the Space Shuttle Columbia (1 February 2003, only 11 months before the Spirit landing). Figures 7 and 8 identify those hills and several other named features. Spirit's lander was designated the Columbia Memorial Station. Following the Space Shuttle theme, on sol 117 two other features were named after Shuttle orbiters. A conical hill later named Von Braun was called Discovery Mount, and the gap in the hills later named Inner Basin was called Endeavour Pass. Those names were only used as camera pointing targets and were not adopted elsewhere. The Promised Land was a possible long-term goal for Spirit after it finished work in the Inner Basin, a depression in the hills which was also referred to as South Basin. A northward extension of the hills, here referred to as the eastern ridge, was visible from the landing site just beyond the rim of Bonneville crater (Figure 7) and may have formed part of an almost buried crater rim 8 km in diameter.

Spacecraft components were seen from orbit by MGS and MRO (Figure 6). The backshell and parachute (Figure 6A) landed 500 m northwest of the landing site, and the heatshield fell on the rim of Bonneville (Figures 6B and 15C). A MOC image (Figure 6C) taken on 19 January (sol 15, MY 26, sol 624) shows a dark streak where the heatshield struck the surface, scattering debris to the southeast before bouncing to its resting place on the rim of the crater. The dark impact point may be still faintly visible in the HiRISE image in Figure 6B, taken nearly three years later. The lander itself, in its airbags, bounced and rolled 250 m from its initial impact point (Figure 5) in a southeasterly direction to its final position (Figure 6D), which is considered the formal landing site. A white annulus of airbags about 3 m across surrounds the darker lander deck in the HiRISE view.

The landing site is identified in Figure 9A, which also shows the entire route followed by Spirit from its landing (upper left corner) to the end of the mission (bottom right). The positions of the rover at 1000 m intervals are also marked for reference. The spacecraft landed near a 200 m wide crater informally named Bonneville. This and other craters near the landing site were named after

lakes on Earth, since Gusev was thought to have held a lake (Cabrol *et al.*, 2003). The plains at its landing site were soon found to consist of basalt, showing that the plains with wrinkle ridges inside Gusev (Figure 4B) were volcanic rather than lacustrine in origin. The rover was driven up to the rim of Bonneville crater to see if its walls had exposed layers of bedrock, or if the crater had penetrated through the lavas to excavate underlying sediments. Since no outcrops or different rock types were found, Spirit was driven to the Columbia Hills nearly 3 km from the landing site. Here, rocks and soils showing evidence of alteration by water were found in abundance, but the hills are not composed of lake sediments. Rather, they appeared to be a complex jumble of impact ejecta layers and volcanic materials modified by water, probably similar to much of the ancient crust of Mars. Spirit climbed the 80 m high Husband Hill from the north side, then descended into a broad hollow called the Inner Basin and spent the remainder of its life examining the complex geology at Home Plate, a small eroded volcanic structure surrounded by salt-rich soils.

A detailed description of Spirit's activities follows, illustrated in Figures 10 to 42 and summarized in Tables 8 to 15. The entire rover route is shown at a standard scale in maps plotted on a HiRISE image base with a 100 m square grid for scale (Figures 10, 16, 17, 25 and 30). Sites of more complex operations are depicted on larger-scale maps based on HiRISE images or reprojected rover panoramas (*e.g.* Figures 11, 12 and 13). In all rover route maps, black dots are end-of-drive locations and white squares are stops for *in situ* observations (ISO), the main science locations. Some scientific observation was done at every stop, but at the ISO locations the instruments on the IDD were also used.

A key to map coverage is shown in Figure 9B. MGS MOC images were used for route planning before the first MRO HiRISE images became available in November 2006. Here MOC images are included in Figures 5D and 6C to illustrate the image quality available to planners at the time, but HiRISE images are used for the route maps. Every stop received a name early in the mission, but this became impracticable after Spirit left Bonneville. The tables and descriptions of Spirit's activities are derived from NASA's Planetary Data System, especially the Analyst's Notebook documents, augmented by monthly summaries of activities compiled

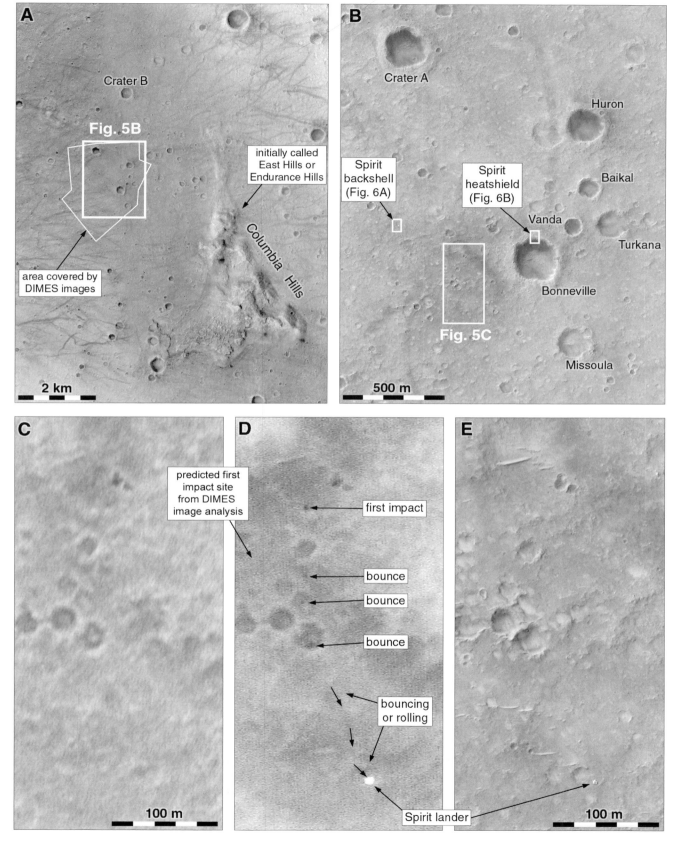

Figure 5. Spirit landing site. **A:** HiRISE mosaic of Columbia Hills region. **B:** Bonneville and nearby craters in HiRISE image PSP_001777_1650. **C:** Part of image 2E126462405EDN0000F0006N0M1 (Spirit Descent Camera image 3). **D:** The same area in MOC image R1303051 (19 January 2004) showing the marks made as Spirit bounced and rolled to its final landing site. **E:** The same area in HiRISE image PSP_001777_1650.

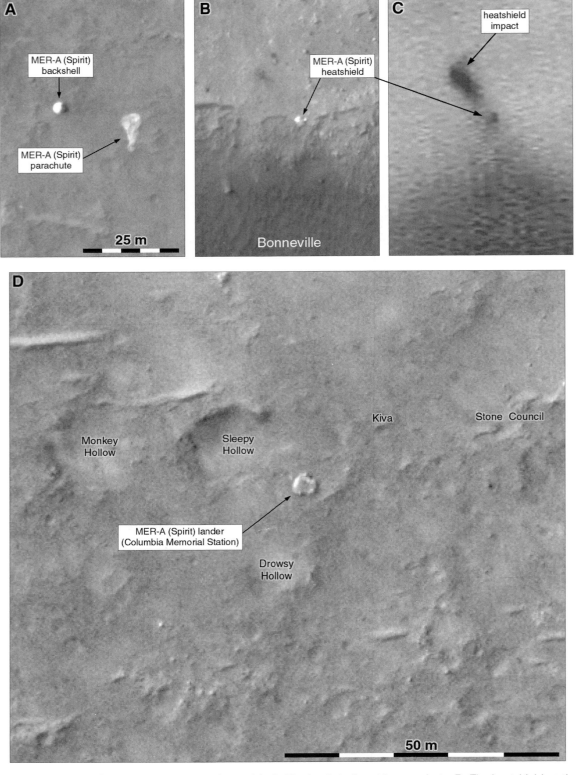

Figure 6. MER-A spacecraft components as seen from orbit. **A:** The backshell and its parachute. **B:** The heatshield on the rim of Bonneville. **C:** The heatshield and a dark mark made by its impact in MOC image R1303051 taken on 19 January 2004, two weeks after landing. **D:** The lander itself. **A**, **B** and **D** are parts of HiRISE image PSP_001777_1650 taken on 12 December 2006. **A**, **B** and **C** are all at the same scale.

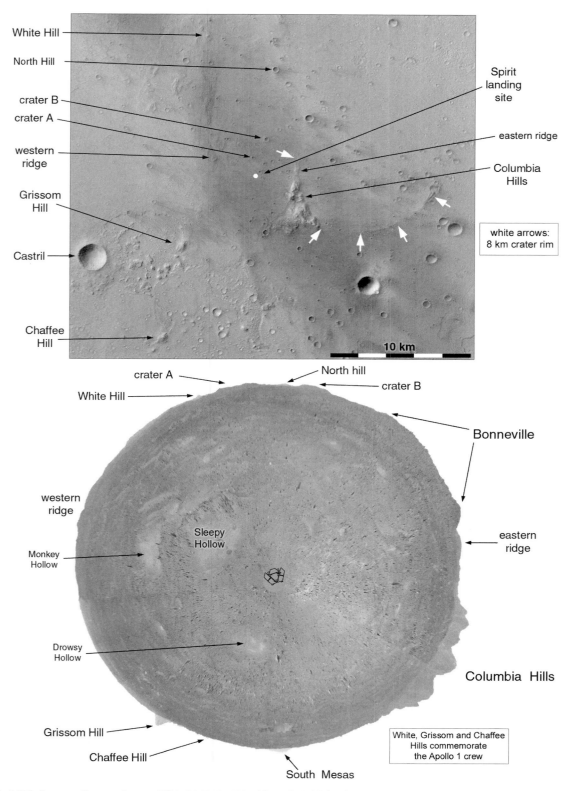

Figure 7. MRO Context Camera image PO1_001513_1654_XI_14S184W (top) and reprojected landing site panorama with exaggerated horizon relief (bottom), showing locations of horizon features. The South Mesas are part of the possible delta shown in Figure 4B. North Hill is probably part of the crater rim indicated here.

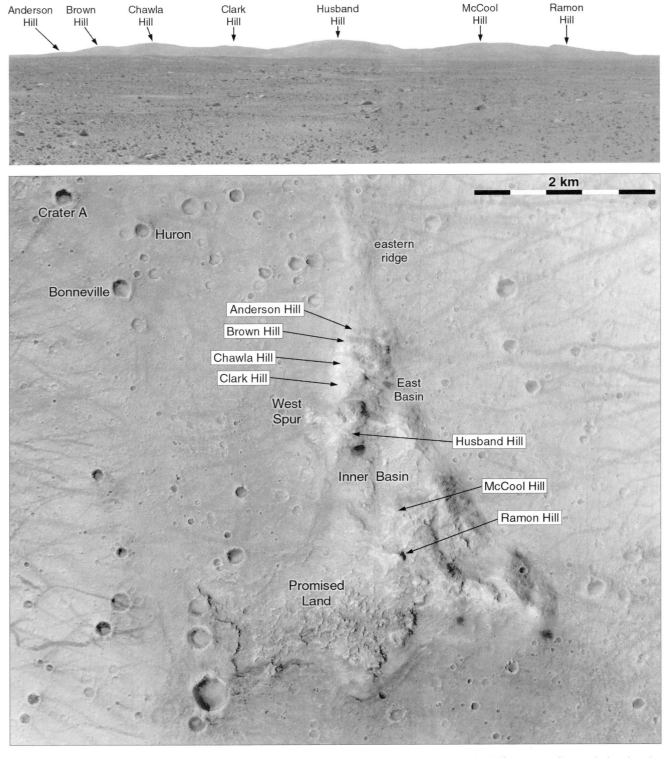

Figure 8. The Columbia Hills. Part of the panorama from the landing site (top) and a mosaic of HiRISE images (bottom) showing the names assigned to commemorate the crew of the Space Shuttle Columbia.

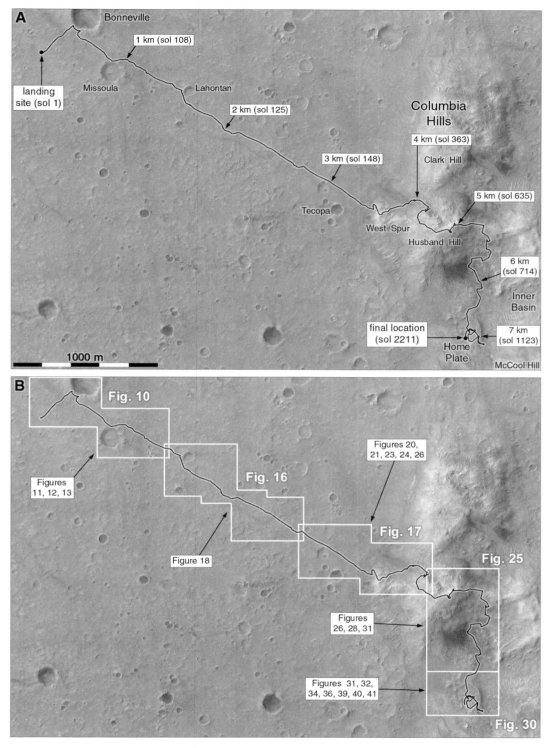

Figure 9. A: Full Spirit traverse with distance markers. **B:** Index of Spirit route maps. White outlines mark sections of the main route map. Text boxes identify larger scale maps with additional details.

for the Planetary Society by A. J. S. Rayl, and by descriptions in Arvidson *et al.* (2006, 2008a, 2010), Crumpler *et al.* (2005, 2011), Golombek *et al.* (2006a) and Squyres *et al.* (2004a).

MER-A, known as Spirit, was launched on 10 June 2003 at 17:59 UT from Cape Canaveral Air Force Station into a parking orbit, then placed on its Mars trajectory. Trajectory corrections were made on 20 June, 1 August, 14 November and 27 December 2003. Several additional corrections in the final days of flight were not required. A severe solar flare on 28 October 2003 affected the attitude control systems of both MER vehicles, requiring use of a backup system for about a week (Erickson *et al.*, 2004). MER-A arrived at Mars on 4 January 2004 (MY 26, sol 609).

The landing sequence was similar to that of Pathfinder. The cruise stage was jettisoned 15 min before atmospheric entry and fell to the surface, its fragments probably striking about 80 km west of the landing site near 14.9° S, 174.3° E, by analogy with Curiosity (Figure 179A). The lander entered the atmosphere encased in its heatshield and backshell. Following its initial atmospheric braking, the spacecraft was slowed further by a parachute, deployed at an altitude of about 7 km. The heatshield was dropped just after the parachute inflated, and the lander was released to drop down a 20 m long cable under the backshell. A radar altimeter on the lander base detected the surface at a height of about 2500 m, and then three images were taken by the descent camera from elevations of 2000, 1700 and 1400 m and analyzed onboard to measure the horizontal motion of the descending vehicle. This system (Descent Image Motion Estimation Subsystem, DIMES) allowed thrusters to reduce horizontal motions to an acceptable level. At a height of about 120 m the airbags were inflated and the thruster system fired to slow and stabilize the descent. Finally, at a height of only 10 m the cable was cut and the thrusters carried the backshell and parachute away while the lander fell to the ground, encased in its airbags.

The cushioned spacecraft bounced and rolled across the surface, finally coming to rest safely at 04:35 UT (Earth-received time) at 14.57° S, 175.48° E. The first impact was 300 m west of Bonneville crater and the lander bounced and rolled 250 m in a southeasterly direction before coming to rest 300 m southwest of Bonneville (Figure 5). The airbags deflated and were retracted by internal cables, the triangular landing stage panels opened, and the rover deployed its solar arrays. The landing occurred a little after noon local time, towards the end of southern hemisphere summer, and the landing sol was called sol 1, as it had been on Pathfinder. The Viking surface missions began on sol 0. Ten minutes of radio silence followed the landing, raising apprehensions on Earth until confirmation of a safe landing arrived. If the tetrahedral lander came to rest on one of its upper faces (petals), it would right itself as the petals opened, but it was already resting on its base petal with the rover facing almost due south.

The Pancam Mast Assembly (PMA) deployed soon after landing and a full Navcam panorama was transmitted, showing a flat plain littered with small rocks (Figures 7, 12A and 15A). The landing site was significantly less rocky than the Viking and Pathfinder sites had been, and it was pocked with numerous small depressions referred to as hollows. One of them named Sleepy Hollow, a few meters northwest of the lander, contained marks made by the airbags as the lander rolled across it (Figure 12A). Monkey Hollow, to its west, commemorates the Chinese Year of the Monkey, which began on 22 January 2004. The DIMES images showed the general area of landing, and the first surface images of the landing site showed a bulge in the airbag, initially thought to be a rock, in front of (south of) the rover. This might block departure from the lander in its initial forward direction, but the airbag could be retracted further or the rover could turn and depart in another direction.

Over the next few sols Spirit was gradually unfolded and deployed, and the rover obtained a full Pancam survey of its site (Figure 15A) to help the science team locate it and plan the initial surface operations. The High Gain Antenna (HGA) deployed on sol 2 but its link direct to Earth was partly blocked by the PMA, slowing progress over the next few days. When roving later in the mission both MER vehicles would normally turn at the end of a drive to optimize communications, but that was not yet possible here. The instruments and systems were checked and were mostly in good condition, but a temperature sensor had failed in the Mini-TES and HGA experienced current spikes.

On sol 3 one of the cables linking the rover to the lander was cut, the southern airbag was retracted more to

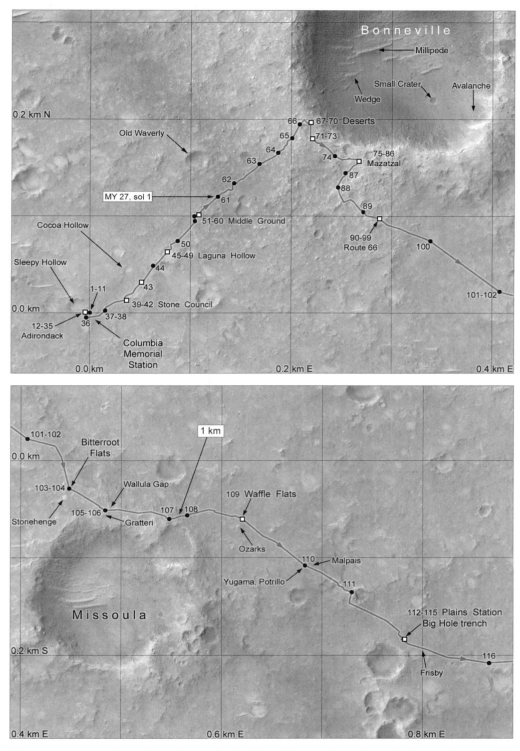

Figure 10. Spirit traverse map, section 1. The grid spacing is 100 m. The map is continued in Figure 16 and additional details are shown in Figures 11, 12 and 13. In all rover route maps, black dots are end-of-drive locations and white squares are stops for *in situ* observations (ISO).

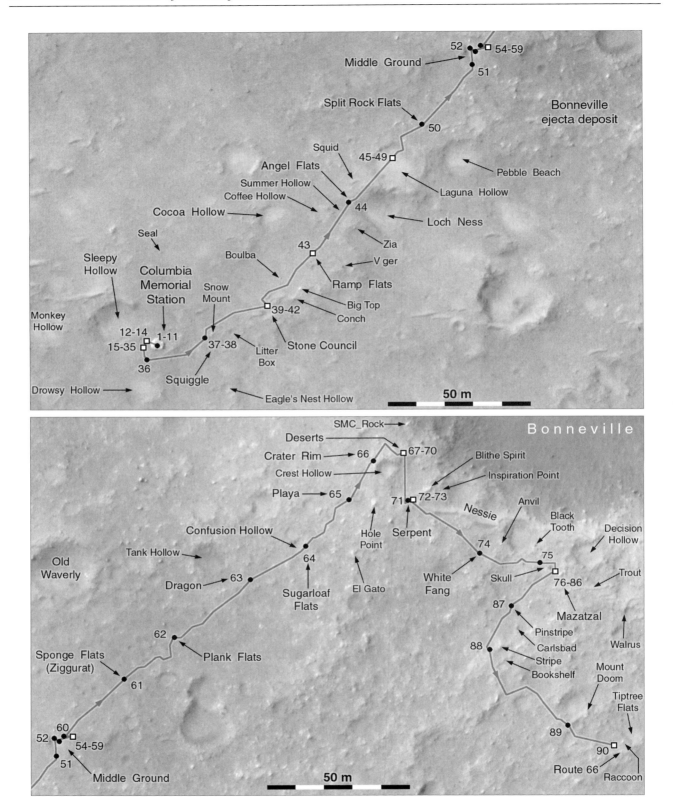

Figure 11. Spirit activities during the 90 sol primary mission, drawn on HiRISE image PSP_001777_1650. Decision Hollow on the southern rim of Bonneville was also referred to as Sandbox and Purgatory Hollow.

try to clear the exit path, and the Pancams viewed the Columbia Memorial patch and the Astrobot. A DVD supplied by the Planetary Society and containing the names of four million people collected by NASA was mounted on the lander body with fasteners resembling Lego blocks and carrying an image of the Astrobot, a Lego character called Biff Starling. Biff and Opportunity's Astrobot Sandy Moondust were part of an outreach program run by the Planetary Society.

Sol 4 was taken up with analysis of the HGA issues and the unexpectedly high temperature of the rover, and to decide how to deal with the airbag, which would not retract enough. Another attempt to pull in the airbag was made on sol 5 and Pancam images were taken of a range of hills to the east (Figure 5A), variously called the East (or Eastern) Hills or Endurance Hills until the team eventually settled on the name Columbia Hills. On sol 6 the southern lander petal was lifted to facilitate airbag retraction, but still it would not pull in enough, so the team decided to turn the rover and depart from a different ramp. Pathfinder had two egress ramps but the MER landers had three, offering more options for a safe route to the surface. The first steps in freeing the wheels and deploying the rover were made on sol 6 and continued on sol 7, and new images showed the area called Magic Carpet where the airbag left crisp impressions in the fine dust (Figure 12A). Rover preparations continued on sol 8 with arm tests and the release of the middle wheels. Mini-TES observed a rock called Sparky south of the lander and Ichabod Flats in Sleepy Hollow on sol 9, and on sol 10 Spirit completed its Pancam panorama, cut the last cable holding it down, backed up slightly and turned to exit the lander on its western side. Another small turn was made on sol 11 as plans for surface operations were finalized.

The initial plan was to examine a rock called Sashimi and nearby soil called Wasabi, to study Sleepy Hollow, and then to drive up to the rim of Bonneville crater, taking some important data before the team had to stand down on sol 21 to focus attention on Opportunity. Spirit rolled off its inert lander at 11:00 local time on sol 12 (MY 26, sol 620, at 8:41 UT on 15 January 2004), and on the next sol it deployed its IDD to obtain MI views of the soil at First Pebble Flats (Figure 12A). It also conducted a coordinated atmospheric observation with Mars Express, looking up at the sky while the orbiter looked

down, and tested RAT operations. On sol 14 it placed the MB on the soil, and on sol 15 the APXS, testing all the instruments and obtaining the first detailed soil analysis of the mission. The rover team had been slow to delete transmitted files from memory and noted now that the flash memory was too full. Then Spirit drove 2 m to the rock Adirondack for the first detailed study of a rock and observed the Eastern Hills and several rocks with Mini-TES. Sol 16 was devoted to imaging Adirondack for ISO planning, and on the next sol the MI, MB and APXS were used on the rock. The plan was to use all three IDD instruments on the rock before and after RAT brushing, and a third time after RAT grinding.

A serious software problem here nearly terminated the mission. On sol 18 the planned imaging and RAT brushing were lost when communications failed. A radio signal on sol 19 showed the rover was still operating, and analysis of the problem continued on sol 20 as Opportunity arrived at Mars. The plan had been for Spirit to analyze Adirondack and search for dust devils in a pre-programmed sequence while communications with Opportunity took precedence. A report on sol 21 noted that the rover was detecting an error as soon as it turned on, and rebooting with a one-hour delay. That sequence repeated every hour so the rover was not sleeping and its batteries were discharging. On sol 24 the team tried to download data and upload a software patch during the brief periods when communications were possible, but the instructions were not received that sol or the next. Finally on sol 26 the process worked and recovery began.

Despite these problems the analysis of data already on the ground was continuing, and by sol 27 the Viking veteran geologist Michael Carr was asking for a labeled panorama to help keep track of rock names. Naming became complicated on these rover missions because so many landscape features or instrument pointing targets were studied by the IDD and remote sensing instruments. Viking had required few names and Pathfinder was static, but rovers would see new features and targets at every stop. Some names were used at more than one place or even at more than one landing site, some objects received more than one name, and variant spellings complicated the issue further. Names given on maps and images in this atlas are selected from mission maps and documents and an online database maintained by

NASA's Planetary Data System, but inconsistencies with some other documents are inevitable.

Files were downlinked and deleted on sol 28, and the first new science was a set of coordinated observations with Mars Express on sol 29. Mini-TES data from the surface were compared with the orbital data to provide "ground truth" for the remote observations. An IDD campaign on Adirondack failed on sol 30 because a command file was corrupted, and on the next sol preparations were made to reformat the flash memory. This was done on sol 32 to resolve the earlier problems and allow the work on Adirondack to resume. It was brushed on sol 33, imaged through all the Pancam filters, and then all three IDD instruments were used on it. On sol 34 the RAT ground into the rock and the two spectrometers analyzed the hole. MI images of the hole were taken on sol 35, and Pancam images of the RAT magnets were taken to see if magnetic dust adhered to them. Adirondack was found to be a basalt fragment, part of a lava flow and not composed of lake sediments. A drive to a flat high-albedo rock called White Boat was lost due to an error on sol 35, but on sol 36 Spirit drove counter-clockwise around the lander and took images of White Boat and the lander.

The long traverse northeast towards the rim of Bonneville began on sol 37 with a 21 m drive ending at a small drift called Squiggle. Typically images would be taken at the end of a drive to assess the surroundings and help plan the next drive, and on the basis of those images additional imaging and ISO might be done on the following morning before the next drive began. During the 90 day primary mission, the team on Earth followed Mars time, living on the 24.6 hour cycle of Mars operations so the team was always available to plan the next sol's activities regardless of the time at which the end-of-drive images were obtained, but this process was physically stressful and very difficult to reconcile with the world beyond the mission. Later the team members reverted to life on normal terrestrial time, so sometimes the end-of-drive images were received in time to plan for the next sol and at other times they were not. If they were, drives could occur on every sol, but if not, drives took place on every second sol with remote sensing or IDD work on non-driving days. These periods of alternating drives were called restricted sols, and to reduce planning time instructions would be prepared in two-day

blocks, or three-day blocks over a weekend and even longer over a holiday period. The rover route maps show these patterns of work where sequences of daily drives alternate with periods of driving on every second or third day.

Stone Council, a group of rocks visible from the landing site, was the next target (Figure 12B). It was approached on sol 39 for ISO on a dune called Arena on sol 41. A flaky rock called Mimi at Stone Council looked very different from Adirondack, but it was also found to be a basalt after ISO on sol 42. On sol 43 Spirit drove to Ramp Flats, for ISO on the soil early on sol 44 and imaging of nearby rocks. The pattern continued with a drive later on sol 44 and ISO on the soil at Angel Flats early on sol 45.

The next stop was at Laguna Hollow, one of the many circular depressions like Sleepy Hollow at the landing site but the first to be examined (Figure 12C). These hollows, floored with fine material, are probably secondary impact craters partly filled with wind-blown sediment. Laguna Hollow was located in a dust devil track visible in orbital images. As Spirit arrived later on sol 45 it stopped, backed up and moved forwards again to leave scuff marks which exposed soil without the ubiquitous dust cover, and on sol 47 it used its wheels to dig a trench called Road Cut in the hollow to look deeper into the soil. The wheels were steered to allow the rover to turn in place, and it made 12 two-way partial turns, but the left front wheel rotated in the opposite direction each time to excavate the trench.

This was Spirit's first trench and the second of the MER mission, occurring three sols after Opportunity's Big Dig trench in Eagle crater on its sol 23. The soil here was harder to excavate than that at Big Dig, as 12 of Spirit's wheel passes dug only about 7 cm deep, whereas six passes had dug down 10 cm in Meridiani. The rover backed up to point Mini-TES at the trench, and then on sols 48 and 49 the IDD instruments were all used on the trench walls and floor. The front Hazcams made a sequence of images of the growing rover shadow as the Sun set on sol 48.

IDD work at Laguna Hollow ended on sol 50 and Spirit was driven to a second hollow, Middle Ground, entering it on sol 52 (Figure 13A). A soil area called Sugar on the rim of Middle Ground was analyzed just before entering the shallow depression. Once in the

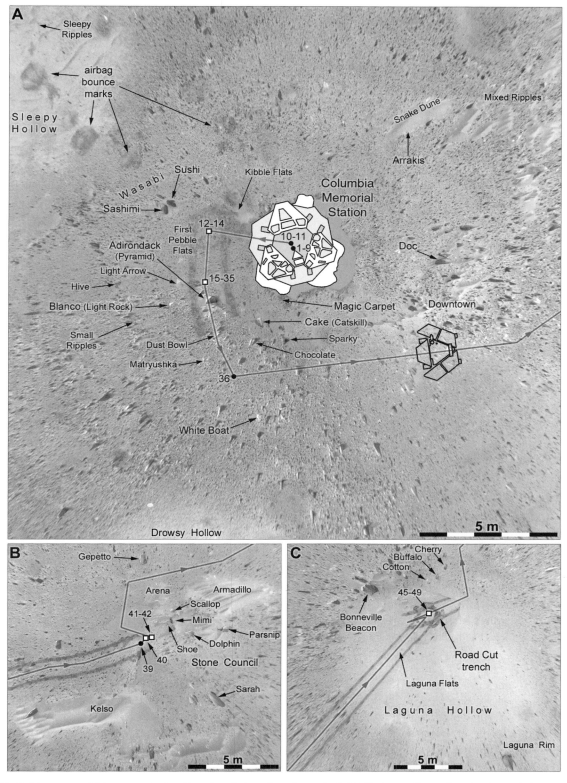

Figure 12. Spirit rover activities near the landing site. **A:** Landing site, sols 0–36. **B:** Stone Council, sols 39–42. **C:** Laguna Hollow, sols 45–49. Sol numbers indicate the end of sol location. Maps are composites of reprojected panoramas. The rover is shown to scale in **A**.

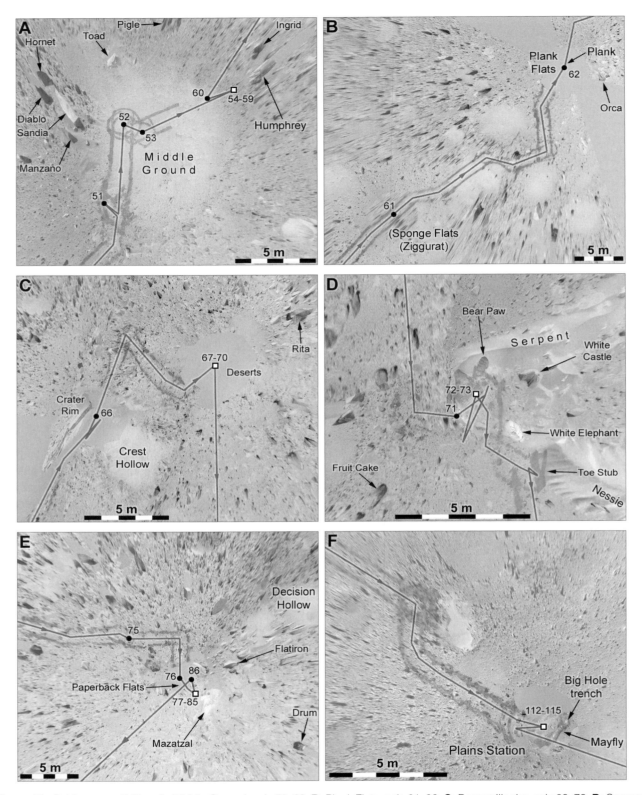

Figure 13. Spirit rover activities. **A:** Middle Ground, sols 51–60. **B:** Plank Flats, sols 61–62. **C:** Bonneville rim, sols 66–70. **D:** Serpent, sols 71–73. **E:** Mazatzal, sols 75–85. **F:** Plains Station, sols 112–115. The images are mosaics of reprojected panoramas, which contain significant distortions caused by relief.

hollow, a dust devil search was made, the MI and APXS were used on the rover magnets on sol 53, and the IDD was used on the soil called Ridge 1 before the rover approached a large rock called Humphrey. On sol 55 the MI and APXS were used on Humphrey before brushing. Then three adjacent brushings were made to clean an area large enough to fill one Mini-TES pixel, and the rover backed up, made the Mini-TES observation and moved back in. On sol 57 the spectrometers were used on the brushed area, and on sol 58 the RAT ground into the hard rock. The grind was cut short and the IDD instruments used on the shallow hole, but more depth was required to cut through any zone of chemical alteration, so on sol 59 the RAT ground deeper and the brush cleaned the hole for further IDD work on that sol and the next. Humphrey was also basaltic, and results to this point suggested that the local rocks had been subjected to a very small amount of chemical alteration by water but nothing like the amount anticipated by the crater lake hypothesis. Also on sols 59, 60 and 61 a full panorama was taken to document the surroundings.

Middle Ground was left behind on sol 61, the first day of a new Mars year (MY 27, sol 1). Spirit now entered the rougher terrain of Bonneville's ejecta blanket (Figure 13B). Astronomical observations, including images of stars and the rising Earth, were made before dawn on sol 63, and Orion was viewed early on sol 67. The sol 63 images also included a streak interpreted as a moving object, initially suspected to be Viking Orbiter 2. Later, Selsis *et al.* (2005) suggested this was a meteor associated with Comet 114/P Wiseman–Skiff, but Domokos *et al.* (2007) concluded it was more likely to be a cosmic ray strike on the camera detector. On sol 68 Deimos was observed crossing the disk of the Sun. Spirit used its IDD on a rock called Plank early on sol 63 (Figure 13B), dug an accidental rut called Serendipity Trench during a back-and-forth maneuver on sol 64, and on sol 65 for the first time it was able to see the far side of Bonneville crater and its heatshield perched on the northern rim. The minimum mission success criterion of 300 m driving was reached on sol 64. The rover reached the crater rim on sol 67 (Figure 13C) and surveyed the walls, looking for rock outcrops, which might reveal details of the local stratigraphy, but none were found. A large panorama made from this vantage point showed the crater interior, the rims of other more

distant craters to the north and east, and the wide plains of Gusev, including Spirit's parachute and backshell to the northwest (Figures 14 and 15C).

Soils at the rim of Bonneville were analyzed on sols 68 and 70, Mini-TES made coordinated observations with Mars Global Surveyor on sol 69, and Spirit looked for water ice halo effects in the sky on sol 70. Most rocks seen so far, including Adirondack, Mimi and Humphrey, had been dark basalts, but the higher-albedo rocks like White Boat had not yet been analyzed. The rover now drove south along the rocky crater rim looking for a good example of the light-toned rocks to study. Spirit skirted the western end of a dust drift called Serpent, one of a chain of drifts that followed the southern rim crest of Bonneville. The main part of the long drift, southeast of Serpent, was named Nessie. Dust drifts in Bonneville itself were dark, but Serpent and other drifts outside the crater were bright. Were they made of the same dark material but covered with brighter dust? The wheels were used to scuff Serpent on sol 72 to reveal its interior, which was indeed darker. The scuff, called Bear Paw, was observed by Mini-TES, multispectral Pancam images and all the IDD instruments. As Spirit left that site on sol 74 it stopped briefly at Nessie to make another scuff called Toe Stub (Figure 13D) for Mini-TES observations. Two large rocks near here had been prominent landmarks during the drive towards the rim of Bonneville, Hole Point just 10 m west of Serpent and El Gato about 25 m southwest of the drift.

A large high-albedo rock called Mazatzal was reached on sol 77 (Figure 13E) after analysis of local soil targets. On sol 78 the IDD instruments studied the rock at a spot called New York, and on the next sol, after some early Deimos imaging, the rock was brushed at two locations, lightly at Illinois and heavily at New York. The spectrometers were used on New York but the APXS door did not open fully. That observation was repeated on sol 80 while the Pancam and Mini-TES looked back at Bonneville crater. The New York site was ground by the RAT on sol 81 (making a new target called Ratted) and cut deeper again on sol 83 (now named Brooklyn, so each unique dataset would have its own name), with IDD analyses at each stage. The final activities on Mazatzal were a RAT brush of six adjacent locations at Missouri and a Mini-TES observation of that large cleaned area on sol 86. A decision about the future route, around the

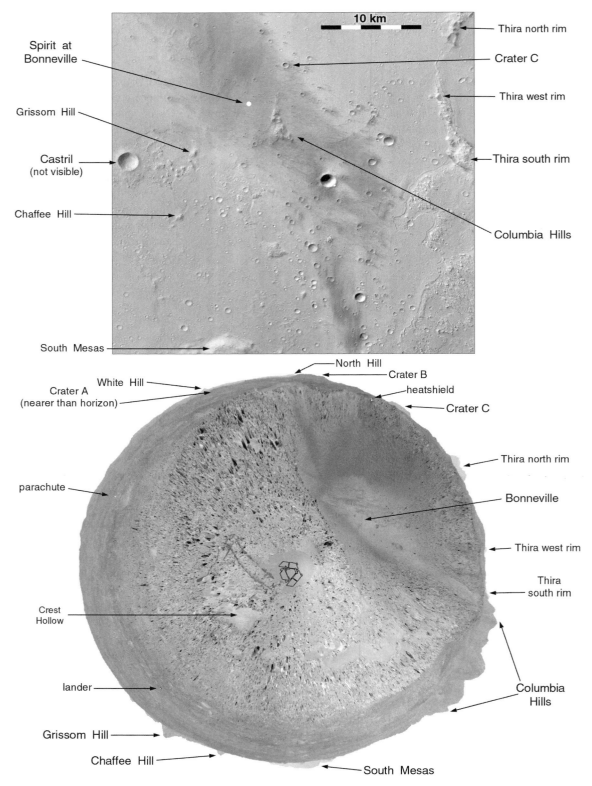

Figure 14. A mosaic of MRO Context Camera images (top) compared with a reprojected panorama from the rim of Bonneville (bottom) with exaggerated horizon relief. Hills on the southern horizon are more visible here than in Figure 7. Craters A and B are shown in Figure 5. The South Mesas are part of the possible delta shown in Figure 4B. The distant rim of Thira is now partly visible.

crater rim or out across the plains, had to be made at Mazatzal, so a nearby hollow was named Decision Hollow, but different versions of mission maps circulated and one of them called the same feature Purgatory Hollow. Since none of these names were official, inconsistencies were not uncommon.

Spirit left the crater rim on sol 87, observing rocks on its way, and on sol 89 as it surpassed 600 m of driving it had achieved all its mission goals. The 90 sol primary mission ended as the rover approached a rock called Route 66 and the first of many extended missions commenced. The primary mission route from the landing site to Route 66 is shown enlarged in Figure 11. Here on sol 92 the rover magnets were analyzed to assess the composition of magnetic dust adhering to them, and Mini-TES and the TES instrument on MGS made coordinated observations. The two spectrometers were calibrated on sol 93 and 94 and then used on Route 66 while the cameras observed sunrise and sunset and saw the distant rim of Gusev, up to 100 km away, for the first time. The atmosphere was usually too dusty to reveal such distant features. The rover's flight software was replaced here, taking several sols to load and check it carefully before its first use on sol 98. This version allowed longer drives and corrected several known problems. On sol 99 a six-spot RAT brush mark was made for ISO and Mini-TES observations before leaving on sol 100 (MY 27, sol 40, or 14 April 2004).

Now the long drive to Spirit's new target, the distant Columbia Hills, began. Moving quickly to the southeast, the rover mounted the rocky rim of Missoula crater, but again no bedrock outcrops were visible in the crater walls. On sol 104 it observed Phobos passing in front of the Sun, causing a partial eclipse, and later watched Phobos setting to test atmospheric refraction. Dust devil searches were made occasionally during the drive, including on sol 104, and on sol 105 the rover viewed the interior of Missoula, but as at Bonneville the eroded rim and inner slopes had no obvious rock outcrops. The drive distance exceeded 1000 m on sol 108. For 50 sols Spirit drove almost every day (Figures 16 and 17), pausing only rarely to examine rocks or regolith in case a sudden failure should end the mission.

The stops included soil targets on sols 110 (Waffle Flats) and 113 (Mayfly), a vesicular basalt block (Potrillo; "vesicular" means containing bubble-like voids) on sol

111, and a trench called Big Hole dug at Mayfly on sol 113 with IDD observations before and after trenching (Figure 13F). The rover looked into Lahontan crater on sol 120 (Figures 16 and 18A). Lahontan's ejecta appeared unusually smooth in MOC planning images but did not look distinctive on the surface. Early on sol 122 MI viewed targets including Owens at Cutthroat, a location in the rover's tracks. Spirit's longest drive in one sol was 124 m on sol 125. The morning after that a soil target called Lead Foot was analyzed but the IDD was held too high above the surface and the results were poor. Figures 8, 9 and 16 show dust devil tracks winding across the plains in later HiRISE images. They appeared in different positions in contemporary MOC images, as also seen in Figures 5D and 5E where the subtle dark streaks are dust devil tracks. Spirit drove over both dark tracks and bright inter-track areas, but they were not noticeably different on the ground. Searches for active dust devils during this period were not successful.

Long drives here were made in two modes. After each drive, Pancam stereoscopic images of the anticipated drive direction were used to make a topographic map for route planning. Out as far as the Pancam could see clearly, a distance very dependent on topography, the drive could be pre-planned from Earth, but beyond that the rover had to watch for hazards itself by analyzing stereoscopic images onboard. If a hazard was detected in this autonomous navigation (autonav) mode, an attempt would be made to bypass it. Pre-planned drives covered about 180 cm per minute and autonav drives only 50 cm per minute. On sol 129 the autonav routine failed to find a path down a ridge, which had not been seen in MOC images, and the drive was cut short. Images taken from the end point enabled planners to direct a safe path down the ridge on the next driving sol.

The last section of the drive to the hills included a stop on sol 135 to use the IDD instruments on the soil and then dig a trench called The Boroughs (Figure 18C). Images revealed darker soil in the bottom of the trench, which was analyzed using the IDD instruments on sols 140 and 141 after several days of computer problems. A full panorama called Santa Anita was obtained here over several sols, and the rover drove away on sol 142 (Figure 18D). A plan for another trench in the plains was dropped so the rover could reach the hills as quickly as possible. The drive distance exceeded 3000 m on sol 148

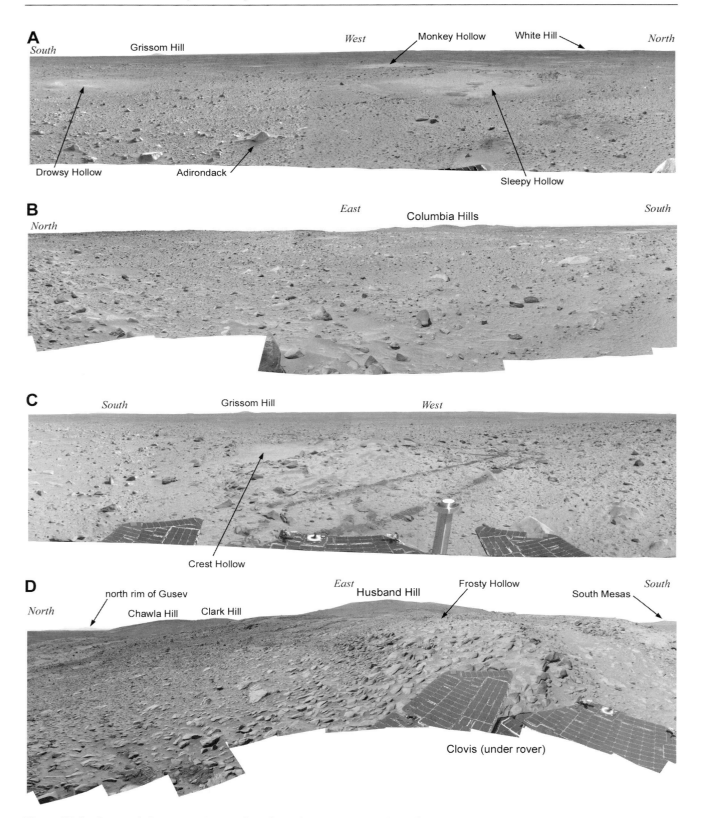

Figure 15 (both pages). Panoramic images from Spirit, from the landing site to Columbia Hills. **A:** Landing site panorama, sols 2–10. **B:** Legacy panorama from Middle Ground hollow, sols 55–60. **C:** Bonneville panorama from the southern rim of Bonneville crater, sols 68–69. **D:** Cahokia panorama from Clovis rock on the western tip of West Spur, sols 219–220.

Figure 15 (continued)

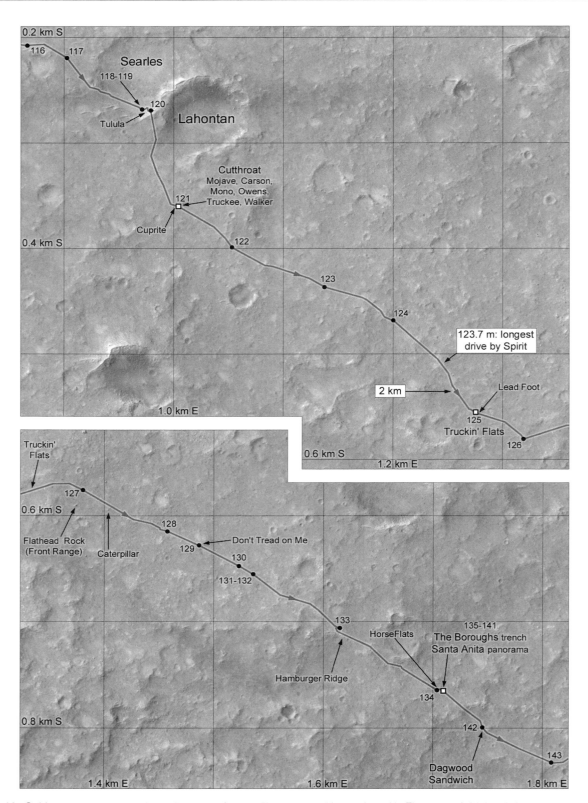

Figure 16. Spirit traverse map, section 2. This map follows Figure 10 and is continued in Figure 17. Additional details are included in Figure 18.

Table 8. *Spirit Activities, Landing Site to West Spur, Sols 1–200*

Sol	Activities
1	Entry, descent and landing, first images (MY 26, sol 609)
2–12	Panoramic imaging (2–11), rover preparation, egress from lander (12)
13–14	Mars Express coordinated experiment (13) and ISO First Pebble Flats soil target
15–17	Drive to Adirondack (15), imaging (16), ISO Blue target on Adirondack (17)
18–28	Computer memory management problems, some imaging during recovery (25–26)
29–30	Mars Express coordinated experiment (29), brush Prospect target on Adirondack (30)
31–35	Computer repair (31–32), brush (33), RAT (34) and ISO (33–35) Prospect target
36–37	Drive to White Boat (36), drive around south side of lander and east to Squiggle drift (37)
38–39	Computer problems (38), drive to Stone Council and image large drifts (39)
40–43	Approach Stone Council (40), ISO Mimi rock (41), Shoe target (42), tracks (43)
44–45	ISO Ramp Flats target (44), Halo target in Angel Flats (45), drive to Laguna Hollow
46–50	ISO Laguna Hollow (46), dig Road Cut Trench (47), ISO targets in trench (48–50)
51–54	Drive to Middle Ground (51–52), imaging, approach Humphrey rock (53–54)
55–59	Brush Humphrey (55), imaging, ISO Humphrey (57), RAT Humphrey (58, 59) and ISO (59)
60–67	Image Humphrey and finish Legacy panorama (60), drive to Bonneville (60–67)
68–71	Take Bonneville panorama (68–69), ISO targets on crater rim (70), drive along rim (71)
72–77	Dig trench in Serpent Drift (72), ISO trench (73, 74), drive to Mazatzal (75–77)
78–80	ISO Mazatzal (78), brush Mazatzal (79), ISO New York target on Mazatzal (80)
81–82	RAT and ISO New York (81), ISO other Mazatzal targets (82)
83–89	RAT Brooklyn (83) and ISO (83–86), imaging (86), drive down crater rim (87–89)
90–93	Drive to Route 66 (90–91), study magnets (92), ISO Route 66 (93), image tracks
94–98	Software update (94–96), ISO Route 66 (94, 95, 97), software test (98)
99–101	Brush Route 66, image Back Lot and Orel targets (99), drive towards Columbia Hills (100–101)
102–106	Photometry, atmospheric studies (102), drive to Missoula crater (103–105), image crater (106)
107–113	Drive to hills (107–112), ISO Waffle Flats (110), ISO Mayfly (113), dig trench (113)
114–121	ISO (114) and imaging (115–116) of Big Hole trench, drive to hills (116–121)
122–126	ISO Owens soil target and imaging (122), drive to hills (122–126), ISO Lead Foot rock (126)
127–134	Drive to hills (127–134), image Flat Head target (128), correct software problem (131–132)
135–138	ISO, dig and image The Boroughs trench (135), problems prevent science (136–138)
139–141	Take Santa Anita panorama (139–140), ISO The Boroughs trench (140–141)
142–148	Image The Boroughs trench (142), drive towards Columbia Hills (142–148)
149–151	Image Columbia Hills, ISO Joshua target and magnets (149–151), drive to hills (151)
152–159	Imaging, drive to hills (152–157), move to End of the Rainbow feature (158–159)
160–165	ISO End of the Rainbow (160) and Pot of Gold targets (161–164), move (164, 165)
166–168	ISO, move (166), ISO Jaws soil (167), move (167–168)
169–175	RAT Pot of Gold (169), ISO Pot of Gold (169–172), ISO Breadbox (175), imaging
176–178	Calibrate front Hazcam (176), ISO String of Pearls (177), driving tests (178)
179–185	Imaging, ISO rover tracks (180–182), wheel tests (183–185)
186–191	Imaging, arm position recalibration (186), wheel tests (187–191), ISO Limping Flats (191)
192–194	Drive to Wooly Patch feature (192, 193), ISO Mammoth (193) and Sabre (194) targets
195–198	RAT Sabre target (195), ISO Sabre (196–198), RAT Mastodon target (198)
199–200	ISO Mastodon RAT site on Wooly Patch (199), image both RAT sites (200)

Note: In this and all subsequent MER activity tables, "RAT" (as a verb) refers to a RAT grinding operation, "brush" refers to a RAT brushing operation, and "ISO" refers to *in situ* observations, which may include the use of any combination of IDD instruments (Microscopic Imager and spectrometers).

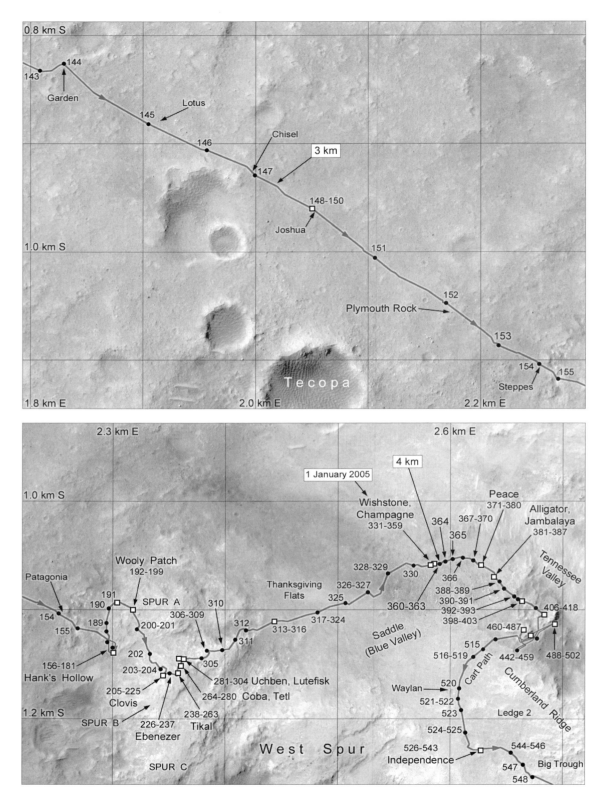

Figure 17. Spirit traverse map, section 3. This map follows Figure 16 and is continued in Figure 25. Additional details are presented in Figures 20, 21, 23, 24 and 26. White squares are main science sites.

(Figure 18E), and a prominent boulder called Plymouth Rock (Figures 17 and 18F), which had served as a target for drivers, was passed on sol 152. A short IDD campaign was conducted on Joshua rock on sols 150 and 151. The geology of the plains is described by Crumpler *et al.* (2005), Golombek *et al.* (2006a) and Arvidson *et al.* (2006). Only basaltic rocks were found in the plains and nothing resembling the anticipated lake sediments was seen.

As the rover approached the hills from about sol 120 onwards, the Science Operations Working Group (SOWG) began considering how to explore them. Their first suggestion, on about sol 127, was to investigate the base of West Spur and then drive south around the foot of Husband Hill before climbing onto a south-pointing spur called Lookout Point (Figure 19). A survey from that vantage point could help plan future operations in the Inner Basin, a depression which was probably a remnant of an old impact crater in the hills. The hills themselves were formed as part of the rim of another crater about 8 km across, indicated in Figure 7, so it was obvious from the start that they must have had a complex history. By sol 136 a more detailed plan had been developed (Figure 19), which would visit a dark deposit later known as El Dorado and several bright patches in the Inner Basin including Home Plate, a polygonal high-albedo feature initially suspected of being an evaporite deposit (Cabrol *et al.*, 2003). From there it would climb McCool Hill and arrive at the edge of the Promised Land, a rough area southwest of the hills and on the edge of the landing ellipse.

Another possible route was suggested on sol 180 (Figure 19). It would take the rover up to the summit of the 80 m tall Husband Hill by traversing West Spur and climbing up the western slope of Husband Hill, a trip taking about a month to complete. It was far from certain at first that an MER rover could climb hills like these, but by this time Opportunity had successfully negotiated steep slopes inside Endurance crater (Figure 53), so climbing now seemed more feasible. The new route was probably too steep, though a sol 230 sketch showed a more realistic zig-zag path up the hill, but this plan and the Inner Basin route were also unsatisfactory from an illumination perspective. Gusev crater, 15° south of the equator, was now experiencing southern winter and Spirit needed to orient itself as much as possible on

north-facing slopes so the solar panels could generate sufficient power for even minimal operations. A better path was suggested on sol 306. The rover would descend West Spur on its northern side and later climb Husband Hill on its northern flank. When the time came to enter the Inner Basin on the hill's south-facing slopes, the season would be more favorable, and the goal for the following winter was to explore targets on the north-facing slopes of McCool Hill. In this rugged terrain, illumination became a major factor for route planning in every winter season.

Spirit arrived at the foot of the low hills comprising West Spur (Figures 17 and 20B) on sol 157 (MY 27, sol 97). The right front wheel had been experiencing elevated wheel currents for several weeks prior to this, and on sol 156 data were gathered to try to diagnose the problem. On sol 158 the IDD instruments were deployed on an area of soil called Shredded which had been disturbed by the wheels, before driving to a cluster of small rocks named End of the Rainbow which lay around a shallow depression called Hank's Hollow (Figure 20A). The name commemorated Henry "Hank" Moore, a USGS geologist on the Viking and Pathfinder missions, who had died in 1998. Their chemistry and weathered appearance showed immediately that the hills were very different from the plains and had experienced more alteration by water (Crumpler *et al.*, 2011).

The rock Pot of Gold on the southern rim of the hollow was characterized over sols 160 to 173, including several sols spent trying to get into a better position. Pot of Gold was too small for the RAT to be used effectively on its very irregular and spiky surface. Other targets here included a deeply eroded rock called Breadbox and a cluster of very bright features called String of Pearls. On sol 178 Spirit backed up to allow Mini-TES observations of recent targets, using a new capability called visual odometry (visodom) for the first time. This used onboard comparison of sequential images to estimate distance traveled, a tool which would become very useful later. It was particularly important for detecting lack of movement if the rover wheels lost traction and spun uselessly in soft soil, and would stop a drive before the rover became badly embedded.

After some last IDD studies of disturbed soils on sol 182 the rover was driven to a level area nearby called Engineering Flats (Figure 20A) to address the wheel

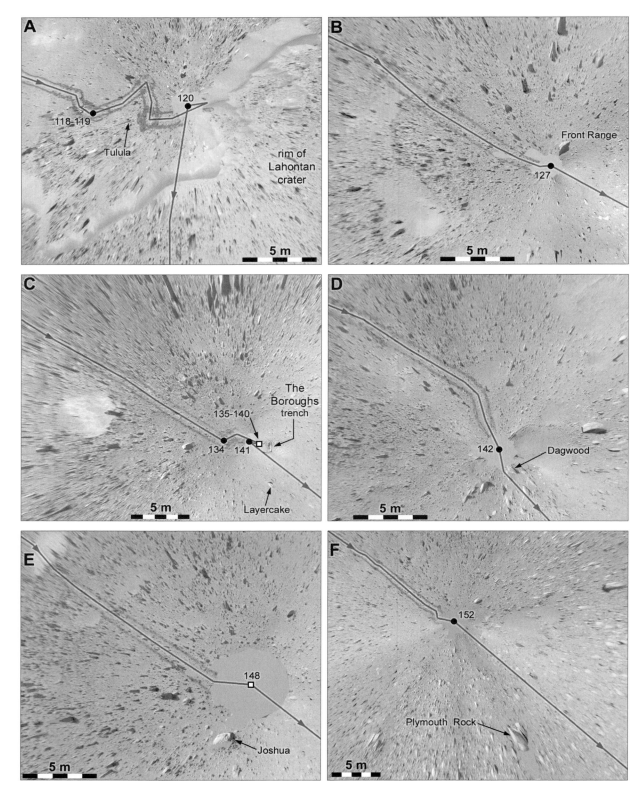

Figure 18. Maps of six stops along the drive to West Spur. **A:** Lahontan, sols 118–120. **B:** Front Range, sol 127. **C:** The Boroughs, sols 134–141. **D:** Dagwood, sol 142. **E:** Joshua, sol 148. **F:** Plymouth Rock, sol 152. The images are reprojected surface panoramas which contain unavoidable distortions, so the scales should be considered approximate.

current issue. On sol 183 the wheel was moved to measure friction, and on sols 184 and 185 the drive system was heated and small movements were made to redistribute lubricant. On sol 186 a set of movements identical to those made on sol 183 showed some improvement due to the heating. Alternating forward and backward driving seemed to help by redistributing lubricant more effectively in the wheels, and another approach was to drive with the right front wheel locked 90 percent of the time. This required backward driving much of the time. Also at this location the positioning of the IDD was recalibrated in Hazcam images. The battery power was very low now, so Spirit moved on sol 187 to a location with a more northerly tilt to recharge, and imaged rock outcrops in the hills, especially Cowlick, a rock later renamed Longhorn. An interesting north–south strip of rocks seen on sol 189 interested the science team, but it was bypassed in favor of north-facing slopes. Sol 191 began with a brief deployment of MI and MB on a target called Jeremiah on a rock named Loofah in front of the rover at a place named Limping Flats.

Spirit now followed a winding route up onto West Spur, aiming for a prominent outcrop called Clovis which was visible from Hank's Hollow and only about 10 m higher. The first stop was a bright rock called Wooly Patch, reached on sol 192. Rock outcrops in the hills made stratigraphic studies feasible but the complexity of the hills made interpretation difficult. The IDD instruments were used on Wooly Patch on sols 193 and 194 and on the next sol the RAT ground a hole (Sabre), showing the rock to be very much softer than the hard basalts encountered earlier. Another hole, Mastodon, was cut on sol 198 after post-RAT analyses of Sabre, and the new hole was analyzed for two sols. Also on sol 198 Mini-TES attempted unsuccessfully to observe Phobos. On sol 200 (MY 27, sol 140, or 26 July 2004) Spirit left Wooly Patch and climbed the hill towards Clovis, observing a small hollow just below Clovis during the approach. The drivers did not expect to enter it but on sol 203 Spirit rolled over its northern rim, ending its drive with a dangerous south-facing tilt. On the next sol it backed up slightly and recharged its batteries, and on sol 205 it skirted the depression, now named sol 203 Hollow, and approached Clovis. Feature names on West Spur were taken from North American archaeological sites (Figures 20B and 21A).

Work began on Clovis with pre-brush IDD data collection, a brush at a spot called Plano on sol 214 and more IDD analysis. The MB instrument was now being used for a full 24 hours as its energy source gradually weakened, though some data could be obtained in shorter sessions. The rock was ground for 2.5 hours with the RAT on sol 216, cutting a hole 9 mm deep, and IDD data were collected again. This time MB data from this interesting rock, extensively altered by water, were collected for 48 hours spread over six sols while the Cahokia panorama, showing the plains from this elevated viewpoint, was taken with the Pancam. The south rim of Gusev near Ma'adim Vallis was visible now with unprecedented clarity (Figure 29D). An area near Plano was brushed for Mini-TES observations on sol 225, producing a flower-like pattern of six circular marks with a seventh in the center (Figure 22). Similar patterns had been made at Mazatzal and Route 66, but here the process was interrupted after the first five brushings, creating a shape similar to the Olympic rings which was imaged to honor the Olympic Games then in progress in Athens. The picture of Clovis in Figure 22 shows the completed pattern.

Spirit now spent 100 sols driving from one north-tilting outcrop to another (Ebenezer, Tikal, Tetl, Uchben and Lutefisk) across West Spur (Figure 21). Soils were also examined, including two sites called Ring of Kerry and Killarney Flats on sol 227. Ebenezer, one of a group of dark rocks in this area, was subjected to the common pattern of pre-brush, post-brush and post-RAT IDD analyses while images were taken to monitor clouds and to estimate atmospheric dust levels by imaging sunset on sols 228 and 230. A RAT brush mosaic (group of brushed areas) was made on sol 236 for Mini-TES observations on the next sol (Figure 22). The rover was driven to a rock called Tikal on sol 238. It was previously referred to as Big Blue Spot for its location in a color-coded illumination map, where blue signified a good northern slope, necessary for sufficient power in the depths of winter. The winter solstice (winter in the southern hemisphere, summer in the northern) was on sol 253 (MY 27, sol 193). This period also coincided with conjunction, when Mars passed near the Sun as seen from Earth, so communications were interrupted from sols 244 to 255 while Spirit sat at Tikal. Spirit could still communicate with

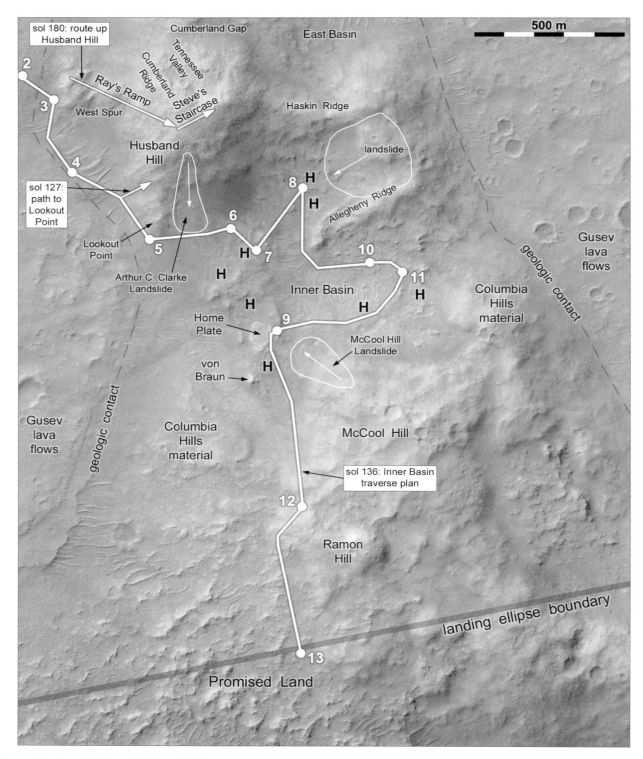

Figure 19. Inner Basin and Husband Hill exploration plans from the Science Operations Working Group between sols 127 and 180. The white numbers are science targets or route planning waypoints, from a map first distributed on sol 136 (site 1 was near Tecopa crater). The H symbols indicate features resembling Home Plate, from the same map and from Rice *et al.* (2010a). Landslides are from Rice *et al.* (2010a). The HiRISE image is PSP_008963_1650.

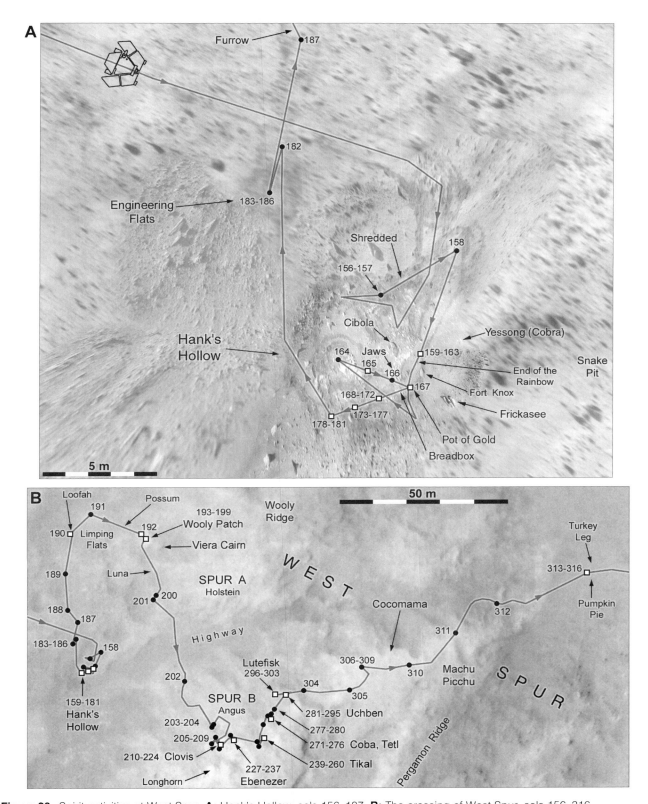

Figure 20. Spirit activities at West Spur. **A:** Hank's Hollow, sols 156–187. **B:** The crossing of West Spur, sols 156–316.

Mars Odyssey and the data were sent to Earth via the orbiter's more powerful transmitter if possible. Rover-to-Earth communication tests were also undertaken during conjunction.

One goal over the conjunction hiatus was to look for surface changes, a difficult thing to do on a rover mission. A Pancam mosaic was made on sol 239 for comparison with post-conjunction images, and, to repeat this at a microscopic scale, the MB ring was pressed into the soil twice on sol 240 (Figure 22), making a very smooth surface, which was then imaged by the MI for comparison with later images. This had unintended consequences because soil adhered to the MB ring which was then placed on one of two rover magnets for a long integration during conjunction. The possible contamination jeopardized future studies of the magnet. On sol 241 Spirit also imaged Pergamon Ridge, part of West Spur on the way to Husband Hill, as the first half of a long-baseline stereoscopic survey for route planning. The large rock Palenque and other targets were imaged from here on sol 243.

Observations after conjunction began on sol 256. The MI observed the contaminated magnet and the soil impressions from sol 240 on sol 258. Sols 259, 260 and 261 were planned together as the team transitioned to the pattern it would follow for much of the rest of the mission, planning for two sols each Monday, two sols on Wednesday and three on Friday, so they could transition back to a five-day work week. Over that weekend the soil impressions were analyzed and the rover moved backwards for more imaging. On sol 263 the second part of the long-baseline stereoscopic survey of Pergamon Ridge was obtained, and it became apparent that the ridge could not be reached without crossing very poorly illuminated areas. On sol 264 Spirit approached a layered rock called Tetl and analyzed it with the IDD instruments over the next few sols. Several driving software problems around here were tackled during a stop over sols 278 to 281, where the IDD instruments were placed on a soil target called Take a Break before driving resumed. The next target was the rock Uchben, where IDD measurements of a target called Koolik were followed by a RAT grind on sol 285 and more IDD work. Then an area of Uchben called Chiikbes was brushed on sol 290 and analyzed.

The last rock target in this area was Lutefisk, only 2 m west of Uchben. It was studied while planners sought a well-illuminated route to Machu Picchu, a ridge northeast of the rover. At Lutefisk several unusual rocks were examined on sols 297 and 298. Herring contained small lumps or nodules, Twins had larger nodules, Flatfish had no nodules and Roe had many. These targets were imaged with the MI, brushed and analyzed until sol 304. Some APXS data on Roe were lost because the door did not open properly on sol 303. They were recovered early on sol 304 and after that the doors were always left open, as was also the case for Opportunity. Sol 300 was MY 27, sol 240, or 6 November 2004.

An observation planned for a long time but put off for lack of power was finally undertaken on sol 304. The Pancam could not be pointed while its actuators were cold, so now the Pancam was pointed at its target on the previous sol and it was ready for imaging very early the next morning. More images of the same target were taken later that morning to look for differences attributable to overnight frost on the surface. No frost was observed by Spirit, though Opportunity saw frost on the rover body on its sol 257. Later that sol the rover began driving northeast along the ridge, moving from one north-facing spot to another (Figure 20B). The greatest danger was a fault that would interrupt a drive in a bad location between these "lily pads." The spectrometers were placed on the magnets again on sol 307 with the MB offset to avoid the soil contamination described earlier. Spirit was driven across the Spur to Machu Picchu, a dome-shaped hill from which it descended the northern side of the Spur on sol 313 (MY 27, sol 253).

The plan now adopted was for the rover to cross a broad "saddle" or apron of debris sloping down from West Spur and Husband Hill towards the plains. The saddle was referred to as Blue Valley on sol 334, taking its name from its color in the illumination maps used in route planning. Spirit would climb into the gap between Husband Hill and Clark Hill and then turn south to ascend Cumberland Ridge, which formed the west side of Tennessee Valley (Figure 19) and offered a north-facing route to the top of Husband Hill. A scuffed soil on the saddle was analyzed on sols 314 and 315.

Around the US Thanksgiving holiday weekend (sols 318–320) images were taken for a large panorama and of targets with Thanksgiving-related names (Corn, Plymouth, Pumpkin Pie). Mobility was compromised as

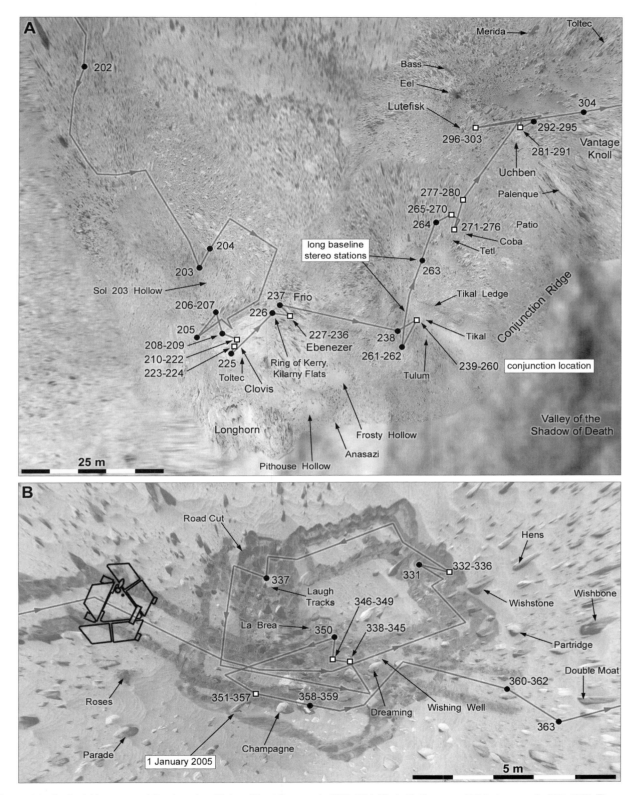

Figure 21. A: Activities around Conjunction Ridge, West Spur, sols 202–304. **B:** Activities near Wishstone, sols 331–360. The name Toltec is used twice in **A**. White squares are main science sites.

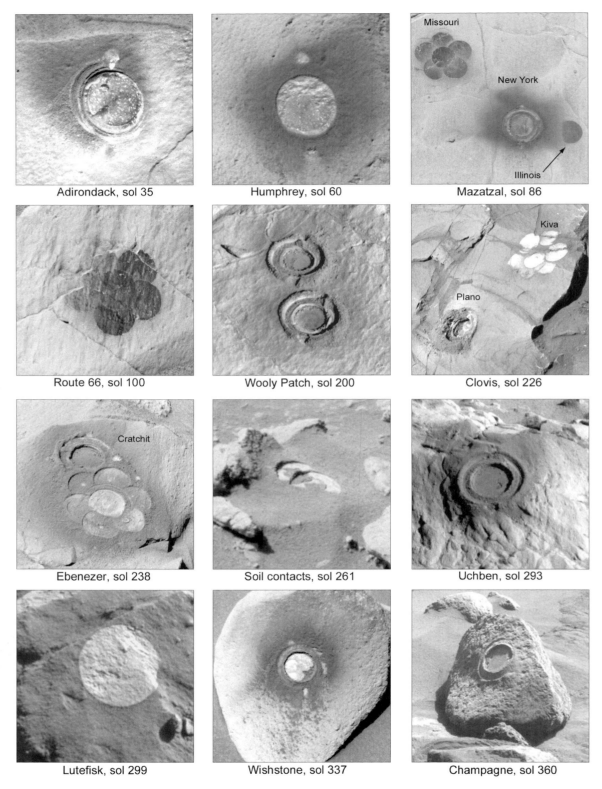

Figure 22. IDD brush and RAT marks on rocks and soil, Adirondack to Champagne, up to sol 358. The dates are of imaging, not necessarily IDD activity (the soil contacts imaged on sol 261 were made 20 sols earlier). Shading, bright or dark, varies with the camera filter used and has no significance here.

the wheels encountered loose soil after sol 332 and even more on sol 339 when a small stone became lodged in the hollow structure of the right rear wheel after studying a phosphorus-rich rock called Wishstone. The stone was eventually shaken loose on sol 345 and the rover made further observations of rocks and soil including a typical IDD and RAT campaign at Champagne (Figures 21B and 22). The rover was at Champagne on sol 354 when Earth saw in a new year on 1 January 2005 (MY 27, sol 294), and left it on sol 360.

Driving was difficult here but the rover started to make good progress on sol 364 and soon turned towards Cumberland Ridge, climbing into Cumberland Gap, a low area between Clark and Husband Hills and examining the Peace and Alligator outcrops on the way. Meanwhile the mission manager was monitoring a dust storm about 700 km southwest of Gusev crater, first noted in a sol 367 report, which had increased atmospheric dust levels and reduced power to the rover. Dust storms typically formed each day in the Ausonia and Eridania regions at about 130° E, and drifted eastwards into Electris (180° E), passing a few hundred kilometers south of Gusev each evening before dissipating at night. Dust levels rose, reducing power to the rover, but this was offset by the northerly tilt on Cumberland Ridge.

Peace (Figure 23A) was a highly eroded or pitted rock. A RAT grind on Peace at a target called Justice was cut short by reduced power on sol 374, but deepened with another grind on sol 377 for IDD analysis. The nearby rock Alligator (Figure 23A) was brushed in a "flower" pattern (Figure 39) on sol 385 and analyzed before moving on. Another software update began with a data upload on sol 395, but the new software was not used immediately since it was still being tested on Opportunity.

The plan around this time was to drive to the Rover Planner's Lookout (Figure 23B) to take a large panoramic portrait of the surroundings, but this was interrupted on sol 398 and again on the next sol when the rover wheels experienced excessive slip in the loose soil and dug up some unusual bright soil, which was given the name Paso Robles. This was the first of several deposits of salt-rich soil to be found by Spirit (Johnson *et al.*, 2007; Wang *et al.*, 2008), and several days were spent analyzing it, finding it to be rich in sulfur. Then the rover drove uphill to Larry's Lookout, an outcrop

about 20 m from Paso Robles which faced into Tennessee Valley about half-way up the hill (Figures 23B and 24A) and was now preferred to the previous lookout position.

Sol 400 was MY 27, sol 340, or 17 February 2005. A significant concern for long-range planning now was that Spirit needed to be back on level ground around sol 500 as the Sun moved south, so the long climb up Cumberland Ridge to the top of Husband Hill had to begin soon. Another three Phobos solar transits, or partial eclipses, were imaged from here on sols 403, 404 and 406, and a Deimos transit was seen on sol 420 (Bell *et al.*, 2005). Meanwhile another small local dust storm was noted about 300 km south of Gusev beginning on sol 413. Mars was now entering the southern spring season in which storms were more frequent in this region. On sol 414 the storm combined with other local dust storms to become a regional event, and over the next two days solar power dropped again.

Spirit reached Larry's Lookout on sol 407. The low power did not stop the busy schedule at Watchtower, the outcrop at Larry's Lookout, which began on sol 409 with a panoramic view of the dramatic scenery. On sol 414 a final approach drive to Watchtower was made, and on sol 415 the outcrop was imaged, brushed, imaged again and analyzed with both spectrometers (Figure 24B). On sol 416 a RAT grind into the rock was made for additional analyses, and this turned out to be the last RAT grind of the mission (Figure 39). The hard rocks, especially on the plains, had worn the grinding head so much that in future only brushing would be possible. The grinder was designed to make three holes and had cut 15.

Despite the need to make progress uphill, Spirit returned to Paso Robles for a more careful examination of that very unusual soil. Power began to climb again as the sky cleared, and on sol 421 a dust devil was seen for the first time by Spirit west of the Columbia Hills. They proved to be common in the spring season, with many seen at once in some searches (Greeley *et al.*, 2010). The wheel tracks made only 10 sols earlier had almost been erased in places by dust erosion or deposition. On sol 424 Spirit arrived at Paso Robles, now referred to as Paso Robles 2, and scuffed the soil to expose bright material again (Figure 24A). Over seven sols several areas of light and dark material were analyzed, and a lump called Ben's Clod was brushed and analyzed.

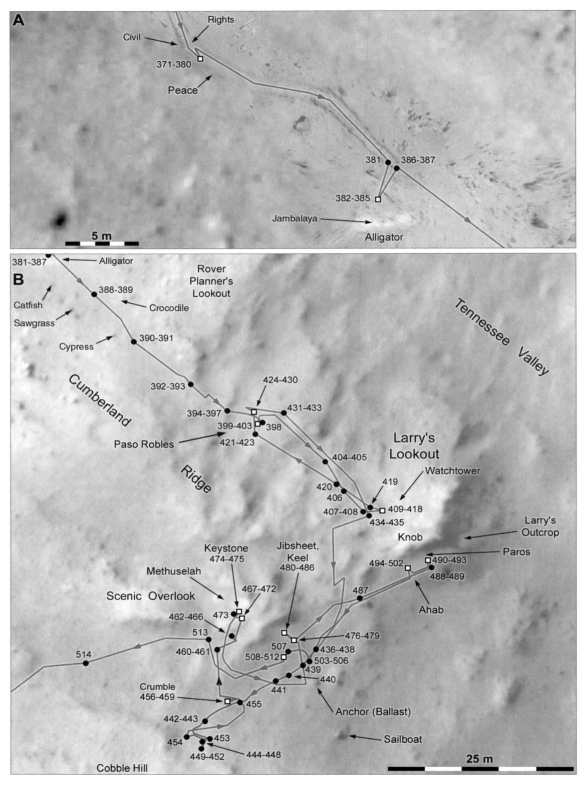

Figure 23. A: Activities around Peace and Alligator, sols 371–387. **B:** Activities at Larry's Lookout, sols 388–514. The HiRISE image in **B** is PSP_010097_1655.

Table 9. *Spirit Activities, West Spur to Larry's Lookout, Sols 201–409*

Sol	Activities
201–210	Tilt stops drives (201, 203), drive to Clovis (202–209), no science (206–207), image Clovis (210)
211–214	ISO Clovis (211), Plano and Cochiti (212–213), brush and ISO Plano (214)
215–219	ISO Plano target (215), RAT Plano (216) and ISO Plano (217–219), imaging
220–224	Take Cahokia panorama (219–220), ISO Clovis (221–223), no science (224)
225–229	Brush and ISO Clovis (225), drive (226–227), ISO soil (227), ISO Ebenezer (228), Cratchit 2 (229)
230–235	Brush and ISO Ebenezer (230), RAT Cratchit 2 (231) and ISO Cratchit 2 (231–235)
236–240	Brush (236) and imaging (237–238) of Ebenezer, drive to Tikal (238), study magnets (240)
241–244	Communication problem (241), imaging (242–243), communication tests (244)
245–257	Atmospheric (245–255) and magnet ISO studies (245–257) during conjunction
258–264	ISO magnets and soil (258) and other targets (259–260), no science (262), imaging
265–271	Failed drive (265), ISO Dwarf target (266–269), imaging (270), drive to Tetl (271)
272–279	ISO Tetl (272–276), no science (277–278), ISO disturbed soil and Take a Break (279)
280–281	ISO disturbed soil (280–281) and Coffee target (281), complete drive to Uchben (281)
282–289	Imaging (282), ISO Koolik (283–284, 286–289), RAT Koolik (285)
290–295	Brush Chiikbes target (290), ISO Chiikbes (291) and Koolik (291–292), imaging
296–303	Drive to Lutefisk (296), ISO Herring, brush (298) and ISO (298–303) Lutefisk
304–309	Images (304), drive to Machu Picchu (304–306), study magnets and image Cocomama (307–309)
310–313	Imaging, drive to Machu Picchu (310), off West Spur (311) and towards Husband Hill (312–313)
314–320	ISO disturbed soil (314–317), imaging (314–320), drive towards hill (317)
321–326	ISO calibration target (321) and magnet (322–324), imaging, drive to hill (325–326)
327–332	ISO solar array (327), drive towards Husband Hill and imaging (328–332)
333–336	Brush and ISO Wishstone (333), RAT Wishstone (334), ISO (334–336) Wishstone
337–340	Image Wishstone (337), drive to hill (337–339), try to dislodge rock in wheel (340)
341–345	ISO disturbed soil (341–342), imaging, try to dislodge rock in wheel (343–345)
346–350	Dislodge rock (346), imaging, ISO Dreaming (348–349), drive to La Brea (350)
351–353	Drive to Dick Clark target (351), ISO Bubbles (352) and Champagne (353)
354–358	Brush and ISO Champagne (354), RAT and ISO Champagne (355–358), drive east (358)
359–371	Imaging and continue drive towards Larry's Lookout (359–371)
372–376	ISO Peace (372), brush (373), RAT (374) and ISO (373–376) Peace
377–381	RAT Peace (377), ISO Peace (378–380), drive towards Larry's Lookout (381)
382–386	Approach Alligator (382), imaging, brush Alligator (385), ISO (385–386) Alligator
387–398	Drive to Larry's Lookout (387–394), update software (395–397), drive to Larry's Lookout (398)
399–409	ISO Paso Robles soil (399–403), drive towards Larry's Lookout and imaging (404–409)

On sol 431 the rover started to drive back up to Watchtower, passing it on sol 436 and moving onto a level area surrounded by rock outcrops which proved to be very interesting. At first, Spirit continued driving, but it was stopped by steep slopes and loose debris on sol 444 and turned back on sol 455 after several more attempts to make progress uphill (Figure 24C). Mars Odyssey, the main communication link between the rover and Earth, was out of operation for a few sols around sol 447, hindering progress further. Spirit would try climbing the hill again, but first the outcrops would be examined to establish their orientations and the overall structure of the hills.

Spirit returned to look at a bright soil called Crumble on sols 456–460 before moving to the outcrops (Figure 23B), first a rock called Keystone on the Methuselah outcrop (sol 474, Figure 24D), then the steep southern edge of Methuselah called Jibsheet (sol 480, Figure 24E) and finally Larry's Outcrop, the south side of Larry's Lookout (sol 490, Figure 24F). Rock targets on Keystone called Pittsburgh and Cleveland (cities in the Keystone State, Pennsylvania) were brushed and

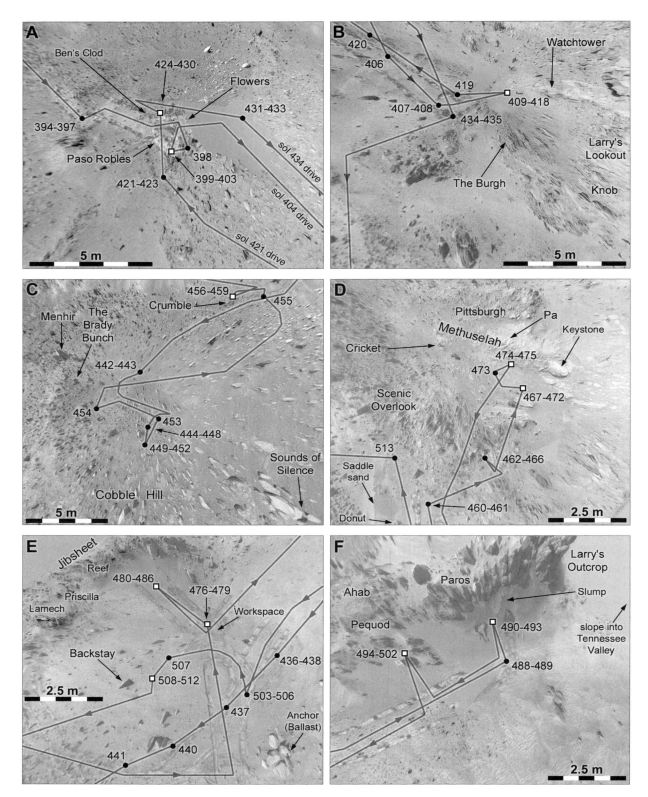

Figure 24. Six sites near Larry's Lookout. **A:** Paso Robles, sols 394–403 and 421–433. **B:** Watchtower, sols 406–420 and 434–435. **C:** Crumble, sols 442–459. **D:** Methuselah, sols 460–475 and 513. **E:** Jibsheet, sols 436–441, 476–486 and 503–512. **F:** Larry's Outcrop, sols 488–502.

analyzed, and then a soil called Bell and the rocks Reef and Davis were examined at Jibsheet. Sol 500 was MY 27, sol 440, or 30 May 2005.

At Larry's Outcrop images revealed a fresh soil slump which resembled the slump features seen near Viking Lander 1 (Figure 74 in Stooke, 2012), and targets called Paros, Ahab and Moby were analyzed. MI images of the rover's solar panels taken on sol 505 showed several coarse grains as well as the expected dust, perhaps raised as high as the rover deck by saltation, a bouncing motion of wind-blown grains. Spirit examined the rock Backstay on sol 509 and found it to be basalt, apparently an ejecta block thrown up from the plains by an impact. On the next sol the rover made a panoramic survey of Tennessee Valley.

Other observations were also made from this area. On sol 464 the evening twilight sky was imaged to look for faint clouds, but none were seen. Clouds were rarely seen at Gusev. Sunset was observed on sol 489 (19 May 2005), helping to reveal how the amount of dust in the atmosphere varied with altitude. A twilight image of Phobos was obtained on sol 497, and dust devils were frequently seen during this period, moving across the plains to the west from this high viewpoint (Figure 29E). The last dust devil of this season was observed on sol 691 from the south side of Husband Hill. Finally, as work concluded in this vicinity, Spirit returned on sol 513 to a "scenic overlook" near Methuselah for another panoramic view of the area (Figure 24D). This location had taken up over 100 sols of Spirit's time and the summit was still some distance ahead.

On sol 515 Spirit began climbing Husband Hill again, skirting the steep slope above Larry's Lookout on its western side to take a more oblique path to the summit (Figure 17). The route followed a terrace called Cart Path, seen earlier from below the hill, which offered superb views across the Gusev plains for regular dust devil surveys. Target names reflected the resumption of driving, including Kerouac, Beatnik and Wanderer. On sol 523 a decision had to be made to climb higher to another rocky ledge, Ledge 2, on the "high road," or to stay on a gentler slope, "low road." The lower path was followed, passing features now given Indian names including Taj Mahal, Santoor and Nilgiri. The rover passed a valley, Big Trough, and rock outcrops at Independence and Voltaire before reaching the summit region

on about sol 580. Route planning was still based on Mars Global Surveyor MOC data and it was often difficult to see exactly where the rover was in those images, so extra pictures were taken on sol 530 to help with localization.

Independence was encountered on the US holiday on 4 July 2005 (sol 528), and targets were given names reflecting US history (Franklin, Penn) and fireworks (Figure 26A). The RAT brush failed to operate several times, so all analyses were done without it, but at Penn the left front wheel was spun on the rock on sol 536 to try to scuff it, with little effect. The Voltaire outcrop was encountered around the date of France's national holiday (Bastille Day, 14 July) and was assigned target names derived from French history, including Descartes, Bourgeoisie and Louvre (Figure 26B). The brush was used again, successfully, on a target called Discourse on Descartes. As Spirit completed work on this outcrop, at a target called Gruyere on sol 571, there were concerns that the launch of Mars Reconnaissance Orbiter might limit communications with Spirit for a few days. Any spacecraft emergency would tie up the Deep Space Network and delay work with other missions.

By sol 576 rover images showed the summit area well enough to identify the highest region (Summit 1), and on sol 581 the horizon east of the hills in the vicinity of Thira crater became visible for the first time. From the top of the hill a full panorama was imaged over sols 583–586 and long-baseline stereoscopic images were taken of the Inner Basin on sols 591 and 593–595 to help plan future operations (Figure 26C). The moons Phobos and Deimos were imaged several times around this time. On sol 585 (25 August 2005) the two moons were imaged six times over 20 min, passing each other in the night sky near Sagittarius. The moons move in opposite directions across the sky, and observations like these helped refine knowledge of their orbits and could also show night-time clouds or hazes. Similar sequences of images were taken on sols 590 and 594, looking towards Aldebaran and the Pleiades, and more on sols 606 and 616. Meanwhile, a soil area called Lambert was analyzed on sols 584–587. A linear feature (Putative Dyke, or Irvine), marking a possible volcanic intrusion, was seen on sol 590, and Spirit returned to it on sol 600 (MY 27, sol 540, or 10 September 2005) for analysis.

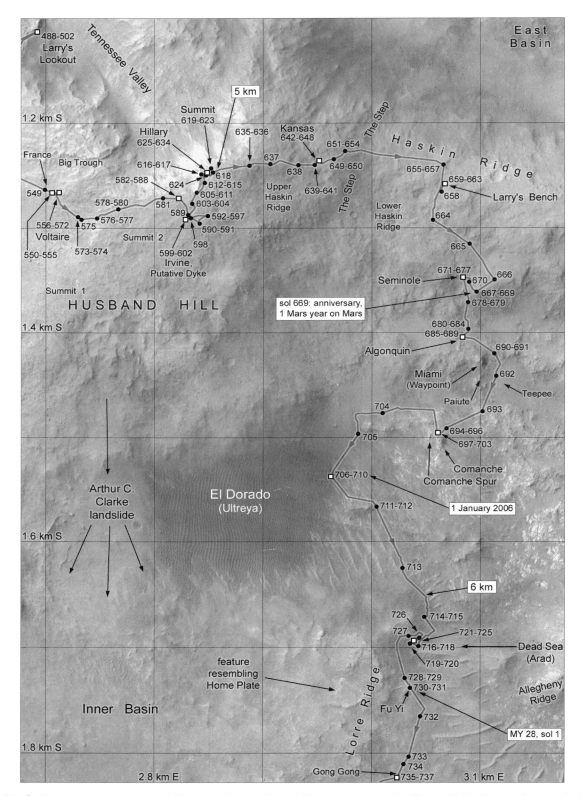

Figure 25. Spirit traverse map, section 4. This map follows Figure 17 and is continued in Figure 30. Additional details are shown in Figures 26, 28 and 31.

On sol 605 Spirit approached the top rim of Tennessee Valley, fringed by a long winding drift called Cliff-hanger, and imaged the view down the valley and out across the broad East Basin, another possible old impact crater. The view extended as far as the rim of Thira crater, 15 km to the east (Figure 27A). Occasionally, now and at other times during the mission when the atmosphere was exceptionally clear, the distant rim of Gusev crater up to 100 km away could also be seen through the dusty atmosphere (Figure 27B). After analysing Cliffhanger the rover was driven up onto a small rise known as Absolute (or True) Summit, the highest point of the summit area, on sol 619 (MY 27, sol 559) for a large imaging survey called the Everest panorama (Figures 26C and 28A). Feature names here referred to the climbing of Mt. Everest.

MB observed its calibration target here, but the rocks on top of the mound were too dusty for reliable analysis, so Spirit descended from True Summit on sol 624 and looped around to examine Hillary, a largely dust-free vertical face on the southern edge of the outcrop (Figure 28A). Some features here took names like Berlin and Nikolaikirche from German Unity Day (3 October 2005). As targets including Khumjung and Namche Bazaar on Hillary were being studied on sol 632, Spirit also made night-time observations of Orion and its nebula to examine camera sensitivity and to look for night-time clouds and meteors. Image exposures as long as 60 s were made shortly after midnight.

The Martian seasons were progressing and once more Spirit would have to move onto a north-facing slope to survive the coming winter, so it now began the long drive to the base of McCool Hill (Figure 25). The summit was left behind on sol 635 and the rover soon descended a slope called The Step to Haskin Ridge, which separated the East and Inner Basins of the hills and was named after team member Larry Haskin of Washington University, who had died on 24 March 2005. On sol 643 (25 October 2005) a night-time search for meteors was undertaken using nine 60 s Pancam exposures while Mars passed through a meteor stream in the orbit of Comet 2001/R1 LONEOS (named for the sky survey that discovered it). Stars in the southern constellations Octans and Pavonis were imaged as well as one possible meteor trail, but the camera detectors were often struck by cosmic rays, which also produced linear trails, making meteor identification uncertain.

A similar meteor search was made on sol 668 (18 November 2005), looking towards the south celestial pole as Mars passed through a meteor stream associated with Comet 1/P Halley. Again, one trail might have been a meteor, but Domokos *et al.* (2007) concluded that both of these trails were probably cosmic ray strikes. Several other meteor observations were attempted over this period. Another astronomical observation was made on sol 675 (27 November 2005) as Phobos was eclipsed. A series of images showed Phobos entering the shadow of Mars during the night. Sixteen Pancam images were taken 10 s apart as the little moon passed from full sunlight to full shadow. Phobos remained faintly visible due to sunlight scattered through the planet's atmosphere. Then on sol 682 (3 December 2005) Spirit imaged Phobos rising and Deimos setting in the west before dawn. Observations like these helped refine the orbits of the satellites.

Spirit was now driven down over a series of broad terraces into the Inner Basin, stopping briefly to examine an outcrop at each level. It brushed and analyzed a target called Kestrel on the rock Kansas on sol 646, and took long-baseline stereoscopic images looking into East Basin on sols 653 and 657 (Figure 25). Early on sol 654 Pancam looked for Mars Odyssey passing overhead, but did not see it. Thresher, a target on Larry's Bench, was brushed and analyzed on sol 660 (Figure 28B), and over the following night the Pancam looked for Mars Odyssey again, also without success. Spirit passed a full Mars year on the surface on sol 669 and reached Seminole two sols later where it brushed and analyzed two targets, Osceola and Abiaka (Figure 28C). Similar activities were conducted at the Iroquet target on Algonquin on sols 687–689 (Figure 28D).

A large rock called Miami or Waypoint had been a target during the zig-zag path down the hill, but now Comanche looked more interesting and became the final stop of the descent. An attempt was made to scuff the surface with the wheels in an area with an unusual polygonal pattern of cracks on sol 694, but it was too hard to be affected. Finally Comanche was approached on sol 697, and attention was directed to the adjacent smaller rock Comanche Spur, which was easier to work on (Figure 28E). A brushing on the target Horseback

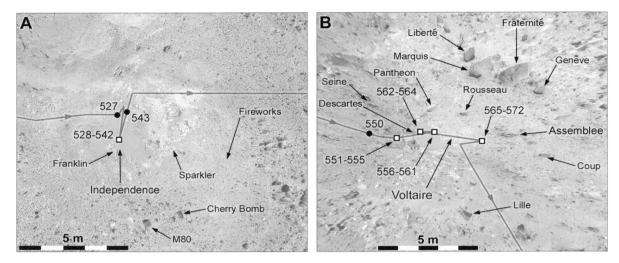

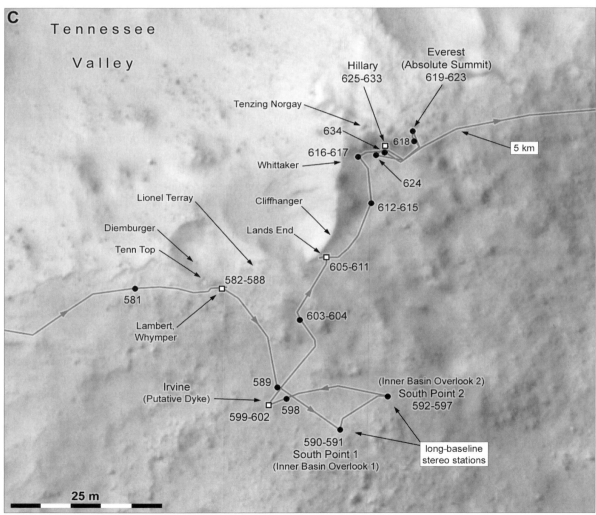

Figure 26. Activities during the climb to the summit of Husband Hill. **A:** Independence, sols 527–543. **B:** Voltaire, sols 550–572. **C:** Summit of Husband Hill, sols 581–634.

failed on sol 698 and was repeated the next sol, followed by IDD instrument studies, and a second target, Palomino, was brushed and analyzed on sols 700–702. Sol 700 was MY 27, sol 640, or 22 December 2005. Comanche was found to contain carbonates, the first such exposure ever found on Mars (Morris *et al.*, 2010), possibly indicating the fate of some of the planet's early, denser, atmosphere. This downward transect revealed details of the complex stratigraphy of the hills (McCoy e*t al.*, 2008; Cole *et al.*, 2012), but it could not be related very easily to the layers of rock seen earlier on the uphill transect.

A final stop before leaving the hills was made on sol 706 on the edge of El Dorado, a prominent patch of dark basaltic sand dunes on the lower flank of Husband Hill (Figure 31A). El Dorado was often named Ultreya in Internet discussions of this mission, a name introduced by Portuguese rover enthusiast Rui Borges, who was also active in promoting naming schemes for features at the Opportunity landing site. The MER missions broke new ground in public outreach, releasing all images as they were received on Earth to encourage the involvement of a wider community. Though some discussions veered into predictably unscientific areas, much of the public involvement was constructive, especially in the area of image processing and panorama construction, and even helped to debunk some of the wilder speculation. The wheels scuffed the dunes at El Dorado on sol 706 to help decide how to dig a trench in them, but the scuff was deemed good enough for analysis. The rover was still here on sol 709 when Earth celebrated a new year (1 January 2006, or MY 27, sol 649 on Mars).

Spirit now began a rapid drive from the dune field across the Inner Basin towards McCool Hill, its survival already somewhat in doubt as the Sun moved further north. The greatest danger to either rover was that inadequate solar power would prevent the charging of batteries needed for overnight heating of critical components. Any north-facing slope would suffice to increase solar power generation, but the rover team hoped to reach McCool Hill where a broad region of suitable slopes and potential science targets would assure a productive winter season.

On sol 714 Spirit approached Lorre Ridge (Figure 25), an irregular group of hills scattered with blocks of vesicular basalt, a distinctive form of lava containing many holes created by gas bubbles. While skirting the eastern side of this ridge, trying to avoid large drifts in the plains east of the ridge, its wheels dug into another patch of bright salty soil on sol 719 (Figure 31B). This deposit, Arad, contained large quantities of ferric sulfate. Spirit backed out of the deposit on sol 721, studied its composition over sols 723–725 and then found a route around the patch of loose soil, slightly higher on the flank of Lorre Ridge. Sol 730 was the start of a new Mars year (MY 28, sol 1), and the rover began it by studying several distinctive vesicular basalt blocks, including Fu Yi on sol 732 and Gong Gong on sol 736, typical of many of the rocks scattered over the ridge. The Chinese names reflected the beginning of the Chinese Year of the Dog on 29 January 2006 (sol 736).

Steering problems on the left front and right rear wheels reappeared now, similar to events on sols 265 and 277 at West Spur. Despite this, the plan now was to drive quickly to Home Plate, the polygonal bright feature identified earlier (Figure 19) and considered by Cabrol *et al.* (2003) to be a possible evaporite deposit. After a rapid characterization of Home Plate, Spirit would proceed to McCool Hill's northern flank to spend the winter among enticing outcrops such as Korolev and Faget, named after prominent rocket and spacecraft designers of the 1960s. Rover planners were always looking ahead to assure suitable illumination for solar power generation, and the current feeling was that any work on Home Plate would have to be finished by sol 760 to allow time to reach McCool Hill, giving very few sols to explore a potentially very interesting feature.

The first close views of Home Plate's northwestern corner on sol 744 (MY 28, sol 15) showed it to be a new type of geological structure consisting of finely layered rocks dipping inwards (Figure 29F). The lowest layers, called the Barnhill unit, were coarser, and in one place a rock had fallen into the layers while they were soft, deforming them to form a "bomb sag" similar to features found in terrestrial volcanic environments (Squyres *et al.*, 2007). This probably required that the layered deposits were damp at the time. The upper layer, the Rogan unit, was finer, with cross-bedding suggesting deposition by wind or as a pyroclastic deposit, a volcanic ash eruption. Precious time was taken to examine rocks at the foot of the outcrop (Gibson, Posey, Barnhill) until sol 754 and on the top (James "Cool Papa" Bell) on sol

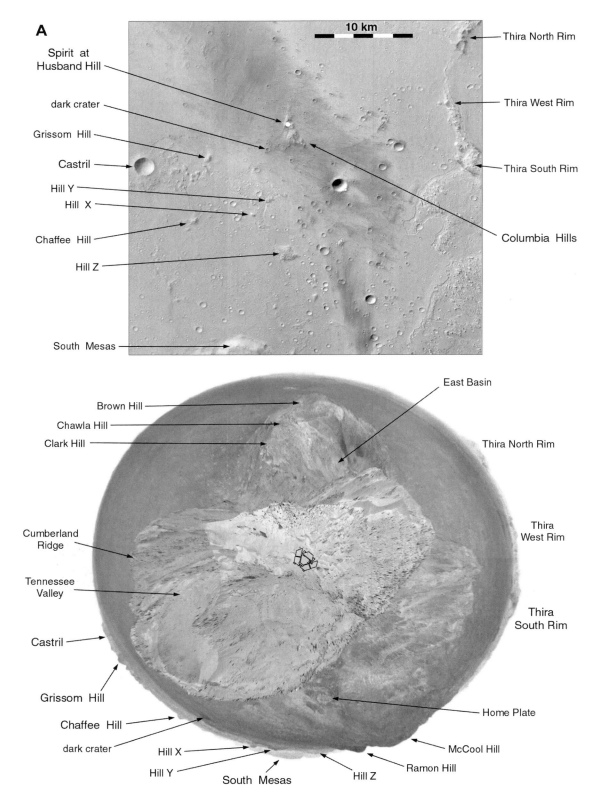

Figure 27 (both pages). Features visible from Husband Hill. **A:** Mosaic of MRO Context Camera images (top) and a reprojected panorama from sols 620–622 taken at the summit of Husband Hill (bottom), with exaggerated horizon relief. The eastern horizon is more visible here than in Figure 14, particularly the western rim of Thira. **B:** THEMIS infrared mosaic with inverted shading (top) and a version of the sol 620–622 panorama processed to show the distant rim of Gusev with 5× vertical exaggeration and extreme contrast enhancement (bottom). Features around the rim are identified by letters.

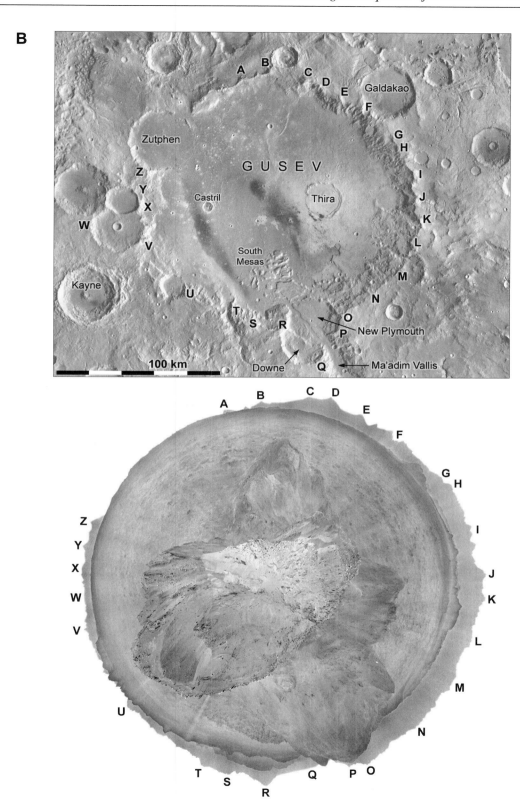

Figure 27 (continued)

Table 10. *Spirit Activities, Larry's Lookout to Husband Hill Summit, Sols 410–600*

Sol	Activities
410–415	Imaging at Larry's Lookout (410–413), move to Watchtower (414), brush and ISO outcrop (415)
416–423	RAT (416) and ISO (416–418) Watchtower, drive back to Paso Robles 2 (419–423)
424–428	Move to and scuff Paso Robles 2 (424), imaging, ISO Paso Robles 2, Ben's Clod (426–428)
429–431	Brush (429) and ISO Paso Robles 2, Ben's Clod (429–431), study solar panel, soil, drive on (431)
432–453	Imaging and drive uphill (432–449), problems prevent science (450–452), imaging (453)
454–460	Slip on slope during drive (454–455), scuff (456) and ISO (457–460) Crumble target
461–468	Drive to Methuselah and imaging (461–466), image Abigail (467–468) and Priscilla (468) targets
469–473	ISO Keystone (469), brush (470) and repeat ISO (470–472) on Keystone, imaging
474–476	Image Pittsburgh (474), brush and ISO Pittsburgh (475), image and drive on (476)
477–480	ISO Liberty Bell soil target (477–480), imaging, image Keel target on Jibsheet (480)
481–487	ISO Reef target (481–483), brush (484) and ISO (484–487) on Davis target
488–493	Drive to Larry's Outcrop (488), imaging, ISO Paros target (491–493)
494–495	ISO Slump soil feature (494), drive to Outcrop_West feature (494–495), imaging
496–501	Brush (496) and ISO (496–498) on Ahab, ISO Moby Dick (499–501)
502–503	ISO Doubloon soil target (502), other soil targets (503), drive to Methuselah (503)
504–508	Imaging, ISO magnets (505–506), drive to rock Backstay (507–508)
509–514	ISO Backstay (509–512), drive to and imaging of Methuselah (513–514)
515–522	Drive to Cart Path (515–516), aborted drive (517), imaging and drive towards summit (518–522)
523–531	Drive to Independence, imaging (523–528), brush (529) and ISO (529–531) Franklin
532–533	ISO Jefferson target (532), Ladyfinger and Pyrotechnic targets (533)
534–535	ISO Rocket and Missile targets (534), Flare and Silver Salute targets (535)
536–543	Scuff Penn target on Independence (536), imaging, ISO scuff (538–543)
544–551	Drive towards summit of Husband Hill and imaging (544–550), move to Descartes (551)
552–556	Brush (553) and ISO (552, 553–556) Discourse target, move to Bourgeoisies (556)
557–559	ISO Bourgeoisies, Gallant, Gentil (557) and Chic (557–559) targets
560–563	ISO Gentil_Matrice (560–562), move to Hausmann (562), ISO Hausmann (563)
564–566	Imaging (564), drive to rock Assemblee (565), ISO Gruyere target (566)
567–572	ISO Assemblee (567–568), Gruyere (569–570), Gruyere_APXS (571), imaging
573–581	Drive towards summit of Husband Hill and imaging (573–580), arrive on summit (581)
582–584	Panorama of Husband Hill summit (582–584), ISO Couzy target on Lambert (584)
585–588	Imaging and ISO Whymper target (585–588)
589–592	Drive to Inner Basin Overlook #1 (589–590), and Overlook #2 (591–592), stereo imaging
593–600	Imaging and studies of magnets (593–597), drive to (598–599) and ISO (600) Irvine

763 (Figure 31C). Feature names around here were taken from the African-American baseball leagues of the 1930s. The composition and structure suggested that Home Plate was an eroded volcanic deposit, possibly draped over an actual vent.

On sol 764 Spirit left this site for the final sprint to McCool Hill, already behind its very tight schedule. It was driven clockwise around the Home Plate plateau, crossing its northeastern corner on sols 768–773, and then turning south through a narrow valley called The Dugout, between the plateau and Mitcheltree Ridge (named after JPL engineer Bob Mitcheltree, who worked on the entry and landing system for the rovers, and had died in an accident on 6 January 2006). An odd rock at the northeastern corner of Home Plate, called Al "Fuzzy" Smith (Figure 31D), was analyzed on sols 769–771 and found to be rich in silica. On sol 775 images of potential targets on McCool Hill were taken, including layered rocks near Oberth. Long-baseline stereoscopic images were taken of this area on sols 776 and 777, separated by an 8 m drive. The winter target site was just above Oberth. Sol 778 was devoted just to atmospheric observations while attention focused on orbit insertion for Mars Reconnaissance Orbiter.

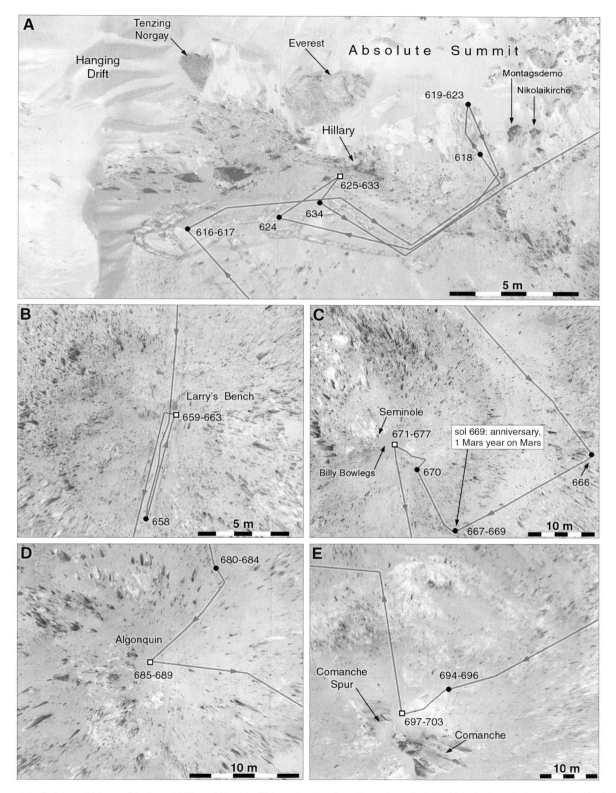

Figure 28. Spirit activities at Husband Hill and Haskin Ridge. **A:** Absolute Summit, sols 616–634. **B:** Larry's Bench, Haskin Ridge, sols 658–663. **C:** Seminole, sols 666–677. **D:** Algonquin, sols 680–689. **E:** Comanche, sols 694–703.

Figure 29 (both pages). Panoramic images from Spirit, from the Columbia Hills to Home Plate. **A:** Lookout panorama, sol 410. **B:** Everest panorama, sol 610. **C:** East half of the Seminole panorama, sol 675. **D:** South rim of Gusev seen from Clovis on sol 210. **E:** Dust devil seen from Larry's Outcrop, sol 496. **F:** Home Plate, sol 751.

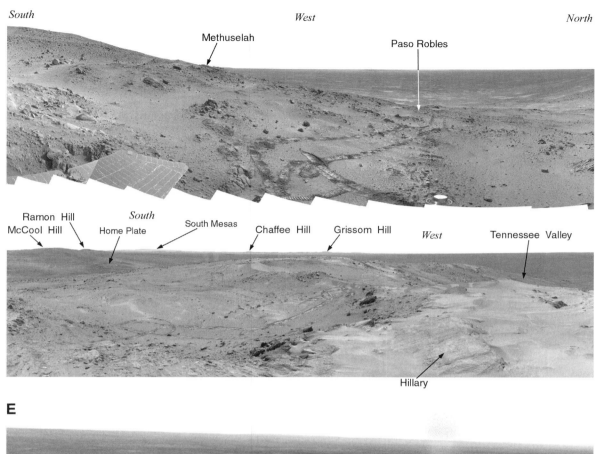

E

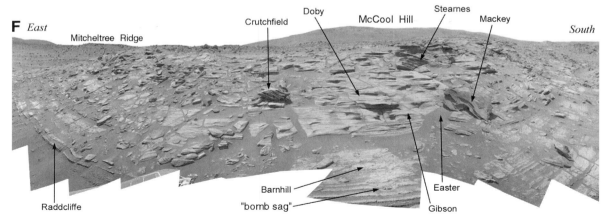

Figure 29 (continued)

Power was very limited but there was still time to reach the safety of McCool Hill.

As Spirit turned east again to cross the plains between Mitcheltree Ridge and McCool Hill, its right front wheel's drive actuator failed on sol 779, locking the wheel so that it had to be dragged behind the rover for the remainder of the mission. Consequently, only backward driving was feasible after this date, except for small maneuvers. On sol 782 bright soil turned up in the tracks again, and on sol 784, because of or exacerbated by the wheel failure, the rover again became stuck in deep, soft, salt-rich regolith at a site named Tyrone (Figure 31E). Irish names were being used now in honor of St Patrick's Day. Power levels were dropping as southern winter approached and Spirit now seemed to be blocked from reaching its preferred winter haven on McCool Hill, but reasonable slopes near Korolev or Faget might still be accessible.

Low on power and delayed further by days devoted only to recharging batteries and by a short safe-mode event on Mars Odyssey, the rover struggled out of the trap on sol 799 onto firmer ground near its sol 782 position. Sol 800 was MY 28, sol 71, or 3 April 2006. Two possible "havens" beckoned, one called Leger Haven just to the south and one a little higher and further west on the northern flank of Crossfield, a possible buried version of Home Plate (Figure 33). Spirit managed to reach the latter area on sol 805 (MY 28, sol 76). This was Low Ridge, named after George Low, the Apollo manager and NASA Deputy Administrator, and the site was referred to as Winter Haven, or Winter Haven 2 if West Spur was counted as the first Winter Haven. Later, it became known as Low Ridge Haven, and the name Winter Haven was transferred to the third wintering site on the north side of Home Plate (Figures 30A and 37).

Spirit remained at Low Ridge Haven with an 11° tilt to the north for 204 sols, passing the southern winter solstice on sol 922. During the winter the rover made many observations, including compositional analyses of several targets within reach of the IDD, change detection, including variable wind streaks on the dark dunes in El Dorado, which was visible above Home Plate, and the acquisition of the largest multispectral panorama of the MER program up to that time, the McMurdo panorama. Work at the haven began with stereoscopic mapping of the IDD workspace on sol 807 and analysis of a target called Halley, a light-toned clod or platy patch near the left front wheel, on sols 809–811 (Figure 32A). Names here were taken from Antarctic explorers and research sites and stations. Undisturbed soils called Mawson and Progress were studied next, including an experiment to explore changes with depth at Progress by brushing the soil a little at a time with analyses at each stage, between sols 822 and 870. Mini-TES and multispectral data around sol 840 suggested that two nearby rocks, Zhong Shan and Allan Hills, were iron meteorites, and a third rock, Vernadsky, might have been another.

The flight software was updated again during this period, with several partial uploads beginning on sol 872, though the new software was not activated until sol 965. Winter temperatures fell as low as about −97° C, so low that on sol 877 the battery heaters turned on for the first time in two years and power became even more limited, but still a few observations could be made almost every day. As the arm was being used, the rover sometimes moved slightly, and targeted remote sensing with the Pancam and Mini-TES might miss its target, so the rover attitude had to be re-measured from time to time, including now on sol 886. IDD work continued on targets called Halley, Halley Brunt Offset and Palmer, an aeolian monitoring target observed to look for wind-induced changes. The last attempt to brush Progress had resulted in a stall, and eventually a small rock was found to be jamming the mechanism. It was dislodged on sol 904 by running the RAT backwards. Sol 900 was MY 28, sol 171, or 15 July 2006. The McMurdo panorama was completed on sol 934 (Figures 33 and 35), and the power now began to increase again after reaching the lowest levels yet experienced. The MER rovers proved more resilient to low power levels than expected, a fact that helped account for their longevity.

Imaging during this period included super-resolution sequences of many targets of interest. Multiple images would be taken separated by slight camera movements, and combined in a way that increased the effective resolution by as much as a factor of 2. Many individual rocks and areas such as the south end of Mitcheltree Ridge were among these targets. The MB spectrometer was used to analyze dust on the filter magnet, eventually accumulating 100 hours of measurements. When the MB was facing downwards, the APXS was pointing

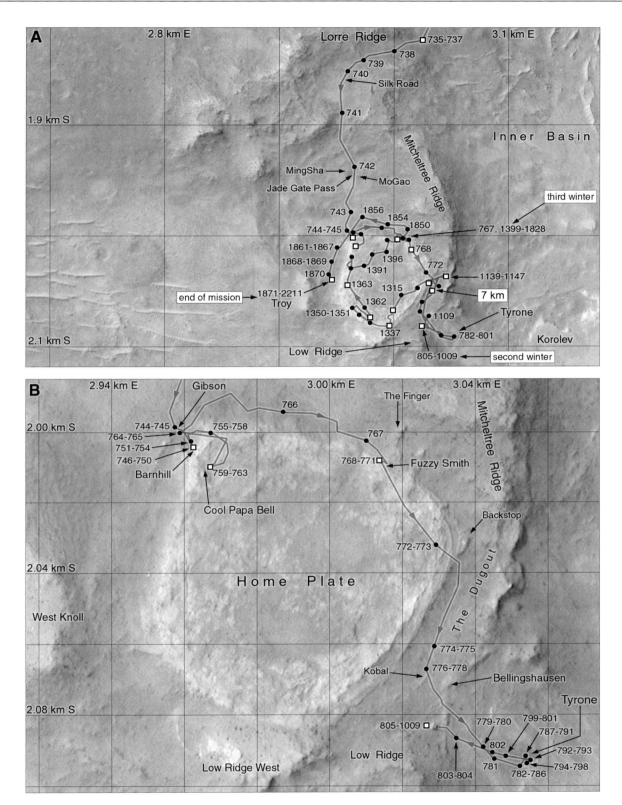

Figure 30. A: Spirit traverse map, section 5. This map follows Figure 25. Additional details are shown in Figures 30B, 31, 32, 34, 36, 37, 40 and 41. **B:** Home Plate activities up to sol 1009. The object immediately to the right of the sol 805 location is Spirit itself seen on sol 1026 in HiRISE image PSP_001513_1655 (22 November 2006 or MY 28, sol 297).

Table 11. *Spirit Activities, Husband Hill to Low Ridge Haven, Sols 601–805*

Sol	Activities
601–606	ISO Irvine target (601–602), drive to drift (603–605), scuff drift (605), imaging (606)
607–624	ISO Cliffhanger scuff mark (607–611), drive to true summit area (612–624), imaging
625–634	Move to Hillary target (625–626), ISO Hillary (628–634), Everest panorama
635–645	Drive to Haskin Upper Ridge, imaging (635–641), drive to Kansas (642), no science (644–645)
646–654	Brush (646) and ISO (646–648) Kestrel, drive east on Haskin Ridge and imaging (649–654)
655–658	Drive down to Haskin Lower Ridge (655), imaging, begin drive down slope (658)
659–663	Turn back to (659), brush (660) and ISO (660–663) Larry's Bench feature
664–671	Drive south down slope (664–670), move to Seminole (671), imaging
672–674	Brush (672) and ISO (672–674) Osceola target on Seminole feature
675–677	Brush (675) and ISO (675–677) Abiaka target on Seminole, imaging
678–686	Drive down slope (678, 680), imaging, no science (683), drive to Algonquin (685), imaging
687–696	ISO (687–689) and brush (688) Iroquet target, drive to Comanche (690–696)
697–699	Approach Comanche Spur (697), ISO (698–699) and brush (699) Horseback
700–703	Brush (700) and ISO (700–703) Palomino target on Comanche Spur
704–710	Drive to El Dorado dunes (704–706), scuff (706) and ISO (707–710) El Dorado sand
711–722	Drive towards Home Plate feature, imaging (711–722), wheels slip in Arad bright soil (719)
723–732	ISO Dead Sea soil target (723–725), drive towards Home Plate, imaging (726–732)
733–735	Drive south along Lorre Ridge, imaging, brake problem (733) and tests (734–735)
736–737	ISO BuZhou and Pan_Gu features on vesicular basalt rocks (736), imaging (737)
738–746	Drive towards Home Plate feature (738–745), imaging, approach Barnhill feature (746)
747–752	ISO Barnhill (747–750), imaging, drive to Posey feature (751), imaging (752)
753–758	Brush (753) and ISO (753–754) Posey, drive up onto Home Plate (755), imaging
759–762	Approach Cool_Papa_Bell (759), imaging, brush (761) and ISO (761–762) Stars
763–764	Brush and ISO Crawfords target (763), drive down off Home Plate (764)
765–771	Imaging (765), drive around Home Plate (766–768), ISO Fuzzy Smith (769–771)
772–778	Drive south towards McCool Hill and imaging (772–778), APXS temperature observations (775)
779–782	Drive south, right front wheel failure (779), tests (780–781), begin backward driving (782)
783–786	Imaging (783), wheels slip in bright soil at Tyrone (784), imaging (785–786)
787–801	Move to try to bypass Tyrone (787–801), imaging of disturbed soil (790–791, 798)
802–804	Abandon McCool Hill goal, drive back towards Low Ridge Haven (802–803), imaging (804)
805	Drive to Low Ridge Haven, reach safe overwintering location (MY 28, sol 76)

upwards, and would often be used to measure the amount of argon in the atmosphere above it. This gave a useful estimate of atmospheric pressure from a rover with no meteorology instruments. Solar conjunction occurred at the end of this period at Low Ridge Haven, as the Sun interfered with radio communication with Earth between sols 991 and 1009. Imaging and MB magnet data were gathered during conjunction for later transmission. Sol 1000 was MY 29, sol 271, or 26 October 2006.

On sol 1010 (MY 28, sol 281) Spirit began moving again with a 70 cm drive to examine unusual soil patches in the tracks, white material called Berkner Island and a blue-gray patch called Bear Island. Allan Hills, the possible meteorite, would be too hard to reach with only five wheels, but King George Island, part of a layered band of rock running through Low Ridge, was accessible and was approached on sol 1022. The wheel was supposed to drive over a thin projecting rock ledge and break it to provide a fresh face, but it slipped and did not reach the rock. The IDD could reach King George Island, however, so it was used over sols 1031–1036 to brush and analyze the rock. While at this location on sol 1026, Spirit was imaged for the first time by HiRISE, the very high-resolution camera on MRO (Figure 30B).

On sol 1039 the rover was driven to Esperanza, one of many dark vesicular rocks on the ridge (Figure 32A), despite initial concerns that it was large enough for the rover's solar panels to strike the rock as it maneuvered.

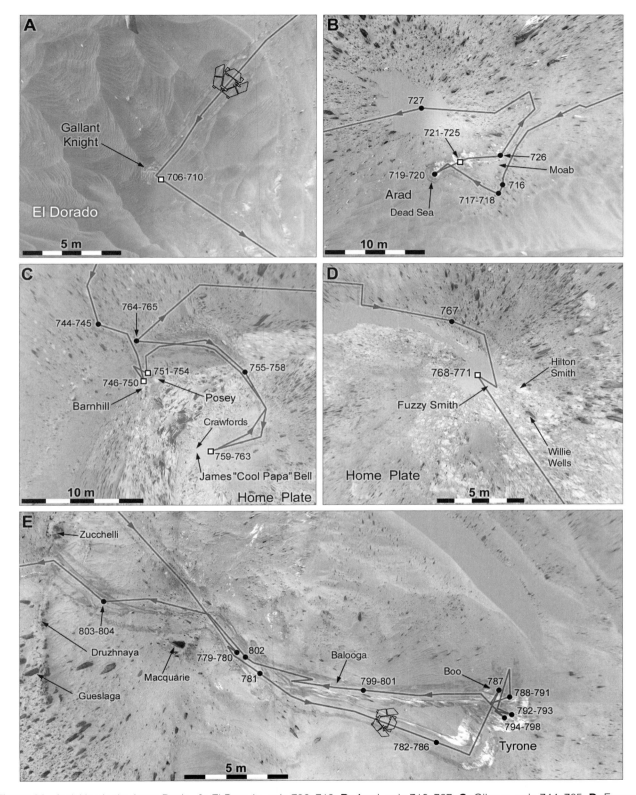

Figure 31. Activities in the Inner Basin. **A:** El Dorado, sols 706–710. **B:** Arad, sols 716–727. **C:** Gibson, sols 744–765. **D:** Fuzzy Smith, 767–771. **E:** Tyrone, sols 779–804. Spirit is driving forwards in **A** and backwards in **E**.

Table 12. *Spirit Activities at Low Ridge Haven, Sols 806–1100*

Sol	Activities
806–811	Imaging (806–808), ISO Halley target on Enderbyland feature (809–811)
812–818	ISO soil (812) and Mawson target (814–818), begin McMurdo panorama (814)
819–829	Imaging (819–821), ISO Progress (822–829), image Tyrone (818, 826)
830–838	Brush (830) and ISO Progress1 (830–831), ISO Halley_Offset (832–838)
839–846	ISO Progress1 soil (839–841), ISO test (842), brush Progress2 (845), imaging
847–856	ISO Progress2 (847, 852–854), no science (850–851), imaging
857–860	Failed brush of Progress3 target (857), ISO Progress3 (858), camera tests (860)
861–867	ISO Halley_Brunt soil (861), imaging, camera test (865), brush Progress3 (866)
868–881	Camera tests (868, 876), imaging, ISO Progress3 (870) and Halley_Brunt (875, 880)
882–890	Camera tests (882, 884), ISO Halley_Brunt (883–885, 889), Mini-TES test (890)
891–897	Imaging, software upload (892), RAT tests (893), ISO Halley_Brunt_Offset1 (897)
898–906	Imaging, ISO Halley_Brunt_Offset1 (901, 904), free jammed RAT (904)
907–912	Software update (907), ISO Palmer ripple (908), clean RAT (910), imaging
913–926	ISO Palmer2 target (913), imaging, camera test (920), image Tyrone (922)
927–931	ISO Halley_Brunt_Offset2 target (927, 929), imaging, finish McMurdo panorama (931)
932–941	Imaging, begin McMurdo rover deck panorama (934), ISO Palmer (937)
942–954	ISO Halley_Brunt_Offset3 target (942, 944, 948)), no science (945–946), imaging
955–964	ISO magnets (955–963), image Tyrone (959), end deck part of McMurdo panorama (962)
965–990	Boot and test new software (965–967), imaging, ISO magnets (968–990)
991–1006	Solar conjunction period (991–1005), ISO magnets (991–1006), imaging
1007–1009	ISO rock clasts (1007–1008), Palmer and Mawson targets (1008), imaging
1010–1016	Move to bright soil tracks (1010), imaging, ISO Berkner_Island_1 soil (1013–1016)
1017–1021	ISO Bear_Island soil (1017–1021), Pancam tests (1021), imaging
1022–1030	Move to layered outcrop (1022), imaging, ISO King_George_Island (1028–1030)
1031–1033	Brush (1031) and ISO (1031–1032) King_George_Island target, imaging
1034	ISO Clarence, Deception and King_George_Island targets
1035–1036	ISO King_George_Island_Offset target (1035), imaging, image Tyrone (1036)
1037–1052	Drive to Esperanza feature (1037–1041), imaging, final move to Esperanza (1051)
1053–1061	ISO Palma target on Esperanza (1053–1061)), imaging, image Tyrone (1062)
1062	Drive to a more favorable energy location, and imaging (1062)
1063–1069	Imaging (1063–1068), move to Montalva target on Troll feature and imaging (1069)
1070–1077	Brush (1070) and ISO (1070–1077) Montalva, imaging
1078–1084	Imaging (1078), ISO Montalva_Offset (1079–1080) and Riquelme3 (1080–1084)
1085	Brush Montalva Daisy targets, ISO Contact and Londonderry, imaging
1086–1094	Imaging (1086), move back (1087), imaging, drive back to Tyrone (1089, 1092, 1094)
1095–1100	Image Tyrone Vista (1095, 1097), move (1096), ISO Mt Darwin (1097–1100)

Another Mars Odyssey safe mode delayed analysis for a week, and when activities resumed a target called Palma on Esperanza was scrutinized with the IDD instruments on sols 1053–1056. The rock was too rough to brush, and orbital observations now showed a large dust storm passing south of Gusev, reducing power to the rover and cutting its work on Esperanza short, so Spirit briefly analyzed a nearby rock called Guselaga before moving on sol 1062 to Troll, a layered outcrop nearby with a northward tilt of about 7°, which improved power generation. For a while the rover would be kept within one short drive of a "safe haven" with a northern tilt, at least until skies cleared or the dust storm season ended.

Sol 1064 was the start of a new year, 1 January 2007, on Earth and was MY 28, sol 335, on Mars. At Troll a light layer called Montalva was analyzed first, over sols 1070–1085, and a darker upper layer called Riquelme was studied on sol 1080 (Figure 32A). A brush mosaic

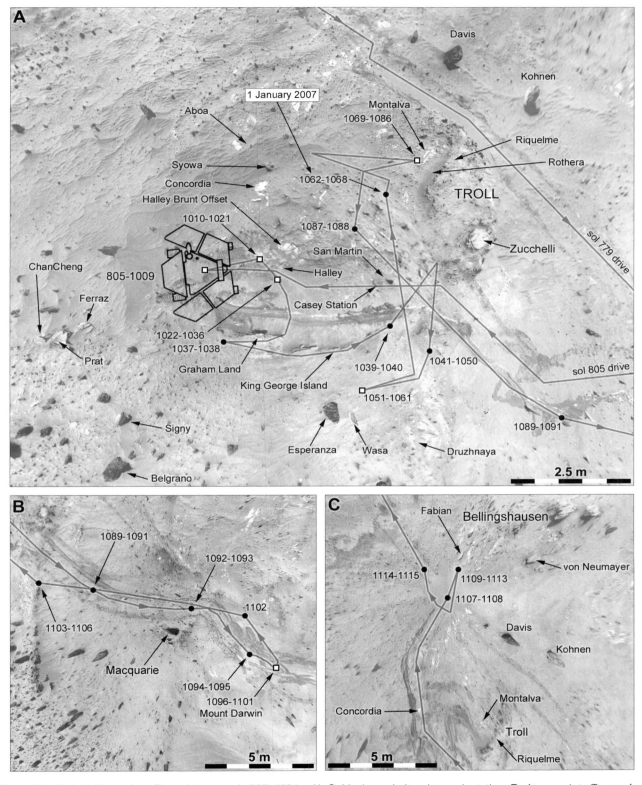

Figure 32. A: Activities at Low Ridge Haven, sols 805–1091, with Spirit shown in its winter orientation. **B:** Approach to Tyrone for further imaging, sols 1089–1106. **C:** Bellingshausen, 1107–1115.

called Montalva Daisy was attempted on sol 1085, though only one of three brushings was effective, to prepare for Mini-TES observations on sol 1088. Mini-TES also measured the surface temperature during Phobos transits (partial solar eclipses) on sols 1074 and 1076–1078 to see if the temperature dropped noticeably. Phobos itself was imaged early on sol 1083, and dust devil movies were often made in the small sector of the Gusev plains visible across Home Plate from this period onwards to watch for the first events of the season.

Now the rover was driven carefully towards Tyrone, approaching closely enough for remote sensing observations on sol 1096 (Figure 32B). Some bright material had been carried out of Tyrone in the wheels and dropped in the tracks, forming a soil target called Mt. Darwin, which was analyzed over sols 1097–1101. Sol 1100 was MY 28, sol 371, or 5 February 2007. The spring equinox occurred on sol 1103 and power was increasing, so now Spirit was directed back towards Home Plate and Mitcheltree Ridge to conduct a thorough examination of this complex area during the summer months. Later it would have to find another north-facing refuge for the following winter.

The ridges surrounding Home Plate contained thin ledges of layered rocks whose stratigraphic relationship to Home Plate was not clear, so these outcrops were examined at two more locations along Mitcheltree Ridge on sols 1109 and 1139. The first area was called Bellingshausen (Figure 32C), but there the ledge was out of reach so a broken slab called Fabian became the target. Some recent IDD observations including APXS at Mt. Darwin had been compromised by poor IDD placement, so the arm position was recalibrated at Fabian with a series of moves called a "Tai-Chi" sequence. This did not leave time for IDD work, so only Mini-TES and multispectral imaging data were collected here.

Spirit moved up into The Dugout, now called Eastern Valley (Figure 34A), after finishing work at Fabian, and imaged Home Plate on the way to find an "on-ramp" or good location to climb onto it later. Mars Reconnaissance Orbiter was in safe mode over sols 1135–1138, and intensive operations with it prevented some communications with Spirit. They shared a radio frequency because when MRO was designed the rover mission was expected to be over long before the orbiter arrived at Mars. Now Spirit could only be commanded when

MRO was hidden behind the planet. APXS argon measurements were being taken once per month around this time to complement weekly observations from Opportunity.

The rover was driven up to Torquas, a rock on the side of Mitcheltree Ridge, on sol 1141 for RAT brushing and analysis of a target called John Carter (Figures 34A and 36A). The names were taken from the Mars (Barsoom) stories by Edgar Rice Burroughs. On sol 1148 Spirit crossed the valley and mounted the edge of Home Plate, but returned to the valley floor two sols later to look for the lowest layer of sediments in this area. A target called Elizabeth Mahon, a nodular deposit rich in silica, was examined around sol 1155, and another rock layer called Madeline English around sol 1170 (Figure 34B). Names in this area commemorated historic female baseball players. On sol 1175 Spirit moved back slightly to look again at silica-rich material in a target called Slide, while the sky became dustier as dust storms swept through Hellas. Another observation made from here was a long-baseline stereoscopic survey of the Arthur C. Clarke landslide on Husband Hill.

Between sols 1086 and 1239 dust devils were again seen on the plains to the west (Greeley *et al.*, 2010). The southern spring and summer seasons were the best time to see these phenomena, corresponding to a six-month period around perihelion. The previous dust devil season had been from sol 421 to sol 691 during the trek across Husband Hill, but the lower vantage point during this Martian spring made viewing more difficult. Very few dust devils were seen by MER-B (Opportunity) at the less dusty Meridiani site. Rover planners now had to choose between driving north to the on-ramp used on sol 1148 or south to look for more silica in the valley, and they decided to go north while continuing to look for more silica deposits.

The drive north began on sol 1184, and it included a stop to examine bright soils dug up by the wheels on sol 1123, a feature called Gertrude Weise containing the targets Kenosha Comets and Lefty Ganote. They were also found to be silica rich. The wheel problem limited mobility but had the unexpected virtue of producing a long trench in the soil whenever it was not crossing solid rock outcrops. Previous deposits of distinctive soils had only been exposed by accident, but now a more thorough study of their extent and nature became feasible. Spirit

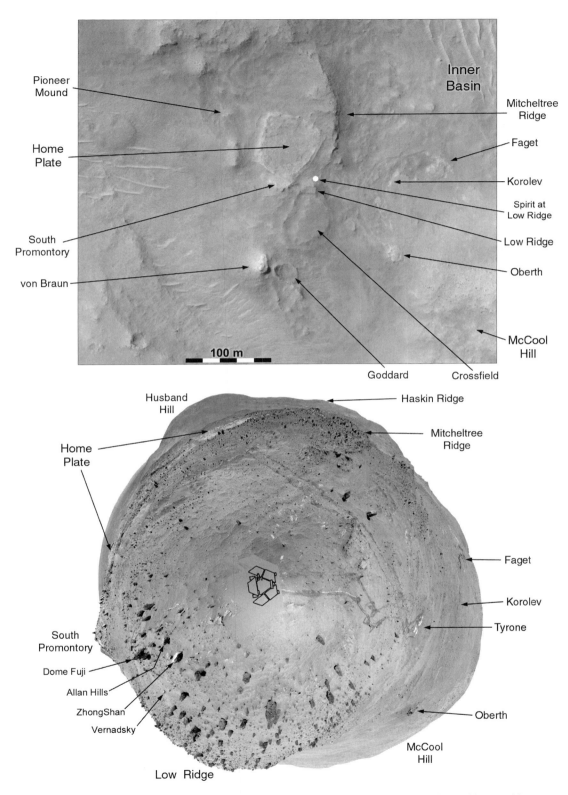

Figure 33. Part of HiRISE image PSP_010097_1655 (top) and a reprojected panorama at Low Ridge Haven with exaggerated horizon relief (bottom), showing locations of horizon features. Low Ridge West has now been renamed South Promontory.

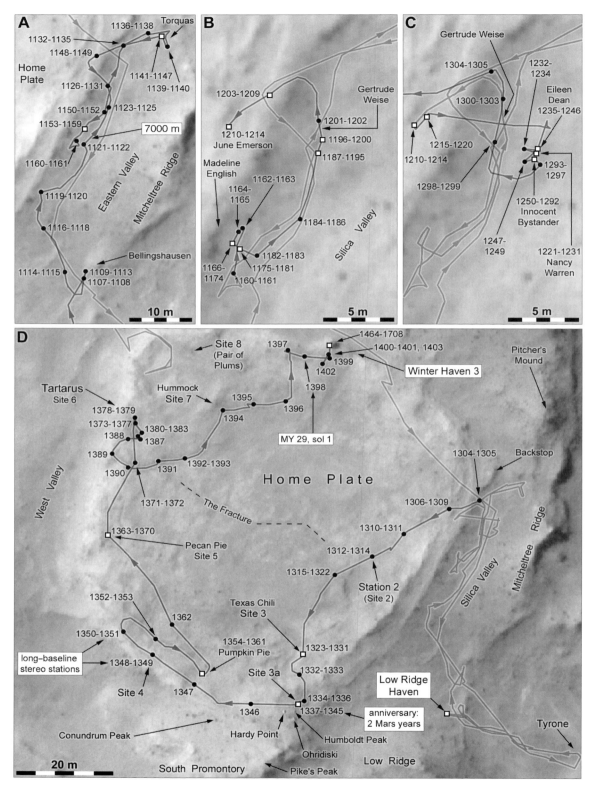

Figure 34. Activities in East Valley and on Home Plate. **A:** Sols 1107–1161. **B:** Sols 1160–1214. **C:** Sols 1210–1305. **D:** Sols 1304–1708. The object under the final location is Spirit itself, imaged on sol 1676 (21 September 2008 or MY 29, sol 279; HiRISE image PSP_010097_1655).

tried to expose more silica by scuffing the soil with its wheels on sol 1201, but was not successful here. Sol 1200 was MY 28, sol 471, or 19 May 2007.

By sol 1209 Spirit was standing on the rim of Home Plate observing rocks called Pesapallo and Superpesis (illustrated on Figure 36A), apparently corresponding to the lower and upper units Barnhill and Rogan seen around sol 750. Targets called June Emerson and Elizabeth Emery were examined here, but this investigation into the structure of Home Plate was soon interrupted. The silica was deemed so significant as evidence for past water in the soil that Spirit returned to the valley on sol 1221 to spend more time at the silica-rich deposits, for which the valley previously known as The Dugout and Eastern Valley had been renamed Silica Valley. In 2008 it reverted to the name East or Eastern Valley when it seemed likely that the rover would drive either east or west of Home Plate on its way to von Braun (Squyres *et al.*, 2008). A target called Nancy Warren was analyzed first, having been identified initially by Mini-TES (Figures 34C and 36A). On sol 1224 a gust of wind cleaned dust off the solar panels, increasing power for a short time, and on sol 1233 Spirit was driven over an outcrop called Virginia Bell to break or scuff its surface, exposing fresh material, as the surface was too rough to brush. It missed its target but broke a piece off a nearby rock, which was consequently named Innocent Bystander.

A dust devil was spotted to the southwest, towards Grissom Hill, on sol 1235 as the rover began studying a hydrated soil called Eileen Dean, but soon a widespread dust storm began affecting both Spirit and Opportunity. The last dust devil of the season was seen on sol 1239 as the storm grew. By sol 1243 Spirit was forced to suspend its observations for a sol but soon took them up again, finishing work on Eileen Dean on sol 1246 before moving the short distance to Innocent Bystander. Deep Space Network (DSN) activities associated with the Dawn spacecraft's launch on 27 September 2007 caused communication delays around sol 1250.

Opportunity's Mini-TES instrument was damaged by dust during the storm, and around sol 1253 plans were devised to use Spirit's MI to image its own Mini-TES to help diagnose the problem. The Opportunity instrument never worked again after this, and soon Spirit had its own problems with dust. The air became dustier on sol

1258 as a dust storm in Hesperia moved towards Gusev, and the MARCI instrument on MRO showed dust over much of the planet. A little work could still be done, including a dust devil movie on sol 1260 and a long MB session on Innocent Bystander from sol 1262 to 1284.

While MB pointed down, MI was facing upwards, and its lens became very dusty despite being fitted with a protective cover, so the Pancam imaged MI to examine its appearance on sol 1279. Winds had not only forced dust under the MI cover, but had also erased rover tracks made only 20 sols earlier and moved small dust ripples on the surface, something rarely observed before on Mars. The rover's survival was uncertain for a time, but it proved resilient again and recovered with no apparent loss of functionality.

On sol 1287 the MI was pointed downwards and the dust cover was repeatedly opened and closed to dislodge some of the dust. It was used again on sol 1291 to observe some of the small ripples that had been seen to move earlier. Finally, on sol 1298 Spirit moved again, this time to Gertrude Weise to repeat its former Mini-TES observations of the soil as a way of recalibrating the instrument with its additional dust contamination. The soil was itself too dusty so a fresh scuff was made by the wheels on sol 1300 (MY 28, sol 571, or 30 August 2007) and the calibration was performed on sol 1303.

Spirit had now spent much longer in the valley than originally anticipated, and in about 200 more sols it would need to reach another north-facing slope to be able to survive the following winter. At this time planners intended to explore Home Plate for several months before moving further south to the complex hill-and-ridge region around von Braun and Goddard, where north-facing slopes and interesting targets promised another productive winter season. Goddard might have been another volcanic feature, perhaps an unburied version of Home Plate, and the spectacular new Mars Reconnaissance Orbiter images showed several other features like these in the vicinity (Figure 19).

On sol 1306 the rover climbed onto Home Plate at the same place where it had done so on sols 1148 and 1203 (Figure 34C), and since previous observations had been made in the northeast and northwest corners, attention was now directed to the southern side (Figure 34D). After some imaging, Spirit conducted a driving test on sol 1310 to see how it would function with five wheels

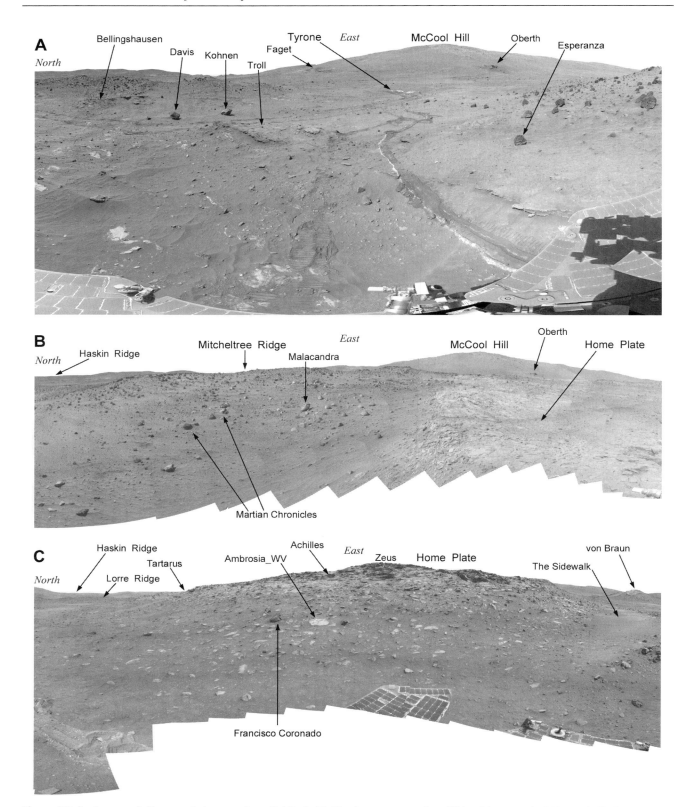

Figure 35 (both pages). Panoramic images from Spirit. **A:** McMurdo panorama, Low Ridge Haven, sols 805–1009. **B:** Bonestell panorama, Winter Haven 3, sols 1477–1696. **C:** Calypso panorama, Troy, sols 1906–1943.

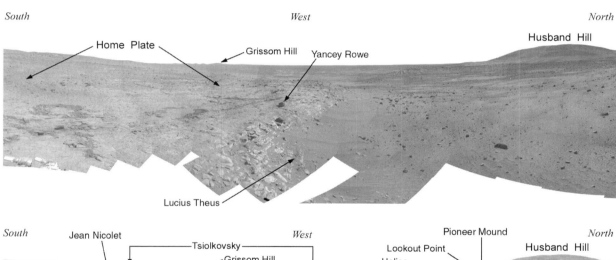

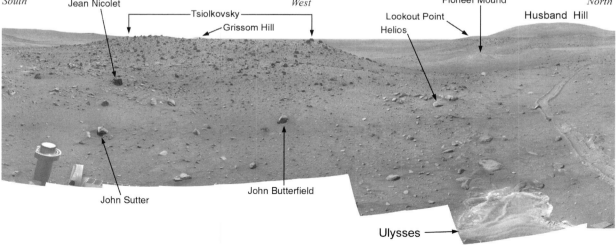

Figure 35 (continued)

Table 13. *Spirit Activities in Silica Valley, Sols 1101–1300*

Sol	Activities
1101–1103	ISO Puenta Arenas target (1101), imaging, drive back to Montalva (1102–1103)
1104–1109	No science (1104–1106), drive to Bellingshausen outcrop (1107), move to Fabien target (1109)
1110–1113	"Tai-Chi" arm calibration movements (sol 1110), imaging, Mini-TES on outcrop targets (1111–1113)
1114–1126	Drive towards Home Plate and imaging (1114–1126), ISO magnets (1125)
1127–1141	Imaging (1127–1140), drive to Mitcheltree Ridge (1132, 1136, 1139), move to Torquas (1141)
1142–1145	Imaging, brush and ISO Torquas2 (1143), ISO Torquas1 (1143–1145)
1146–1149	ISO John Carter target (1146), imaging, drive towards Home Plate (1148)
1150–1154	Drive to Madeline English outcrop (1150), imaging, drive to Elizabeth Mahon target (1153)
1155–1166	ISO Elizabeth Mahon (1155–1159), drive to Madeline English (1160–1166)
1167–1170	Imaging (1167), ISO Belles target (1168–1170), start Ballpark panorama (1170)
1171	ISO Beaches, BlueSox, Madeline English and Everett targets, imaging
1172–1175	ISO Everett (1172–1174), imaging, move to Slide (1175), end Ballpark panorama
1176–1181	ISO and brush Slide (1176–1178), ISO GoodQuestion (1179–1181)
1182–1186	Drive to Gertrude_Weise feature (1182), imaging, drive to White Soil (1184), imaging
1187–1197	Drive (1187), ISO Kenosha_Comets target (1189–1195), imaging, drive (1196)
1198–1204	ISO Lefty_Ganote target (1198–1199), imaging, move back (1201), imaging
1205–1209	ISO and brush Pesapallo target (1205–1208) and Superpesis (1209), imaging
1210–1214	Drive to June Emerson target (1210), ISO and brush June Emerson (1211–1214)
1215–1216	Drive to Elizabeth Emery target (1215), ISO and brush Elizabeth Emery (1216)
1217	ISO and brush Jane Stoll target and ISO Elizabeth Emery target
1218	ISO and brush Mildred Deegan and Betty Wagoner targets
1219	ISO and brush Betty Wagoner's Daughter target and ISO Elizabeth Emery target
1220–1225	Imaging, drive to Nancy Warren target (1221, 1223), ISO Nancy Warren (1225)
1226–1229	ISO and brush Nancy Warren (1226), ISO Nancy Warren (1227–1229)
1230–1231	ISO soil near Nancy Warren and solar arrays (1230), and Nancy Warren (1231)
1232–1234	ISO Darlene Mickelsen and approach Virginia Bell target (1232), imaging
1235–1237	Move to (1235) and ISO (1235–1237) Eileen Dean target, imaging
1238–1245	ISO solar arrays, magnets (1238) and Eileen Dean target (1238–1245), imaging
1246–1250	ISO Eileen_Dean2 target (1246), imaging, drive to Innocent_Bystander (1247, 1250)
1251–1260	ISO Innocent_Bystander (1251–1256) and Innocent_Bystander_offset_2 (1257–1260)
1261–1277	Imaging, ISO Innocent_Bystander target (1266, 1273, 1275, 1277)
1278–1279	ISO Innocent_Bystander, Stealing_first and Stealing_second (1278), imaging
1280–1281	APXS dust study (1280), ISO Innocent_Bystander (1281), imaging
1282–1287	Microscopic Imager calibration (1282), ISO Innocent_Bystander (1284–1287)
1288–1292	ISO Norma_Luker (1288, 1291), imaging, Microscopic Imager calibration (1290)
1293–1299	Move back from Innocent_Bystander (1293), imaging, drive to Gertrude_Weise (1296, 1298)
1300	Move wheels to scuff and drag soil at Gertrude Weise target, and imaging

on solid rock, and then drove to the first science site, Site (or Station) 2. Station 1, the first area seen on the top of Home Plate, was the location examined on sol 760. Imaging goals here were a long-baseline stereoscopic survey of the Arthur C. Clarke landslide on Husband Hill (Figures 19 and 25) and observations of a large fracture crossing Home Plate. An Odyssey safe mode caused a delay here before Spirit could move on, but it reached Station 3 at a layered outcrop called Texas Chili on sol 1323.

The outcrop was brushed and analyzed over sols 1325 to 1329. Here the broad level surface of Home Plate sloped down slightly to the south and was covered by debris, forming a prominent ridge called South Promontory. Large dark boulders were scattered over the ridge, and one called Humboldt's Peak was examined here at Station (or Site) 3b. The drive to the rock was slowed when the jammed wheel caught on a rocky ledge but Spirit arrived on sol 1337. The next sol was Spirit's second anniversary on Mars, in Mars years. The plan here was to study the natural surface and then a brushed surface, but the brushing failed at the start when a procedure called "seek-scan" failed to locate the rock surface properly. RAT diagnostic tests were performed on sol 1346 before Spirit moved on to Station 4 near the southwest corner of Home Plate. Dust devils were seen fairly frequently during the spring seasons each Mars year and rarely between them, but one was spotted now on sol 1340.

Spirit was now looking for routes off the generally steep-sided plateau. At Site 4 (Figure 36B) a long-baseline stereoscopic survey of von Braun and its surroundings was undertaken to plan future activity. Two panoramas were taken from points 8 m apart for the survey on sol 1348, with the idea that, after taking a look into the valley on the western side of Home Plate, the rover would descend its southern edge to reach the von Braun Winter Haven as soon as possible. This plan very soon evolved into an immediate descent, giving a little extra time to inspect the stratigraphy of Home Plate on the way down, but although this steep south-facing slope might give the best available cross-section of Home Plate strata it would be very bad for power generation.

Spirit returned to the off-ramp area on sol 1353 as planners contemplated four potential winter havens: von Braun, West Knob (a hill later named Tsiolkovsky, just west of Home Plate), South Promontory and the northern edge of Home Plate itself. West Knob was imaged on sol 1356, and on sol 1360 MI data were collected from a disturbed soil target called Pumpkin Pie and the Pancam imaged Candy Corn, as the Earth date was Halloween, 31 October. Finally the decision was made. Spirit would overwinter on the northern edge of Home Plate, which it could reach without a south-facing traverse and where slopes were steeper than at South Promontory. The other two options were still inadequately mapped.

Spirit was driven northwards just inside the western rim of Home Plate, stopping briefly for IDD observations at Site 5 (Pecan Pie) facing West Knob, and then moved north to Site 6, only 15 m south of the area studied on sol 759. West Valley, the low area between Home Plate and West Knob, was surveyed from the top of the slope along this drive. At Site 6 the wheels slipped in loose material in a shallow depression called Tartarus and Spirit had trouble climbing out of the hollow. It was trapped from sol 1375 to 1387 with very low power and a dangerous south-facing tilt, but managed to escape on sol 1388 from the south side of Tartarus, passing very close to the steep drop into West Valley (Figure 36C). The rover now turned to the east, moving rapidly towards the northern edge of Home Plate to search for the best location for Winter Haven 3. The slope had to be steep enough, but negotiable, and the departure route for the following summer had to be free of rocks or deep dust. Spirit approached the edge on sol 1397, the last day of MY 28, and on sol 1399 the rover moved to a promising location and transmitted images to confirm its suitability over the next two sols.

Sol 1400 was MY 29, sol 3, or 11 December 2007. Between sols 1403 and 1406 the rover was carefully maneuvered into its first position, tilted about 13° towards the north, for a prolonged season of remote sensing and *in situ* studies (Figure 39A). With the planning team working on Mars time again to take full advantage of the limited time available, Spirit moved incrementally down the slope, facing backwards to enable its IDD to study targets on the top of Home Plate. The IDD was extended on sol 1409 to test the rover's stability, and was then used to analyze a flat rock surface called Chanute. A high-resolution color view of the top of Home Plate was taken from the upper part of the slope (sols 1412–1428) and called the Tuskegee panorama. Feature names here commemorated the Tuskegee Airmen, the first African-American military pilots.

As the Sun dropped lower, Spirit was able to increase its tilt by moving further down the slope. Sol 1419 was 1 January 2008 on Earth and MY 29, sol 22 on Mars, and by now the power generated by the dusty solar panels was so low that the rover's survival was more in doubt than it had ever been before. Power was still sufficient to do a small amount of science on most days, including dust devil searches, change detection sequences at El Dorado and IDD work. Mini-TES was used on many targets, including Fuzzy Smith on sol 1439. A slight downslope movement on sol 1429 increased the tilt to about 16° while keeping Chanute within reach. It had been brushed on sol 1427 but some areas were not

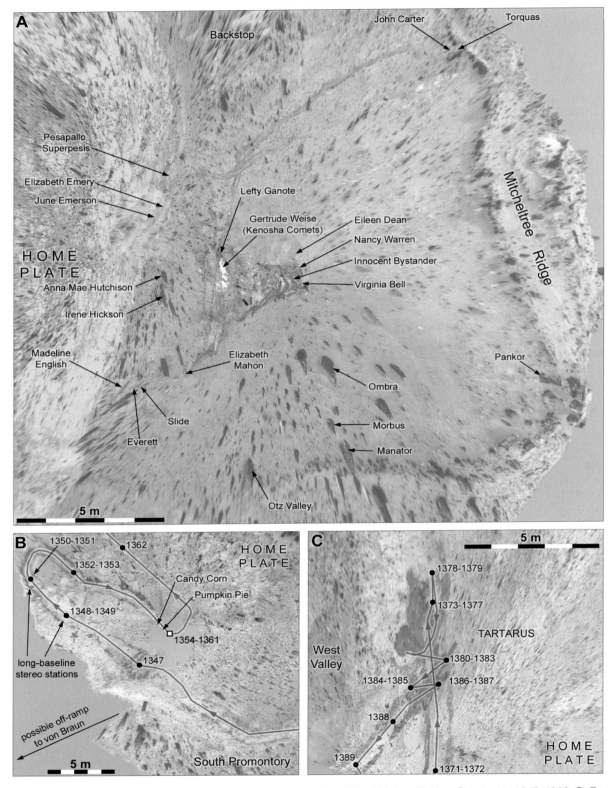

Figure 36. Activities near and on Home Plate. **A:** Feature names in East (Silica) Valley. **B:** Near Site 4, sols 1347–1362. **C:** Tartarus, sols 1371–1389. Severe topographic distortions are present in all three maps.

Table 14. *Spirit Activities, Home Plate and Winter Haven 3, Sols 1301–1800*

Sol	Activities
1301–1309	Imaging (1301–1309), drive towards Home Plate (1304), drive onto Home Plate (1306)
1310–1320	Drive to Home Plate Site 2 (1310, 1312), imaging, drive towards Home Plate Site 3 (1315)
1321–1324	Instructions not received (1321), drive to Home Plate Site 3 (1323), imaging
1325–1330	ISO Texas Chili target (1325), brush (1326) and ISO (1326–1330) Texas Chili
1331–1336	Panorama of Home Plate Site 3 (1331), drive to Home Plate Site 3a (1332, 1334), imaging
1337–1340	Drive to Humboldt Peak (1337), imaging, ISO Humboldt Peak (1340)
1341–1345	Brush Humboldt Peak (1341), ISO Humboldt Peak (1341–1345), imaging
1346–1350	Drive to Home Plate Site 4 (1346–1348), stereo position 1 imaging (1349), drive (1350)
1351–1354	Stereo position 2 imaging (1351), drive to south off-ramp (1352–1354), drive fault (1353)
1355–1360	ISO magnets (1355–1358), drive (1359), ISO Pumpkin Pie (1360)
1361–1367	Imaging, drive to Home Plate Site 5 (1362–1363), RAT scan Pecan Pie target (1366)
1368–1374	Brush (1368) and ISO Pecan Pie (1368–1370), drive to Home Plate Site 6 (1371–1373)
1375–1383	Drive towards Home Plate Site 7, wheels slip in Tartarus (1375, 1378, 1380), imaging
1384–1387	Drive towards Home Plate Site 7 (1384), drive to Site 7 with wheel slip (1386–1387)
1388–1389	Escape from Tartarus (1388), drive towards Home Plate Site 7 (1388–1389)
1390–1403	Drive towards Winter Haven 3 (1390–1403), failed drives (1393, 1395, 1400), imaging
1404–1408	Drive to edge of Home Plate (1404), descent to tilt north (1405, 1406), imaging
1409–1426	Begin Tuskegee panorama, ISO Chanute rock (1409–1422), RAT scan of Chanute (1424)
1427–1434	Brush Chanute (1427), move downslope (1429), ISO Chanute (1431–1434)
1435–1448	Imaging, move downslope (1436, 1440), RAT tests (1442), RAT scan Freeman (1445, 1448)
1449–1456	Imaging, brush Freeman rock target (1450), ISO Freeman (1452–1455)
1457–1472	Drive downslope to increase tilt (1457, 1459, 1463, 1464) and imaging
1473–1475	ISO solar panel, magnets (1473), RAT scan of Wendell Pruitt rock target (1475)
1476–1477	Imaging, begin Bonestell panorama (1476), RAT scan of Wendell Pruitt (1477)
1478–1485	Imaging, brush Wendell Pruitt (1479, 1484), RAT scan of Wendell Pruitt (1482)
1486–1498	ISO Wendell Pruitt (1486, 1489, 1491), imaging, ISO magnet (1498)
1499–1509	Imaging, super-resolution sequence (1503), ISO magnet (1502, 1504) and solar panel (1506)
1510–1591	ISO Arthur C. Harmon soil target (1510, 1512, 1529, 1532), imaging, recharge (1564–1591)
1592–1705	Imaging (1592–1640), recharge and imaging (1641–1656), finish Bonestell panorama (1657–1705)
1706–1724	Imaging, stow IDD (1707), attempt drive up slope onto Home Plate (1709–1724)
1725–1760	Imaging (1725–1736), recharge (1737–1741), conjunction (1746–1760)
1761–1769	Imaging, check Mini-TES (1761, 1763, 1766, 1769), failed drives (1763, 1768)
1770–1776	Recharge (1770), check Mini-TES (1771, 1773), failed drives (1772, 1775), imaging
1777–1793	Imaging, no science (1779–1781), drive downhill (1782), drive to Stapledon target (1793)
1794–1797	Image Thunderbolt (1794), ISO Stapledon (1796), failed drive to Home Plate on-ramp (1797)
1798–1800	Drive towards possible on-ramp (1798), anomaly precludes science (1800), imaging

cleaned properly. Further moves were made on sols 1436, 1440, 1459 and 1463, raising the tilt to 29°. These moves put Chanute out of range of the IDD, so a new target, Freeman, was selected and brushed on sol 1450, and after some analysis a third target was chosen. This was Wendell Pruitt, brushed on sols 1479 and 1484 and analyzed later.

One major goal during the long stay at this site was to compile a large multispectral panorama, the Bonestell panorama (Figure 35B). Imaging began on sol 1477 and built up very slowly. Sometimes only one frame per sol could be taken among other observations. Sol 1500 was MY 29, sol 103, or 23 March 2008. On sol 1503 a super-resolution sequence was obtained for a target called Roger Zellazny, and on sol 1512 another was made of the Arthur C. Clarke landslide. A new IDD target was examined beginning on sol 1510, a soil target called Arthur C. Harmon, but most analysis had to be

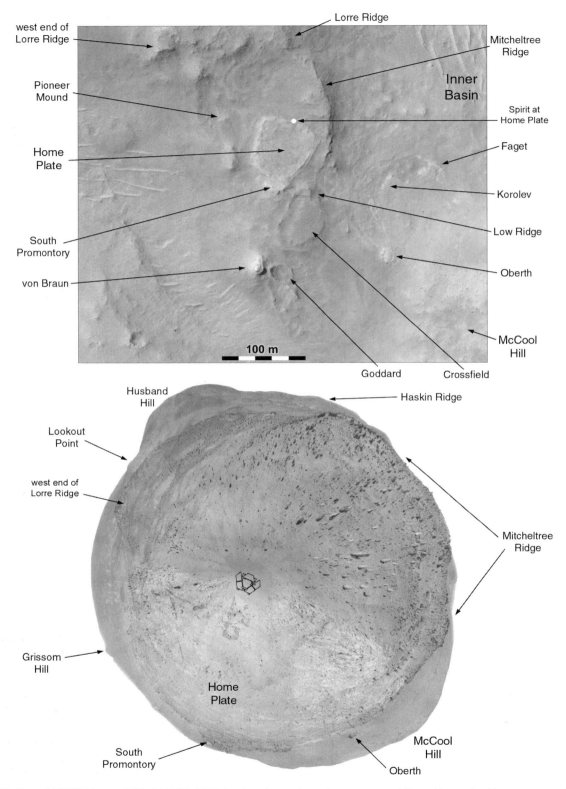

Figure 37. Part of HiRISE image PSP_010097_1655 (top) and reprojected panorama at Winter Haven 3 with exaggerated horizon relief (bottom), showing locations of horizon features.

postponed until after the solstice owing to the very low power. Very little work except occasional imaging was done from sols 1533 to 1656 so all available power could be used to keep the rover alive through the coldest period. The Mini-TES heaters were still used since the instrument might fail if they were shut off, and the rover was woken briefly each day to generate a little extra warmth. Even the Solar System complicated matters, since instructions for sols 1547–1549 were not received because planners had failed to notice an occultation of Mars by the Moon.

Sol 1600 was MY 29, sol 203 and 4 July 2008. The solstice fell just before that on sol 1591 and Spirit survived its third winter, resuming slowly paced imaging for the Bonestell panorama around sol 1637 as the science team planned its future operations. Sol 1700 was MY 29, sol 303, or 14 October 2008. Mini-TES observed an unusual rock called Stapledon on sol 1697 and the big panorama was completed on sol 1705. On sol 1709 (MY 29, sol 312) Spirit made its first move after the long winter, just a few centimeters uphill to reduce its tilt slightly, to improve its solar power generation as the Sun climbed higher again. Another small move was made on sol 1713. Since these moves were successful, the likelihood of being able to drive back up the slope seemed to increase, but later attempted drives rotated the rover rather than climbing higher. The goal was to achieve a tilt of about 20° for the conjunction period.

After several small moves which reduced tilt, a drive on sol 1716 produced a slip to the right. The jammed right front wheel made moving forwards (uphill) difficult, and if it slipped below the top of the slope the rover would have no choice but to drive downwards instead. The intention had been to climb onto the plate and cross it, driving off the south side towards von Braun, or if this was not possible to descend and drive around the plate, probably on its western side (Figure 40A). A cleat on the jammed wheel appeared to be caught on an obstacle, so on sol 1724 Spirit was commanded differently. Of Spirit's six wheels only four could be steered, those at the front and rear. Three of them were set to drive sideways and the jammed wheel, which could be steered but would not rotate, was turned with its cleats parallel to the direction of motion. The rover moved 30 cm sideways (Figure 39A) and could now try again to reduce its tilt.

On sol 1725, before another drive could take place, a significant local dust storm again cut power to levels that seriously compromised the rover. Spirit was commanded to shut off its heaters and waited for power to increase. This occurred only a week after the Phoenix mission had been brought to a sudden end by a similar dust storm, exacerbated by seasonal illumination changes at its high-latitude site. By sol 1730 power had increased to usable levels again and the Mini-TES had survived another dust event. Then, at the end of November, Mars entered superior conjunction and communications were largely blocked over sols 1746 to 1760. The last attempt to drive up onto Home Plate was on sol 1768, but still the movement was sideways rather than upwards, so a new approach was tried. Spirit would drive down the slope and turn east, trying to climb back up onto Home Plate again a few meters further east, close to where it had ascended successfully on sol 766.

Sol 1775 was 1 January 2009 or MY 29, sol 378, and Spirit marked the new terrestrial year a week later by driving downhill on sol 1782. After a pause at the foot of the slope to image the Winter Haven 3 site and the tracks made during the descent, including super-resolution images of a target called Jack Williamson, Spirit moved east to look for a new route up onto Home Plate. The dragging wheel was used on sol 1791 to scuff the small rock called Stapledon (Figure 39B), which had been observed on sol 1697, after which it was analyzed on sol 1796 and found to consist mostly of silica. This find helped to reveal the full extent of the deposit, which was one of Spirit's most significant discoveries. White soil was also seen in the tracks here. As the rover prepared to climb back onto Home Plate on sol 1800 (MY 29, sol 403, or 25 January 2009), a computer anomaly stalled the drive, which resumed on sol 1806. While waiting here the Inertial Measurement Unit (IMU) was monitored as the rover deployed and stowed the IDD to see if any induced motions could be detected. This was done to test the IMU, but it might also be done in future as a seismic experiment during non-driving periods, attempting to observe seismic events or "marsquakes."

The chosen ramp up to Home Plate was not as easy as expected, though Spirit had climbed onto the plate at the same location on sol 767 when it had six functioning wheels. Attempts to climb the slope on sols 1809, 1811 and 1813 failed owing to the very soft dusty soil.

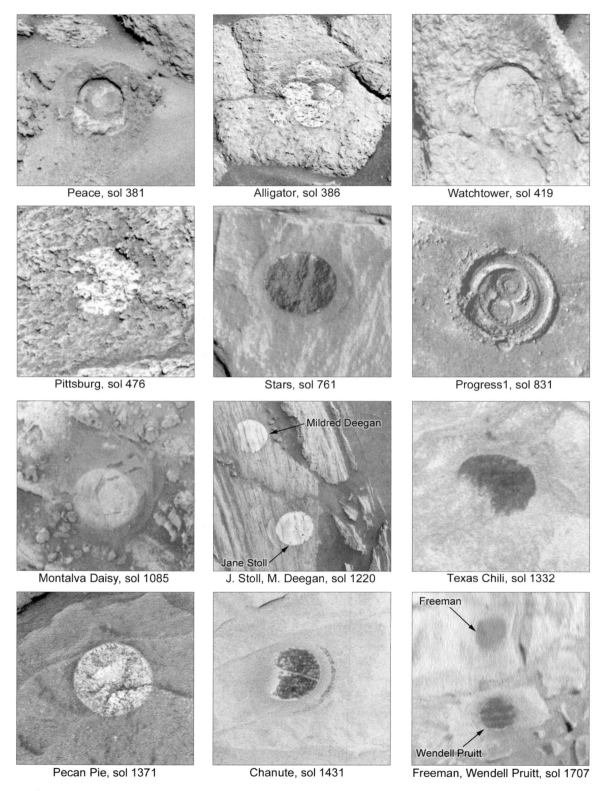

Figure 38. Selected RAT grind and brush marks from Husband Hill to Home Plate. The circular markings are 4 cm across. Dates are those of the images, not the RAT events.

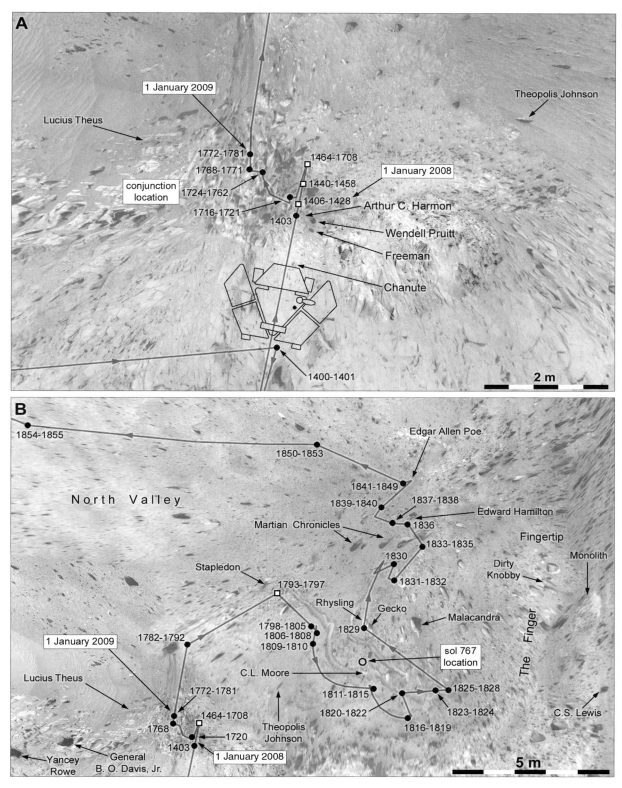

Figure 39. Activities at Winter Haven 3. **A:** Sols 1400–1782 on the northern rim of Home Plate. **B:** Departure from the winter study site, sols 1758–1854. Mosaics include images from multiple viewpoints with significant topographic distortions. North is rotated about 20° counterclockwise from the top in both maps. Many small rover movements occurred between illustrated sites at Winter Haven 3.

One result of this was the discovery of another bright soil patch at C. L. Moore on sol 1813. Spirit's bleak power situation was aided by wind gusts, which cleaned the dusty solar panels a little on sols 1812 and 1820. Dust devils had been seen occasionally in the plains from sol 1784 onwards, another sign of the changing seasons. Despite increasing power, the generally very dusty panels would make for a shorter active season than in the past, and planners were eager to start moving towards von Braun in time for the following winter.

Spirit was driven back west on sol 1820, out of the area of soft soil, and made another attempt to climb onto Home Plate five sols later and slightly further north (Figure 39B). When that failed, the rover was directed north again to skirt a finger of rock protruding from the plateau, hoping to pass north of its "fingertip" and return to the East Valley (Silica Valley) where access to Home Plate was assured. The dragging wheel had uncovered more bright soil at a site called Edmund Hamilton on sols 1836 and 1837. No IDD work could be done in the hurry to drive south, but multispectral images suggested the soils at C. L. Moore and Edmund Hamilton resembled those at Arad and Tyrone and were probably rich in sulfates. After several more drives failed to make any progress in that direction, Spirit finally turned west on sol 1850, to cross the area north of Home Plate, which was now called North Valley (Figure 39B), and another software upgrade was installed on sol 1859. Dust storms were observed by MARCI near the equator around sol 1854 but did not spread into the Gusev region.

Spirit rounded the Gibson area first seen on sol 746 (Figure 40B) and turned to the south to enter West Valley on sol 1861, taking images of other potential science targets in case the route south towards von Braun was impassable. These pictures included a long-baseline stereoscopic survey of an area to the northwest around the western extension of Lorre Ridge, and a small hill called Pioneer Mound to the west (Figure 40B). More white soil rich in sulfates was turned up by the dragging wheel on sol 1861, suggesting it might be found all around Home Plate. The soil targets here included John Wesley Powell and Kit Carson, and the area was sometimes referred to as the Santa Fe Trail, taking names from iconic figures of the American West. Spirit moved slightly uphill on sol 1866 to try to avoid getting caught

in what might have been another area of poor traction, and then continued south.

Good progress was made until sol 1872, when a five-day sequence of computer problems halted operations. The problems included failing to wake up in the morning, computer resets and failure to store science data in flash memory. Diagnostic and recovery work occupied several sols, and the next short drive was not made until sol 1886. On a more positive note, a wind gust cleaned some dust off the solar panels on sol 1881, easing the power situation. Several more cleaning events in the next few weeks brought power levels up close to those at the start of the mission, but driving became harder.

A smooth flat shelf called the Sidewalk at the base of the Home Plate cliff (Figure 40C) looked like a safe driving area, so Spirit was directed towards it, but between sols 1886 and 1892 four drives only covered about 30 cm and on sol 1892 (MY 29, sol 495) the wheels became embedded again. They had broken through a crust overlying very soft dusty material, digging up several layers of different color and composition as the drivers tried to break free. To complicate matters, memory problems recurred on sol 1896 and the left middle wheel stalled on sol 1899. Sol 1900 was MY 29, sol 503, or 8 May 2009. The wheel turned successfully on sol 1916 and might have caught on a buried rock as it was deeply embedded in the soil. This location was named Troy and the soil trench was called Ulysses, as names were now being chosen from the Trojan War as well as the American West.

The soft soils occupied a very subdued crater called Scamander (Figure 40C) and were found to be the most sulfur-rich materials ever seen on Mars. A tan-colored soil at Sackrider (Figure 41A) contained sulfur, a white soil at Olive Branch was rich in silica, and the yellow soil Penina contained ferric sulfate. Cyclops Eye was a red-colored undisturbed soil which was the site of repeated RAT brushing from sol 1965 onwards to explore variations with depth (Figure 41B). Each brushing was followed by microscopic images and APXS analysis of the soil composition, and after bright soil was found under the surface a new brushing site called Polyphemus Eye was started nearby on sol 1981 to see how far the brighter material extended (Figure 41C).

Dust devils were monitored over the plains to the west during this enforced stay at Troy while extrication

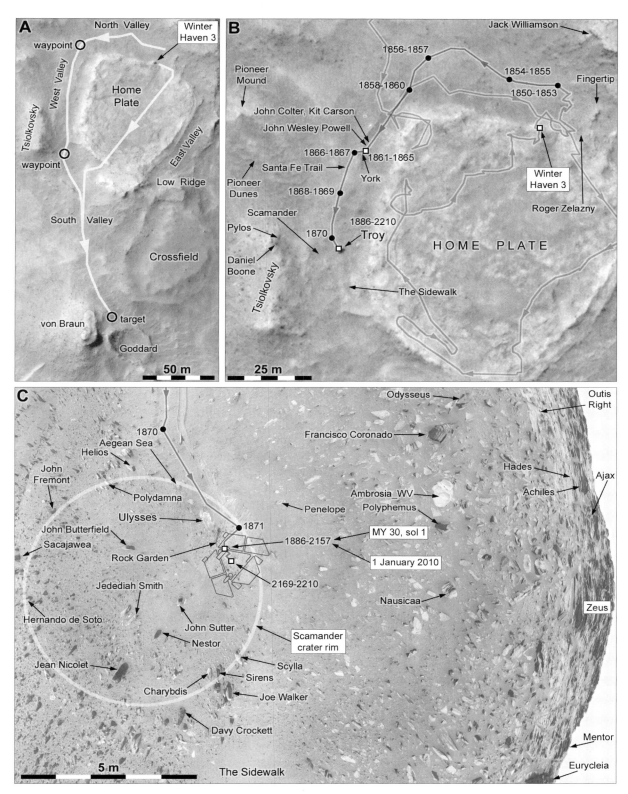

Figure 40. The drive to Troy. **A:** Route plans from SOWG planning meetings. **B:** Traverse around Home Plate, sols 1850–2210. The object at the end of the track is Spirit itself, imaged on 15 February 2010 (MY 30, sol 109; HiRISE image ESP_016677_1650). **C:** Troy, sols 1870–2210. Spirit is located and oriented as it was at the end of the mission. The work at Ulysses was done when the rover was 30 cm further north and rotated to face the sol 1870 position.

methods were tested exhaustively at JPL. Spirit was now in its third dust devil season, which extended from sol 1785 to sol 2058, and attempts were made on several sols after 1968 to take color images of a dust devil. This involved taking red and blue filter images simultaneously through the two Pancams, followed by yellow and infrared images, followed by a Navcam dust devil movie sequence and more Pancam images. If nothing appeared in the movie, the Pancam images were not transmitted. Successful two-color images were obtained on sols 1983 and 2025.

Planners now felt that they needed to reach good north-facing slopes by about sol 1995 if they were to survive, so a speedy extraction was required. The rover drivers were so busy developing extrication methods with the test rover that other work was delayed for a few sols around sol 1960.

Science during this period included compositional studies of the unusual soils, Mini-TES observations of many targets in the valley and nearby slopes, and a panorama, which included a good view of the stratigraphy of Home Plate. This Calypso panorama was started on sol 1910 and completed on sol 1932 (Figure 35C). As Spirit now had more power than it needed, it could also make overnight observations, as it had on Husband Hill about 1300 sols earlier. Observations were made on sol 1941 to measure the opacity of the night sky by imaging faint objects, and the Andromeda galaxy was just discernible. Another session on sol 1943 (22 June 2009 on Earth) imaged the night sky around the bright star Canopus for evidence of night-time clouds, fog or hazes.

Efforts were made to image Earth and Venus setting close together in the west after sunset, but the bright twilight sky compromised the observations. Pictures were taken over several evenings after sol 1943 to show the motion of the planets relative to each other and to the stars. More twilight observations on sols 1934, 1935 and 1947 provided information on the vertical distribution of dust in the upper atmosphere. So much power was being generated that the mission engineers wanted these overnight or morning observations to help drain the batteries.

The deadline for reaching safe north-facing slopes was looming, and an extraction plan was devised after much testing on Earth using a rover in Pasadena. Another source of information to help with this was a novel use of the MI on the IDD, carefully positioned to look under the rover to see if its body was in contact with the ground. Images taken from the top of Home Plate around sol 1365 and during the approach to Troy on sol 1870 showed a cluster of rocks called the Rock Garden that was now underneath the rover, and if the spinning wheels lowered the rover body onto them escape might become impossible. The first of these images, badly out of focus but still useful, were taken on sol 1922 after the procedure was tested with Opportunity. This was repeated frequently over the next 200 sols, especially very late in the mission in conjunction with the final attempted drives. A campaign of Navcam and Pancam imaging at different Sun angles was undertaken around this time to find the best illumination conditions for identifying safe driving areas. That might prove useful for future drives in this difficult terrain.

The first attempt to extract Spirit was planned for sol 1993 but the team was not yet ready. As work continued on Polyphemus Eye and Olive Leaf, a white soil clod or rock (Figure 41A), a regional dust storm arose north of Gusev on sol 2000 (MY 29, sol 603, or 19 August 2009) and began to reduce power to the rover. The recent very high power levels might have allowed Spirit to survive without reaching a north-tilting haven, but by sol 2002 the dust spread over the landing site and power dropped rapidly. Another test for extraction planning was undertaken on sol 2006 (the third anniversary of landing, in Mars years) by pressing the RAT hard into the soil at a location called Ulysses Spear to see if it could help support the rover's weight. This was called the RAT Penetrometer Experiment, and it slightly distorted the RAT brush, which had to be checked to be sure it was still safe to use. Then a series of external and rover issues affected the mission. MRO entered a safe mode and the DSN was needed on sol 2009 to support its recovery, and at the same time severe forest fires near JPL interfered with operations. JPL itself was closed for several days after sol 2011 (31 August 2009), preventing planning for sol 2013.

The problems continued as the HGA experienced faults on sol 2027 and the rover's flash memory became dangerously full by 2037, preventing further observations until enough data could be downlinked to free space. Data collection resumed on sol 2044 with a five-sol analysis of the rock Stratius followed by analysis of undisturbed soil at Thoosa. Everything was operating

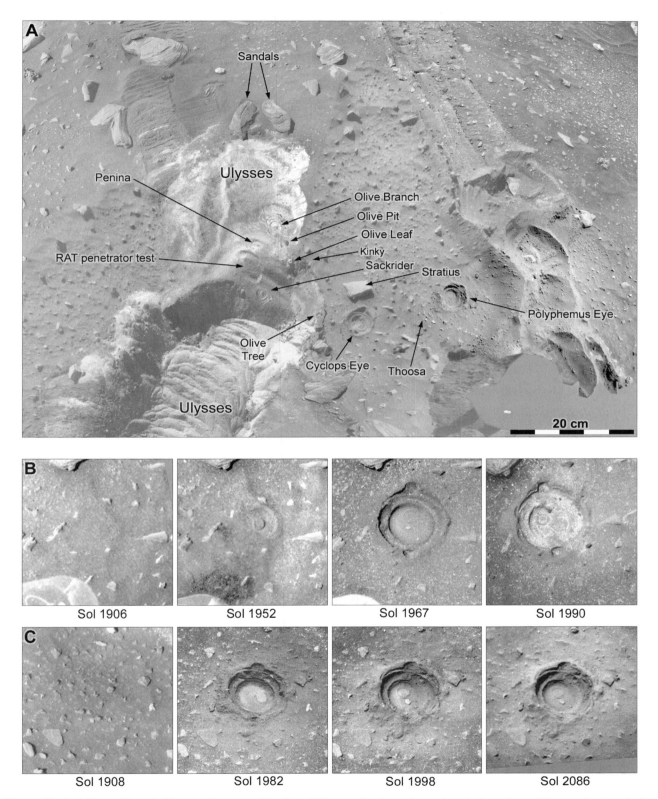

Figure 41. Activities at Troy. **A:** Mosaic of images of the Troy IDD area showing science targets in and near Ulysses, the trench dug by the left front wheel. The scale is only approximate. **B:** Evolution of the Cyclops Eye feature. **C:** Evolution of the Polyphemus Eye feature. Each "eye" is 4 cm across. Dates are those of the images, not the brushing events.

Table 15. *Spirit Activities in West Valley, Sols 1801–2211*

Sol	Activities
1801–1819	Drive stopped by rock (1806) and wheel slip (1809–1818), image C. L. Moore target in tracks
1820–1828	Imaging, drive downhill to escape soft soil (1820), try new route to Home Plate (1823–1826)
1829–1843	Image Edmund Hamilton soil pile in tracks, drive north towards East Valley (1829–1843)
1844–1856	Abandon eastern route, turn (1844), drive towards West Valley (1845–1856)
1857–1861	Imaging, drive west (1858) and into West Valley (1861), excavate Kit Carson bright soil (1861)
1862–1864	Imaging, ISO John Wesley Powell bright soil in tracks, image Pioneer target (1863–1864)
1865–1871	Imaging, drive south (1866–1871), image York target (1867)
1872–1891	Computer problems (1872–1885), imaging, short drives in soft soil (1886, 1889, 1891)
1892–1921	Imaging, drive out of embedded position (1892–1899), dust cleaning events (1899, 1918)
1922–1926	Imaging, MI images of rover underside (1922, 1925) and Sackrider target (1922–1926)
1927–1933	Imaging, ISO Olive Branch1 (1927) and Olive Branch2 (1928–1932)
1934–1939	Imaging, ISO Olive Branch3 (1934), Olive Branch4 (1935) and Penina1 (1936–1939)
1940–1944	Imaging, ISO Penina2 (1940), Penina3 (1941) and Penina1 (1942–1944)
1945–1946	Imaging, ISO Cyclops Eye1 (1945) and Cyclops Eye2 (1946), dust cleaning event (1946)
1947–1965	Imaging, ISO Cyclops Eye3 (1947–1952, 1963, 1965), brush Cyclops Eye3 (1965)
1966–1967	Imaging, ISO Cyclops Eye3 (1966) and Olive Branch5 (1967) targets
1968–1972	Imaging, ISO Olive (1968) and Cyclops Eye4 (1968–1971) targets
1973–1974	Imaging, RAT Cyclops Eye4 and ISO Cyclops Eye5 (1973), DSN problems (1974)
1975–1979	Imaging, ISO Cyclops Eye5 and Olive Branch6 (1975), ISO Olive Branch7, Olive Pit (1979)
1980–1981	Imaging, ISO Polyphemus Eye1 (1980), RAT and ISO (1981) Polyphemus Eye2
1982–1985	Imaging, ISO Sackrider (1982) and Cyclops Eye5 (1982–1985), illumination study
1986–1990	Imaging, ISO Penina (1986), RAT and ISO Cyclops Eye5, image under rover (1990)
1991–1996	Imaging, RAT Polyphemus Eye2 (1993) and ISO Polyphemus Eye3 (1995–1996)
1997–2005	Imaging, ISO Olive Leaf (1997) and Polyphemus Eye3 (1998–2005)
2006–2016	Imaging, RAT Ulysses Spear (2006), ISO Olive Leaf (2006–2016), RAT test (2016)
2017–2033	Imaging, ISO magnets (2017–2023) and Penina4 target (2024–2029)
2034–2044	Computer problems (2034–2035, 2040), image under rover and ISO Stratius (2044)
2045–2060	Imaging, ISO Stratius rock target (2045–2051) and Thoosa soil target (2054–2060)
2061–2064	Imaging, image under rover (2061) and ISO Thoosa (2061–2062)
2065–2070	Computer problems (2065–2067, 2069), imaging, ISO Thoosa (2068)
2071–2074	ISO Thoosa (2071), image BellyRock (2072), problems prevent science (2073–2074)
2075–2082	Imaging, microscopic images of BellyRock (2076, 2081), drive to straighten wheels (2078)
2083–2088	Computer reboot (2083–2084), MI images of BellyRock and failed drive (2088)
2089–2098	Imaging, microscopic images of BellyRock and drives to escape Troy (2090, 2092, 2095)
2099–2103	Imaging, microscopic images of BellyRock and drive stopped by right rear wheel stall (2099)
2104–2116	Imaging, wheel diagnostics and microscopic images of BellyRock (2104, 2109, 2113)
2117–2125	Imaging, microscopic images of BellyRock and drives to escape Troy (2117, 2118, 2120, 2122)
2126–2135	Imaging, microscopic images of BellyRock and drives to escape Troy (2126, 2130, 2132)
2136–2142	Imaging, microscopic images of BellyRock and drives to escape Troy (2136, 2138, 2140)
2143–2149	Imaging, microscopic images of BellyRock and drives to escape Troy (2143, 2145, 2147)
2150–2152	Imaging, microscopic images of BellyRock and drives to escape Troy (2150–2152)
2153–2173	Imaging, recharge batteries (2153), drive to escape Troy (2154, 2156, 2158, 2161, 2165, 2169)
2174–2210	Imaging, last microscopic images of BellyRock (2174), UHF downlink only (2203, 2210)
2211	Spirit enters hibernation, effective end of mission (MY 30, sol 145)

normally by sol 2059, but on sol 2065 a computer reboot caused a safe mode and affected the flash memory again. Data could still be collected but could be stored only temporarily before transmission, and would be lost when the rover shut down. The rover would make observations and transmit them via Mars Odyssey before shutting down for the day, an arrangement that allowed continued science if managed carefully. Work on Thoosa continued for now, but only a flash memory reformatting would correct the problem, as it had previously on sol 32. During this period Spirit saw in a new Martian year, since sol 2067 was sol 1 of MY 30.

The flash memory reformat procedure was supposed to happen on sol 2075 but was postponed until sol 2083 due to a computer error. Now everything was ready for Spirit to try to break free of its sand trap at Troy, where it had remained immobile since sol 1899. In preparation for extraction, Spirit's wheels were correctly oriented on sol 2078 and the final images of Scamander were added to the Calypso panorama before the computer reboot. On sol 2088 the driving began, this time northwards to escape Troy, but halted immediately without any motion after a tilt check failed. On sol 2090 the rover was commanded to drive 2.65 m but slipped so much that it only moved 1 cm, and during the next drive on sol 2092 the right rear wheel stalled, stopping the drive. It resumed working for the next drive but stalled again on sol 2099. Sol 2100 was MY 30, sol 34, or 29 November 2009.

During testing to deal with this problem on sol 2113 the right front wheel moved briefly and unexpectedly, its first motion since sol 779, and the rover experienced electrical behavior suggestive of grounding and other problems. A drive on sol 2117 was attempted with all six wheels, but little progress was made. The illumination was deteriorating so much now that drives would be needed every day if the rover was to escape and survive the winter. By sol 2118 the right rear wheel had stopped working altogether, the right front wheel's intermittent operation was not enough to be useful, and the left middle wheel was lifting out of the soil and not providing any traction. The cumulative motion since 2088 had only been a few centimeters northwards and the tilt was now 5° to the south, whereas 5° to the north was needed.

Driving problems aside, Spirit was still making useful observations. A new wind or dust devil streak was seen on Husband Hill on sol 2122, and the final dust devil movie of the season was made on sol 2129. Many targets were observed by the Pancams, including frequent atmospheric observations between sols 1991 and 2149 as well as multispectral images of Scamander Plains on several sols between 2049 and 2089, von Braun on sols 2114–2117, Pioneer Mound on sol 2114 and the Ulysses trench tailings on sols 2118, 2122 and 2126. Super-resolution sequences were taken of South Valley on sol 2124 and Pioneer Mound on sol 2125. Power was very low and the Mini-TES heaters were now turned off. On 5 January 2010 (sol 2136) the rover planners decided to spend only a little more time attempting to escape from Troy, improving the tilt if nothing else, before transitioning into survival mode with a small amount of science whenever possible.

Beginning on sol 2145 the driving direction was reversed with initially promising results. The rover had been trying to move northwards but it was now driven south towards the Sidewalk again, and over the next few sols Spirit moved about 20 cm southwards and rotated counterclockwise about 35° (Figure 40C). Although this progress was promising the tilt worsened, reaching 11° to the south on sol 2154 before improving by 1.5° on sol 2163. Power dropped rapidly after this, and after a last futile attempt to improve tilt on sol 2169, all driving ceased. The IDD was left deployed over Ulysses after taking some final images underneath the rover on sol 2174, positioned so it could take argon measurements and drive in the spring even if the arm was no longer operational. Imaging continued when possible, including multispectral observations of Circe, a disturbed soil which appeared to be changing color since it was first exposed. The last cloud movie was taken on sol 2185 and the last color imaging of soil on sol 2190. Sol 2200 was MY 30, sol 134, or 12 March 2010.

As the power situation worsened the team on Earth devised a plan for operating with very low batteries. The rover would wake each morning, quickly check the dust levels in the atmosphere and shut down again, allowing the batteries to recharge as much as possible. Once each week a brief radio session would relay the dust data and report on rover system health. The last successful transmission was on 22 March 2010, sol 2210 (MY 30, sol 144), and the next expected signal on sol 2218 was not heard. The rover had probably entered a low-power safe

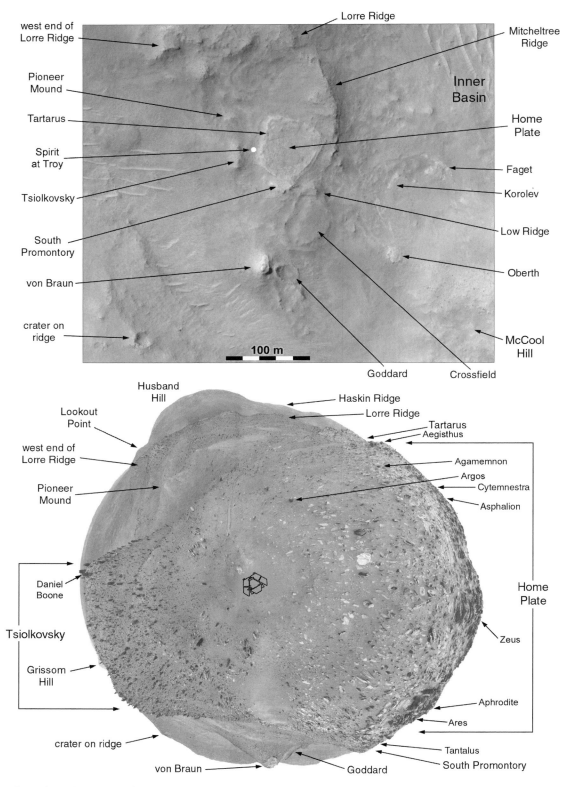

Figure 42. Part of HiRISE image PSP_010097_1655 (top) and reprojected panorama at Troy with exaggerated horizon relief (bottom), showing locations of horizon features.

mode, in effect hibernating until its batteries could recharge. The lowest power generation level previously experienced by Spirit had been during a dust storm around sol 1250 in November 2008, but that had been brief and the batteries had been more fully charged at the time. The southern winter solstice was still 48 sols away on Spirit's last day of operation.

NASA now assumed that Spirit might never leave this location. If it successfully resumed operations late in 2010, it would operate as a static lander, observing any changes in its surroundings and taking part in a unique experiment that required immobility. This would be a six-month period of precise radio monitoring to track the planet's rotation more precisely than had ever been done before. The data could permit a sensitive measure of the precession rate, which in turn would reveal whether the planet's core was liquid or solid. After that was finished, attempts to move small distances might continue in order to explore soil variability in the Troy area. The radio monitoring experiment was eventually undertaken by Opportunity at Greeley Haven during the next winter season.

Several efforts to hear from or transmit to Spirit were made in July 2010, around sol 2320 of the mission, with no success. This was about the earliest that a signal could have been expected. A more likely time to make contact would have been in October 2010, but still nothing was heard. A dust storm observed in the area late in October further diminished the chance of survival. If Spirit had survived until sol 2245 (MY 30, sol 179) it would have matched Viking Lander 1's lifetime, but in the absence of a signal the record for survival on the surface passed to Opportunity on 20 May 2010 (MY 30, sol 200). Illumination increased to a maximum in March 2011 but the radio silence continued. Eventually, on 25 May 2011 NASA abandoned regular efforts to contact Spirit and the mission was declared over. The final resting place of Spirit at 14.60° S, 175.53° E was sometimes referred to as Spirit Station after the mission was over.

Spirit had seen enough over two Mars years to elucidate the complex geology of Home Plate and its surroundings (Lewis *et al.*, 2008; Crumpler *et al.*, 2010, 2011; Ruff *et al.*, 2011). The lowest visible layer was a platy material suspected of being altered volcanic ash, seen at Halley (sol 809). Above that was a nodular silica-rich material seen at Nancy Warren (sol 1225), Stapledon

(sol 1798) and elsewhere around Home Plate. The next layer was the coarse-grained layered deposit of volcanic ash seen at Barnhill (sol 750), including the possible "bomb sag" (Figure 29F). That was topped by a finer-grained wind-deposited sandstone of basaltic composition comprising most of Home Plate, and above that was a thick layer of vesicular lava seen at Lorre Ridge, Mitcheltree Ridge and Low Ridge.

8 July 2003: MER-B (Opportunity)

Opportunity, the MER-B rover, was identical in its design and goals to MER-A. After several delays, the spacecraft was launched from Cape Canaveral Air Force Station at 03:18 UT on 8 July 2003 (7 July, local time), placed in a parking orbit and then set on its Mars trajectory by its upper stage 72 min after lift-off. Trajectory corrections were made on 18 July and 8 September 2003 and on 17 January 2004. MER-B arrived at Mars on 25 January 2004 (MY 26, sol 630) and landed successfully at 5:05 UT, in the early afternoon local time. The landing ellipse as defined in 2002 (Figure 43A) was 119 km long and 17 km wide, centered at 2.07° S, 6.08° W (353.92° E) and oriented at 88° azimuth (almost east–west). Figures 43B and 44 show the site in increasing detail.

The landing area, Meridiani Planum, was known as the Hematite site (Table 3) because orbital data from MGS had found evidence of coarse-grained hematite, an iron oxide mineral usually associated with water (Christensen *et al.*, 2000). Imaging revealed a site unlike any other landscape yet seen on Mars, with outcrops of layered rock only a few meters from the lander in the wall of a small crater. The rover explored that crater and several others of varying size and age during a drive of 39.5 km, lasting 10.5 years, and was still operating as of sol 3700 (MY 32, sol 318, or 23 June 2014). Opportunity became the longest-surviving lander on Mars, surpassing Viking Lander 1's operational lifetime on 20 May 2010 (sol 2245). Asteroid 39382, a member of the Hilda family of asteroids in a 3 : 2 orbital resonance with Jupiter, was named Opportunity in October 2004 to commemorate the rover.

Figure 43A shows the regional setting of the Opportunity landing site. Some locations nearby had been considered as landing sites for Mars Pathfinder, Mars-98 and Mars Surveyor 2001 (Tables 66 and 74 in Stooke, 2012). This area of the cratered southern highlands is situated on a broad regional slope running downwards to the northwest, to Chryse Planitia. The cratered terrain south of the site is cut by many small channels, evidence of the presence of water in the past. The northeast corner of Figure 43A is smoother, clearly delineating another geologic unit superimposed on the cratered terrain and partly filling the 150 km diameter Miyamoto at 3° S, 353° E (called Runcorn in some contemporary documents). These plains correspond with the hematite deposit identified from orbit by the TES instrument on MGS (Christensen *et al.*, 2000). Bopolu, a fresh 20 km diameter crater at 3.0° S, 353.7° E, covered the hematite unit with ejecta consisting of cratered uplands material. Blocks from this crater were scattered over the Opportunity landing site, and one might have been examined early in the mission (Bounce Rock, Figure 51B).

The landing ellipse (Table 7), initially 119 by 17 km, extended across the hematite-rich plains unit named Meridiani Planum, north of Bopolu and the older 20 km crater Endeavour (Figure 43A). The dark area around and east of the ellipse in Figure 43A is the classical dark marking Sinus Meridiani (Figure 1 in Stooke, 2012). Figure 43B shows the landing ellipse in more detail. Several craters show bright wind streaks extending to the southeast. Many irregular bright patches scattered across the ellipse were not understood when the site was selected. They were found to be outcrops of wind-deposited sandstones cemented by sulfate salts that had precipitated out of ground water.

Components of the landing system fell 20 km east of the target (Figure 43B). The lander in its protective air-bags touched down between the backshell and heatshield within a region brightened by the blast of its thrusters (Figure 44D). It then bounced and rolled northwards before coming to rest inside Eagle crater (Figures 44D and 45D). The parachute and backshell fell 400 m southwest of the landing site (Figure 45B). The bright patch of disturbed soil near the backshell was seen to fade over the course of several years in successive HiRISE images. The heatshield fell 750 m southeast of the landing site and about 200 m south of Endurance crater (Figure 45C). The cruise stage which accompanied the rover to Mars probably fell about 80 km west of the landing

site, near 2.1° S, 353.0° E, by analogy with Curiosity (Figure 179A).

Figure 47 shows the full route of Opportunity from its landing site (top left) to the end of coverage in this atlas on sol 3700 at lower right. The rover landed in Eagle crater on 25 January 2004 (MY 26, sol 630) and spent two months studying the small outcrop of rock inside it. Then it was driven about 700 m east to Endurance crater to examine the much larger stratigraphic section in its walls. After six months in Endurance it undertook a drive of over 6 km southwards to Victoria crater in the hope of finding even larger outcrops. The long drive began on a very smooth surface which became covered with progressively larger sandy drifts, some of which caused serious problems with mobility. From Erebus crater southwards the drifts were superimposed on Etched Terrain, a broad pavement of the same evaporite-cemented sandstones seen in the walls of the crater.

Opportunity reached Victoria on sol 952 (MY 28, sol 244) and entered it on sol 1291. After surveying the stratigraphy of Victoria for 12 months the rover emerged on sol 1634 and began a long drive south and east to reach the much larger crater Endeavour. Smooth surfaces were encountered again around Victoria itself and near Endeavour. Opportunity reached the rim of Endeavour at Cape York on sol 2681 (MY 30, sol 636) and examined the very old rocks underlying the Meridiani plains, including exposures of clay indicating a wet environment early in Martian history. After spending almost one Mars year at Cape York, the rover was driven south to another segment of the rim of Endeavour called Solander Point, arriving there on sol 3398 (MY 32, sol 16). The mission was still in progress, examining outcrops on Murray Ridge south of Solander Point, as of sol 3700 (MY 32, sol 318, or 23 June 2014). The total distance driven by sol 3700 was 39.5 km, breaking the previous record for driving distance on another world set by Lunokhod 2 on the Moon in 1973.

The International Astronomical Union (IAU) naming rules for small craters on Mars specify that small craters should be named for terrestrial towns. Large craters are named after scientists with some connection to the study of Mars. Landing site feature names are mostly unofficial and not governed by these rules, so all crater names at the Opportunity landing site were taken from famous ships and spacecraft involved in exploration, and many rocks

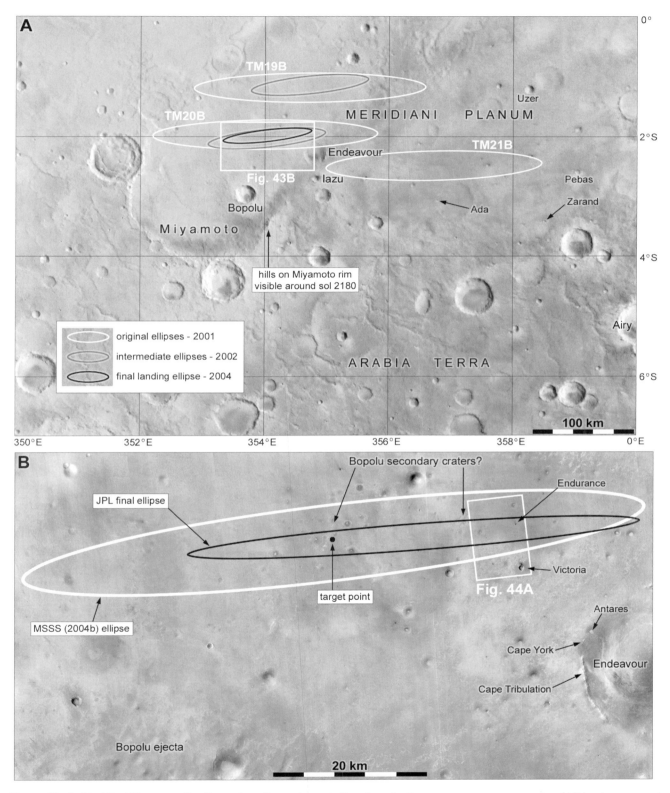

Figure 43. A: Meridiani Planum, with ellipses from three stages in the site selection process, on a composite of Viking images and Mars Orbiter Laser Altimeter (MOLA) topography. **B:** The final Opportunity landing ellipse on a composite of MGS MOC images (Malin Space Science Systems, 2004b) and MRO CTX images. The target was the center of the larger ellipse but final predictions suggested Opportunity would land further east, in the 60 by 4 km ellipse shown in black.

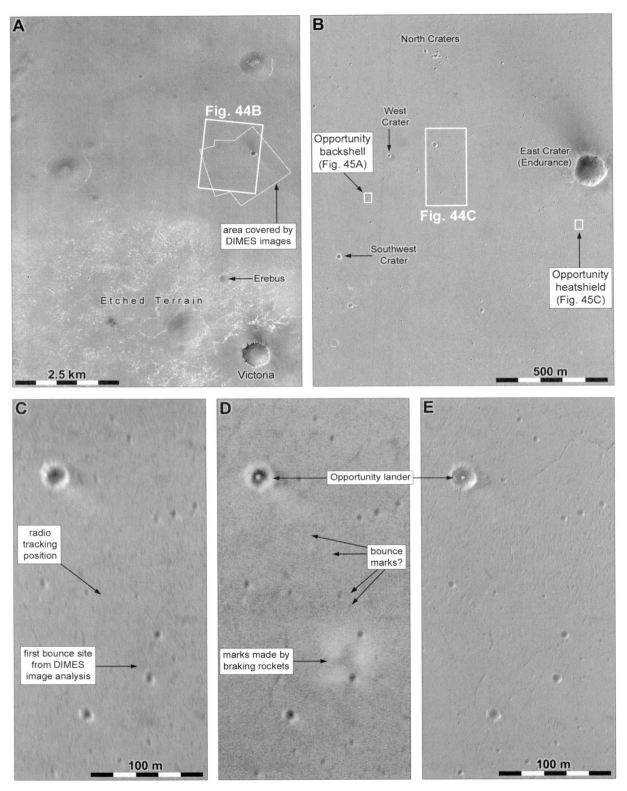

Figure 44. Opportunity landing site. **A:** CTX image B02_010486_1779_XN_02S005W showing the landing region. **B:** HiRISE image PSP_009141_1780 showing nearby craters used to locate the lander. **C:** DIMES image 1e128278513edn0000f0006n0m1 showing first estimates of the landing location. **D:** MOC image R1602188 showing rocket exhaust effects on the surface, possible bounce marks and the lander on the surface. **E:** Detail of **B** for comparison with **C** and **D**.

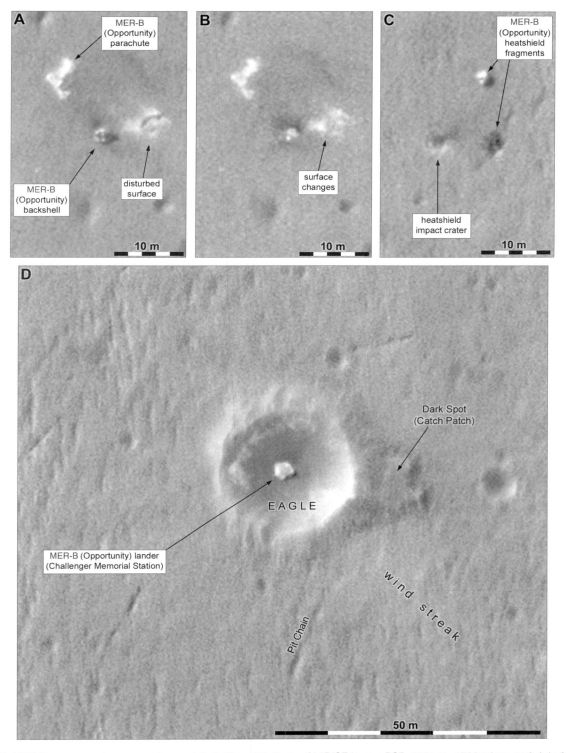

Figure 45. MER-B components as seen from orbit. **A**, **C** and **D:** Parts of HiRISE image PSP_009141_1780 taken on 8 July 2008 (MY 29, sol 206). **B:** Part of image ESP_013954_1780 taken on 18 July 2009 (MY 29, sol 571), showing changes to the backshell impact surface markings.

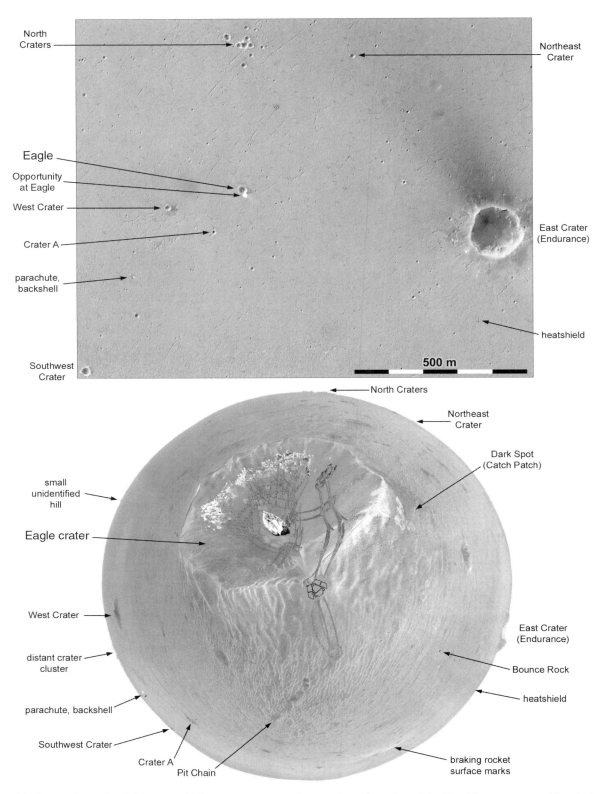

Figure 46. Comparison of HiRISE image PSP_009141_1780 and a reprojected version of the Lion King panorama. The dark streak extending northwest of Endurance was not present at the time of the landing, illustrating the variability of surface markings.

south of Victoria were named after islands. The crater names Victoria and Endeavour, initially commemorating the ships of Magellan and James Cook, respectively, were eventually made official as they are also the names of towns on Earth.

A detailed description of Opportunity's activities follows, illustrated in Figures 49 to 115 and summarized in Tables 16 to 27. The entire rover route is shown at a standard scale in maps plotted on a HiRISE image base with a 100 m square grid for scale (Figures 49 and 58 and succeeding maps, each one identified in the caption to the previous map). Sites of more complex operations are depicted on larger-scale maps based on HiRISE images or reprojected rover panoramas (*e.g.* Figures 50, 51 and 52). A key to map coverage is shown in Figure 48. MGS MOC images were used for route planning before the MRO HiRISE images became available in November 2006. Here an MOC image is included in Figure 44D to illustrate the image quality available to planners at the time, but HiRISE images are used for the route maps. The tables and descriptions of Opportunity's activities are derived from documents provided by NASA's Planetary Data System, especially the Analyst's Notebook documents, augmented by status reports from the Jet Propulsion Laboratory, monthly summaries of activities compiled for the Planetary Society by A. J. S. Rayl, and by descriptions in Squyres *et al.* (2004b, 2006a, 2006b, 2009, 2012), Arvidson *et al.* (2011, 2014a), Golombek *et al.* (2006b, 2014) and Farrand *et al.* (2014).

Early images showed several low ridges on the horizon, which Tim Parker (JPL) identified with nearby crater rims to help locate the lander (Figures 44B and 46). The prominent rim of Endurance crater, 700 m to the east, was hidden by Eagle's own rim in the first images. Figures 50 and 51A show a mosaic of approximately rectified panoramic images from Opportunity, showing the lander in Eagle crater, which is 20 m in diameter. West and north of the lander an outcrop of layered rock, first called Great Wall and later Opportunity Ledge, was exposed in the low crater wall, the first outcrop ever seen by a lander on Mars apart from some possible bedrock patches at the Viking 1 site. Long narrow drifts ringed the crater rim on the northwest and southeast sides.

The lander had entered Eagle crater from the east and rolled around the floor before coming to rest at 1.95° S,

354.47° E, lying on one of its side petals. It was righted as the petals opened. Opportunity's solar panels unfolded, several images were taken to document the landing, and then the camera mast was deployed to permit the first survey of the site. The initial images showed an elevated horizon, suggesting that the rover was sitting in a small crater. The mast would not block the HGA as Spirit's had done, allowing better communications and a faster deployment, and the front egress route appeared clear of airbags and ready to use. The airbags had left imprints in the soil of the crater floor, which were nicknamed lily pads. On sol 2 the descent images were transmitted, helping to locate the landing area. The overnight electricity use by a heater on the arm was worrying. It turned on at night and was off during the day, but it used as much energy as a typical day of science activities, about one-third of daily usage, and it could not be turned off. Despite the shock of the airbag landings, this was the only serious fault experienced by either rover. Opportunity's Astrobot, called Sandy Moondust, was imaged on sol 2.

The HGA was deployed on sol 3 and extensive imaging began as Opportunity was made ready. The rover stood up on sol 4 and Mini-TES was prepared for work, and an atmospheric APXS measurement was taken. Then on sol 5 the rear petal of the lander was moved to tilt the platform for easier departure. More images were taken for topographic mapping and planning, and then on sol 6 the middle wheels were released and the arm was deployed after its heater turned off during the morning. Tracking during landing suggested the site was further north than initially expected, about 3 km north of an area of bright rock outcrops called Etched Terrain (Figures 44A and 47), 2 km west of a degraded crater with bright dunes on its floor and east of a young crater which might be accessible to the rover. This location appears to be about 3 km east of Endurance.

The rover rolled onto the ground on sol 7, drove 3 m and paused to examine the soil, which Mini-TES showed to be rich in hematite. Images showed the surface was covered with small round spherical nodules nicknamed blueberries, which turned out to be the hematite-bearing component of the surface. They had been pressed down into the soil in the lily pads and in rover tracks. The steering was tested on sol 8 with several small turns, and on sol 9 the Pancam imaged the IDD instruments and MI

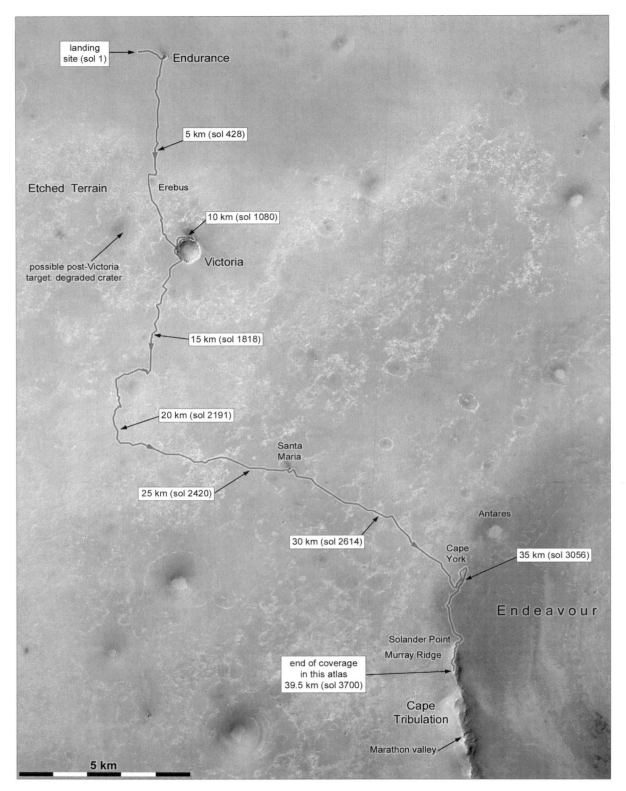

Figure 47. Full Opportunity traverse, showing drive distances at 5 km intervals. The base is CTX image B02_010486_1779_XN_02S005W.

viewed a soil target called Sidewalk. On the next sol MI viewed a site called Tarmac before putting MB on it for 24 hours. These targets were in an area called MER-lot (parking lot) or Merlot. APXS was placed on Tarmac on sol 11, while other targets were examined by the Pancams and Mini-TES, including an outcrop called Chani, later named Snout. The science team wanted to dig a trench in the soil here but the drivers had not finished practicing the activity.

After another steering test on sol 12 and observations of the test tracks (Fishtail), Opportunity drove towards Snout, also called Stone Mountain at the time. The backshell and parachute now became visible on the plains to the southwest. On sol 13 the rover approached Snout, and on the next sol MI viewed the soil at its foot before positioning itself for IDD operations. The plan now taking shape would have Opportunity make a circuit of the outcrop before digging a trench in an area rich in hematite in the Mini-TES data. Sol 15 was taken up with MI observations of the soil at Piedmont below the outcrop and Robert E. on Stone Mountain. APXS was placed on Robert E., finding it to be sulfur rich, and MB was placed on the same spot overnight. The finely layered rocks contained blueberries, both embedded in and eroding out of the outcrop.

Over sols 16–21 the rover drove counterclockwise around the crater, taking stereoscopic images and Mini-TES data covering the whole outcrop from several locations to help plan its future activities. On sol 16 Opportunity stopped at Alpha to survey the outcrop, and early on sol 17 the MI and MB were used on blueberry-rich soil called Berry Flats before a drive to the next survey location, Bravo. The area surveyed here was called Beta, and on sol 18 MI took images here, but they were out of focus and a fault prevented the next drive. MI and MB were used at a target at Bravo called Dark Nuts on the next sol, before a successful drive to the next location, Charlie, which included a test of visual odometry (visodom).

Opportunity looked to the southwest to view sunset in five sets of color images on sol 20. The rate of dimming of the Sun near the horizon revealed the amount of dust in the sky, which was found to be nearly twice that measured by Pathfinder, causing the Sun to appear much fainter. The sky above the Sun had a blue tint, as Pathfinder and Viking had also seen, because fine dust in the atmosphere scatters red light sideways and blue light on towards the observer. Elsewhere in the sky we see only the red part of the light, but near the Sun we see the bluer part of the spectrum. Also on sol 20 a set of MI and MB observations of a target called Flashy 2 were spoiled because the arm was held too high, and the same fault prevented a drive later that sol. Mid-drive images of Delta, the next survey point, were taken as if the drive had occurred but did not show the desired targets. The rover arrived at the hematite-rich area on sol 21, passing but not imaging the outcrop at Delta and Echo, the last survey point.

The trenching location was examined before digging on sol 22, with Pancam, MI, MB and APXS, and Mini-TES, which also conducted surface observations as MGS flew overhead, looking down with its TES to help calibrate the two instruments. The next sol began with MI observations of the trench area and tracks, and then the wheels were spun to dig a trench called Big Dig, the first soil trench dug by either rover, to see if the hematite was concentrated on the surface or extended deeper. The dig was designed to preserve layering in the trench walls. On sol 24 the MI and MB examined the trench floor, with APXS used overnight, and on sol 25 all three instruments looked at the far wall of the trench. Meanwhile the Pancam and Mini-TES looked at the Delta area to recover the observations lost earlier. These outcrop images were used to identify targets for detailed analysis.

Opportunity backed away and imaged the trench on sol 26 before driving around it, preserving it in case it needed to be re-examined, and skirted the lander to approach the El Capitan area. Power was severely limited by the arm heater problem, which reduced the scientific work on every sol. The next sol was taken up with IDD work on a rock called Stone Tablet and a soil patch called Red Sea, and images of the outcrop and the distant heatshield, ending with a small move to position new targets for analysis. These were Guadalupe and McKittrick, the upper and lower parts of El Capitan, respectively, and MI showed linear pits called vugs on Guadalupe, probably formed where soluble crystals had been removed.

IDD work on sols 28 and 29 characterized the targets before Opportunity's RAT was used for the first time on sol 30. It ground 4 mm into McKittrick, MI viewed the results and APXS was immediately applied to the hole.

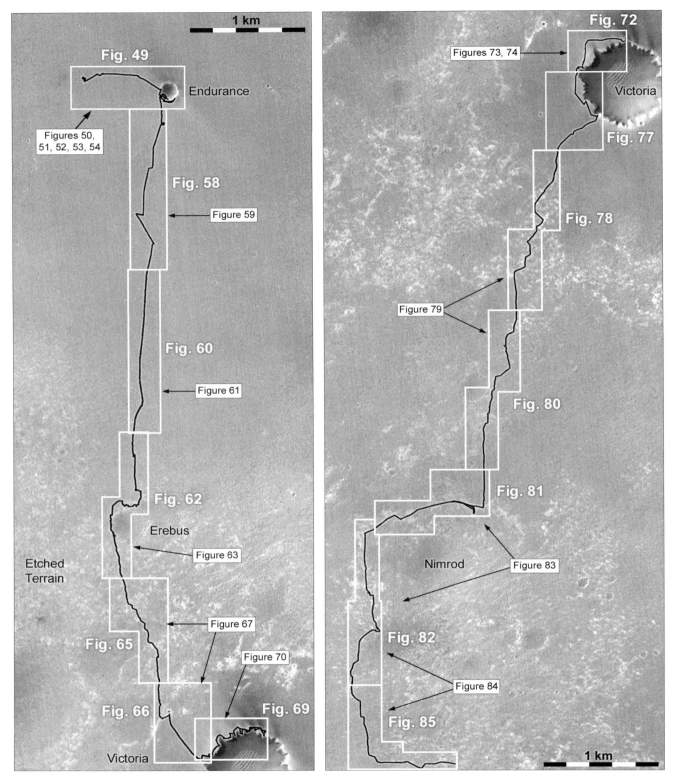

Figure 48 (both pages). Index of Opportunity route maps. White outlines mark sections of the main route map. White text boxes identify larger-scale maps with additional details. The background CTX images are B02_010486_1779_XN_02S005W and B02_010341_1778_XI_02S005W.

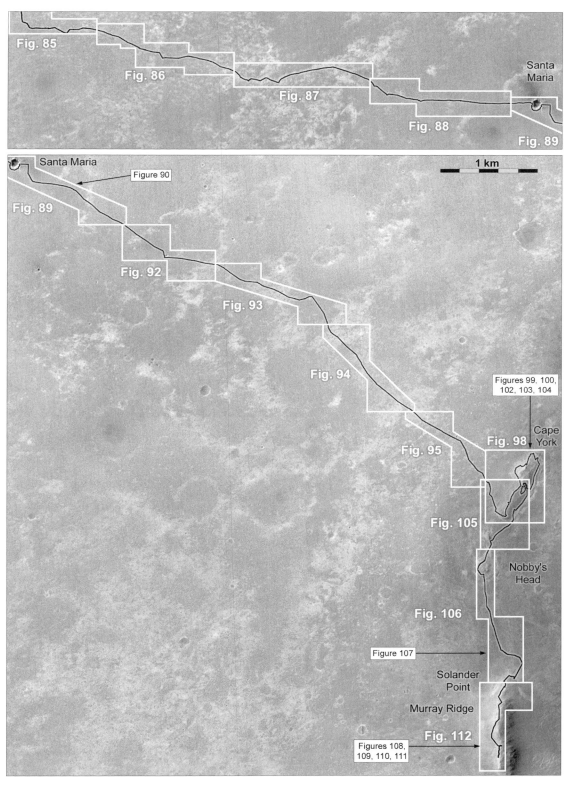

Figure 48 (continued)

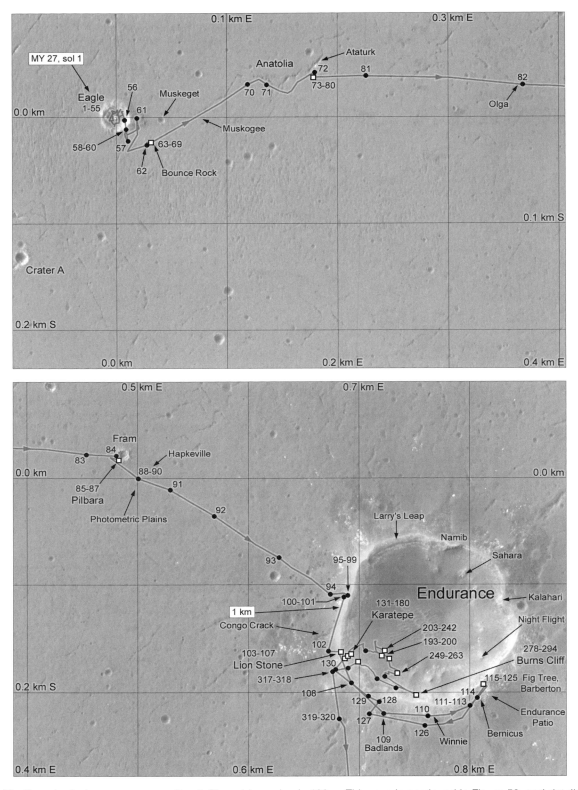

Figure 49. Opportunity traverse map, section 1. The grid spacing is 100 m. This map is continued in Figure 58, and details are shown in Figures 50 to 54. In all rover route maps, black dots are end-of-drive locations, white squares with dark borders are main science stops.

Power was very low after these activities, so sol 31 was mainly devoted to recharging the battery and using MB on the RAT hole. The MB analysis continued on sol 32, and on sol 33 a small move placed Guadalupe for analysis. On sol 34 MI viewed the target, the RAT ground into it, MI looked at the new hole and MB was placed on it. This new hole was called King, and APXS was used on it on sol 35. Remote sensing with Pancam and Mini-TES continued throughout these activities, including views of the horizon east and west to plan future drives.

Opportunity finished the APXS observations on sol 36 and backed away from the holes for Pancam multi-spectral imaging. Pancam and Mini-TES were used on the magnets on the next sol, and on the RAT holes, before the rover moved eastwards to a new target called Last Chance, not far from Stone Mountain. The magnet study continued on sol 38 with MI observations, followed by IDD work on a soil target called Pay Dirt. MI viewed the soil, MB was placed on it, and then MI looked at the MB impression in the soil, before a short drive put Last Chance within reach. Finally, on sol 38 a coordinated observation was made with Mars Express. This was for photometry, with images taken looking north, south, east and west to look at surface brightness variations with illumination direction while the site was imaged simultaneously by Mars Express.

Deimos was seen crossing the Sun on sol 39, the first satellite transit seen on Mars, and another coordinated observation was made as well as stereoscopic images of Last Chance and IDD work on a target called Makar. Sol 40 saw the end of an overnight APXS session and more MI observations, and early on sol 41 more ISO at Wave Ripple, a nearby dust drift. Later on sol 41 the rover used visodom for the first time during a drive to the Slick Rock area (Figure 50B), for ISO at Flat Rock. During this traverse on sol 41 the Mars calendar registered the start of a new year, sol 1 of MY 27.

These study areas had been planned from the initial survey of the outcrop. Sol 42 included a RAT grind at Mojo on Flat Rock and a practice imaging sequence for the Phobos transit on sol 45. The grind failed because the arm was poorly placed but the problem was diagnosed on sol 43, and MB and MI were used successfully at the target now called Mojo 2. On sol 44 a grind was attempted at that location, but it stalled during grinding when an embedded blueberry was dislodged from the

rock. Despite this, MB and APXS were used on the hole, and on sol 45, after a Phobos transit observation (Bell *et al.*, 2005), Mojo 2 was brushed and imaged with MI and then viewed with Pancam after Opportunity backed away. APXS measurements were compromised by a partly open door, so the instrument was viewed by Pancam to diagnose the problem. The door was opening properly but closing unreliably.

The Pancams imaged Berry Bowl, a small hollow in the outcrop filled with blueberries, and Briar Patch, an adjacent surface free of berries, on sol 42, and Mini-TES observed them on sols 43 and 45. The precise composition of blueberries was still uncertain because all targets so far had been mixtures of berries and surrounding soils. Here the IDD spectrometers could be used on a target containing rock and berries and on adjacent rock without berries, allowing the blueberry composition to be isolated. On sol 46 all IDD instruments were used on Berry Bowl, and on the next sol they were used on Empty, part of Briar Patch. Then on sol 48 Empty was brushed for ISO to confirm that blueberries were hematite nodules, before a drive to Shoemaker's Patio, the last stop along the outcrop.

The Patio took its name from Eugene Shoemaker, the renowned USGS planetary geologist. First, on sol 49, the IDD instruments were used on Raspberry Newton, a crack filled with red material also referred to as Filling. Then a small move on sol 50 brought Opportunity to Shark Tooth for more ISO. This was a dark red rock, which probably formed in a fracture like Raspberry Newton but was now exposed at the surface. Sol 51 began with a wheel scuff and imaging of the rock Carousel to test its hardness and the appearance of freshly scraped surfaces, and then Opportunity began a traverse around the south side of Eagle crater to inspect several soil targets.

The first drive on sol 51 was down the slope, counter-clockwise around the lander and back up the slope to examine small dusty ripples on the crater wall. Sol 52 began with ISO of the soil target Punaluu, and then Opportunity moved to the next location, Neapolitan, and examined its magnets with APXS. Sol 53 similarly began with ISO of the soil targets Vanilla and Cookies&Cream at Neapolitan, followed by another drive and Mini-TES observations of parts of the crater not yet completed. Sol 54 began with ISO on the targets Coconut 2 and

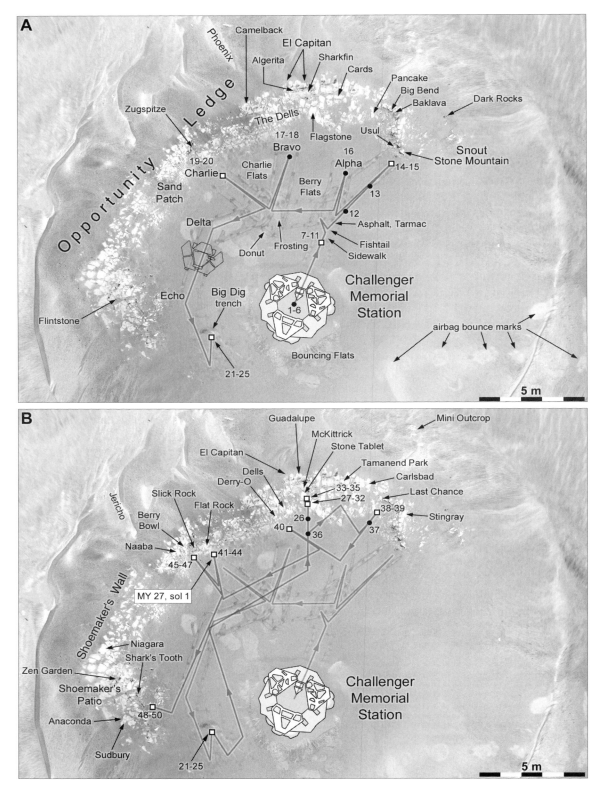

Figure 50. Opportunity activities in Eagle crater. **A:** Sols 1–25. **B:** Sols 26–50. The base images are reprojected surface panoramas from Opportunity. The lander, designated the Challenger Memorial Station, and rover are shown to scale. The name Big Dig was used again in Figure 51A.

Chocolate Chip at Mudpie, after which the rover was driven to a small patch of wind ripples called Meringue very close to the rover egress position beside the lander. The wheels were spun to cut a trench in the ripples and the rover backed up to image the result. MB and MI were used on the trench on sol 55, and then Opportunity drove east to climb the crater slope to its last stop, a group of small rocks or cobbles called Black Forest.

The cobbles were not accessible so the soil nearby was investigated with images, MB and MI on sol 56 before a first attempt to leave the crater later that sol. The plan was to drive up the slope, out onto the surrounding Meridiani plains and then a little to the south, but the upslope drive failed due to excessive slip and the turn to the south moved Opportunity along the crater's inner wall (Figure 51A). The wheels slipped in the loose soil on the upslope part of the drive, cutting a deep rut, which, like the sol 23 trench, was named Big Dig after a large highway excavation in Boston. Another drive south on sol 57 (MY 27, sol 17) brought the rover out onto the plains at last, after some imaging of targets including Fluffer Nutter, a bright patch on the crater rim disturbed by tracks, and Cool Whip, which was undisturbed. The activities inside Eagle crater are described by Squyres *et al.* (2004b).

The plains were very smooth and were covered with miniature ripples of dusty material. Linear depressions and chains of pits suggested drainage of surface material into fractures, including one called Pit Chain just south of Eagle. Sols 58–60 were spent back on the rim, imaging targets including Pit Chain and the rock Scoop and conducting ISO on the bright material at Mont Blanc. On sol 60 a full Pancam panorama called Lion King (taking its name from a popular animated film, which included a vista from a prominent viewpoint) was made to document activities in the crater and to view its surroundings. Batteries were recharged on sol 59 and flash memory was carefully monitored as it was nearly full.

The first targets on the plains were a patch of dark soil east of Eagle and a bright wind streak on the downwind (southeast) side of the crater. The dark soil target Munter at Catch Patch was reached on sol 61 and analyzed on sol 62, followed by a zig-zag drive into the wind streak, and on sol 63 Opportunity analyzed the brighter soil target Cleo. Then the drivers directed the rover to Bounce

Rock, so named because the airbags had bounced on it as the lander rolled towards Eagle crater. A software error required diagnosis on sol 64, but on sol 65 MI and MB was used on the rock, and on sol 66 APXS made its analysis. Later that sol the RAT ground 7 mm into Bounce Rock and MB was placed on the hole, a target named Case, while Pancam made a super-resolution sequence of images of the distant backshell, the best view of it that would be obtained. MI and APXS were used on Case on sol 67, and other spots on the rock were analyzed as well.

Bounce Rock was found to resemble EETA79001, one of the Martian meteorites called Shergottites found on Earth, and was very different from the rocks in Eagle crater. It was interpreted as ejecta from a distant crater, possibly Bopolu (Figure 43), whose secondary craters were seen to cross the landing ellipse only 8 km from Eagle crater. On sol 68 the rover repositioned itself, and on sol 69 soil targets called Luna and Grace near the rock were examined. This was the last sol on Mars time for the MER crew on Earth, the last time they would have to follow Mars time rather than terrestrial clocks, which was necessary to extract maximum results from the rovers but came at a physical and psychological cost.

After finishing the overnight APXS observation on sol 70, Opportunity was driven over Bounce Rock to try to fracture it, but without effect, followed by a record drive of 100 m over the hazard-free plains. The destination was Anatolia, a group of linear trenches possibly developed over fractures or karst-like solution hollows, and skirted on sols 71 and 72 (Figure 52A). After using MI and MB on a soil patch just east of Anatolia, another trench was dug on sol 73, the deepest so far, intended to expose the inner structure of a dust ripple. The battery was recharged on sol 74 as MB was placed on the trench debris, and then new rover flight software was uploaded on sols 75 to 77. Operating software was always called "flight software" on these missions even if the vehicle was no longer flying through space. After another MB measurement of the soil on sol 78, the new software was put into operation on sol 79 and APXS took an atmospheric measurement. Activities here concluded with the use of all IDD instruments on the trench on sol 80, and on sol 81, after a brief approach to the nearby Anatolia trough for imaging, the rover headed east towards a small, fresh, blocky crater called Fram. Craters at the

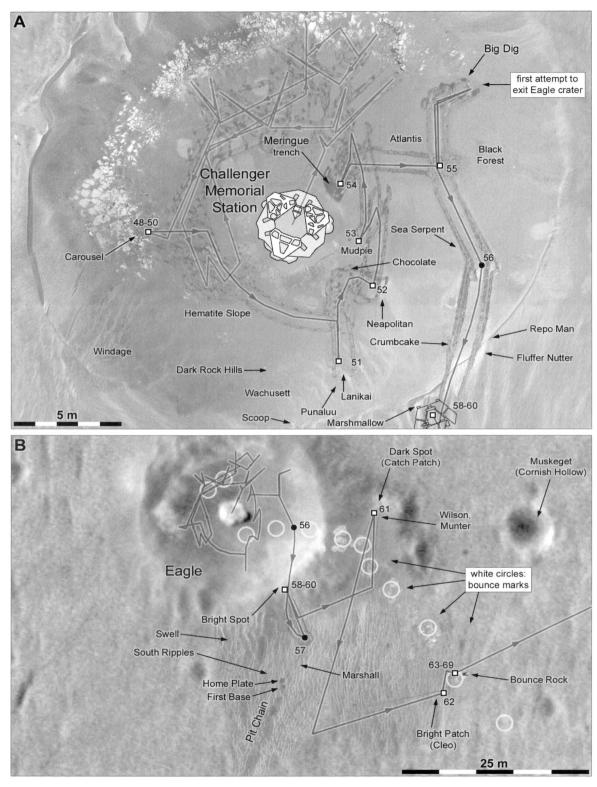

Figure 51. Activities in and around Eagle crater. **A:** Completing work in the crater, sols 51–60. The rover is at the location where it obtained the Lion King panorama. **B:** Leaving the crater, sols 56–69. The base is HiRISE image PSP_001414_1780, taken on 14 November 2006 (MY 28, sol 289), with additions from surface images.

Table 16. *Opportunity Activities in and Near Eagle Crater, Sols 001–101*

Sol	Activities
1	Entry, descent, landing, first images (MY 26, sol 630)
2–6	Panorama from lander and remote sensing, rover preparation for egress
7–10	Egress from lander (7), remote sensing (8) and ISO Tarmac soil target (9–10)
11–15	Remote sensing, drive to outcrop feature Snout, ISO Snout and Robert E. target (14–15)
16- 17	Drive around outcrop, image rocks from Alpha (16) and Bravo (17) locations
18–21	No science (18), ISO Dark Nuts soil target at Bravo (19), imaging (20), drive to trench (21)
22–25	Pre-trench ISO (22), dig Big Dig trench (23), ISO floor and walls (24–25)
26–29	Drive to El Capitan outcrop area (26), imaging (27), pre-RAT ISO at Guadalupe (28–29)
30–33	RAT McKittrick target (30), ISO McKittrick (31–32), move towards Guadalupe (33)
34–36	RAT Guadalupe target (34), ISO Guadalupe (35), then drive towards Big Bend (36)
37–39	Approach Last Chance area (37), imaging and final approach (38), ISO soil (39)
40–45	Drive to and image Wave Ripple (40), Berry Bowl (41), RAT and ISO Flat Rock (42–45), drive (45)
46–50	ISO Berry Bowl (46–47), drive to Shoemaker's Patio (48), ISO Shark's Tooth (49–50)
51–52	Scuff Carousel rock and drive to south wall of Eagle (51), ISO Punaluu, drive to Neapolitan (52)
53–54	ISO Vanilla, Cookies&Cream, drive, Mini-TES crater (53), ISO Coconut, drive to Meringue (54)
55–57	Dig and ISO trench, drive to eastern wall of Eagle (55), ISO soil and failed exit (56), leave crater (57)
58–59	Image plains and Pit Chain outside Eagle, return to crater rim (58), ISO Bright Spot on crater rim (59)
60–62	Obtain Lion King panorama (60), drive around crater (61, 62), ISO and image Dark Spot (62)
63–64	Drive to Cleo (62), ISO Cleo, drive to Bounce Rock (63), science prevented by computer error (64)
65–69	ISO Bounce Rock (65), RAT rock (66), ISO Bounce Rock and Maggie soil (67–69)
70–79	Drive towards Anatolia (70–72), dig trench (73), ISO trench (74), software update (75–79)
80–84	Complete Anatolia trench ISO (80), drive towards Fram crater (81–84)
85–87	Approach Fram crater, ISO and imaging (85), RAT Pilbara target (86), ISO Pilbara (87)
88–91	Drive to plains (88), photometry experiment, soil scuff and ISO (89–90), drive and remote sensing (91)
92–101	Drive towards Endurance crater, arrive on sol 95 (MY 27, sol 55), remote sensing of crater (96–101)

Note: In this and all subsequent MER activity tables, "RAT" (as a verb) refers to a RAT grinding operation, "brush" refers to a RAT brushing operation, and "ISO" refers to *in situ* observations, which may include the use of any combination of IDD instruments (Microscopic Imager and spectrometers).

Meridiani site were named after vessels of exploration, including spacecraft like Eagle and ships like Fram.

Sol 82 began with cloud imaging, followed by another record drive of 141 m, which exceeded any drive of Spirit's mission. Opportunity stopped near Fram on sol 83 (Figure 52C), and on sol 84 it used MI and MB on a soil target called Nullarbor before rolling up to the crater's rocky rim. A small move on the next sol brought the rover to a rock named Pilbara, where MI observed a target on sol 86 before the RAT ground 7 mm into it, the deepest grind yet made. After Opportunity backed up for better access, the IDD instruments were used on the hole on sol 87. Pilbara displayed vugs like those seen in Eagle crater, and Mini-TES suggested the presence of clays in another rock called Hammersley on the east side of Fram.

The Pilbara MB measurement concluded on sol 88 and Mini-TES viewed the target after backing up.

Fram was too rocky to enter, so the rover drove southeast into the plains again and scuffed the soil with its front wheels. Images taken there and just prior to stopping gave a long-baseline stereoscopic view of Endurance crater, the next target, whose rim formed low hills in the distance. This site was called Photometric Plains as Opportunity made photometric observations over sols 88 to 90, imaging the surface in different directions at different times of day as the IDD instruments were used on Nougat in the scuff and Fred Ripple nearby.

Sols 91 to 95 were taken up with daily drives towards the 160 m diameter Endurance crater while planners contemplated options. Opportunity might enter the crater if slopes permitted, explore the ejecta, or turn south. The rover traversed a cluster of small craters (Figure 49) and reached a low point on the rim of Endurance on sol 95 (MY 27, sol 55), making a large stereoscopic panorama

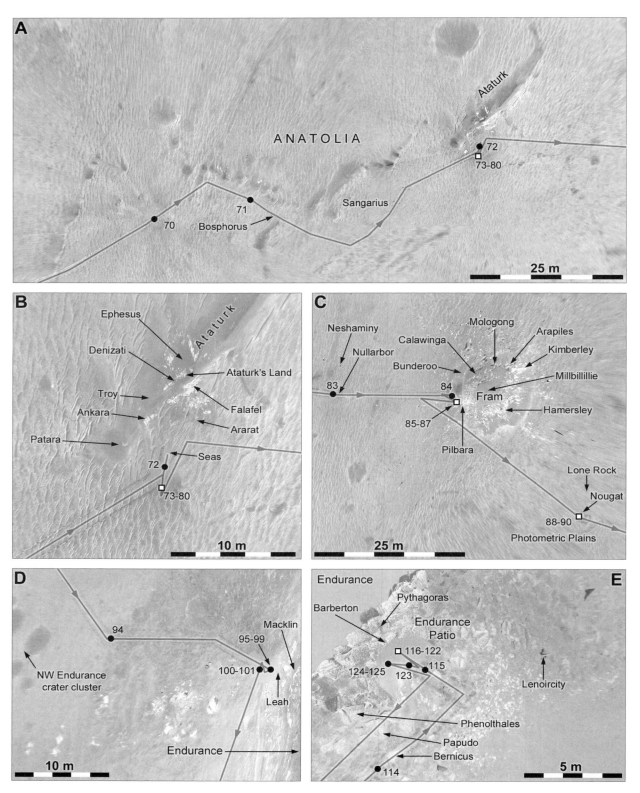

Figure 52. Opportunity activities on the approach to Endurance crater. **A:** Anatolia, sols 70–80. **B:** Anatolia trench site, sols 72–80. **C:** Fram, sols 83–90. **D:** Endurance rim panorama site 1, sols 93–101. **E:** Endurance rim panorama site 2, sols 114–125. These reprojected panoramas contain distortions due to surface relief.

of the dramatic view over the next four sols, the "B. B. King" panorama. On sol 100 long-baseline stereoscopic images were taken from points 1.5 m apart of two possible entry points to the crater, Larry's Leap and Karatepe, north and south of the rover, respectively. APXS and MI were used on the soil at the rim and

Mini-TES was pointed at Burns Cliff, a vertical exposure of rock with strata clearly visible on the south wall of Endurance, but it pointed too high and the data were spoiled. This outcrop was about 3 m high, exposing much more than Eagle's roughly 30 cm deep section, so it was an important target.

Table 17. *Opportunity Activities in Endurance Crater, Sols 102–315*

Sol	Activities
102–108	Drive to Lion Stone (102–104), ISO (105–106), RAT Lion Stone (107), ISO Lion Stone (108)
109–122	Drive around Endurance, remote sensing (109–122), examine Barberton pebble (121–122)
123–124	ISO McDonnell_Hilltop_Wilson soil target (123), ISO Pyrrho (124)
125–131	ISO Diogenes (126–127), drive around Endurance, remote sensing (128–131)
132–137	Approach rim (132), test drive (toe-dip) into crater (133), drive down to Tennessee (134–137)
138–142	RAT Tennessee (138), ISO Tennessee (139–141) and Bluegrass, Siula Grande and Churchill (142)
143–147	RAT Cobble Hill (143), ISO Cobble Hill (144), RAT Virginia (145), ISO Virginia (146–147)
148–151	RAT London (148), ISO London (149) and Tennessee (150), RAT Grindstone (151)
152–155	ISO Grindstone (152), RAT Kettlestone (153), ISO Kettlestone (154–155)
156–159	Recharge (156), drive to 6th Layer (157), imaging (158), atmospheric studies (159)
160–162	ISO Millstone (160), RAT Drammensfjorden target on Millstone (161), ISO target (162)
163–168	Remote sensing (163), ISO My_Dahlia soil target (164–167), magnet study (168)
169–174	Drive to Razorback (169–172), ISO Arnold Ziffel target (173–174)
175–180	Move to Diamond Jenness (175–176), RAT Diamond Jenness (177–178), ISO (179–180)
181–183	ISO Razorback (181), drive to Mackenzie (181), RAT Mackenzie (182), ISO Mackenzie (183)
184–185	Image Mackenzie, drive to Inuvik (184), approach Tuktoyaktuk, image Arctic Islands (185)
186–193	RAT Tuktoyaktuk target on Inuvik (186), ISO target (187–188), drive to Axel Heiberg (189–193)
194–197	RAT (194) and ISO (195–196) Axel Heiberg, ISO Sermilik target on Axel Heiberg (197)
198–200	ISO fragment broken off Sermilik (198), and Jiffypop (199), no science activity (200)
201–206	Drive to and image Dune Tendril (201–202), drive to Ellesmere, image Shag, Auk, Escher (203–206)
207–210	Imaging and atmospheric study (207), drive to Escher (208–209), ISO Escher (210)
211–214	Mini-TES dunes, ISO Kirchner target on Escher (211–213), brush EmilNolde and Kirchner (214)
215–217	ISO EmilNolde and Kirchner_RAT (215–216), brush and ISO Otto Dix (217)
218–221	RAT and ISO Kirchner (218, 220), conjunction sequence software upload (219), drive (220)
221–222	Microscopic imaging to choose placement of MB over conjunction (221), remote sensing (222)
223–236	Conjunction, ISO Spherules near Auk (223–235), drive to Auk soil target near Ellesmere (236)
237–243	ISO Auk (237–238), No Coating (239–240), Barbeau (240–241), move to Miro (242–243)
244–245	ISO Lyneal, Llangollen and Platt Lane targets on Welshampton (244), imaging (245)
246–249	ISO Void target on Rocknest (246–249), make cloud movie (248)
250–257	Drive to Wopmay (250–256), image magnets (253), ISO soil targets and reach Wopmay (257)
258–259	ISO Otter, Jenny, Hiller, Jet Ranger 2 and Twin Otter targets on Wopmay (258–259)
260–278	ISO Otter (260–261), image Wopmay (261–263), difficult drive to Burns Cliff (264–278)
273–279	Study magnets (273, 279), arrive at Burns Cliff (278), drive east along cliff and imaging (279)
280–285	Drive east, ISO Wanganui (282), imaging, continue drive to east along Burns Cliff
286–294	Remote Sensing of Whatanga and cliffs from eastern end of drive
295–304	Imaging, abandon plan to exit east of Whatanga, drive west towards Karatepe feature
305	Brush Paikea, ISO Wharenhui, Paikea and Contact targets on Black Cow
306–309	RAT Paikea, ISO Paikea (306), RAT Wharenhui, ISO Wharenhui (307–309)
310–312	ISO Karatepe (310), RAT Wharenhui again and ISO (311–312), drive uphill (312)
313–315	Remote sensing (313–314), exit from Endurance crater (315) (MY 27, sol 275)

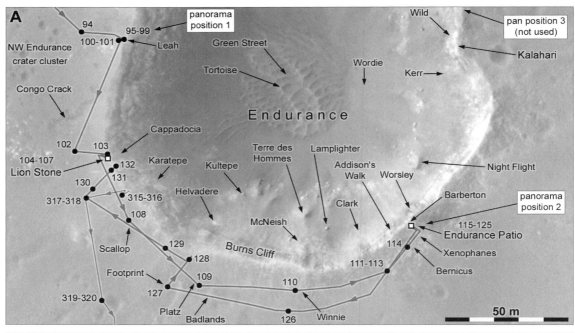

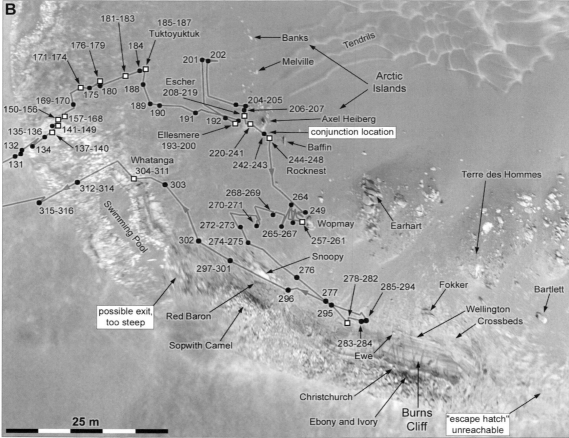

Figure 53. A: Activities on the rim of Endurance, sols 94–132 and 315–320, drawn on HiRISE image PSP_001414_1780. **B:** Activities inside Endurance, sols 131–316, on a mosaic of reprojected rover panoramic images. Some features received two names for different observations (Helvadere and Snoopy, Lamplighter and Bartlett, Kultepe and Wopmay). White squares are main science stops.

The arm heater, which drained power all night, was limiting science activity, but a partial solution had been uploaded with the new flight software. At night the rover could be placed in an even less active mode called deep sleep in which even that heater was disabled. This was first attempted late on sol 101 after a very busy day of remote sensing, which generated heat in the electronics inside the rover body. The danger was that something might fail overnight but Opportunity awoke unscathed on sol 102. Deep sleep was used throughout the mission after this, not every night but whenever it was needed to conserve power.

The Karatepe entry point was chosen for investigation of Endurance because a large rock near it could be examined. The rim of Endurance had very few large ejecta blocks because the friable sandstone of the plains was rapidly sandblasted to a smooth surface, leaving flat-lying slabs and blueberries on the surface. Opportunity drove south on sol 102, exceeding 1 km of distance driven, and on sol 103 it imaged a broad depression called Congo and then approached Lion Stone. After a final approach on sol 104, it used MI and MB on Lion Stone and MI on a nearby soil target on sol 105, and APXS on Lion Stone on sol 106. The rover moved slightly to position the arm correctly on sol 106 and imaged the heatshield to the south of Endurance. On sol 107 the RAT ground into the rock, making a useful hole despite its rough surface. APXS was placed on it overnight and MB on sol 108, followed by MI images of the hole. The team considered that it may have been an ejecta block derived from a dark layer in Karatepe, or possibly thrown there from another impact.

Before Opportunity was sent into Endurance, the planners wanted to characterize the crater interior from two or three points on its rim. On sol 109 the rover drove south around the sloping crater rim while the team was still considering when to begin regular use of deep sleep. The rover survived its first test of the procedure but the danger of losing Mini-TES was considered very real, and it would be essential for this survey of the crater. The team decided to delay regular use of deep sleep until after the survey was completed.

Sol 110 saw the rover move south of the crater where the outward-sloping rim now faced south, limiting solar power and science activities still further. The drive ended with a wheel scuff, which moved a subsurface rocky plate. Sol 111 was power limited again but another 40 m drive was made, and a Deep Space Network problem prevented the next sol's instructions being transmitted. The rovers always had standard "runout" instructions for these situations, so sol 112 was devoted to imaging and sol 113 was spent sleeping and recharging the battery. Short drives continued on sols 114 to 116 as Opportunity reached a new viewpoint to image the crater again. A potential third crater survey from a point further north on the eastern rim was now dropped (Figure 53A). Power was limited, but a full panorama and Mini-TES survey were obtained over sols 117 to 120. The base image for Figure 53B was constructed from a combination of this panorama and the one from sols 96–99, reprojected to fit HiRISE images for geometric control.

A 3 cm pebble called Barberton, examined with the IDD instruments on sol 121, may have been a meteorite but was too small for thorough analysis (Schröder *et al.*, 2008). It was rich in magnesium and nickel, contained olivine, and may have been a mesosiderite meteorite of a type possibly derived from the asteroid 4 Vesta. The first regular deep sleep occurred over the night between sols 121 and 122 at Barberton, with temperatures dropping to $-52°$ C, but Mini-TES survived. A test rover at JPL was being used now to practice driving on rocky slopes as the drivers prepared for a descent into the crater. A target called McDonnell was examined with the IDD instruments on sol 123, and another called Pyrrho on sol 124, followed by a small move to another target called Diogenes. Deep sleep was not used here, to allow an overnight data transmission, so on sol 125 only MI images of Diogenes could be taken.

On sol 126 MB examined Diogenes and the rover tested driving methods to prepare for crater entry. The front wheels were locked and the other wheels pulled the rover to measure friction on a rocky surface; each wheel was driven forwards and backwards while all the others were locked, scuffing the rocky surface. Then Opportunity drove 72 m back around the crater towards Karatepe, and another 50 m on sol 127, stopping to image the crater and take Mini-TES data. Deep sleep was used about every other sol now. On sol 128 the rover drove up to the rim to investigate a possible entry point, but it was too steep, so on sols 129 and 130 Opportunity returned to the Lion Stone area and the preferred entry point at Karatepe.

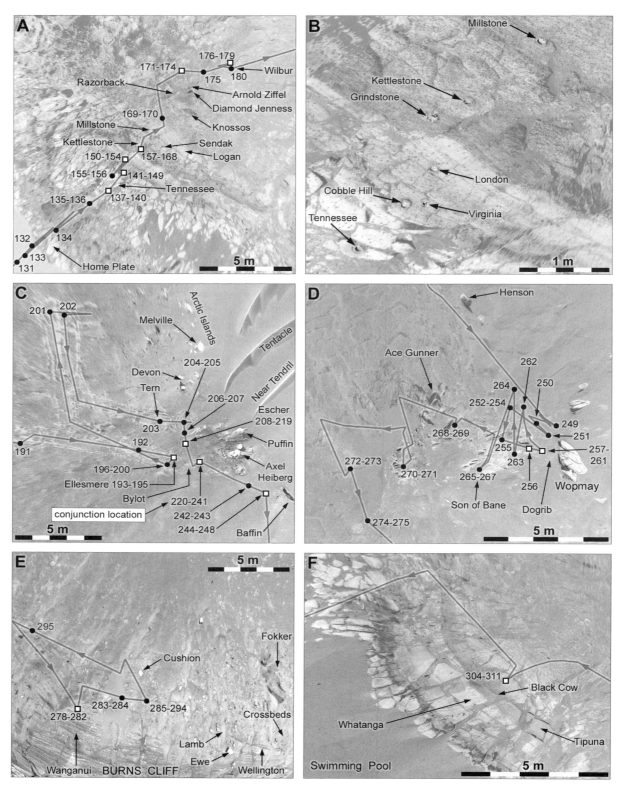

Figure 54. Activities inside Endurance crater. **A:** Karatepe, sols 131–180. **B:** Karatepe RAT holes. **C:** Arctic Islands region, sols 191–248. **D:** Wopmay area, sols 249–275. **E:** Burns Cliff, sols 278–295. **F:** Whatanga, sols 304–311. All maps include severe relief distortions and scales are approximate.

After imaging the entry route on sol 131 the rover backed up to improve the data link and transmitted its images to help with planning. Sol 132 (MY 27, sol 92) was the first move onto the slope, with just the front wheels over the rim. There was no guarantee that Opportunity would be able to climb back out of Endurance, so this might be its final resting place. The rover entered the crater carefully, gradually moving down the 28° slope to examine multiple layers of rock (Figures 53B, 54A and 54B). On sol 133 it drove down to place all six wheels on the slope, and then backed out to test driving on the slope. On the next two sols it descended a little more before pausing on sol 136 to choose targets for analysis. The slope faced east, which was good for morning power generation and warming but less satisfactory for data transmission.

The first target, Tennessee, was reached on sol 137 and the RAT ground 8 mm into it on sol 138 to allow analyses on sols 139 and 140. A short drive on sol 141 brought Opportunity to a contact between two distinct rock layers, for ISO on sol 142 at targets called Bluegrass, Siula Grande and Churchill, and a RAT grind at Cobble Hill on sol 143. Stability was a concern for a grind at a slope of 23°, so the grind was only 3 mm deep, but it was enough to analyze the target on sols 144 and 145. Later on sol 145 the RAT ground 4 mm into a new target called Virginia (Figures 54B and 56), which was analyzed on sols 146–147. The process was repeated at the target London, with a grind on sol 148 and analysis overnight with APXS (often used at night because it gave better data at colder temperatures) and MI and MB on sol 149.

On sol 150 Opportunity drove uphill slightly and then down to a position slightly below London, crossing a small step to reach a new rock layer. The tilt here was nearly 26°. After taking multispectral Pancam images of a target called Grindstone on sol 151, the MI was used on it and then the RAT ground into the rock. Analysis extended from late sol 151 to early sol 153, after which Kettlestone was ground into. Analysis of Kettlestone continued until sol 155, followed by a calibration procedure to ensure the expected IDD placement matched Hazcam images of the arm. Lastly on sol 155 Opportunity drove 1 m up the hill, mounting the low step again. The next sol's instructions failed to reach the rover because of a timing error so some runout images were taken and existing data were transmitted.

Pancam and Mini-TES viewed the recent RAT holes on sol 157 and then Opportunity was driven downhill to the next target area, ending on a slope of nearly 29° with one wheel off the ground, an unstable position that would not permit use of the RAT. On the next sol, images of the sky showed clouds above the crater, and MI images were taken of the next target, Millstone, after which a small move set the wheels firmly on the sloping ground. Sol 159 included more atmospheric observations, which were frequently interspersed with surface studies throughout the mission, and the Mini-TES mirror pointing was calibrated. A target called Drammensfjorden on Millstone was viewed with MI on sol 160 and the MB was calibrated by observing a rock slice of known composition which was mounted under the rover, the first use of that procedure. More cloud images were taken as Drammensfjorden was ground into and analyzed on sols 161 to 163, and later on sol 163 the APXS doors were tested again and Opportunity observed the slope above it.

My Dahlia, a soil patch near Millstone, was another ISO target on sols 164–167, and images to support future drives into the crater were taken on sol 166. APXS was used on the rover magnets to measure the composition of the fine dust adhering to them on sol 167, and MB did the same on sol 168. After finishing work here, Opportunity drove downhill towards a rock called Knossos on sol 169. The visible layering in Karatepe had been left behind now and the surface was composed of debris shed from the upper parts of the slope. Analysis of the rock layers showed that the lower layers were deposited as sand dunes, the middle layers as sheets of sand, and the upper layers were formed in an environment with occasional surface water (Grotzinger *et al.*, 2005). A test later on sol 169 involved taking an image during a data transmission to see if observations and downlinks could run simultaneously. The process seemed to work well. New targets now appeared in the images. A debris-free slab of rock called Flatland and a protruding ridge called Razorback, possibly a cemented fracture fill material, were imaged on sol 170 and approached on sol 171. The 3.7 m drive downslope included significant slipping on the loose material and the rover ended the drive with a broken fragment of Razorback called Arnold Ziffel in the IDD workspace. A crushed section of Razorback was named Hogshead. Sol 172 was devoted to engineering,

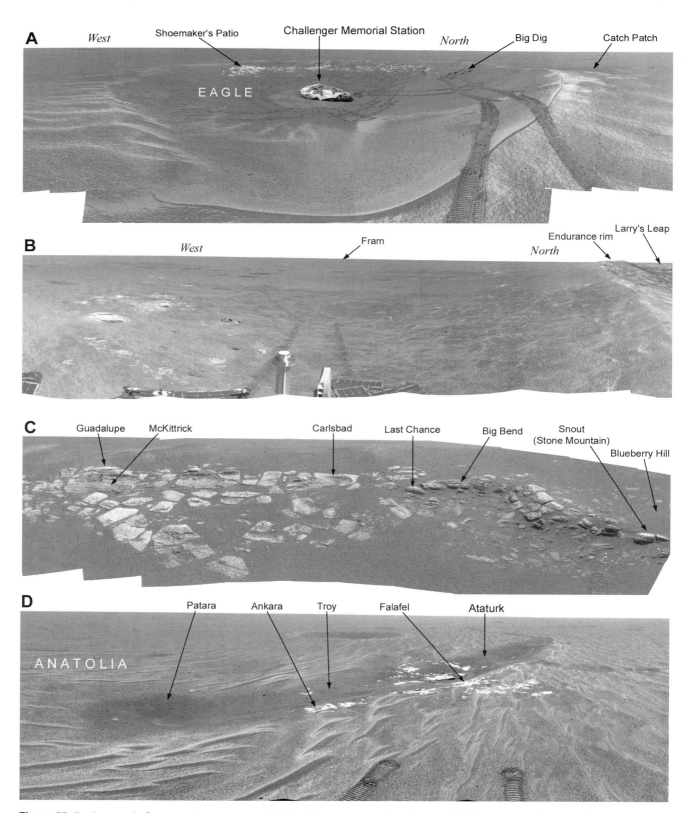

Figure 55 (both pages). Opportunity panoramas. **A:** Lion King panorama from the rim of Eagle crater, sols 58–60. **B:** Endurance rim panorama, sols 95–99. **C:** Opportunity Ledge outcrop in Eagle crater, sol 17. **D:** Anatolia, sol 81. **E:** Burns Cliff, Endurance crater, sols 287–294. **F:** Heatshield area, sol 330.

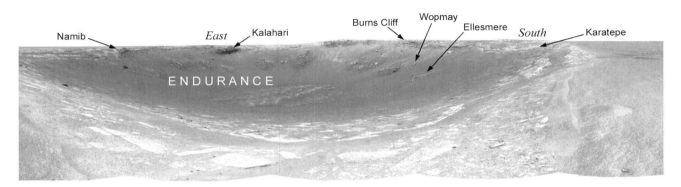

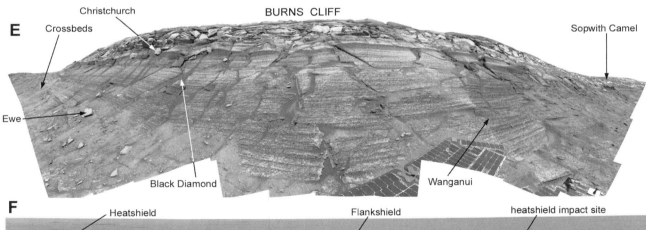

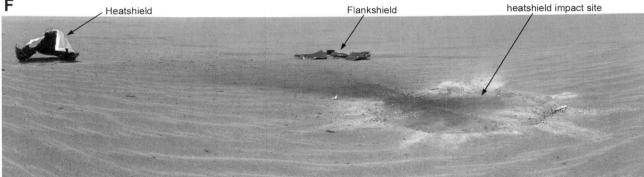

Figure 55 (continued)

with Pancam and Mini-TES views of the solar panels and more tests during data downlink. Arnold Ziffel was analyzed on sols 173 and 174, and on sol 175 the rover backed up a little to image the surroundings and then drove further downhill.

The next target was Diamond Jenness, reached on sol 176 and ground into on sol 177. A problem commanding the MI was encountered here. APXS was used overnight on the shallow RAT hole, and then it was deepened with an additional grind on sol 178 for further analysis with MB and APXS on sol 179. The two APXS observations might reveal variations in composition with depth. Opportunity now drove around a sandy area and down to a new target called Mackenzie on sols 180 and 181. Mackenzie was ground into on sol 182 and images were taken with the Pancams during the RAT activity, again testing simultaneous activities to try to achieve greater operational efficiency in future. Analysis of Mackenzie extended through sol 183 and MI observed the hole on sol 184 before a short drive back for imaging and an 8 m downhill traverse to a rock called Inuvik (Figure 53B). This drive made use of visodom to track its progress.

The slope here was only 15° but the wheels slipped badly in the loose soil. On sol 185 Opportunity moved to reach a target called Tuktoyaktuk on Inuvik (rover documents spell this as Tuktoyuktuk). After taking cloud images on sol 186, the rover conducted ISO on Tuktoyaktuk, including a RAT grind, until sol 188. Data return was slowed now by intense DSN activities during the MESSENGER Mercury orbiter mission launch. Sol 188 included driving tests to understand wheel slip in this environment of rocky plates with a thin soil cover. Over the next three sols Opportunity drove across the 17° slope towards a new target called Axel Heiberg, one of a line of rocks running north to south near the foot of the slope which took their names from Canada's arctic islands. Beyond them lay the striking dune field in the center of the crater.

Axel Heiberg may have represented a deeper stratigraphic level in the crater, possibly related to a series of rocky mounds extending around half the crater circumference at this level, with Axel Heiberg as a small outlier (Figures 53B and 54C). The rock was approached on sol 192 as the rover cleared old data from its flash memory, and after a cloud search and a small move on sol 193 a target on the rock was ground into on sol 194 and analyzed until early sol 196. Then a small move was made to place a light-toned vein called Sermilik within reach of the IDD, and it and other nearby targets were investigated. On sol 198 MB and APXS were placed on a part of the vein broken by the wheel, and an attempt to grind a target called Jiffypop on sol 199 was prevented by a stall. RAT diagnostic tests on sol 200 failed, leading to suspicion that a rock fragment was stuck in the RAT bit.

The dune field lay across a very smooth surface which may have been too soft to risk driving on, but a cluster of narrow tendrils or tentacles extended almost to the northern rocks of the Arctic Islands group and might be reachable (Figure 54C). On sol 201 Opportunity drove back to reach that area, and on sol 202 it drove closer in the hope that images would reveal hints of rock plates under the soil which would improve traction. No signs of buried rocks were seen, so the rover backed up and returned to the previous area, reaching a rock called Ellesmere on sol 203. On sol 204 the drivers tried to reach a target called Shag, but excessive slip in the loose soil brought Opportunity close to the edge of the safe area, so on sol 205 and 206 they directed the rover to move to another target called Auk. A rock called Escher with an unusual polygonal pattern on it now looked more interesting so the rover stayed there. Cracks in the rocks here suggested desiccation of once-wet sediment.

Sol 208 began with a small move to Escher and imaging of the rock and the disturbed soil around it. The rover recharged on sol 209 and on sol 210 viewed the RAT, finding that the troublesome rock had fallen out since sol 200. Pancam and Mini-TES studies of the dunes spanned sols 210 and 211, after which MB was placed on a target called Kirchner on Escher. The RAT was declared fit for use on sol 212 and MI viewed Escher, but a test of simultaneous use of Mini-TES and IDD operations failed. Mini-TES data were corrupted when the IDD moved. Escher was then analyzed until early on sol 214, and later that sol two RAT brushings were attempted on targets EmilNolde and Kirchner RAT. MI viewed both brushed areas and MB was placed on Kirchner RAT late on sol 214. APXS was placed on the same target on sol 215, and, though its doors failed to open completely, the data were still acceptable. MB was used on EmilNolde on sol 216, and a new target called Otto Dix was brushed on sol 217 for MI viewing.

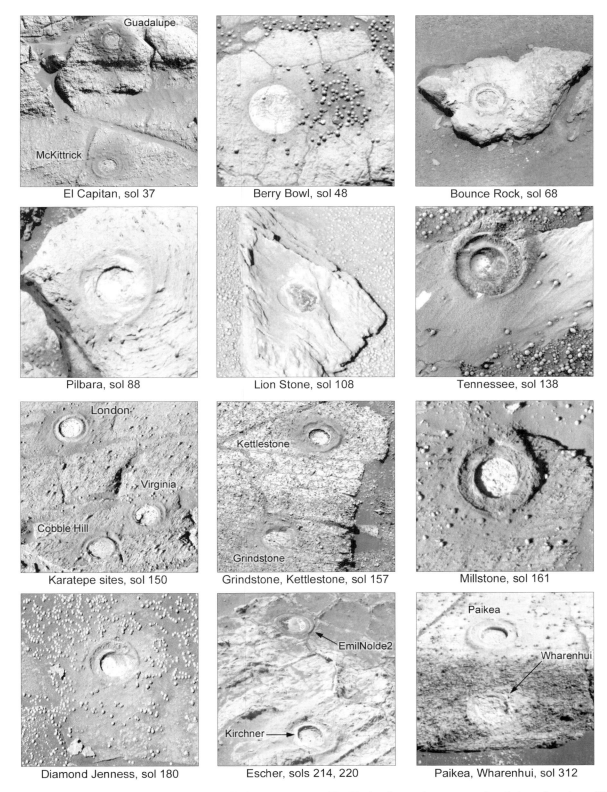

Figure 56. Selected RAT activities from Eagle to Endurance craters. The Escher image is a composite of views from two different sols and does not represent the simultaneous appearance of the two RAT sites. The dates are of the images, not the RAT events.

Later on sol 217, with the APXS doors now fully open, the instrument was used on a poorly cleaned part of EmilNolde and then on a properly brushed part of it. The complex operations continued with MI views of the APXS target and a RAT grind on sol 218, which cut 7 mm deep in 100 min of operation. All IDD instruments were now used on that target, with APXS use continuing overnight. The door problem had been resolved with a decision to leave the doors open from now on. Commands to operate the rover during conjunction were transmitted on sol 219, and after APXS use concluded on sol 220 the rover drove 5 m to reach a patch of blueberries.

Those berries were viewed by MI on sol 221 and Mini-TES made atmospheric observations, and then on sol 222 Pancam multispectral images were taken of the berries and the MB was placed on the target Auk for use over conjunction, supported by good power generation on this north-facing slope. Sol 223 had been set aside for conjunction preparations if problems had delayed other actions, but it was used instead for imaging and to calibrate APXS with a sky observation. The conjunction break spanned sols 224–237, during which MB ran a very long integration and communication tests were conducted to see what signals Opportunity could receive as it passed near the Sun.

On sol 238 regular communications resumed, the MB integration on Auk ended, MI viewed a soil target, Pancam imaged a rock called Ellesmere and the rover backed up a short distance. MER operations now transitioned to five-day weeks, cutting out weekend work to reduce costs. Three days of activities would be planned at once to cover the weekends. The IDD instruments were used on a target called No Coating on sol 239, and on Barbeau on sol 240 and 241. Then on sol 242 Opportunity was moved to reach a group of small chunks resembling popcorn, a target called Miro. The Navcams made a long-baseline stereoscopic survey of the area ahead for planning purposes on sol 243 and a final move brought Miro within reach. MI viewed three targets on Welshampton on sol 244, and on sol 246 MB was placed on a target called Void. MB operated until sol 248, taking longer now to get good results as its radioactive source weakened, and APXS took over on sol 249.

Later on sol 249 the next part of the trek began with a 20 m drive towards a 2 m long rock called Wopmay. The wheels slipped badly in the loose soil. After another drive on sol 250, a final approach was made on sol 251 using visodom to measure progress, as the slip was unpredictable, but even so the rover ended up too close to the rock for the arm to deploy properly. After multispectral imaging of Wopmay on sol 252, Opportunity moved back and used Mini-TES on the rock on sol 253, as well as some magnet imaging, and sol 254 was taken up with atmospheric observations. The north end of the rock was chosen as the site for IDD work, so over two drives on sols 255 and 256 the rover moved to the new site. Wopmay had a massive but cracked texture like Escher. Sol 257 began with a Pancam search for frost, as described for Spirit on its sol 304. Images were taken just after sunrise and again three hours later. No sign of frost was apparent on the soil, but it was seen on the rover deck and calibration target (Landis *et al.*, 2007). Then MI observed the soil near the rock and a last drive brought the arm within reach of its targets. MI viewed targets called Otter, Jenny and Hiller on the rock on sol 258 and Jet Ranger 2 and Twin Otter on sol 259 before placing APXS on Otter. That observation continued until sol 261, when MB took its place. The target names here are taken from northern Canadian aviation history.

After some final imaging, Opportunity tried to reach the southern, upslope, end of Wopmay with drives on sols 262 and 263, but the wheels slipped badly in the loose soil and on sol 263 a wheel was caught on a buried rock. The planners decided to leave Wopmay and climb up to the major rock exposure at Burns Cliff, backing away from Wopmay on sol 264. A drive on the next sol was stopped after a steering motor began drawing too much current and the following two sols were spent imaging to determine the rover's exact location. Driving resumed on sol 268 as Opportunity tried to avoid a buried rock named Son of Bane. Memory was nearly full as well and images of Wopmay that duplicated earlier views had already been deleted on sol 263 without sending them to Earth, a very rare occurrence, but a big downlink on sol 269 improved the situation. The north-facing slope was good for power generation, reducing the need for deep sleeps and permitting some night-time data transmissions. The wheel slip was so bad here that driving became a series of zig-zag movements to try to climb out of the loose soil, and the team was just about to give up and backtrack to Karatepe when a drive on sol 272 was successful.

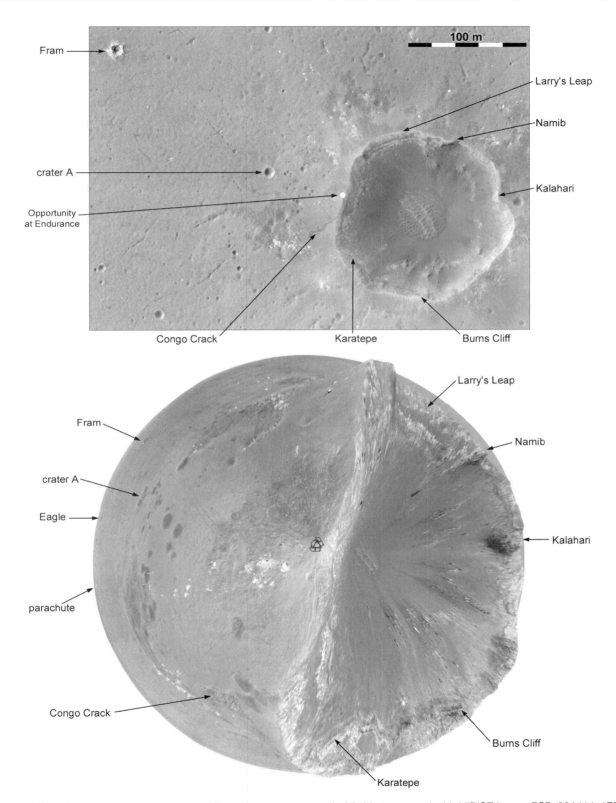

Figure 57. Circular panorama from the rim of Endurance crater on sols 95–99, compared with HiRISE image PSP_001414_1780.

On sol 273 the IDD was placed over the rover magnets for MI and APXS observations. Then on sol 274 Opportunity reached a rocky area beneath the cliffs which might allow a traverse along the outcrop, though a sandy strip east of the rover might be difficult to cross. After imaging on sol 275, several drives took Opportunity up to and around the top of the sandy area, reaching it on sol 278 and driving past it on sol 283 without difficulty despite the steep slope, as much as 30° in places. Mini-TES was used on the cliffs on sol 279 and imaging followed, with MI and APXS work on a target called Wanganui on sols 282 and 283. An interesting unconformity or cross-bedded outcrop called Whatanga, where higher layers of rock cut older, lower layers at an angle, lay ahead and would make a good target for study, after which it might have been possible to leave the crater by a route called escape hatch, slightly further east (Figure 53B).

Unfortunately, careful analysis of the terrain suggested the slope was too dangerous to leave by that route, so a small move forwards was made later on sol 283 to reduce the tilt a little, and another on sol 285 allowed the best possible observations of the unconformity, multispectral and super-resolution imaging and Mini-TES over the next few sols. The eventual geological interpretation of this area was that the sulfate-rich sandstone layers seen at Eagle crater and in Burns Cliff were underlain by thick cross-bedded sandstones, evidently old sand dunes cemented by salts precipitated from fluctuating ground water (Squyres *et al.*, 2006a).

Remote sensing continued until sol 294, including dramatic cloud images on sol 290, long Mini-TES observations of the cliff and imaging of many targets around the area, including the rocks Bartlett and Cushion (Figures 53B and 54E). The drive west to Karatepe began on sol 295, and by sol 297 another potential escape route called Early Egress Chute was examined, but it was too steep and was partly blocked by a rocky obstacle. Sols 298–301 were taken up with imaging over the US Thanksgiving holiday, and further driving brought Opportunity to a site with an interesting contact between rocks of two different colors on sol 304. This was called Black Cow, and two targets called Warenhui and Paikea were examined over several sols.

On sol 305 MI imaged both targets and the contact area, and Paikea was brushed for APXS to analyze it overnight. Opportunity took images of a dark rock called Tipuna on sol 306, viewed Paikea with MI, and then ground into Paikea for two hours, imaged the hole and placed APXS on it overnight. Warenhui was ground into on the next sol and APXS used on it, and the Pancams and Mini-TES made coordinated observations with MGS as it passed overhead. MB was placed on the Warenhui hole on sol 308, but the grind was not very satisfactory, so after two days of imaging a second grind was made to dig deeper into the rock on sol 311. MI and APXS were used on the new hole, and on sol 312 the rover backed up for imaging and then departed, driving to the northern end of the Swimming Pool feature.

The "Pool" was a smooth patch of soil which had to be bypassed so Opportunity could climb out of the crater on a rocky slope. The rover reached the crater rim just south of its entry point at Karatepe on sol 315 (MY 27, sol 275) and spent sol 316 imaging clouds and its surroundings, as planners noted a significant power drop due to the reduced tilt. The exploration of Endurance confirmed preliminary observations in Eagle crater that water had flowed over the surface or in depressions between sand dunes to produce characteristic rippled layering (Grotzinger *et al.*, 2005). Opportunity had now demonstrated beyond doubt, with this and the blueberries, the presence at some time in the past of water on the surface of Mars.

The old tracks from sol 130 had been degraded by 186 sols of wind action and dust movement, so on sol 317 Opportunity drove to the tracks for Mini-TES analysis on sol 318 while also imaging its next target, the heatshield. On sol 319 MI made mosaics of the old and new tracks to help understand surface processes and particle motions before the rover began driving towards the heatshield. Engineers wanted to inspect the heatshield to investigate its behavior during entry, though some images might not be released due to ITAR regulations (International Traffic in Arms Regulations, a set of very tight restrictions on the release of technical information with possible military applications). Then Opportunity would cross the Etched Terrain, an area of rocky outcrops very different from the smooth soil of the landing area (Figure 47), hoping to reach the 800 m diameter crater Victoria, 5 km to the south, which offered rock exposures far better than Burns Cliff.

Opportunity was now directed southwest, down off the elevated rim of Endurance, and then south towards

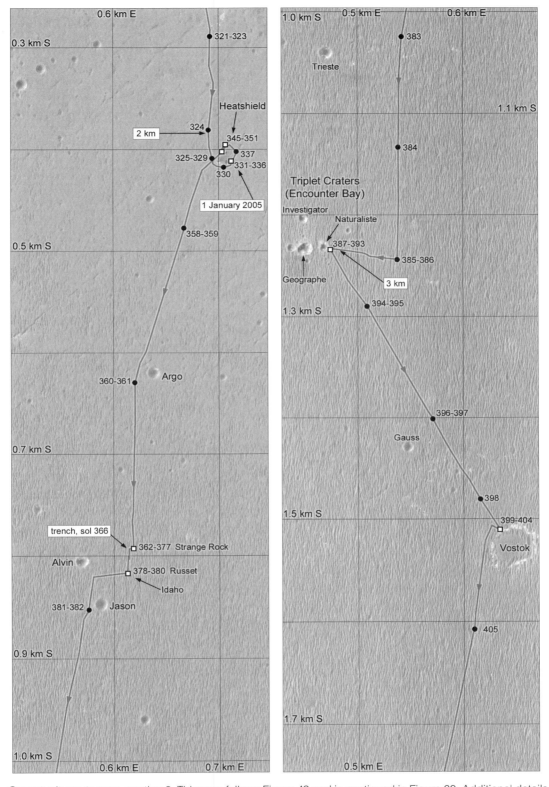

Figure 58. Opportunity route map, section 2. This map follows Figure 49 and is continued in Figure 60. Additional details are shown in Figures 59 and 61. White squares are main science sites.

the heatshield. Deep sleeps were needed again, the first on sol 320, and the rover drivers alternated forward and backward driving and used heaters on the drive actuators to reduce wear, as they were also doing with Spirit. MI viewed the solar panels on sol 321, and would investigate a cobble, a small rock, while planning heatshield images, if one was found. On sol 325 Opportunity exceeded 2 km of driving, and a map shown at a SOWG meeting suggested a route to Victoria. This might pass craters later called Triplet and Vostok, then drive down the east side of Erebus, pass Beagle and reach Victoria near Bottomless Bay (Figures 62–68).

The heatshield debris field was imaged from West Point on sols 326–329 as MB was calibrated. The heatshield had broken into two pieces, scattering small fragments in the process, and had bounced, leaving a prominent but shallow pit and a spray of dark ejecta (Figure 59A). Opportunity moved to South Point on sol 330 for more imaging as the sky darkened. A dust storm seen by MGS was increasing sky opacity now and reducing available power. On sols 331 and 332 the rover moved to the first large heatshield section, named Flankshield, as the storms worsened. Several local dust clouds were now visible from orbit in the Meridiani area. The landing had occurred shortly after the previous dust storm in this area, so Opportunity made observations to characterize this first storm of the mission. The rear Hazcams became mottled here as wind gusts blew dust and possibly heatshield contamination onto the lenses.

Inspection of the heatshield might have yielded insight into the effects of entry and heating, possibly aiding the design of future hardware. The specific targets for observation were the stagnation point, the part of the curving surface facing directly forwards during entry, and the thermal protection system to estimate the depth of charring. The best targets were out of reach on Flankshield but MI viewed parts of it on sol 334 as a new terrestrial year, 2005, began, and early on sol 337 APXS was used on one of the rover magnets over sols 334–336. On sol 337 Opportunity drove to East Point for more imaging, thoroughly documenting the large heatshield sections and the debris field around them, and then drove further, but now it experienced data management problems caused by over-filling its flash memory. Some data were lost as files were automatically deleted to avoid repeating Spirit's early problems.

On sol 338 Opportunity recovered from the previous sol's situation, established knowledge of its attitude again by imaging the Sun, and took some more images from North Point. The next sol included imaging from two points along a drive, examining the charred side of the heatshield here. Planners thought they might also inspect the "divot," the impact pit, especially on its western edge, the "incision" where the heatshield first hit the surface, where soil layering might be preserved. On sol 340 the rover moved forwards to place MI on a target on the heatshield for use that sol and the next. MB was used on the rover magnets on sols 341 to 343. MI was experiencing errors, first seen on sol 176, which suggested wires in the IDD might be degrading, possibly limiting future IDD use. On sol 344 the rover backed up to image the heatshield and downlinked data to allow more file deletes and clear memory.

Unfortunately, the shield had been turned inside out by the impact and the most useful material was hard to see. Contamination of the IDD instruments was also a factor in abandoning the study of the heatshield, but another unusual target lay nearby, a large rock with an unusual appearance. Opportunity drove to the rock on sols 345 and 346, imaged it and downlinked more data to help clear the flash memory. Meanwhile a future drive path was being sketched out, hopping from crater to crater to keep track of the rover's location in the otherwise almost featureless plains. These waypoints included Argo, Alvin, Jason, Trieste and Triplet, and other points of interest identified in MOC images, leading to Vostok, which might be a buried crater or a low mesa.

Heatshield Rock was pocked by many large holes and had shiny patches reflecting a spectrum closely resembling the sky. It was eventually identified as an iron meteorite, unofficially named Heatshield Rock by the MER team, and later officially named Meridiani Planum by the Meteoritical Society. Meteorites are officially named after the location in which they are found, and this was the first to be identified on another planet (Schröder *et al.*, 2008). Barberton, observed on sols 121 and 122, was not interpreted as being a possible meteorite until later. In 2014 Meridiani Planum was still the only meteorite on Mars that had been given an official name.

MI viewed Heatshield Rock, briefly referred to as Spongebob after a children's cartoon character, on sol

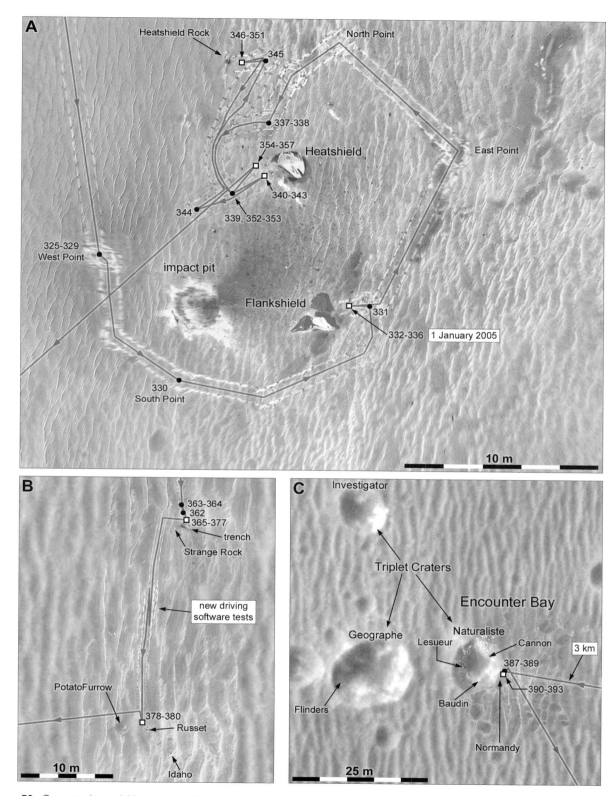

Figure 59. Opportunity activities south of Endurance. **A:** The heatshield area, sols 325–357. **B:** Strange Rock and Russet, sols 363–380. **C:** Triplet craters, sols 387–393.

347, then used MB on it for 19 hours into sol 348, and replaced it with APXS overnight into sol 349. The RAT brushed part of the rock named Squidworth and MI viewed it on sol 349 before using APXS on that spot overnight and MB on sol 350. The APXS observation on the brushed area was repeated on sol 351, and after final MI observations on sol 352 the rover backed away to collect additional images and Mini-TES data and then returned to the heatshield.

Atmospheric observations took up much of sol 353 and a final approach to the heatshield took place on sol 354. MI viewed it on sol 356 with its dust cover closed to prevent contamination if it accidentally touched the charred area, and with the cover open early on sol 358. The dust storms were abating now and the sky had cleared noticeably. On sol 357 Mini-TES viewed Heatshield Rock from a distance at different times of day and night to measure another diagnostic property, its thermal inertia, from the rate at which its temperature changed. Finally the rover drove away on sol 358 to resume its path south.

The long drive south to Victoria would cross the bright-toned region called Etched Terrain that had been seen from orbit (Figure 47). The plains near Eagle and Endurance had been very smooth and easy to drive over, but further south they became covered with progressively larger north–south oriented drifts. Because this orientation coincided with the drive direction, the drifts did not present much of an obstacle at first, and drives of 150–200 m per sol were common. Route planning was still being done with MGS MOC images, which barely resolved the drifts in most areas. The rover also stopped occasionally to examine rocks lying between the drifts (Figure 58 and Table 18). The craters showed various stages of infilling and degradation, reflecting their different ages.

Driving records fell several times during this period, first a distance of 154 m on sol 360, then 156 m on sol 352, including a reversal of direction after 90 m to relieve stress on the wheels. The sol 363 drive was prevented by a software fault, so the rover made atmospheric observations instead, and the next sol was lost as a DSN problem prevented transmission of instructions. A rock called Strange Rock was studied with multispectral images and Mini-TES on sol 365, followed by a drive of a few meters. Strange Rock appeared to be a typical fragment of the sulfate-rich sandstone seen in Eagle and

Endurance, thrown here by a recent impact. On the next sol Opportunity used its wheels to cut a trench through a drift crest to examine its inner structure, and on sols 367–371 the IDD instruments examined its walls, including a target called Caviar. On sol 371 the wheels made another soil scuff and Opportunity drove backwards to use the IDD instruments on the scuff on the next two sols. A change in APXS was noted now, requiring a recalibration. Between sols 250 and 368 a chamber in the instrument full of nitrogen had leaked, filling instead with Mars atmosphere.

On sol 373 the Pancams viewed the RAT cutting head to estimate wear of its teeth. About 65 percent of the teeth remained. Opportunity's RAT wore down much more slowly than Spirit's because most rocks at Meridiani were very soft. New flight software was uploaded to improve future mobility on sols 374–376 and used from sol 376 onwards. The new mobility tools, autonav and visodom, were tested on sol 378 (Figure 59B). Then Opportunity turned to face a rock called Russet on sol 379 for ISO on sol 380 at targets called Eye and Bridge of Nose, the latter just above a site named Nose which was difficult to reach. Russet was another locally derived ejecta block. MI viewed Russet early on sol 381, and then the rover moved south again, imaging the small craters Alvin and Jason on sol 382. Sols 383–385 were the first set of three consecutive drives all planned at once. Sol 383 set another record with a 105 m directed drive and a 72 m autonav drive in which internal analysis of successive images was used for hazard detection and avoidance. So many images were being taken for this that the flash memory was almost full by sol 385 and only a little atmospheric imaging was possible on sol 386 while data were transmitted.

The small crater Trieste was bypassed on sol 384 instead of being visited as suggested earlier. Then on sol 387 Opportunity drove across the north–south drifts, which were larger now but still easy enough to cross safely, to examine the Triplet crater cluster, an area now named Encounter Bay. The Triplet craters, which receive individual names from the voyage of Nicolas Baudin to Australia in 1800, were fresh (Figure 59C), probably a cluster of secondary craters formed by blocks ejected from a large distant primary impact crater.

After a survey with Pancam and Mini-TES, a final approach on sol 390 put the rock Normandie in the arm's

Table 18. *Opportunity Activities From Endurance to Erebus, Sols 316–631*

Sol	Activities
316–326	Imaging, drive off crater rim, microscopic images of old tracks (319), study solar panels (321)
327–330	Instrument calibration (327), magnet study (328–330), drive to South Point near heatshield
331–339	Drive to Flankshield, ISO Flankshield and magnets (334–337), drive to heatshield (337, 339)
340–344	Imaging, ISO fractured edge of heatshield (341) and magnets (341–344)
345–352	Drive to Heatshield Rock, ISO rock (347–348), brush rock (349), ISO rock (349–352)
353–364	Imaging, return to heatshield, ISO heatshield (356, 358), drive south past Argo crater
365–366	Remote sensing of Strange Rock and approach trench site (365), dig trench on ripple crest (366)
367–373	ISO trench (367–371), make wheel scuff (371), ISO trench and wheel scuff (372–373)
374–379	Software update (374–376), imaging, drive to test software (378), drive to Russet target (379)
380–382	ISO Russet ejecta block (380–381), drive to Jason and Alvin craters (378), image Jason (382)
383–393	Drive south (383–386), east (387), approach Naturaliste (390), ISO Normandie (393–393)
394–405	Drive to crater Vostok (394–399), ISO Laika (400), RAT and ISO Gagarin (400–405)
406–417	Drive past James Caird, longest MER drive, 220 m (410), ISO Mobarak trough target (415–417)
418–421	Approach ripple (418), ISO Norooz and Mayberooz targets on ripple (419–420), reach Viking (421)
422–432	Image Viking (422), drive to Voyager (423), image Voyager (424–427), drive south (428–432)
433–440	Wheel steering problem appears (433), mobility tests (437), software reset (440)
441–445	Anomaly recovery precludes science (441–442), ISO Cure soil feature (443–445)
446–462	Drive, wheels buried in Purgatory Ripple (446), imaging while planning escape strategy (447–462)
463–484	Short drives to escape Purgatory Ripple, and remote sensing
485–496	Remote sensing (485–490), drive north from ripple (491), return for imaging (496)
497–504	Imaging, ISO North Ripple (498), approach Purgatory Ripple (504)
505–514	ISO Purgatrough soil feature (505–506) and wheel tracks (507–510), imaging
515–537	Drive east around Purgatory Ripple and south towards Erebus crater, actuator test (525)
538–542	*In situ* study of magnets and solar array (538), drive south towards Erebus crater
543–544	Approach Ice Cream feature (543), brush and ISO OneScoop target (544)
545–549	RAT OneScoop (545), ISO OneScoop (545–549), continue southward drive (549)
550–554	Approach Outcrop551 (550), ISO Reiner Gamma soil target and Arkansas cobble (551–554)
555–557	Drive to FruitBasket (555), brush LemonRind (556), ISO LemonRind (556–557)
558–560	RAT and ISO Strawberry (558), ISO LemonRind (559), RAT and ISO LemonRind (560)
561–562	ISO Strawberry, image FruitBasket (561), drive towards crater Erebus and remote sensing (562)
563–577	Computer reset (563), recovery testing (564–576), short test drives (569, 571, 576) and imaging
578–582	Drive east to Erebus Highway (578, 580), imaging, drive south on Erebus Highway (582)
583–591	DSN transmitter failure (583), imaging (584–587), drive south towards Erebus (588–591)
592–593	Approach South Shetland (592), RAT and ISO Penguin target on Elephant feature (593)
594–598	ISO Kendall target on Deception feature (594–595), computer anomaly (596–598)
599–605	Drive west around Erebus outcrops, wheels slip in Telluride Drift (603), back out (605)
606–608	Image Telluride tracks (606–607), drive west around Erebus and look for cobbles (608)
609–624	Detour to west around large drifts, problems prevent science (610–613 and 623)
625–631	Drive south towards Olympia outcrop and rim of crater Erebus

workspace for analysis on sol 392–393. Initial results suggested that the target was not important enough to use the RAT on, so on sol 394 the rover backed up for imaging and began a drive towards Vostok. Opportunity viewed a crater called Gauss on sol 396 (the name was reused at Victoria on sol 1663). Mini-TES developed a fault, which precluded its use for a while, so the drive to Vostok continued until it was reached on sol 399 (Figure 61A).

Unlike Triplet, Vostok was clearly old, eroded and nearly filled with drifts. Jason was intermediate in age. At Vostok two targets called Laika and Gagarin were examined with MI on sol 400, and Gagarin was the target of APXS on sol 400, a brush and ISO on sol 401 and a RAT grind on sol 402, followed by more ISO over sols 402–404. Sol 405 began with a set of "Tai-Chi" moves to calibrate IDD movement with the cameras, followed by Mini-TES testing and a drive south, and driving records of 183 m and 190 m were set on sols 406 and 408. A Deimos transit was not observed on sol 408 because it fell during a much-needed communications pass via one of the orbiters. Sol 409 was used for battery charging and data transmission, and then the longest drive on a single sol by either MER, a 220 m traverse, was made by Opportunity on sol 410 just south of James Caird crater. At the same time Opportunity finally exceeded Spirit's driving distance to date.

Sol 411 was devoted to recharging batteries, and a large file delete took place to clear flash memory on sol 413. Driving occurred on most days as the distant Victoria crater beckoned. An interim target here was a study of the difference between a drift crest and the flat area between drifts. Drifts were growing in size now and constrained driving more than before. On sols 415–417 *in situ* observations were made at Mobarak, a trough floor target, and on sol 418 Opportunity mounted a drift to image the nearby blocky crater Vega (also called Knarr) and to use the IDD instruments on drift crest targets Norooz and Mayberooz on sols 419–420 (Figure 61B).

The next targets were two craters, Viking and Voyager, imaged on sols 422 and 424, respectively. The drive on sol 424 was stopped by excessive tilt as the rover mounted the rim of Voyager. Mars Odyssey entered a safe mode now, curtailing communications with the rovers over sols 425–427, though a small direct transmission from Opportunity was made on sol 426. On sol 428 the rover exceeded 5 km of driving, but the next drive on sol 431 was interrupted by a fault, and a 150 m drive on sol 433 was halted as the right front wheel steering actuator failed. On the next sol the rover backed up to see if its wheel was stuck on anything, but the actuator failed again and never recovered. Luckily the wheel was facing nearly forwards at the time, and could still drive but would not turn for steering, so this had little impact on mobility. Steering diagnostic tests occupied much of sols 435 and 436, and a test drive on sol 437 was fine except that a commanding error caused it to curve rather than head straight.

Over the next few sols the rover reversed driving direction periodically to distribute lubricant evenly in its wheel bearings. On sol 439 a multispectral image was taken of the trench cut by the turn at the end of the last drive, and then driving resumed. Sol 440 saw the first test of Mini-TES since a fault on sol 394, and then a software problem halted rover operations. Opportunity recovered on sol 441, transmitted data back to Earth, and performed ISO at a soil target called Cure over sols 443–445. Driving resumed on sol 446 (MY 27, sol 406) with an interim target, a north–south patch of rocky outcrop called Erebus Highway, which might offer a less troublesome path to Erebus crater and the Etched Terrain. After 40 m of a planned 90 m drive the wheels became embedded in a large drift, but they kept turning as commanded, digging even more deeply into the very soft soil.

This site became known as Purgatory (Figure 61C). Opportunity imaged its surroundings and the tracks on sols 447 and 448, and looked ahead to Erebus Highway on sol 449 with a super-resolution sequence. Those images were somewhat degraded, so they were retaken on sol 453. Also on sol 449 Earth was imaged in color about one hour after sunset, appearing elongated because it moved during the 15 s exposures. On sol 455 Navcam images of the rover's shadow were taken near sunset, to see how much light passed under the rover and assess clearance between the rover belly and the drift. The use of MI to look under the rover, as done by Spirit late in its mission, had not yet been developed. Meanwhile the "sandbox" at JPL was prepared to test extraction methods, and computer reset and other rover issues were investigated. Over sols 446–460 a large panorama called Rub-al-Khali was compiled to document the area and help plan an escape route, and on sol 446 a wind gust cleaned dust off the solar panels, boosting the power available for future activities.

A plan to free Opportunity was devised at JPL and tested with a duplicate rover. The rover would drive backwards out of the trap, so the wheels were straightened on sol 461 and the last panorama frames were taken on sols 462 and 464. The first drive on sol 463 consisted of ten short sessions, which would have moved a free rover 2 m, but in fact moved Opportunity only about 3 cm. This was repeated with slightly longer drives on sols 465 and 466 to test the procedure, and

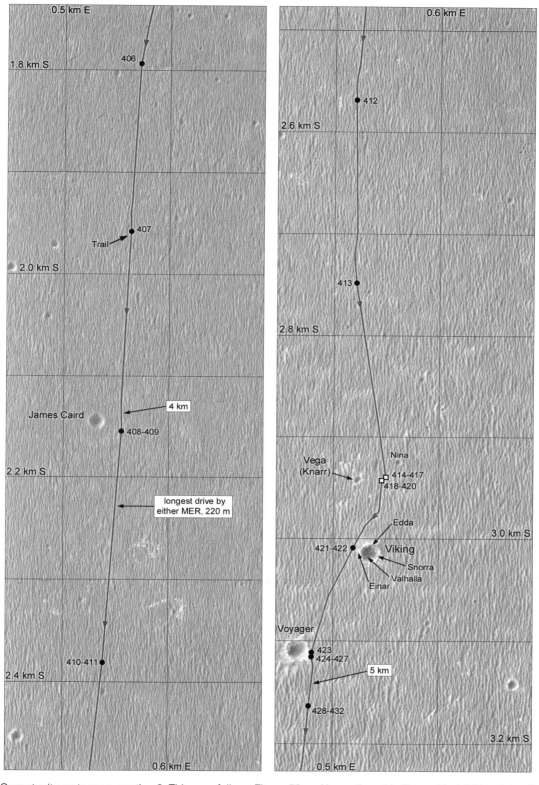

Figure 60. Opportunity route map, section 3. This map follows Figure 58 and is continued in Figure 62. Additional details are shown in Figure 61.

again commanded drives of a few meters produced movement of a few centimeters. Visodom was used to measure actual movement. Then daily drives of longer duration were tried, while cloud imaging and magnet observations continued to ensure some science data during this hiatus.

Sol 473 was spent imaging the area while the rover team wondered whether it was safe to continue this long drive to Victoria crater, or if another target should be sought in safer terrain. The southward path still seemed the best alternative. Daily driving resumed on sol 474 and continued until 483, by which time drives of 20 m were commanded each day, producing movement on the order of 10 cm each time, but good progress was clearly being made as the cleats on the wheels gradually pushed through the fine soil. Suddenly the 20 m drive commanded for sol 484 (MY 27, sol 444) produced a leap of 14 m out of the drift and into a broad trough to the north (Figure 61). Opportunity now studied the problem area briefly to help identify and avoid future traps like Purgatory. The tracks were carefully studied on sols 487 and 489 with Pancam and Mini-TES, the first use of the latter since sol 440.

The rover now executed a turn to approach Purgatory for a careful study, making several small moves over sols 494–506 and ISO of North Dune on sol 498 and of the tracks in Purgatory on sols 505–507. Mini-TES was also useful here, and MB collected data for 24 hours around sol 508. APXS and a night-time Mini-TES observation on sol 509 also contributed to the study, and MI completed the job on sol 510 as a route to the south was selected. Paths east and west of Purgatory were considered and an eastern route was chosen, so after some atmospheric observations over the US holiday weekend on 4 July (sols 513–514) the drive commenced.

Opportunity's route now took it north on sol 515, east on sol 516 into a wide trough and south again on sol 517. The flash memory was almost full again and the downlink was limited as the drivers were cautious about the rover's usual "turn for comm" routine, turning to ensure a good link with the relay satellite, because of the steering problem. For two sols the rover recharged and downlinked data, before driving resumed. Now Opportunity would check for wheel slip as it drove by processing images onboard. If the rover became embedded

again, the lack of motion would be detected and the drive would stop. The sol 520 drive was cut short at 10 m by a software error during a slip check, then two sols of 15 m drives were successful, but the sol 523 drive was again cut short at 16 m of 23 m commanded. Rear-looking images of the tracks were used to estimate movement but sometimes changes were not detectable, stopping the drive. The process took time and was subject to error, so a new strategy was used on sol 524, making several 5 m drives interspersed with short slip checks, which should allow greater distances per sol. It worked well and Opportunity drove 27 m.

Wind gusts were cleaning the solar panels again around this time, so power was increasing. On sol 525 the right front wheel steering was tested again without success, and on sol 526 Opportunity drove 32 m. The sol 528 drive was cut short by one of the slip checks in a region with many cobbles, which were interesting enough to image but not enough to stop for. Cobble targets called Apollo and Scotty were imaged on sols 530 and 531, sol 530 being the anniversary date of the Apollo 11 lunar landing and the death of James Doohan, "Scotty" on Star Trek. Other cobble targets around here included Glasgow (sol 532) and Sconset (533), and a large Mini-TES raster was made on sol 535. Drives on the order of 30 m continued every few sols around now, much shorter than before Purgatory but gradually increasing as drivers became more confident. The changes in driving distances are very apparent in Figures 60 and 62.

All this driving was planned using MGS MOC images, but it was proving difficult to locate the rover in them. The location estimate on sol 536 was 75 m north of the estimate after the previous drive. Only when HiRISE images became available 500 sols later was it possible to fix these old locations precisely (Stooke, 2013). Another large Mini-TES raster was made on sol 537. The instrument was working well now and the thermal data measured a surface temperature of 300 K as the low-albedo surface was warmed by the Sun.

After a long stretch without seeing any rocky outcrops, the sol 539 drive brought Opportunity to an area called Parking Lot, with small rock exposures in hollows between the drifts, the first sign of the Etched Terrain and the first outcrops not associated with an impact crater. The drifts south of it looked difficult so the planners decided to turn east to reach the Erebus Highway, a

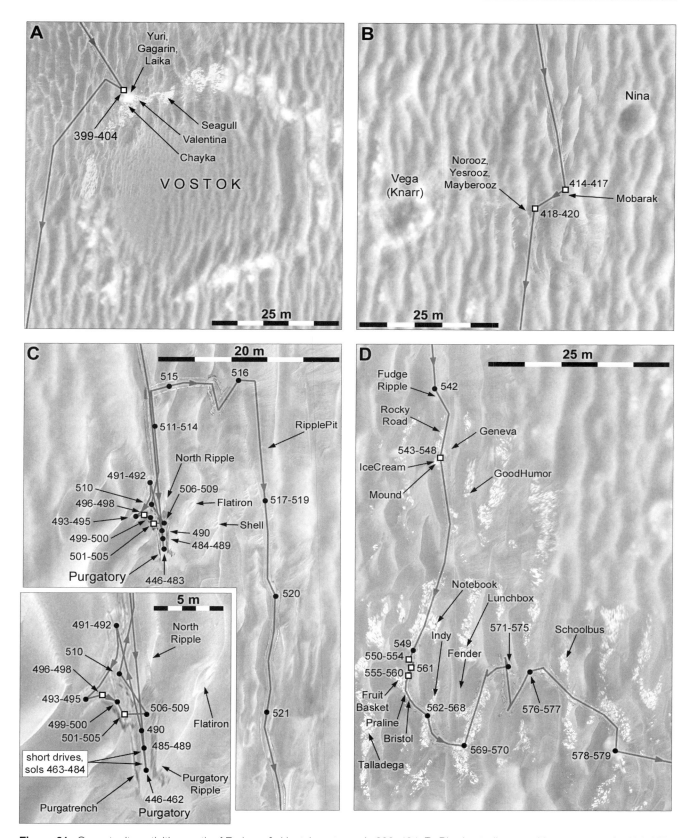

Figure 61. Opportunity activities north of Erebus. **A:** Vostok crater, sols 399–404. **B:** Ripple studies near Vega crater, sols 414–420. **C:** Purgatory, sols 446–521, and inset map showing details of Purgatory operations. **D:** Fruit Basket area, sols 542–579.

Table 19. *Opportunity Activities, Erebus to Victoria, Sols 632–951*

Sol	Activities
632–638	Approach Olympia (632), brush and ISO Kalavrita (633–637), brush and ISO Ziakas (638)
639–648	Drive to cobbles (639), ISO Anistasi (641–644), move, calibrate instruments (647–648)
649–653	ISO Jerome and Heber targets (649), drive south, arrive at Rimrock feature (653)
654–670	ISO Rimrock, arm joint failure (654), no science (655–656), imaging and tests (659–670)
671–685	Unstow arm (671), ISO Williams (671–678), brush and ISO Ted (679–685)
686–701	ISO Hunt (686–690), RAT and ISO Ted (691–696), make Fenway panorama (699–701)
702–708	Imaging (702–706), move to (707) and ISO Scotch target on Lower Overgaard (708)
709–717	Imaging (709–712), ISO Scotch_doublemalt (714), Branchwater (715), Bourbon (716), move (717)
718–725	Imaging, ISO Upper Overgaard (719, 721, 723), move to Roosevelt (724–725)
726–731	ISO Rough_Rider (726–730) and Fala (730) targets on Roosevelt, move to Bellemont (731)
732–739	ISO Bellemont feature (732–733), more arm problems (734), tests (735) and imaging (734–739)
740–756	Drive to Payson outcrop (740–742), drive through Halfpipe, imaging Payson (745–756)
757–760	Descend into Second Halfpipe, image Hokan (757), back to Payson (758–759), leave Erebus (760)
761–790	Drive south towards Victoria crater (761–790), image Red River Station (764) and Bosque (783–784)
791–802	Brush and study Buffalo Springs target on Rawhide (791), drive towards Victoria (792–802)
803–809	Approach Brookville (803), brush and ISO Brookville (804–806), drive south (807–809)
810–820	Approach Pecos River (810), ISO it (811), drive (812–817), brush and ISO Cheyenne (818–820)
821–832	Drive, imaging (821–824), ISO Alamogordo Creek soil (825–827), drive south (828–832)
833–851	Stuck in Jammerbugt drift (833), back out of drift (836–841), drive towards Victoria (842–851)
852–878	Imaging, load new software (852–876), drive south towards crater Beagle (853–878)
879–885	ISO Westport (879–881) and Fort Graham (880), imaging, drive to Jesse Chisholm (882–885)
886–890	ISO Joseph McCoy target on Jesse Chisholm (886–890) and Haiwassee target (890)
891–896	Scuff soil (891), drive towards Beagle, RAT and ISO Baltra (893–895), drive to Beagle rim (896)
897–908	Image Beagle (897–906), system fault (902) and recovery (903–904), drive onto Annulus (905–908)
909–916	Approach (909) and ISO Isabella and Marchena ripple targets (910–911), drive (912–916)
917–923	Move to trench location (917), dig trench (919), arm problems (920), diagnostics (923), imaging
924–927	ISO trench targets including Powell (924–927) and Powells_Brother (928), arm tests (925–926)
929–936	Drive to Victoria (929), approach crater Emma Dean (931), image it, approach Cape Faraday (936)
937–943	Imaging, ISO Cape Faraday (939, 941), arm fault (939) and tests (941), drive to Victoria (943)
944–951	Flight software boot (944, 945), imaging, mobility tests (948), drive to Victoria rim (950–951)

larger expanse of outcrop. On sol 543 Opportunity drove over a drift called Fudge Ripple, down a trough called Rocky Road and stopped at an outcrop named Mound.

This was the first chance to study the Etched Terrain, so on sol 544 MI observed the layered outcrop, now called Cone (it resembled an ice cream cone in shape) and a target called OneScoop, then brushed OneScoop and put APXS on it. On the next sol MI viewed the brushed area, used APXS on it again, ground into it and used APXS again. On sols 546 and 547 MI and MB data were gathered on the same target, and on sol 548 the RAT bit was imaged to assess wear again. The cutting teeth were considered a very limited resource, only to be used when really necessary, even as early as this, though the

device was still in limited use 3000 sols later. After overnight APXS operations and a last MI view of the hole early on sol 549, the rover moved back slightly for Pancam imaging and drove south again for 26 m.

Several cobbles were found here, including a target called Arkansas, approached with a short drive on sol 550. It was viewed with MI on sol 551, and MB data were taken, but the operators were concerned that the 3 cm cobble could become wedged in the instrument. On the next sol MI observed a soil target called Reiner Gamma near a cobble called Perseverance and APXS was used on Arkansas. A dust devil search was made on sol 553, though no dust devils were seen at Meridiani until sol 2310, and the ISO at Arkansas continued until

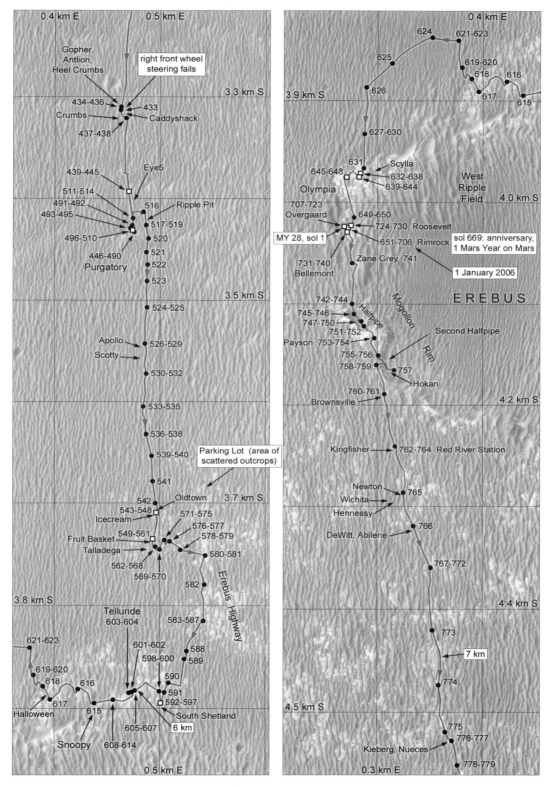

Figure 62. Opportunity route map, section 4. This map follows Figure 60 and is continued in Figure 65. Additional details are shown in Figures 63 and 64. White squares are main science stops.

late on sol 554. Then Opportunity drove south to Fruit Basket, another part of this first patch of outcrop, for ISO over sols 556 and 557 at a target called LemonRind, which resembled some of the crack-fill material seen at Shoemaker's Patio in Eagle crater but better exposed. It was viewed by MI, brushed clean, imaged with MI again and then analyzed by APXS and MB. On sol 558 similar procedures were used at Strawberry, this time with MI, a RAT grind, MI again and APXS. Attention then reverted to LemonRind with MB on sol 559 and overnight, followed by a RAT grind and APXS on sol 560. After more MI work on Strawberry, the rover backed up for Pancam imaging and Mini-TES of the Fruit Basket targets on sol 561.

Now Opportunity was driven on a path referred to as Tim's Southern Route on sol 562, but after only 7 m the drive was stopped by a fault. On sol 563 the rover powered off just after waking up again. It was restarted on sol 564, and on the next sol it recovered knowledge of its attitude and imaged targets called Crash and Burn on the outcrop and Tent Seam on a drift crest. No new work was performed until sol 569 as the rover team tried to identify the problem. On sol 569 a test drive was made, and on sol 570 Mini-TES and APXS were tested. The next drive was a zig-zag around ripples on sol 571, and the next on sol 576 used visodom to test that as well. Everything seemed normal again, but Mini-TES was suspected of being involved in the problem. It was tested again on sol 577, and then left unused until the situation was resolved.

Driving continued from sol 578 on, with dust devil searches on sol 579 and 581. Drives were typically about 30 m long now, both forwards and backwards, carefully skirting drifts until Opportunity reached the Erebus Highway on sol 582 and drove south on sol 583. On sols 583 and 584 data could not be transmitted to Earth due to a DSN failure, but it was received on sol 585. On sol 586 stereoscopic imaging of a target called Washboard was taken as the science team developed plans to visit Vermilion Cliffs, the eastern side of Erebus crater (Figure 64), to examine stratigraphy. The rock outcrops north of Erebus would be explored first to investigate rinds, cobbles, polygonal fractures and fracture fill materials. By sol 589 it seemed preferable to drive to the north rim and image the east and west sides of the shallow crater to see which might offer safer driving.

On sol 591 (MY 27, sol 551) Opportunity reached an outcrop called South Shetland and on sol 592 it moved 2 m to a target called Deception, darker than the surrounding rocks, which included a second target, Elephant. On sol 593 Elephant was brushed and the MI and APXS were used on Penguin, the name given to the brush mark. Sols 594 and 595 were devoted to APXS analysis of Deception and imaging of numerous other targets nearby, including Clarence, Snow, Nelson, Robert and Bellows, a laminated target near Snow. A sudden fault, a software timing issue, caused the rover computer to reboot on sol 596, and as it recovered from the anomaly on sol 598 planners decided to take the western route around Erebus.

On sol 601 Opportunity moved forwards 34 m to examine that western route, and on sol 603 it began a 45 m drive to the west, which ended after only 5 m in a large drift called Telluride. The rover backed off the drift on sol 605 and was able to bypass it at a low point on sol 608. Another computer reset occurred on sol 610, similar to the sol 563 event. Mini-TES had been used again on sol 599 despite the previous concern, and was on during this reset, so now its use was suspended again until the problem could be solved. The rover recovered on sol 611 and was supposed to drive on sol 612, but the instructions were not received. On sol 614 multispectral images of targets Chinook, Scirocco and Santa Ana were taken. Snoopy, also targeted here, was a bedrock patch misinterpreted as an impact crater from the MOC images.

The path now crossed patches of bedrock and meandered between drifts along the northern rim of Erebus, a broad shallow depression 330 m across, presumably a heavily eroded old impact crater, cut into the west side of an even larger and older crater, the 600 m diameter Terra Nova (Figure 64). Particularly large drifts blocked a direct path, so Opportunity turned north and west again into a patch of small drifts, and then south again onto another large outcrop area called Olympia, at the northwest corner of Erebus. Targets called Caramel and Cider were imaged on sol 619, and a dust devil search was made without success on sol 620. A software error on sol 622 prevented transmission of data, but Opportunity was driving again by sol 624. A different hazard caused problems on sol 628, when a nearby dust storm raised so much atmospheric dust that the rover had too little power to wake up from deep

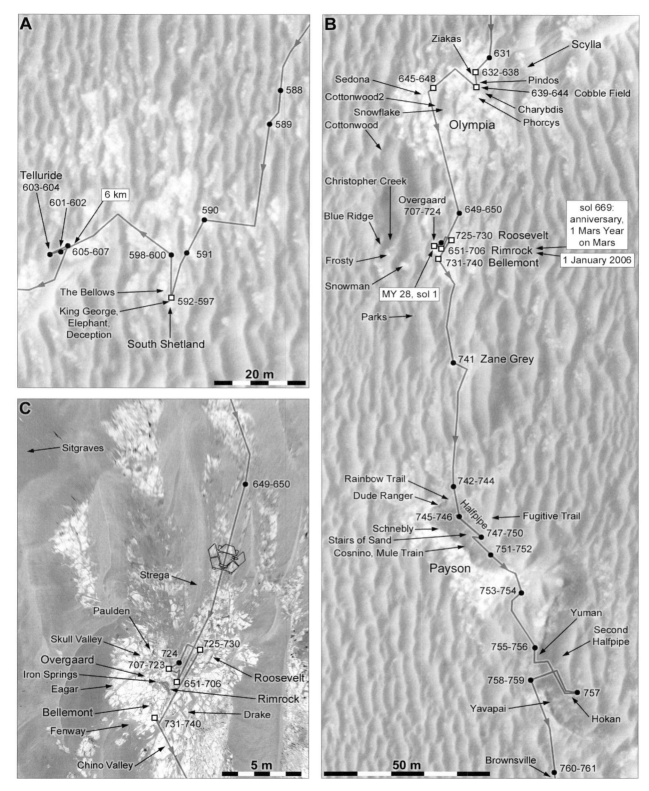

Figure 63. Opportunity activities near Erebus. **A:** South Shetland, sols 588–607. **B:** Erebus western rim, sols 631–761. **C:** Olympia, sols 649–740.

sleep, and when it did wake up later in the sol the computer failed again.

After arriving at Olympia on sol 631 (MY 27, sol 591), the first task was to characterize the outcrop at a target called Kalavrita. MI viewed it before and after a RAT grind on sol 633, and MB and APXS were used on it over sols 633 to 637. Care had to be taken to avoid filling flash memory again. Mini-TES was used again on sol 636, this time without problems. Another target here called Ziakas was investigated on sol 638 with MI, a brushing and APXS. Then Opportunity backed up to image Ziakas and drove 6 m to a cluster of cobbles on sol 639. On 641 the MB began a 40 hour integration of a cobble called Anistasi, spread over four sols, with some use of APXS on sol 642 and imaging of the cobble area Charybdis and an outcrop called Phorcys on sol 644. A large eroded drift called Scylla, on the eastern side of the outcrop, was another target.

On sol 645 Opportunity moved back to image Anistasi and then drove south, imaging layered outcrops called Show Low and Sedona on sol 647 and calibrating all of the IDD instruments. Pancam imaged an outcrop patch called Winslow on sol 648 as well as a cobble named Snow Flakes and Cottonwood, a drift, and on the next sol MI viewed outcrop targets Jerome and Heber before the rover was driven south, looking for a site to stop over the US Thanksgiving holiday. A promising site was found on sol 651 at a small rocky ledge named Rimrock. On sol 653 the rover observed sunset, and on sol 654 *in situ* observations were planned on a target on Rimrock called Prescott. The usual sequence of MI pictures, RAT brushing and APXS analysis was prevented when the arm failed to deploy from the folded, stowed position it always adopted when not in use.

The MB was supposed to collect data over sols 655–656, but could not with the arm still stowed. Diagnostic tests were performed on the IDD on the next two sols, and on sols 659–661 very small movements of the arm were successfully commanded while targets including Chino, Bellemont and Camp Verde were imaged. The fact that the arm moved at all suggested that a wire may have broken but the pieces sometimes made contact. As troubleshooting continued, many other targets were imaged until sol 665, including the rover deck, a super-resolution sequence of Bellemont and targets called Young, Cherry and Paulden. The arm moved as commanded on sol 666, and on the next sol test movements were attempted in colder and warmer times of day. The colder test failed. Mini-TES examined a rock called Parks, which might be a target on the way to Payson. Then the sol 668 imaging plans were prevented by an uplink failure.

Operations resumed on sol 669, the first anniversary of landing, but a planned coordinated observation with Mars Express failed. Another, early on sol 670, succeeded. The engineers had concluded that a wire in the IDD had cracked but that the arm could still be used with different software settings. On sol 671 the arm was unstowed and an MI mosaic of the target Williams was started. It was interrupted by another fault. After some atmospheric observations and another Mars Express coordinated observation on the next sol, the rover continued with the MI mosaic on sol 673, but it was interrupted again. On sol 674 the MI dust cover was closed and APXS was positioned for an overnight analysis of Williams. The analysis continued on sol 675 as engineers devised a way to drive with an unstowed arm, possibly resting on the deck, to reduce the use of the arm's shoulder joint. Meanwhile images were taken of Roosevelt, Forest Lakes and Paulden. MB was used on Williams over sols 676–678, and the MI mosaic was finished on sol 677.

The ISO continued on a target called Ted on sol 679 with an MI mosaic and an APXS analysis, while Navcam viewed the front of the deck to help plan an IDD resting spot. Williams and Ted were both parts of Rimrock, Ted being a blueberry-free bedding plane in the rock. On sol 680 Ted was brushed and APXS was placed on the brushed area, after which MB was used on it over sols 681–683. The terrestrial year 2006 began on sol 681. MI viewed the brushed area of Ted on sol 684, and then MB was returned to it until sol 685 while the Pancam was imaging targets and building up a multispectral panorama of Erebus. Another target here was Hunt, which differed in color from Ted. MI and APXS work began on Hunt on sol 686 and MB over sols 687–690. On the morning of sol 687 (29 December 2005) Earth and Jupiter were seen rising in the pre-dawn twilight. Images taken at 60 s intervals over a 30 min period showed wispy high-altitude clouds and the rising planets, Earth followed by Jupiter. Earth occasionally faded as it passed behind the clouds as they drifted towards the east.

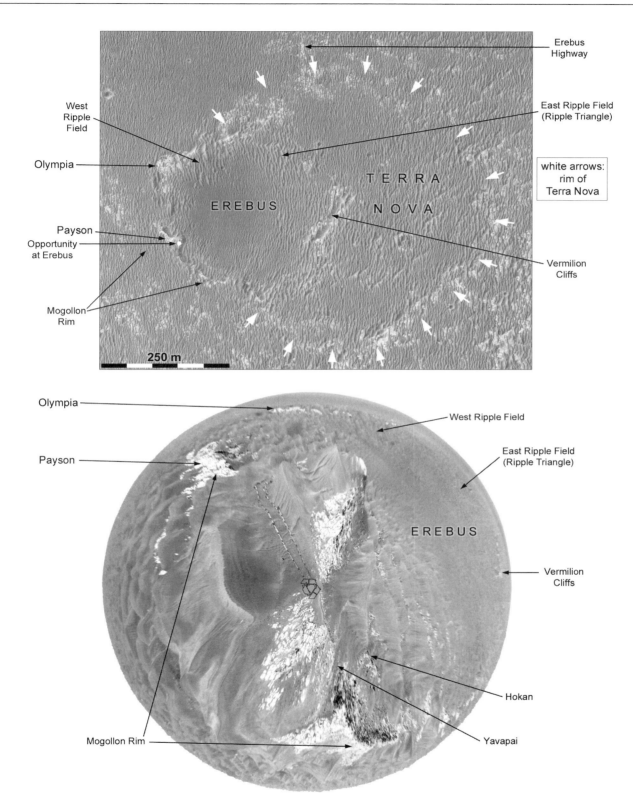

Figure 64. Circular panorama from the rim of Erebus crater on sol 755, compared with HiRISE image PSP_001414_1780. The rim of the ghost crater Terra Nova is indicated in the HiRISE image.

On sol 691 the RAT ground into Ted for 28 min, cutting nearly 3 mm deep, MI viewed the hole and MB analyzed it until sol 695 while the Erebus panorama was completed and sunset was imaged on sol 695. IDD work here concluded on sol 696 with APXS on the Ted hole.

On sol 697 the MI, Pancam and Navcam were calibrated by viewing the sky, and on the next sol and on sol 701 the IDD performed a set of "Tai-Chi" moves to calibrate its position in Hazcam images for future operations. A low Sun mini-panorama called the Fenway panorama was compiled on sols 699 and 701, and photometric observations using Navcam to take images at different times of day were performed over sols 699–701. The next IDD site would be a spot first called Iron2a but now named Overgaard, so multispectral images of it were made on sol 702. Imaging on sol 703 was prevented by an uplink failure, and on sol 704 the first attempt to stow the arm using new settings also failed. Mini-TES made coordinated observations with Mars Express on sol 705, and on the next sol the arm stowed successfully, APXS pointing upwards, as Pancam made coordinated observations with Mars Express and viewed a Deimos transit of the Sun. Phobos was observed during similar transits on sols 707, 708 and 709. Sol 709 was also the first sol of Mars year 28. Earth and Jupiter were seen rising again in the morning of sol 718 (29 January 2006).

On sol 707 Opportunity backed up 1 m and moved forwards again 1.7 m to Overgaard, intending to reach the upper part of it but falling short with Lower Overgaard in range of the arm. Overgaard was an outcrop with distorted bedding in the finely layered rocks, interpreted as ripple marks caused by shallow surface water flow (Lamb *et al.*, 2012). The sediments here lacked blueberries, suggesting that these layers were stratigraphically higher than those of Eagle crater. The iron-rich ground water from which the hematite concretions had precipitated did not extend this far up the sequence of layered rocks.

MI was used on a Lower Overgaard target called Scotch on sol 708, but the arm stalled again and the Scotch images were out of focus. It was tested on sol 709 and more coordinated observations with Mars Express were made with Pancam and Mini-TES on sols 710 and 711. Pancam made super-resolution images of targets Loupp and Dewey on the next sol. A sol 713 MI

sequence on Bourbon was canceled because of the arm stall, but Pancam imaged a target called Mariah, a projecting rock layer named (via the song *They Call the Wind Mariah*) after the violent Santa Ana winds then (24 January 2006) affecting JPL. The arm was back in operation on sol 714 and MI repeated its observations of Scotch. More mosaics were made on the next two sols, covering the targets Branchwater and Bourbon, and then Opportunity moved forwards slightly to reach Upper Overgaard on sol 717.

Sol 719 coincided with the birthday of Wolfgang Amadeus Mozart (b. 27 January 1756), so the targets imaged by MI were given names associated with the composer: Don Giovanni, Nachtmusik and Salzburg. Another arm fault occurred during the Salzburg imaging, but that work was completed on sols 721 and 723. The arm was stowed and Opportunity made short drives on sols 724 and 725 to work at Roosevelt, a possible fracture fill with blueberries below it and in it but not above it. APXS and MB analyzed a target called Rough Rider, which was too rough to brush, on sol 726, and MI made a mosaic of it on the next sol. MB returned to it on sols 728–730 as Pancam observed the nearby targets Delano, Eleanor, Teddy and Franklin, with names associated with US Presidents, and Panama Canal, a soil area just west of Rough Rider. On sol 730 APXS was placed on a nearby ripple called Fala overnight.

The next target here was Bellemont, reached on sol 731. The planners were looking ahead, anticipating stops at an outcrop called Zane Grey and the Payson area before commencing the long drive to Victoria crater, stopping along the way only for particularly compelling targets such as playa deposits or exotic cobbles. MI viewed three targets on sol 732, Vicos, Tara and Chaco, before work was stopped by another arm stall. A fourth target, Verdun, was imaged on sol 733 but another stall prevented imaging to fill a gap in the Chaco mosaic. The rover team wanted to move before winter reduced their power and mobility, but a drive to Zane Grey on sol 734 was prevented when an arm stall prevented it from stowing safely. Sol 735 was devoted to imaging and IDD diagnostics as planners contemplated whether to drive on top of Payson or in the trough below it. Imaging from Zane Grey would help them decide.

The sol 736 drive to Zane Grey was postponed as a DSN fault delayed the command uplink until the next

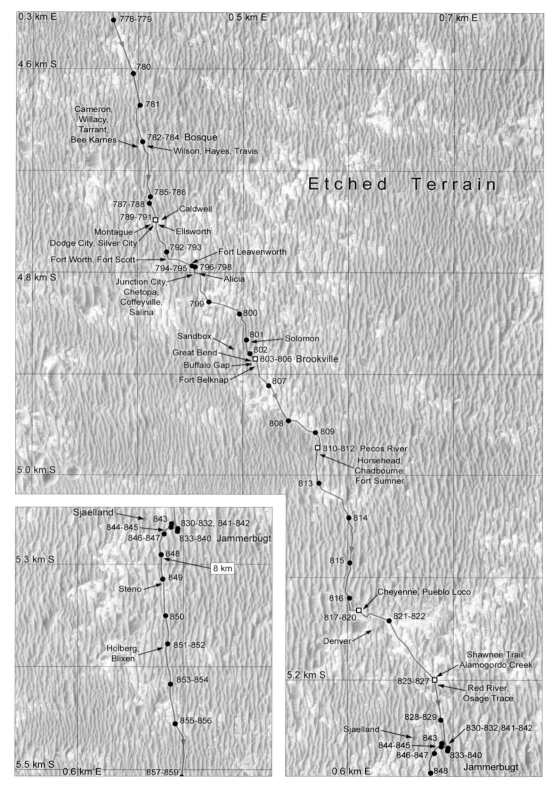

Figure 65. Opportunity route map, section 5. This map follows Figure 62 and is continued in Figure 66. Additional details are shown in Figure 67.

sol. After more imaging, the rover drove on sol 740, but was stopped after only 21 cm by a high wheel current reading. Finally, on sol 741 the 35 m drive was successful, and on the next sol Opportunity crossed a drift and approached Payson. This 1.5 m high rock face was part of the western rim of Erebus, the Mogollon Rim (Figures 63B and 64), and the drive would take place at its base to permit images of its stratigraphy. A wide trough between the drifts and the rock face was named Halfpipe (Half Pipe in some sources), and Opportunity viewed targets called Dude Ranger, Mysterious Rider and Rainbow Trail with Pancam or Mini-TES on sol 744. After more driving, another section of the outcrop was examined on sols 748–749, this time at targets including Wilderness Track, Deer Stalker and Fugitive Trail. The names were taken from books by Zane Grey. Most of the imaging had been done after noon with the rocks of Payson in shadow, so morning imaging was done on sol 750. The rocks were clearly cross-bedded (Arvidson *et al.*, 2011).

Opportunity climbed out of Halfpipe on sol 753 onto the top of Payson where a spur extended into Erebus, and on sol 755 it drove across the spur to the top of the next "half-pipe" area. After imaging the area and looking for clouds on the next sol, the rover descended into the next depression on sol 757 to look at the rock face. Unfortunately it was obscured by loose rubble in places, and this "half-pipe" was blocked by a drift at its far end. Several rocks here, called Yuman, Hokan and Yavapai, were observed with Pancam and Mini-TES, and on sol 758 the drivers brought Opportunity up out of the depression again.

After recharging and imaging clouds on sol 759, the long drive south began on sol 760 (MY 28, sol 52). The rover crossed the outcrop and entered a trough between drifts. The general orientation of drifts was still north–south, roughly the desired driving direction, but eastward moves would be made on outcrop patches as needed to reach Victoria, 2 km further south. Deep sleep was used every few nights as needed.

Mini-TES observations and imaging, including cloud searches and movie sequences, were made on sol 761, as at many stops along the way. The sol 762 drive brought Opportunity to Red River Station for imaging of that outcrop and other targets named Kingfisher and Rush Springs. Western and cowboy names continued to be used here as at Erebus. Drives on the order of 30–60 m were common now, with more frequent slip checks where outcrop patches were absent.

Mars Odyssey went into safe mode on sol 768, causing a delay until driving resumed on sol 773, when the rover team was no longer in restricted sols and Opportunity could be driven almost every day. The rover passed the 7 km point of the drive on sol 774 and stopped for recharging on sol 777, where it observed targets named Nueces, Kleberg and San Patricio. After several more drives another outcrop called Bosque was reached on sol 782, providing remote sensing targets as the rover moved back into restricted sol planning.

A "monster ripple," a particularly large drift, lay ahead and the rover had to cross from one trough to another to avoid it, climbing onto the drift to its left on sol 787 and down into the next trough on sol 789, stopping early because of a visodom error. An outcrop here was examined with MI and a brushing on Buffalo Springs on sol 791, and imaging of the cobble Caldwell, the outcrop Silver City and a drift called Dodge City. Driving resumed without using APXS as originally planned.

On sol 796 images of a target called Fort Leavenworth were taken, and another drift had to be crossed, but excessive tilt stopped that drive. Further imaging at that location included targets named Junction City and Salina. A drive on sol 800 also crossed drifts, cutting a trench unintentionally. It was imaged on sol 801 before continuing southwards. An outcrop beckoned and Opportunity reached it on sol 802, but the rock was too fragmented and dusty for useful IDD work, so the rover drove another 8 m to Brookville on sol 803. Brookville was examined on sol 804 with MI, then brushed, and MB was placed on it as another target named Great Bend was viewed with Mini-TES. APXS was used on Brookville overnight starting on sol 805, then replaced with MB as the Pancams observed Great Bend and a banded ripple called Fort Belknap. Names from the Chisholm trail, a nineteenth-century US cattle trail, had been used up, and names were now being taken from the Goodnight–Loving trail, another path used to take Texas cattle to various markets. Opportunity drove away on sol 807 after imaging the brush target.

On sol 808 the first view of Victoria crater was glimpsed. This was a bright outcrop sometimes referred

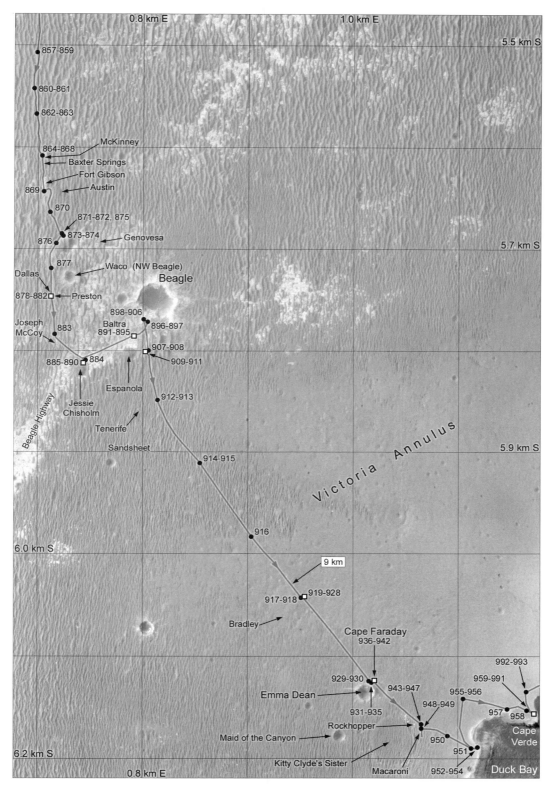

Figure 66. Opportunity route map, section 6. This map follows Figure 65 and is continued in Figure 69. Additional details are shown in Figure 67. White squares are main science stops.

to in Internet discussions as "the Beacon," part of Cape St. Mary (Figure 68) but often misinterpreted as part of the far side of Victoria's indented rim. On sol 810 another banded ripple target was approached for MI work on sol 811. This target was Pecos River, and other remote sensing targets nearby were called Horsehead and Chadbourne. On sol 812 Opportunity transmitted via Mars Odyssey while it was low in the sky, testing procedures for the forthcoming Phoenix mission, and on sol 816 it imaged the setting Sun. Then on sol 817 it drove towards a small depression, and on the next sol it made an MI mosaic of the outcrop Cheyenne before brushing it and placing APXS on it. MI viewed the brushed area on sol 819 and then placed MB on it for two sols while another outcrop called Pueblo was also imaged. Sol 821 began with Pancam imaging of that outcrop before the drive resumed, and another Phoenix communication test was made on sol 822.

The sol 823 drive exceeded 70 m on an expanse of outcrop. The scientists were looking for a soil target, as they had not analyzed one for 200 sols and wanted to see one before they reached the Victoria ejecta. The season was moving deeper into winter, Spirit was now almost at Low Ridge Haven, and Opportunity was targeting north-tilting lily pad surfaces again when possible. On sol 825 MI and APXS were used on a soil patch called Alamagordo Creek, followed by a 24 hour MB integration over sols 826 and 827. Pancam imaged Grenada, and then they drove south again. A nearby blocky crater called Sjaelland was imaged from a distance of 25 m on sol 830, and Matt Golombek suggested that the rover should move closer for better pictures, but it was directed south instead. As it turned south on sol 833 (MY 28, sol 125) its wheels became embedded in a large drift, but unlike the Purgatory event this drive was stopped very quickly by the slip check procedure. The drift was named Jammerbugt, after the "Bay of Wailing," site of many shipwrecks on the northwest coast of Denmark. Danish names were being used here, as 5 June (sol 841) was Denmark's Constitution Day.

Opportunity recharged on sol 834, looked for clouds and imaged the area. On sol 836 it drove north to escape the trap, now driving forwards since it had been driving backwards on sol 833. A commanded drive of 5 m moved it only 9 cm due to the excessive slip, and similar drives on sols 837–840 produced similar results, but on sol 841 it escaped the drift, this time stopped by visodom when it was seen to be moving properly again. The trap was imaged on sol 842, and on the next sol the rover skirted the large drift and headed south on its west side. A drive on sol 844 was cut short by excessive slip again, and after imaging some targets here on the next sol, Opportunity crossed to another trough on sol 846 and made better progress. The MB instrument was causing concern, as its data were becoming degraded, possibly due to cable deterioration in the IDD.

Another Phoenix communication test on sol 847 was canceled to save power. The rover was still fairly close to the blocky crater Sjaelland, and on sol 848 the Pancams took images before driving south. Opportunity imaged a layered outcrop target called Steno on sol 850, and as the drive continued a new version of flight software was slowly uplinked over several weeks, beginning on sol 851. Targets called Blixen and Holberg were observed by Pancam and Mini-TES on sol 852 before the rover drove south, skirting a large drift called Pitcher's Mound on the next sol. Images of the soil on sol 858 showed very few blueberries, which had almost covered the ground at Eagle crater and helped reduce soil erosion. A ripple or drift at that location, known as Boggy Depot, was imaged on sol 859, and after several drives a sunset was viewed on sol 864. Atmospheric observations were made frequently along this traverse, looking for clouds or measuring levels of dust in the air.

A three-day plan was uplinked to carry the rover over the US 4 July holiday. Sols 867 and 868 were taken up with imaging and Mini-TES observations of targets called Austin, McKinney, Baxter Springs and Fort Gibson, with a drive on the third sol. The flash memory was filling up again, so many files were deleted on sol 871. On the next sol Opportunity imaged a target called Rock Bluff, and on sol 873 it tried to cross a drift and was stopped by excessive slip. The rover backed off the drift on sol 875 as the Pancam and Mini-TES examined a distant rock called Genovesa, thought to be a possible meteorite.

Now Opportunity was nearing a large expanse of outcrop containing an impact crater named Beagle, after Darwin's ship. It performed a coordinated observation with Mars Express on sol 877, and on sol 878 the rover scuffed the soil with its wheel unintentionally while the planners sought an IDD target, and the scuff became the

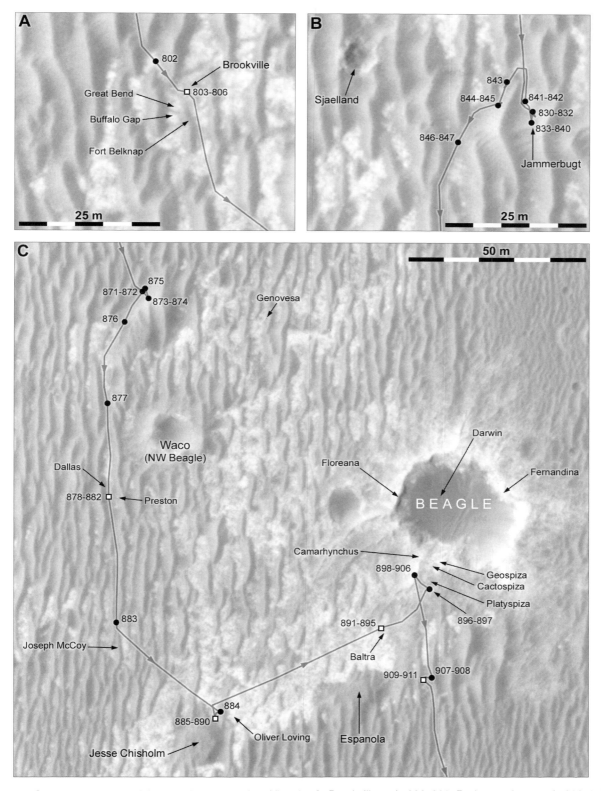

Figure 67. Opportunity rover activities on the approach to Victoria. **A:** Brookville, sols 802–806. **B:** Jammerbugt, sols 830–847. **C:** Beagle crater, sols 873–911.

target. A berry-free soil in the scuff was named Westport, and it was examined with the Pancam, MI and APXS on sol 879 along with Dallas, a soil target in the tracks behind the rover, and Waco, a nearby crater. An undisturbed soil called Fort Graham was looked at with MI on sol 880 and MB was used on Westport over two sols, while Mini-TES examined Preston, an ejecta block. The soils here would be compared with the Victoria ejecta blanket. Sol 882 was lost to a DSN problem, but before leaving this location on sol 883 the Pancams imaged Fort Graham and the nearby banded ripples, which might not be seen again as Opportunity headed for Victoria.

On sol 884 Opportunity drove to and imaged a dark mound of coarse-grained sediment named Jesse Chisholm, making a final approach on sol 885. The target here, a cobble, was called Joseph McCoy. These were the last names associated with the old Texas cattle trails referred to earlier. MI and APXS were used on Joseph McCoy as Jesse Chisholm was imaged by the Pancams on sol 886, and then MB was placed on the cobble for three sols as the Pancams viewed Sandsheet and Mini-TES examined Jesse Chisholm. On sol 890 the MB integration ended and APXS was placed on Ignatius, another target here. Then on sol 891 the rover scuffed the dark mound to examine its mechanical properties, imaged the scuff and drove 55 m towards Beagle. Jesse Chisholm may have been a clod of ejecta from Victoria, or an old crater filled with drift material that has been preserved as the surroundings were scoured by wind.

A target called Baltra was investigated on sol 893 with the MI, a 3 mm deep RAT grind and APXS, followed on sol 894 with more MI views and the use of MB for two sols. Meanwhile Beagle and targets called Espanola and Bartolome were imaged, Espanola being another dark, coarse-grained mound. Names of targets here were taken from Darwin's famous voyage. Then on sol 896 Opportunity backed up to image Baltra and drove up onto the rim of Beagle to image the crater. On the next sol the rover recharged and Mini-TES was calibrated, and on sol 898 it moved to a different viewpoint for additional imaging. The cooler winter nights threatened Mini-TES, so its heaters were enabled on sol 899, and the instrument observed Beagle and other targets over the next six sols while the Pancams imaged the crater and targets including Fernandina, Camarhynchus, Cactospiza, Floreana and

Darwin. This imaging was interrupted by a fault which put Opportunity into a safe mode on sol 902.

After another Mini-TES calibration on sol 906, the rover was driven to a new banded ripple target on the next sol. On sol 908 a test was done for NASA's future Mars Science Laboratory (MSL) rover, imaging the Sun at noon with the left Navcam. A later session would observe a sunset as well, to test the ability of the Navcams to measure atmospheric dust levels, a task usually done with the higher-resolution Pancams. MSL would carry the same Navcams but very different high-resolution cameras. After a large data deletion on sol 909, Opportunity moved closer to the ripple to investigate the light and dark bands running along its sides. MI and APXS investigated targets Marchena and Isabella while Mini-TES examined targets called Pinzon and Pinta over the next two sols.

Opportunity drove onto Victoria's ejecta deposit, or annulus, on sol 912, and characterized it and a rock called Tenerife with Mini-TES and multispectral imaging on the next sol. The annulus rose slightly above the Etched Terrain, sloping gently up to the crater rim, and was as smooth and easy to drive over as the plains outside Eagle crater had been. Like those plains, these were covered in blueberries, the small hematite nodules first seen at Eagle. Blueberries were lacking in the rocks and on the surface at Erebus, suggesting that those seen here had been excavated from deeper layers by the Victoria impact. On sol 914 the rover drove 72 m over the smooth surface, and the Navcam imaged the Sun near sunset as part of the test for MSL.

Two more long drives on sols 916 and 917 brought the driving distance over 9 km. Then on sol 919 a target for trenching was imaged and after moving 2 m the trench was dug by the wheels to further characterize the annulus. Mini-TES examined the trench on sol 920 but an IDD stall interrupted further analysis. After imaging on sols 921 and 922 and IDD diagnostic tests on sol 923, the arm was fixed and MI and APXS were deployed on sol 924. Diagnostic tests continued to confirm procedures for stowing the arm, and MB collected data over sols 925–927. Targets for observations in and around the trench included Powell on the trench floor, Sumner on the wall, Hall on soil outside the trench and Hawkins, a nearby ripple. A distinctive cobble was named Bradley and a crater nearby was called Canonita. Finally on sol

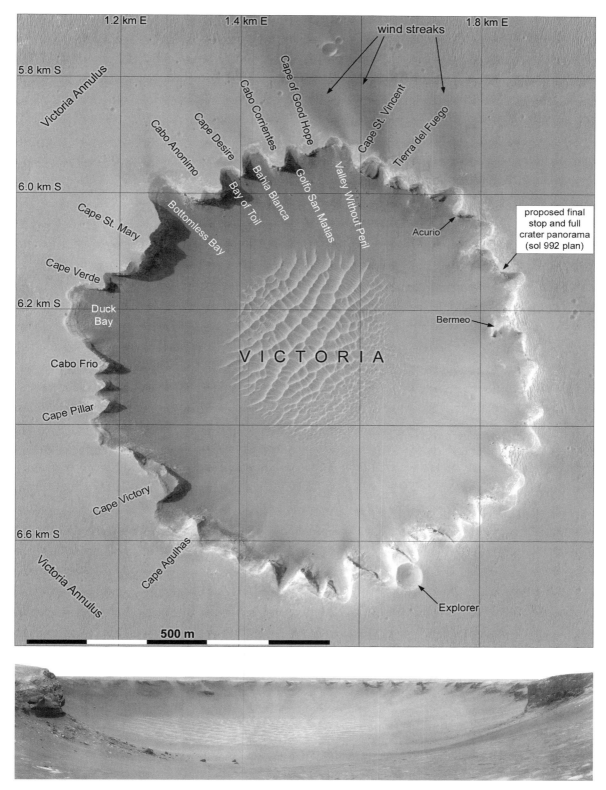

Figure 68. Victoria crater showing named features and prominent wind streaks (top). The background is HiRISE image PSP_001414_1780. Panoramic view across Victoria from Duck Bay, sol 1332 (bottom).

928 MI observed Powell, and MI and APXS both examined the nearby target Powell's Brother.

On sol 929 Opportunity moved back to image the trench and then drove 100 m to its next stop near the crater Emma Dean. Two sols later it moved 5 m to an ejecta block called Cape Faraday. Hazcam images of the RAT bit helped the planners estimate how many more RAT grinds might be possible. After more imaging the rover made a small move on sol 936 to put Cape Faraday within reach of the arm. Mini-TES collected data on targets Thompson and Jones while the rover team considered Cape Faraday as a target. Was it stable enough to grind and free of blueberries? The team was also contemplating operations in Victoria crater, thinking of driving to Cabo Frio (Figure 68) to survey the crater, using Pancam and Mini-TES to examine Cape St. Mary remotely before deciding whether to drive to Cape St. Mary or Point St. Anthony (not identified, but probably Cape Victory or Cape Agulhas in Figure 68) to make a high-resolution panorama of the crater.

Opportunity used MI on Cape Faraday on sol 939, but an IDD fault prevented a planned RAT grind and post-RAT MI observations, as well as the use of MB on sol 940. Imaging and Mini-TES observation of a target called Beaman replaced the MB analysis. On the next sol APXS was used on Cape Faraday after some IDD diagnostics and imaging of Dellinbaugh, a target in the center of Emma Dean. Pancam images of Beaman and Mini-TES analysis of Dellinbaugh followed on sol 942, and on sol 943 Opportunity drove away, stopping mid-drive to image a shallow 50 m diameter crater called Kitty Clyde's Sister.

After moving 62 m the rover stopped in an open area to boot its new flight software on sol 944 and test it on sol 945. The next sol was spent taking images while a new plan was developed. Now Opportunity would drive to a broad recess called Duck Bay in the crater rim, use Pancam and Mini-TES on Cape Verde and then decide whether to move to Cape Verde or Cabo Frio to make a big panorama. Pancam and Mini-TES data were collected here on sol 947 at targets called Macaroni and Rockhopper, and more images were taken of Kitty Clyde's Sister. Driving tests on sol 948 checked the new software as the sky darkened, decreasing power, and many clouds were seen.

Drives on sols 950 and 951 brought Opportunity close to Duck Bay, and the rover arrived on the rim of Victoria on sol 952 (MY 28, sol 244), greeted by a dramatic vista across the 800 m diameter crater. Names of features around Victoria were taken from places visited during the first circumnavigation of Earth by Ferdinand Magellan (Fernão de Magalhães), between 1518 and 1522, in the ship *Victoria*. This scheme was suggested by Portuguese space enthusiast Rui Borges. The deeply indented rim of Victoria lent itself to the names of capes and bays seen on Magellan's voyage (Figure 68).

The first viewpoint at Duck Bay was chosen to avoid a large drift fringing the rim, whose stability was suspect. Sol 953 was devoted to Pancam and Mini-TES observations, looking for an area of bare rock on the rim for analysis over the approaching conjunction. Cape Verde looked better than Cabo Frio for that purpose, so over sols 955 to 958 Opportunity drove northwards around Duck Bay and out onto the narrow promontory. A final approach on sol 959 put the rock Fogo within reach of the arm, and by driving too far and then backing up the wheels left tracks in front of the rover to be examined before and after conjunction for any changes. Some imaging was done here, and the flash memory was cleared to leave space for conjunction observations.

The target Cha on Fogo was investigated on sol 961 by MI, followed by a brush, MI again and APXS while Mini-TES examined a target called Elcano. Cha was ground into on sol 962, viewed by MI and analyzed by APXS again. Then on sol 964 MB was placed on the hole to collect data over conjunction, and Mini-TES examined a soil target called Axio (Figure 70A). Memory was cleared again over several sols, the conjunction instructions were uplinked on sol 966, and Pancam viewed Pigafetta on sol 967.

Conjunction was on sol 975, but Opportunity was out of communication between sols 970 and 984, during which atmospheric and other images were taken and MB analyzed Cha. Those observations continued for two additional days after communications were restored, and as the flash memory had filled up more than expected two sols of large downlinks and file deletions followed. MB continued collecting data on Cha until sol 990 and many sites were targeted by Mini-TES, including Zubileta, Pigafetta, Arriata, Huelva and Antonio. A large panorama was compiled over sols 985–991, and on the next sol Opportunity imaged the area in front of it to observe any changes over conjunction.

The first MRO HiRISE image of this site was taken on Opportunity's sol 997 (MY 28, sol 289), greatly helping the process of navigating and planning entry into the crater. A plan for the exploration of the crater was developed from the first image while Opportunity paused during conjunction around sol 980. The partial circumnavigation of the crater was designed to allow imaging of the exposed rock faces of the capes, often from both sides, and to inspect possible entry routes. HiRISE suggested that the best exposures of layered rocks would be found in the area around Tierra del Fuego. Opportunity had made high-resolution panoramas of Endurance from two points far apart (Figure 53A) and the same would be done here, from Cape Verde and a point just beyond Tierra del Fuego (Figure 68). In the end the second panorama was not taken.

The cliff exposures revealed local variations in stratigraphy, but generally followed a simple pattern. The lower layers consisted of cross-bedded sandstone originating as sand dunes which had been cemented by minerals precipitated out of ground water. An upper layer of this material had been exposed at the surface in the past and weathered, showing up now as a bright band visible all around the crater. Above that the bright layer was covered with a deposit of ejecta containing jumbled blocks of the sandstone. An unusual deep stratum of soft layered sandstone, exposed only at the foot of the cliffs of Cape Verde on the north side of Duck Bay, became a target for Opportunity later.

The drive around Victoria began on sol 992, including a test of target tracking, imaging a target as the rover passed it to estimate distance traveled. On sol 994 Opportunity drove up onto Cape St. Mary, the highest part of the crater wall, and viewed the crater from this vantage point. Downlink was very limited now as MGS had suffered a spacecraft emergency on Opportunity's sol 986 (MY 28, sol 277, or 2 November 2006) and DSN was occupied with attempts to recover it. The long-lived orbiter's mission ended with this failure of a solar panel steering joint, fortuitously just as MRO was beginning its operations at Mars. On sol 999 the rover was driven to the edge of Cape St. Mary to image Cape Verde from the north and also performed a mid-drive test for future long-baseline stereoscopic imaging across the crater. Sols 1000 and 1001 were taken up with imaging and atmospheric observations.

Opportunity took the second set of long-baseline stereo images of the opposite wall of Victoria on sol 1001, and on the next sol it moved to the edge of Cape St. Mary to see if the prominent rock layer on the south side of Cape Verde was also visible on the north side. It was not obvious here and could not be traced around the crater, suggesting that the layers were not regional strata but local structures. Nevertheless, each projecting rock face would be imaged in color stereo to map local depositional variations. The status of MGS was still uncertain, so commands were sent to the orbiter to transmit to Opportunity on sols 1005 and 1006 in case the rover relay on MGS was still working though its main transmitter was not. Opportunity received nothing and efforts to contact the orbiter were soon abandoned.

Driving resumed on sol 1009, moving north towards Bottomless Bay, or Bahia Sin Fondo. MRO and Opportunity conducted coordinated observations on sol 1013, the rover viewing the sky and ground as the orbiter looked down. New hazard-avoidance software called D-star (D*), developed at Carnegie Mellon University, was used on sol 1014. Later it would help control drives but here it was only tested. On sol 1016 the rover drove closer to the possible entry point at Bottomless Bay, and after an argon measurement on sol 1018, two Pancam mosaics of the "Bay" were made on sols 1019 and 1021, separated by a 2 m drive to provide enough offset for good stereoscopic viewing. Mars Odyssey was in a safe mode following a charged-particle event, limiting rover communications, and was only recovered on sol 1027.

The rover was driven around Bottomless Bay, imaging a potential entry point and nearby targets called Malua and Timor on sol 1029. After moving carefully around the rim of Bottomless Bay, Opportunity imaged the south side of the narrow depression on sols 1033 and 1034. Tests of IDD placement were done on sol 1033, including some moves that had caused faults, to prepare for future ISO. Then a target called Rio de Janeiro was brushed for ISO and multispectral imaging over sols 1035–1038 and Mini-TES analyzed targets named Catalonia, Valencia, Murcia and Navarra. A cobble target was needed, so the Pancam observed a field of small rocks near Cabo Anonimo on sol 1038. Cabo Anonimo was so named because it had not been labeled on an earlier planning map, rather than from a terrestrial location. A 14 cm cobble, large and stable enough for brushing, was selected for study and named Santa Catarina, so after collecting atmospheric and argon data on sol 1040, Opportunity approached the rock on sol 1041. A new Earth year, 2007, began on sol 1043.

RAT bit imaging on sol 1045 to assess wear showed that the grinder motor did not move as commanded, so the brush was not used here as intended. Diagnostic tests on sols 1045 and 1047 showed that the bit would move, but had lost the ability to tell when it was in contact with a target. Santa Catarina was examined by all of the rover instruments over sols 1045–1053 and interpreted as another possible meteorite (Schröder *et al.*, 2008), maybe part of the same fall as Barberton, seen on sol 121. Those other cobbles were also viewed by Mini-TES and Pancam, and all appeared very similar, so a possible drive to another cobble called Igreja was dropped. The total time MB spent gathering data on Santa Catarina was 80 hours, reflecting its diminishing source strength.

On sol 1055 Opportunity backed up to view Santa Catarina with Pancam and Mini-TES, and then drove across Cabo Anonimo towards the Bay of Toil. It made an argon measurement on sol 1057 and used Mini-TES on nearby targets, including Pacific, before reaching the next bay on sol 1060. On arrival it imaged the opposite wall of the Bay of Toil and moved for stereoscopic imaging, taking the second set of images on the next sol before moving on. An outcrop called Guam and cobbles named Gallego, Vasco and Gomes were imaged on sol 1062.

Opportunity tried to image Comet McNaught early on sol 1063, APXS took an argon measurement and Mini-TES viewed several targets, including Gomes and Santandres. Then on sol 1064 another test of the visodom software was attempted. The surface here was often featureless, so the visodom software used images of the tracks to measure distance traveled, but this was easier if the tracks themselves were less uniform. In the first test a wheel was dragged for 5 cm of every 60 cm of progress, and in the second two wheels were spun after every 60 cm. That second test was stopped early by a fault, and the next sol was devoted to imaging, including views of the sunset. Then on sol 1066 more visodom tests were made, this time using a variety of scuff patterns and wheel turns to find the most effective method for marking the tracks.

Opportunity drove towards Cape Desire on sol 1067, imaged Guam again, and reached Cape Desire on sol

1069 to look into Bahia Blanca. Another software test was carried out, indicative of the constant efforts to improve the rover. The front Hazcams took images of the surface with the IDD to one side, then it was swung back, the MB was touched to the surface to confirm its location and MI images were taken, all automatically. This would allow IDD work without a delay while planning operations on the ground. The rover team were also recommending now that Mini-TES be used while it was still available, as it had suffered a temporary failure one Mars year earlier due to very low temperatures and its fate was unpredictable.

Imaging on sol 1070 included Ceuta, the IDD autoplace test area, Bahia Blanca and Cabo Corrientes. Cabo Corrientes and Cabo Anonimo were observed in stereo and multispectral images over the next few sols as Opportunity moved around the rim. The rover exceeded 10 km of driving on sol 1080, tested D* again on sol 1082, and observed Phobos transits on sols 1082 and 1083. As it drove onto Cabo Corrientes on sol 1084 the drive stalled, possibly because of the visodom wheel movements, but further drives on sols 1087 and 1089 brought Opportunity to the rim for imaging of the surroundings. A cobble called Santiago had been viewed on sol 1085, as well as an outcrop on the rim called Extremadura on sol 1094, and long-baseline stereo imaging of Cape Desire followed on sols 1091 and 1095.

Cape of Good Hope, sometimes just called Cape Hope, was imaged on sol 1096, and then Opportunity drove towards it, viewing targets Madrid and Alava on the cape on sol 1097 as Mini-TES examined dunes at the base of the next promontory and also collected data on Madrid and another location called Coslada. The rover conducted more visodom test drives on sol 1100 (MY 28, sol 392, or 27 February 2007) and imaged the surroundings on sol 1101, including an odd cobble called Donut. On sol 1102 Pancam viewed Cabo Corrientes and MI was used on Donut, but an arm joint stalled and the images were out of focus. Tests on the next sol showed no problem and MI successfully imaged Donut. Then multispectral and long-baseline stereoscopic images of Cabo Corrientes were taken over the next few sols.

On sol 1108 Opportunity drove towards Valley Without Peril, another potential crater entry location. The RAT was tested again on sol 1109 to prepare for use inside Victoria, but the test was interrupted by a fault and

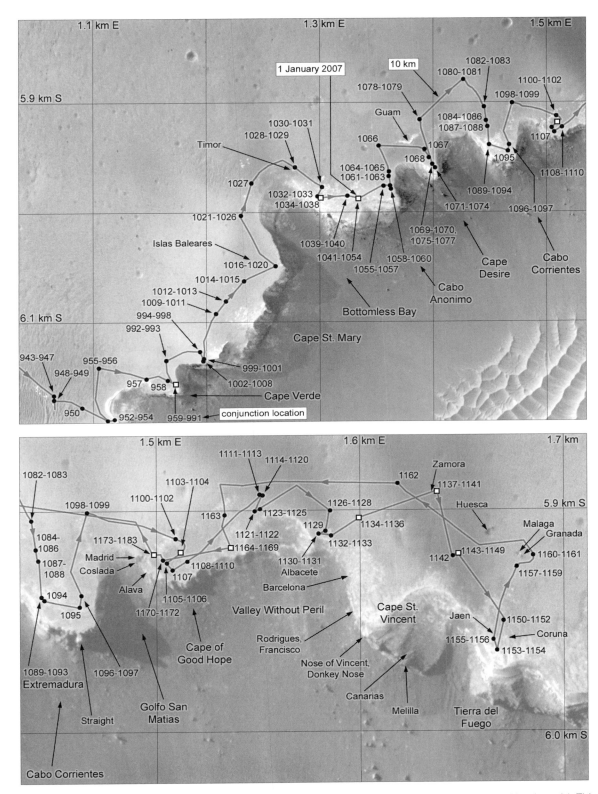

Figure 69. Opportunity route map, section 7. The lower section is enlarged for clarity, with scale indicated by the grid. This map follows Figure 66 and is continued in Figure 72. Additional details are shown in Figure 70.

Table 20. *Opportunity Activities on the Rim of Victoria, Sols 952–1290*

Sol	Activities
952–960	Arrive at Duck Bay (952), imaging (952–954), drive to Cape Verde (955–959), image Fogo (960)
961–970	Brush (961), RAT (962) and ISO (966–990) Cha, start of conjunction period (970)
971–991	Take Cape Verde panorama during conjunction period (970–984) and after (985–991)
992–1008	Drive towards Cape St. Mary (992–999), to St. Mary Point (1002), image Cape Verde (1002–1008)
1009–1034	Drive to Bottomless Bay (1009–1016), image cliffs (1017–1020), drive around bay (1021–1034)
1035–1038	Brush (1035) and ISO Rio de Janeiro target and imaging (1036–1038)
1039–1058	Drive across Cabo Anonimo (1039–1058), ISO Santa Catarina cobble (1045–1053)
1059–1069	Image Bay of Toil (1059–1060), drive around bay (1061–1069), Victoria crater panorama (1063)
1070–1094	Drive to tip of Cape Desire and imaging (1070–1077), drive to Cabo Corrientes (1078–1093)
1094–1097	Move to Extremadura (1094), imaging (1095), move (1096), image Cape of Good Hope (1097)
1098–1103	Drive towards Cape of Good Hope and imaging (1098–1102), ISO Donut (1103)
1104–1116	Imaging and drive to Cape of Good Hope (1104–1107), drive to Valley Without Peril (1108–1116)
1117–1134	Fault prevents science (1117–1120), drive to Valley Without Peril and imaging (1121–1134)
1135–1138	ISO Salamanca (1135) and Sevilla (1135–1136) targets, move (1137), imaging (1138)
1139–1142	ISO Palencia (1139–1140) and Pontevedra (1139) targets, imaging, move (1142)
1143–1156	ISO Alicante target (1143–1149), drive to Tierra del Fuego and imaging (1150–1156)
1157–1163	Begin return to Duck Bay (1157), imaging, drive around Granada (1160), drive west (1162–1163)
1164–1167	Drive to Cape of Good Hope (1164), RAT test and ISO Viva La Rata (1166), imaging
1168–1171	Brush and ISO Viva La Rata (1168), drive, image Madrid (1170), Pedriza (1171)
1172–1183	Imaging, move to rock Cercedilla (1175), brush, RAT and ISO Penota (1176–1183)
1184–1204	Drive west and imaging (1184–1197), image Paloma target (1194), drive south (1198–1204)
1205–1232	Image Cape St. Mary and Duck Bay from Cape Verde (1206–1214), move towards Duck Bay (1232)
1233–1281	Very limited activity during dust storm (1233–1271), small moves and imaging (1272–1281)
1282–1288	Self-portrait (1282, 1284), drive to entry point (1285–1286), instrument tests (1287, 1288)
1289–1290	Drive to Paolo's Perch target (1289), imaging, prepare to enter Victoria crater

had to be repeated on sol 1112 as engineers sought a way to work around the failed encoder. This test was still not successful. After an argon measurement later that sol, Pancam and Mini-TES viewed the nearby "normal" soil for comparison with the later dark streaks (Figure 69). Sol 1116 was supposed to see a drive to a good vantage point for imaging the nearby valley and Cape St Vincent on its far side, but a computer reset prevented it and it took two sols to recover operations. The rover then rested and recharged over the terrestrial weekend, resuming activities on sol 1121 with a drive to the western rim of Valley Without Peril and several sols of long-baseline stereo imaging of the valley.

On sols 1126 and 1127 the rover approached the eastern side of Valley Without Peril for further imaging, the latter drive stopping early when the high-gain antenna blocked the view in Navcam images used to monitor progress. On sol 1128 the RAT diagnostic tests

were successful, allowing the RAT to bypass the failed encoder, and Pancam and Mini-TES viewed targets Gerona and Burgos. The rim was finally reached on sol 1129 and on the next sol the targets Albacete, Toledo and Cadiz were examined.

Planners had been looking carefully at the terrain below the various capes around this part of Victoria and decided now that it would be impossible to enter the crater in one bay and depart from another. The rover would enter and leave at the same place in a bay with well-exposed layering, after inspecting the dark wind streaks. One possibility was a short toe-dip into Valley Without Peril to inspect cross-bedding in its eastern wall as closely as possible. Long-baseline stereoscopic images to support activities in the valley were taken on sols 1131 and 1133 from points about 5 m apart, separated by Mini-TES observations of Barcelona and an argon measurement.

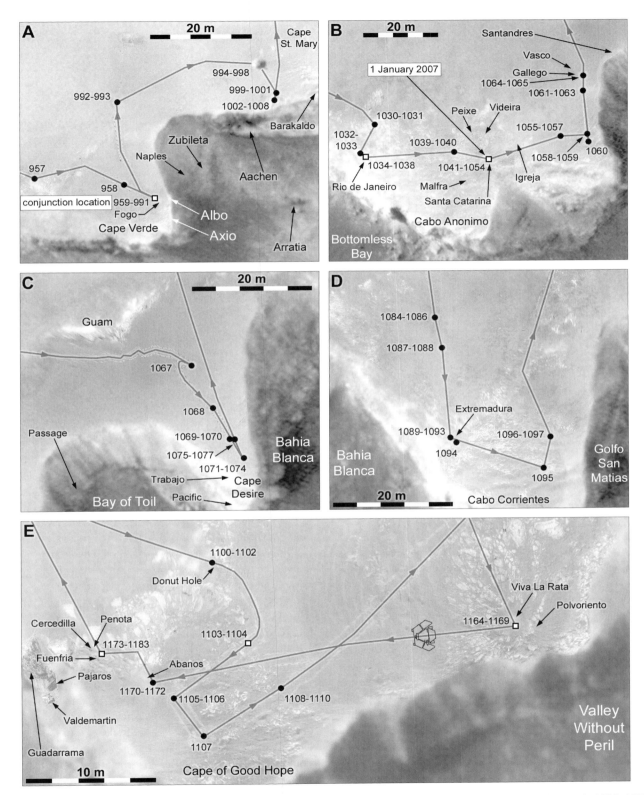

Figure 70. Opportunity activities at sites around Victoria crater. **A:** Cape Verde, sols 957–1008. **B:** Cabo Anonimo, sols 1030–1065. **C:** Cape Desire, sols 1067–1077. **D:** Cabo Corrientes, sols 1084–1097. **E:** Cape of Good Hope, sols 1100–1110 and 1164–1173. **A** shows Opportunity as seen on sol 997 in HiRISE image PSP_001414_1780 (14 November 2006 or MY 28, sol 289).

Opportunity drove to a point between the two most prominent dark streaks on sol 1134 and made photometric observations to the east and west at three times of day. MI viewed a soil target called Salamanca on sol 1135 but work on a target called Sevilla was stopped by an IDD fault. It was done instead on sol 1136 as Mini-TES characterized the soil and Pancam viewed the tracks. Then on sol 1137 the rover moved into the middle of a dark streak and repeated the photometry observations over two sols. Stereoscopic MI views of Palencia and Pontevedra were taken on sol 1139, and, as APXS measured soil composition on the next sol, Pancam and Mini-TES observations were obtained of the target Zamora in the tracks and a nearby undisturbed area.

The rover was driven to a second dark streak location closer to the crater rim on sol 1142 and approached a target called Alicante on sol 1143 to deploy APXS on it. MI and APXS studies of Alicante on sol 1145 were unsuccessful because a previous surface contact by the MB to establish the IDD location failed, but the process was repeated successfully on sol 1148. Now both light and dark areas had been thoroughly characterized. Other nearby targets called Avila, Cordoba, Colomenero and Grenada were viewed by Pancam, and Avila by Mini-TES, and MB was used on Alicante on sols 1146 to 1149. Cordoba and Grenada were small craters and Colomenero was a cobble target. The results of this survey of the rim of Victoria are described by Grant *et al.* (2008) and Squyres *et al.* (2009).

On sol 1150 Opportunity drove to Tierra del Fuego, the last point it would examine along the rim of Victoria. An earlier plan to take a full panorama from a point further southeast was abandoned to save time. Several targets, including Jaen, Castellon and Coruna, were examined here on sol 1152 before the rover moved to the rim for long-baseline stereoscopic imaging of Cape St. Vincent on sols 1155 and 1157. Other observations here included Mini-TES views of the sky and ground at four perpendicular azimuths on sol 1154 and of targets Melilla and Canarias on sol 1156. A sunset was observed on sol 1156, and now the mission planners decided to enter Victoria at Duck Bay, where they had first arrived. It had the best combination of negotiable slopes and interesting targets. Opportunity drove to Grenada on sol 1157, and a wind gust cleaned the solar panels on sol 1158, increasing the available power.

On sol 1160 Opportunity imaged Grenada and performed a test of its D* hazard-avoidance software. It was told to drive to a point beyond a group of rocks called Segovia, and that the rocks constituted a hazard. The rover successfully chose a path around the rocks to the desired location near Grenada, where it imaged its tracks and the little blocky crater and used Mini-TES on soil named Malaga in the crater. Two long drives on sols 1162 and 1163 returned the rover to Cape of Good Hope, and Opportunity used MRO for a large data downlink. Here some further imaging was undertaken, including views of Cape St. Vincent on sol 1166. The RAT was also tested on that sol, on a target called Viva la Rata. The loss of an encoder prevented the procedure called "seek-scan" which established the position of the surface relative to the RAT, but a new procedure would achieve the same result. The RAT would grind with the lowest current limit setting until it stalled, demonstrating that contact had been made, and then brushing and grinding could follow. MI stereoscopic images were also obtained on the target.

Pancam super-resolution images of cross-bedding at Cape St. Vincent were obtained on sol 1167, as the rover would not be entering Victoria here. Then a RAT brushing of Viva la Rata and MI stereo imaging were followed by APXS on that target on the next sol. Mini-TES examined targets called Rodrigues and Francisco but not a dust target called Polvoriento because its position was uncertain. The IDD failed to stow properly, preventing a drive on sol 1169, but after testing on sol 1170 it stowed properly and was driven across the broad cape to view Madrid on the crater rim, very close to its sol 1105 location. Madrid and another target called Guadarrama were imaged on sol 1171, and then Opportunity moved to examine a dark cobble called Pedriza, but its drive was cut short when a wheel struck a rock. Argon was measured later that sol, as it was typically every ten sols at this time.

Pancam and Mini-TES examined other targets here, including Cercedilla and Fuenfria. Pedriza now appeared to be dangerously close to the edge of the slope, so Cercedilla was chosen as a dark cobble target instead. Opportunity approached it on sols 1173 and 1175 and conducted ISO on Penota, a target on the cobble, on sol 1176. Mini-TES viewed targets including Hierro and Matabueyes on sol 1176 and Pancam imaged many targets on sol 1177 as the IDD used its new "seek-grind" procedure to prepare for RAT work on Cercedilla. Penota was brushed on sol 1178 for a full range of ISO over three sols, and then the RAT ground the same target on sol 1182 for further ISO of the ground surface. The RAT was now fully usable for its operations inside Victoria. Cercedilla was unlike any other rock seen here and may have been ejecta from deep within Victoria.

Work here finished with multispectral imaging of Penota and the nearby target Abanos, and Opportunity left the area on sol 1184. Several aspects of the D* and visual target tracking software still had to be tested and the cobble field around Santa Catarina would be good for that, so Opportunity was driven to that area over several sols. One D* test occurred on sol 1188 and a cobble called Paloma was chosen for target tracking and imaged on sol 1190. On the next sol the target tracking was conducted as the rover moved away from and towards Paloma, measuring its progress with and without autonav and visodom use. Another part of the test occurred on sol 1194 and then Opportunity moved on towards Duck Bay, exceeding 11 km of driving on sol 1196. The fifth and last D* test was carried out on sol 1200 (MY 28, sol 492, or 10 June 2007) and the procedure was now certified for use.

Opportunity drove rapidly back to Duck Bay, stopping on sol 1204 to make new long-baseline stereoscopic images from Cape Verde to plan the entry. The images were taken on sols 1205 and 1210, interrupted by two driving faults. More stereo imaging separated by a short drive was done on sols 2012 and 2013, this time of Cape St. Mary, with additional views under different lighting conditions. Mini-TES and super-resolution Pancam observations of a target called Zaragosa followed on sol 1213 before driving away on sol 1214. A cleaning event boosted power on sol 1214, but this advantage would prove very short-lived.

On sol 1216 the sky darkened with dust as Opportunity conducted some final target tracking tests. MI was used to image the solar panels and magnets on sol 1217, and Hazcam images of the RAT bit were taken. Mini-TES experienced a low-signal problem on sol 1217, possibly caused by its mirror. Then the IDD stalled, and diagnostic tests were run on sol 1218 before driving towards Duck Bay again. A dust storm was building up and the sky had darkened considerably, but the solar panels were still clean on sol 1219. By sol 1220 the dust

storm, part of a system widespread in the southern hemisphere of Mars, had reduced power to the lowest level ever seen by either rover. The survival of Opportunity was suddenly very uncertain.

Very little could be done now as the rover focused on survival between sols 1221 and 1231. Dust levels were monitored daily and MI made a test image of the magnets before its dust cover was closed on sol 1223. On sol 1224 an image across the crater was made to compare with a clear pre-storm view. The launch of the Dawn asteroid mapping spacecraft, which eventually occurred on 27 September 2007, was repeatedly delayed through June and July, around Opportunity's sol 1220. It might have needed DSN resources usually available to Opportunity if a problem arose after launch, so many sols of runout commands were transmitted on sol 1229. Then on sol 1232 Opportunity was driven nearly 40 m to the west to take advantage of slightly improved power, but a second drive on sol 1235 was canceled because a DSN problem prevented communications.

MRO's MARCI instrument now showed that the dust storm was almost global in extent, and by sol 1237 power was so low that no heating was possible on the rover and its radio receiver frequency was drifting, preventing the receipt of instructions. On sol 1238 commands were received after the transmission frequency was adjusted, and Opportunity's low-power settings were modified to help it survive. Dust levels were monitored with daily images of the Sun, but only downlinked every few sols as conditions allowed. By sleeping much of the time, the rover was able to survive, and a tau (atmospheric opacity) measurement on sol 1246 showed the dust was clearing slightly, but this was offset by increased dust on the solar panels.

By sol 1251 the rover was able to wake for long enough to warm its electronics, and over the next few sols conditions gradually improved, including some imaging starting on sol 1254. New instructions were transmitted on sol 1257, and on sol 1261 Opportunity took images of a target called Soria where old and new tracks from sols 957 and 1232 crossed. Now they were inspected for signs of change due to the storm. Pancam and Hazcam images of MI were taken to assess dust coverage on sol 1268 and transmitted via Mars Express. The dust cover was closed, but care would be needed when it was opened to keep dust off the lens. A small

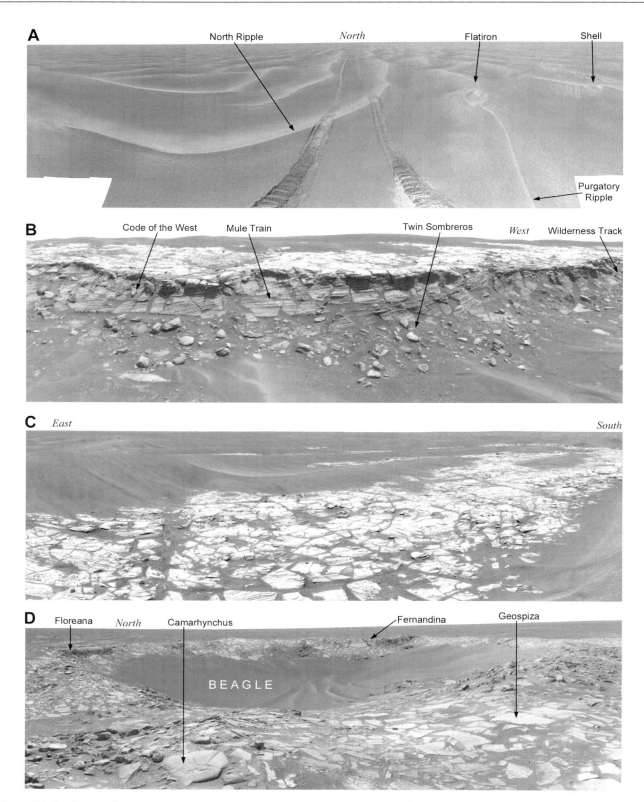

Figure 71 (both pages). Panoramas from Purgatory to Victoria. **A:** Purgatory, sol 456. **B:** Payson outcrop, Erebus crater, sol 747. **C:** South end of Brookville outcrop, sol 807. **D:** Beagle crater, sol 898. **E:** Duck Bay, Victoria crater, sol 952. **F:** Looking across Bottomless Bay to Cabo Anonimo, sol 1016. **G:** Golfo San Matias and Cabo Corrientes, sol 1106. **H:** Looking over Cape of Good Hope and across Valley Without Peril to Cape St. Vincent, sol 1170.

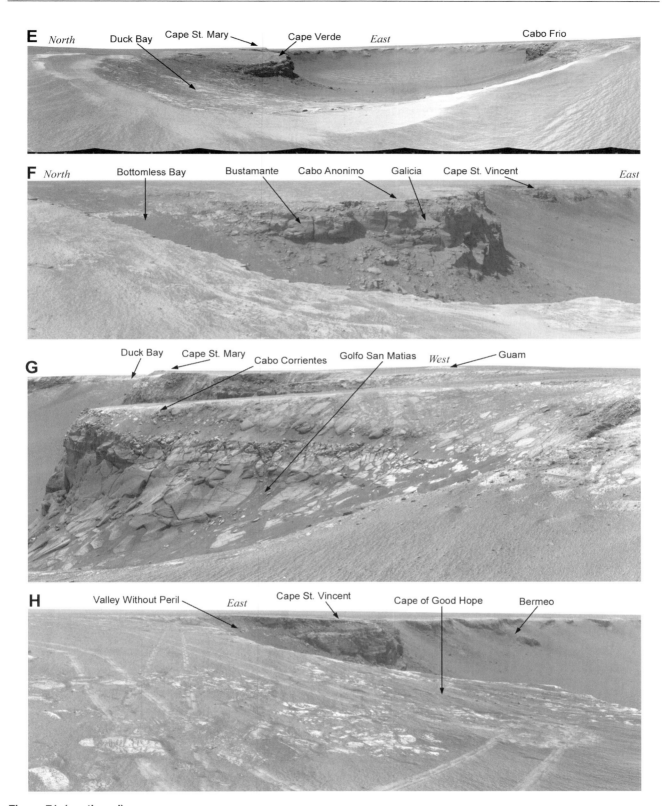

Figure 71 (continued)

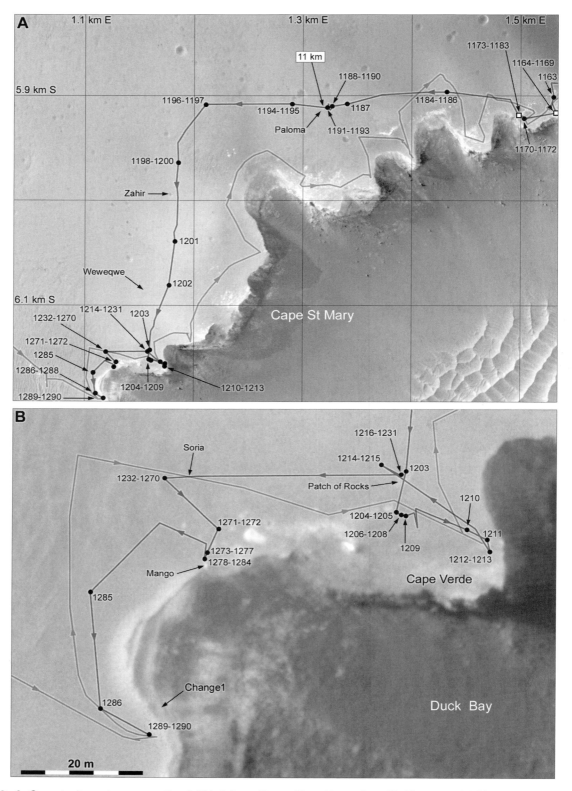

Figure 72. A: Opportunity route map, section 8. This follows Figure 69 and is continued in Figure 77. Additional details are shown in Figures 73 and 74. **B:** Enlarged view of the return to Duck Bay.

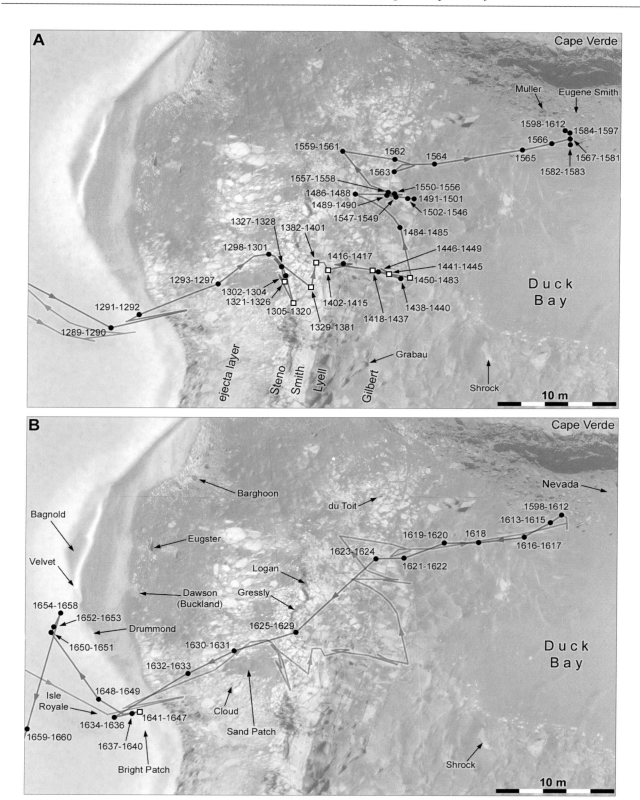

Figure 73. Opportunity activities in Duck Bay. **A:** Entry and descent, sols 1289–1612. **B:** Departure, sols 1598–1660. The background image is a composite of panoramas taken on sols 952–953 and 991, projected onto a HiRISE base image for geometric control.

drive on sol 1271 left the rover with a better tilt to the south to improve its power levels slightly, and it was allowed to recharge on sol 1272. A fortuitous wind gust blew some dust off the solar panels on that sol but power was still low.

Pancam imaged MI again on sols 1274 and 1277, and also viewed a rock called Mango (also referred to as Tarragona) on sol 1275. Late in the evening Opportunity conducted a communications test with Mars Express on behalf of the Phoenix mission. Then on sol 1278 the rover moved closer to the crater rim using visodom to track its position, and measured argon on sol 1279. Pancam viewed Duck Bay on sol 1281, and on sols 1282 and 1284 a mosaic of images of the solar panels (called a deck panorama) was taken to assess dust coverage. MI was imaged again on sol 1283 to help plan its dust cover opening as preparations continued for a descent into Victoria. The rover team had considered entering at this location, but now decided to enter where the first views of Duck Bay had been made on sol 952, as more rock outcrops were available there.

Opportunity was driven around Duck Bay on sols 1285 and 1286. The MI dust cover was cleaned on sol 1287 by pointing it downwards and fluttering the cover. It may not have helped very much, but now MI could be used, and on the next sol it imaged Mini-TES on the mast to see if its cover was opening. Mini-TES could not see anything and it now appeared that its mirror had been coated with dust despite its cover having been closed during the storm. Mini-TES never recovered despite prolonged efforts to revive it, a serious blow to mission science.

On sol 1289 the rover was driven to Paolo's Perch, the very rim of Duck Bay, where a long fringing dust drift was lowest. It made a toe-dip drive in and out on sol 1291 (MY 28, sol 583) to test traction, and was driven carefully down the rocky slope after a delay caused by a Mars Odyssey safe mode over sols 1294–1297. A patch of loose soil was bypassed on sol 1298 as traction was much better on bare rock. On sol 1300 (MY 28, sol 592, or 20 September 2007) the software protections put in place during the dust storm could be relaxed, and Opportunity conducted another Phoenix communication test with Mars Express.

The rocky slope resembled that seen earlier at Karatepe in Endurance crater. It was topped by broken rocks representing the ejecta of Victoria, planed smooth by wind-blown sand, below which several distinct bands of rock were exposed. From the top down they were called Alpha, Beta and Gamma, or Steno, Smith and Lyell, in different weekly mission reports. The first science target was a bright band of rock, visible around most of the crater in the "bays" and as a notch or terrace on the "capes." This was originally interpreted as the pre-impact surface of the Etched Terrain, but analysis showed it marked the top of the zone altered by interactions with ground water before the impact (Edgar *et al.*, 2012).

Opportunity was using deep sleep to save power roughly every other sol around this time. It drove carefully down to the Alpha layer, reaching it on sol 1305, and on sol 1307 it conducted a stability test to ensure the rover could use its IDD on the rocky hillside. The arm was moved around the area where the RAT would be used at the target called Steno RAT, but it stalled during the test and again on sol 1309. The problem seemed to be due to the 25° slope and it was solved by increasing current to the arm motors.

After a successful IDD test on sol 1311, ISO began with MI and APXS deployed at Steno RAT. Blueberries and vugs like those in Eagle crater were seen in the MI images. A RAT grind scan located the surface on sol 1312, and on the next sol Steno RAT was brushed. MI and APXS were used on it, and then the site was ground into on sol 1316 and analyzed again, this time including MB over sols 1317–1320. Pancam viewed the nearby targets Dolomieu and Arduino, and MI viewed the RAT hole on sol 1320 using multiple identical images, all subdued in contrast by dust, to improve the quality of contrast-stretched versions. Then Opportunity made a small move on sol 1321 to a second location on the Alpha layer, moving back, turning and moving forwards, allowing the expected slip to position the target where it was needed. This was called Hall, and MI and APXS analyzed it on sol 1322. Mini-TES testing ended now pending further analysis of its condition.

Sols 1324–1326 were lost due to a DSN problem, and on sol 1327 the rover imaged Cape Verde, but inadvertently aimed too high. Then it made multispectral images of Smith and Hall and drove towards Smith, stopping early due to a visodom error. Sol 1328 imaging included a novel new kind of image. On almost every sol of the mission, tau (atmospheric opacity caused by dust) was measured by imaging the Sun with Pancam, usually followed by transmitting only the small part of the image containing the Sun. Now the Sun was imaged 64 times on one image, offset each time to position the Sun in many different locations. In this way the pattern of dust on the camera itself could be mapped.

Opportunity imaged a target called Sedgwick on sol 1329, and then drove to Smith. On the next sol it measured argon and conducted another Phoenix communication test with Mars Express. MI and APXS viewed the target, using multiple MI frames again to improve the image quality. The APXS results showed Smith to have an unusual composition rich in zinc and silica. Pancam viewed Lyell and an area of thin layers called Sharp at several times of day on sol 1333 to see when viewing conditions were best. Then a RAT grind scan was attempted at Smith RAT, but it failed. Pancam imaged numerous targets in the vicinity as RAT diagnostics were undertaken on sols 1336 and 1339 and MB was placed on Smith over sols 1336–1340. While here, the rover passed its second anniversary of landing, in Mars years, on sol 1338. The grind scan was repeated on sol 1341 but failed again, and the mistargeted Cape Verde images were retaken on sol 1342. RAT diagnostics resumed on sol 1343, and it soon became apparent that a second RAT encoder had failed.

MB returned to Smith for sols 1344–1346 as the Pancams imaged more targets and a Navcam cloud search was conducted on sol 1344. More diagnostic tests were made on sol 1356 and another RAT scan was attempted using a new method on sol 1347, followed by a brushing of Smith and ISO on sol 1348. Pancam viewed Cape Verde again on sol 1349 to help plan routes to its base. The RAT was tested again on sol 1350 to try to assess whether it had ground a little as well as brushing, and Pancam observed the brushed area. More tests on sol 1352 showed that the new RAT operational method was running the brush backwards, and imaging with the Pancam revealed that the brush bristles had been bent. The brush would not be used while the issue was studied. MB resumed work on Smith on sol 1352.

Other work continued during the pause in RAT operations. On sol 1353 Cape Verde was imaged at different times of day, APXS was used on the rover magnets on sol 1354 and communications with MRO were tested to assist the Phoenix mission on that sol and the next.

Super-resolution images of Cape Verde and Cabo Frio were taken over several sols, and a new spot was selected for analysis, a clean rock area called Smith 2 or Smith's Brother. MI stereoscopic images were obtained on sol 1359 but the IDD stalled again as APXS was placed, causing it to measure only the atmosphere. While the team waited for more information, the Navcams were used on sol 1360 for a Sun-finding test for the benefit of the future MSL mission. MER usually used the Pancam to locate the Sun but MSL would not carry Pancams.

After tests on sol 1361 APXS was placed on Smith 2 for analysis, followed by several sols of MB analysis, and a Pancam mosaic called the Pettijohn panorama was compiled. Another Phoenix communication test was conducted with MRO on sol 1368 and a RAT grind scan was successfully performed. This allowed a grind at Smith 2 on sol 1370, followed by MI stereo imaging and Pancam multispectral imaging. More MI images were taken on sol 1373 followed by APXS analysis and several sols of MB use and a new deck panorama. That work concluded on sol 1379. Mars year 29 began on sol 1378, and back on Earth 2008 began on sol 1399, both events occurring while Opportunity was on the Smith layer in Duck Bay.

The science team hoped to drive to a target in Lyell called Ronov, but the path was too steep, so Newell was chosen as a target instead. Opportunity drove to it on sol 1382 and used MI and APXS on Newell, also called Lyell 1 or Lyell RAT, on sol 1384. A RAT grind scan on sol 1386 allowed a grind over the next two sols and six sols of MB integration ending on sol 1394, over the Christmas period on Earth. Two MSL Sun-finding tests were also performed on sols 1388 and 1390, as the first one failed. MI and APXS examined Lyell 1 on sol 1395, and then APXS was moved to a nearby target called Lyell 3 for use on sol 1397. Then MB was redeployed at Lyell 1 on sol 1399 for three more sols. Pancam targets here included Walcott, Graham and Smith RAT. Sol 1400 was MY 29, sol 23, or 1 January 2008.

On sol 1402 Opportunity was driven downslope to place the Smith–Lyell contact area within reach of the IDD. MI stereoscopic images were taken of the contact area on sol 1404, but an IDD stall prevented APXS from use at Smith 3 and Lyell 4, targets on either side of the contact. After diagnostic tests on sols 1407 and 1409 the work resumed with ISO at Smith 3 on sol 1410, MI imaging of Lyell 4 on sol 1411 and ISO of Lyell 4 on sols 1411–1414. Then the rover was driven down to reach a target called Buckland at the bottom of the Lyell band on sols 1416 and 1418.

Buckland, also called Lyell B, was examined by MI on sol 1421, but again the APXS placement failed. An attempt to place MB on Lyell B also failed when sol 1424 commands were not transmitted due to a DSN failure. A RAT grind scan succeeded on sol 1427 and the target was ground on sol 1429. The familiar pattern of ISO was repeated here with MI on sol 1430, APXS on the next sol and MB over sols 1432–1434. A frost search was made by imaging the grind tailings near noon on sol 1435 and 10 min before sunrise on the next sol. Changes might reveal the presence of frost, but none were seen. The IDD stalled again on sol 1436 but on the next sol it was able to place MB on Lyell B for one more sol of data.

Opportunity backed up to image Buckland and then drove towards Gilbert, the layer beneath Lyell, on sol 1438, but stopped short. On sol 1441 it drove to Gilbert, scuffed the surface and backed up to image the scuff. The right front wheel had dug deep into a patch of soil between rock plates. The RAT bit was inspected on sol 1443 and then MI viewed a new rock target called Lyell Exeter on sol 1443. APXS analyzed Lyell Exeter on sols 1443–1445 as the Pancams viewed the left scuff in front of it and a communication test was performed with Mars Express, for once not on behalf of Phoenix. Another test with MRO on sol 1446 failed because that orbiter had entered a safe mode.

On sol 1446 Opportunity backed up to reach Gilbert by a route that would avoid the deep soil patch. The rover measured argon and performed a cloud search, and then on sol 1450 (MY 29, sol 73) it moved down to Gilbert a little to the north of the scuffed area. APXS measured argon again on sol 1451, and the Pancam was measuring atmospheric dust levels several times a day around this time in ongoing efforts to understand how much was on the camera rather than in the sky. MI and APXS examined the rover magnets on sols 1452–1454, and more cloud movies were taken early in the mornings. A target called Gilbert A was selected for analysis and imaged on sol 1455, but ISO later that sol were prevented by another arm fault. They were recovered on sol 1456 after some testing.

Adjacent to Gilbert A at the edge of the flat slab of rock was a thin vertically projecting ridge, apparently a hard filling in a crack now exposed by movement and erosion. A small flake broken off it and now lying horizontally was called Dorsal, and it became the next target. MI stereoscopic images were taken on sol 1457 followed by APXS observations, and MB analysis followed on sol 1461, but the instrument position was not ideal. This had to be done before using the RAT on Gilbert, which would contaminate Dorsal. APXS was used again on sol 1462 followed by additional MI stereoscopic images, and the MB was used at a location called Dorsal Tail on sols 1463–1467. Dorsal Tail was misplaced again, mostly on soil, so another target called Dorsal New was chosen for MI on sol 1468 and MB over four sols.

Other observations were made through this period, beginning with cloud images on sol 1458, imaging of Cabo Frio on sol 1461, views of the Lyell RAT hole on sol 1463 and a super-resolution panorama of the entire visible rim of Victoria over sols 1463–1472. That was known as the Rimshot panorama. Another communications test with Mars Express on behalf of Phoenix took place on sol 1461. The Pancams imaged a nearby cobble called Jin on sol 1463 and a site called Lyell Oxford on sols 1464, 1480 and 1481.

Finally, on sol 1475 the RAT ground into Gilbert A, also called Gilbert RAT after this, and Pancam took multispectral images of the hole. MI viewed it on sols 1477 after a sol of cloud imaging, and MB then took data for three sols followed by more MI views on sol 1482. Then attention was shifted to a slightly offset position as MB data were collected on sols 1482 and 1483. The ISO data now looked good, so Opportunity left Gilbert on sol 1484, moving away to image the RAT site before turning towards Cape Verde. The target, a band of layered rock near the base of the cliff (Figures 73 and 74), was observed on sol 1487, and the rover made cloud movies and an argon measurement. The approach to the cliff base would involve crossing loose soil below the area of rocky outcrops and plates traversed so far, and it might also put the rover in shadow at certain times, so pictures of the shadow at different times were collected on sols 1489 and 1490. A drive on sol 1489 included a toe-dip and backup to assess soil conditions, and the scuffed soil made by that drive was viewed on sol 1491.

This caution was warranted, as drives on sols 1493 and 1495 both ended early, curtailed by excessive slip, and an attempt to retreat on sol 1496 was also cut short. Cloud and argon observations were made as planners examined options here, and some progress was made on the retreat on sols 1499 and 1502 (sol 1500 was MY 29, sol 123, or 13 April 2008), but the first was stopped by excessive tilt and after the second the arm failed to unstow. Since the arm joint fault on sol 654 the arm had been unstowed after every drive in case it should fail in the stowed position. This was the same arm joint that had also failed on sol 1404 at Lyell. Many IDD tests followed, often ending in another fault, and the problem seemed restricted to the joint which had caused problems so often before. During these tests the trenches dug by the left and right wheels, named Williams and Harland, were observed several times to watch for changes, and the cobble Jin was targeted again. The last set of change detection images, taken on sol 1540, revealed many changes caused by wind. A Pancam panorama of Cape Verde and Duck Bay, the Garrels panorama, named after the geologist Robert M. Garrels, was compiled over sols 1529–1545 after an attempt to begin it on sol 1526 was lost to a communication fault. It was taken using only half-frames to avoid the worst of the dust on the camera, and only when the Sun was behind a nearby ridge to prevent glare.

Diagnostic tests on the arm joint at warm and cool times of day showed it worked best during warmer periods. On sol 1531 the arm was unstowed using the maximum available power setting, and after this it was never stowed again. Various tests were conducted on it over sols 1538–1544 as Phoenix landed on Mars on Opportunity's sol 1543. The rover drove again on sol 1547, continuing its retreat from the soft soil area. The strategy now was to hold the arm vertically in front of the rover with the instrument turret hanging down, not resting on the deck as once considered. As the arm now blocked the front Hazcam view, it was moved out of the way for imaging after every drive, a maneuver called a salute. After measuring argon and viewing clouds for three sols, Opportunity looked again at Cape Verde on sol 1551 to plan lighting for its close observations.

Driving was still difficult during this final approach to Cape Verde, with a lot of slip on sol 1552 as a wheel caught on the edge of a rock. More shadow images were taken on sol 1553, but Mars Odyssey entered a safe mode and no Opportunity data could be transmitted. Cloud movies, argon measurements and imaging of the cobbles Agassiz, Barnes and Wilson occupied several sols here, until the rover freed its wheel on sol 1557 and moved to examine the wheel trap area on sol 1562. Rover driver Paolo Belluta nicknamed the trap Tartar Sauce, a play on the more serious trap Tartarus where Spirit was caught around its sol 1380, 200 sols earlier.

The target for driving now was a place called the Staging Area or Safe Haven, from which close images of Cape Verde and its lowest strata could be taken. Shadow imaging continued on sols 1562–1565, and small drives continued with a toe-dip on sol 1563 to test traction and four daily drives after that, ending on sol 1567 about 7 m from the base of the 6 m high cliff. Power dropped slightly as the cliff blocked part of the sky. A Pancam tau mosaic, Sun images taken at many places across the frame to map dust on the camera, was taken on sol 1570, which was the southern winter solstice. With the noon Sun so far to the north, staying out of the shadow of the cliff was very important. Imaging the cliff from this location was so important that part of the Garrels panorama would be deleted without being transmitted to Earth to make room for new data. Monochrome stereo images were retained; color coverage was deleted.

The Cape Verde images were taken over sols 1570–1573, with a few frames retaken on sol 1577 to remove Sun glints. Then several specific targets on the cliff named Alpha, Bravo, Charlie, Delta, Echo and Foxtrot were shot in super-resolution over sols 1574–1578. The RAT was tested under warm and cold conditions to prepare for future operations, and afternoon and dusk images of Cape Verde were collected. The Charlie and Delta super-resolution frames were retaken on sol 1581 as the previous frames were overexposed. After this intensive study of Cape Verde the rover began its final approach to a target called Hutton on the rock Nevada, thought to be an *in situ* exposure, not a fallen block.

Drives on sols 1582 and 1584 experienced a lot of slip, and on the latter sol the left front wheel looked as if it was about to pick up a potato-sized rock which might jam it. Driving was halted to plan the next move. Cloud images, argon measurements and a Pancam tau mosaic were taken over sols 1585–1590. On sol 1591 a drive avoided the problem rock but slipped badly. The rover team now began to consider alternatives to Nevada, including leaving Victoria. A variety of Pancam targets were imaged to document variations in weathering, including Mawson, Murchison, Mackay, King, Playfair and Eugene Smith, and short drives made a little headway in a zig-zag manner. A computer outage at JPL caused some difficulties for the team on sol 1598.

Suddenly on sol 1600 (MY 29, sol 223, or 25 July 2008) the motor in the left front wheel began drawing excess current, as Spirit's right front wheel motor had done before failing on its sol 779. Slip was very bad and the wheel might just have been stuck, so tests were conducted on sols 1602 and 1604. The motor seemed to be fine, but in case it was not, the team decided to leave Victoria after some additional imaging. If a wheel failed now, the rover might never get out of Duck Bay. On sol 1607 Opportunity turned and made a very small move away from the cliff as planners considered their post-Victoria options. They wanted to look at cobbles and blueberry-free soils on the plains, and then drive to an outcrop area south and a little west of Victoria. Plans to image Cape Verde were halted by a computer reboot on sol 1608, which prevented a weekend of work. On sol 1611 the rover recovered and imaged Cape Verde in stereo at dusk.

Driving was still difficult, as the right front wheel was slipping in deep soil, but moves on sols 1612 and 1613 freed it, and after imaging clouds and targets called Dawson and Barrell on sol 1615, the rover was able to make 3 m of progress on sol 1616. A target called Du Toit was viewed on sol 1620, and by sol 1623 all wheels were back on rock. The sol 1625 drive was over 9 m long, and was cut short by tilt as Opportunity drove over a small step in the slope. A target called Barghoon was imaged on sol 1628 and two more called Dawson and Eugster on sol 1631. Opportunity was now taking images frequently during drives to create a movie sequence of its departure from Victoria. The final drive and exit took place on sol 1634 (MY 29, sol 266) when Opportunity rolled up over the fringing dust drift and onto the surrounding plains, at the same spot where it had entered on sol 1291 and approximately one Mars year after first reaching the crater's rim on sol 952.

The first acts on leaving the crater, apart from continuing the almost daily cloud movies, were to image the old tracks on sol 1636, to assess changes caused by wind over a Mars year, and a cobble called Isle Royale (Figure 76). The troubled wheel would now be heated before each drive to help redistribute lubricant and prolong its life. The current plan was to drive to an outcrop 1000 m south of Victoria after testing IDD operations, preferably on a cobble, and if the arm worked satisfactorily the team would look for more cobbles. As no cobble was available without additional driving, the arm was tested on a nearby ripple instead. A spot called Bright Patch 2 was chosen and approached on sol 1641 for the IDD test.

MB was successfully placed on the dust on sol 1642 and data collected for five sols, but the target was not pure enough to distinguish fine bright dust from the ubiquitous coarser material. A communication test with MRO was conducted on sol 1644 and MI viewed a target called Victoria Ripple Field on sol 1645 as the Pancams imaged the target Schuchert. Mini-TES was tested on sol 1646 but was still unusable. Schuchert was imaged again as APXS was placed on the ripple field target, but a better dust target was now seen in a hollow on the crest of the large drift north of the rover. On sol 1648 Opportunity drove towards it, imaging Bagnold, the largest dust-filled hollow, and some possible cobbles south of the rover.

Two post-Victoria plans had been considered. One would involve a further search across the Etched Terrain for good exposures of rock, perhaps at different stratigraphic levels. The other would take the rover back along its previous traverse to examine numerous cobbles, or small rocks that had been seen scattered across the surface. These might include meteorites and ejecta from distant craters, but despite their inherent interest most of them had been bypassed in favor of Victoria, the exception being Santa Catarina on sol 1045. In the end the two plans were combined, and Opportunity would examine any cobbles it encountered during a long drive across previously unseen parts of the Etched Terrain.

One suggested target for that drive was a highly eroded crater about 2000 m west-northwest of Victoria and 1000 m in diameter (Figure 47). The more ambitious target eventually chosen was the much larger crater

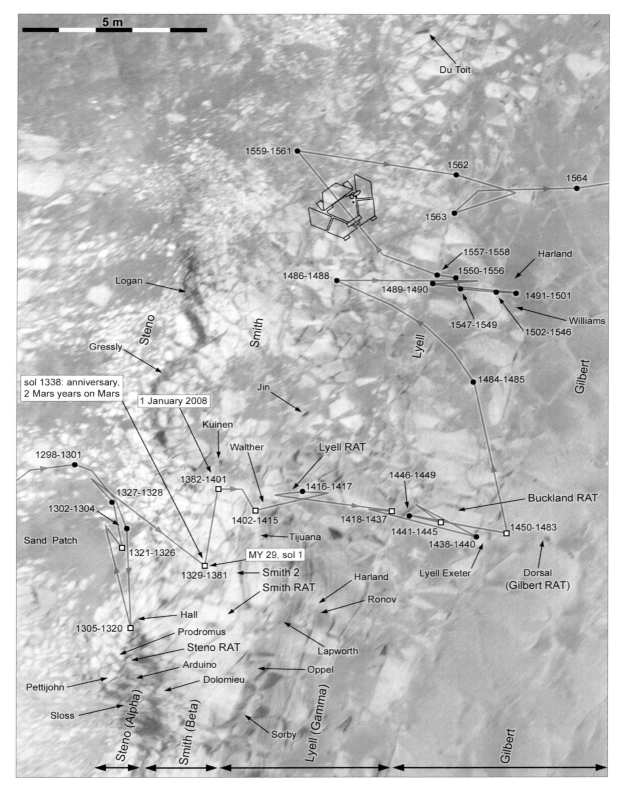

Figure 74. Stratigraphic studies in Duck Bay, sols 1298–1564. The background is from Figure 73. RAT holes shown in Figure 75 are located precisely on the mosaic. The rover, shown to scale, was driving backwards at that location on sol 1559. White squares are main science stops.

Table 21. *Opportunity Activities in Victoria, Sols 1291–1634*

Sol	Activities
1291–1306	Toe-dip, drive in and out of Victoria (1291), drive to Alpha layer and imaging (1292–1306)
1307–1320	ISO Steno (1307–1312), brush (1313) and RAT Steno (1316), ISO Steno (1317–1320)
1321–1331	Move, ISO Hall target (1322), imaging, drive to Smith (1327–1329), imaging
1332–1346	ISO Smith (1332–1340), RAT tests (1339), RAT Smith (1341), ISO Smith (1343–1346)
1347–1358	Images, brush (1348) and ISO Smith (1348–1358), magnets (1354), RAT test (1350)
1359–1370	ISO Smith 2 (1359–1366), ISO Stall (1359), brush (1368) and RAT (1370) Smith 2
1371–1383	Imaging, ISO Smith 2 target (1371–1381), drive to Newell (1382), imaging (1383)
1384–1396	ISO Lyell 1 (1384–1395) and Lyell 2 (1395), RAT Lyell 1 (1388–1389), imaging (1396)
1397–1403	ISO Lyell 3 (1397) and Lyell 1 (1399–1401), drive to Tijuana (1402), imaging (1403)
1404–1408	ISO Tijuana, arm shoulder joint problem (1404), imaging (1405–1408)
1409–1422	ISO Tijuana (1409–1415), drive to Buckland (1416–1418), ISO Buckland (1421–1422)
1423–1437	Imaging (1423–1428), RAT Buckland (1429), ISO Buckland (1430–1437)
1438–1445	Drive towards Gilbert (1438, 1441), scuff soil (1441), ISO Lyell Exeter (1443–1445)
1446–1454	Drive to Gilbert (1446, 1450), imaging, ISO magnets (1452–1454)
1455–1462	ISO Gilbert_A (1455–1456), ISO Dorsal (1457–1462), DSN uplink problems (1458–1460)
1463–1474	ISO Dorsal_Tail (1463–1468), Dorsal (1463) and Dorsal_New (1468–1472) and imaging
1475–1490	RAT (1475) and ISO (1477–1483) Gilbert_RAT, drive towards Cape Verde (1484–1490)
1491–1547	Difficult driving (1491–1547), imaging, arm shoulder joint failure (1502) and tests (1504–1544)
1548–1582	Imaging, drive towards Cape Verde (1548–1567, 1582), Cape Verde panorama (1571–1581)
1583–1604	Imaging, drive towards Nevada target (1583–1600), wheel problems (1600) and tests (1602–1604)
1605–1634	Imaging, drive back up across Duck Bay (1605–1633), exit from Victoria (1634)

Endeavour (Figures 43 and 47), named after Captain James Cook's ship during his first global voyage (1768–1771), and the only feature at this site other than Victoria whose name became official. Endeavour, first mentioned in team documents as a potential target on sol 1649, was older than the Meridiani Planum deposits and its rim would contain materials characteristic of the ancient cratered highlands to the south. It was 12 km away in a straight line, about as far as Opportunity had already driven at this point, but in fact a substantial detour was required to bypass a region of large drifts similar to Purgatory and the total distance from Victoria would be at least 20 km. The drive was to take 1050 sols or nearly three Earth years.

Opportunity drove to the large drift, reaching it on sol 1654, but the wheels slipped too much in the drift and the dust target was not accessible. Similar material was eventually examined at North Pole on the rim of Endeavour (Figure 99C). Mini-TES made a test observation of a target called Velvet without useful results, and after some imaging and argon measurements the rover left the area on sol 1659 and began its long trek to the south. A young rocky crater called Sputnik on the rim of Victoria was imaged on sols 1661 and 1663 as well as a small crater called Gauss, while cobble targets were sought in the vicinity. A 152 m drive on the smooth terrain of the Victoria Annulus later on sol 1663 was the longest since Purgatory, and the next on sol 1666 was 129 m long. Opportunity was skirting the crater rim, still making cloud movies and measuring argon, and the Pancams were becoming cleaner and its images better.

Two last points of interest on the rim of Victoria were still to be visited. On sols 1668 and 1670 the rover drove to an unnamed viewpoint at the northern edge of Cape Victory to observe the cliffs of Cape Pillar on sol 1671. Then on sol 1673 it drove south to examine Cape Victory itself from Cape Agulhas, the last imaging site it would visit here. A drive on sol 1674 was stopped by excessive tilt, closer to the steep bank than expected. The foreground outcrop, Savu Sea, was imaged on sol 1676 and the rover moved to a safer spot before driving onto Cape Agulhas on sol 1679. The first of many unsuccessful attempts to recover use of Mini-TES was made on sol 1680 by shaking its mirror to try to remove the dust

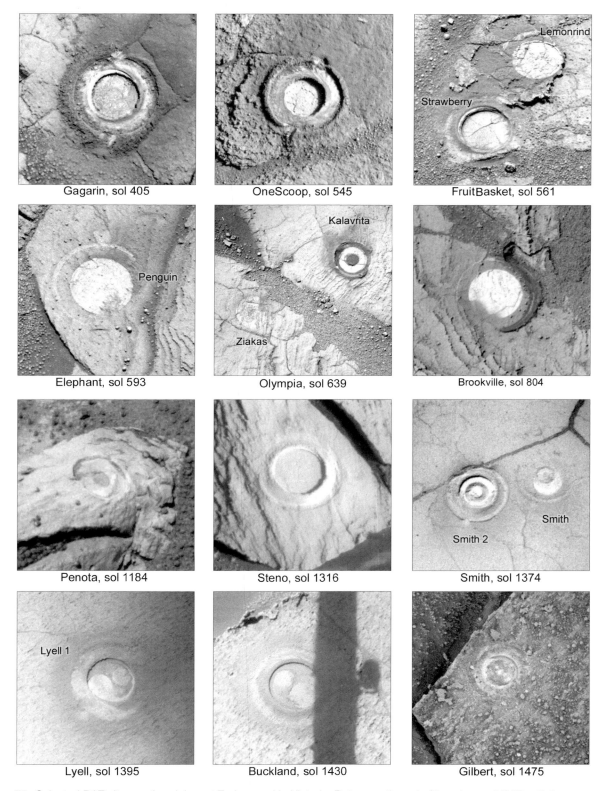

Figure 75. Selected RAT sites on the plains, at Erebus and in Victoria. Dates are the sol of imaging, not RAT activity.

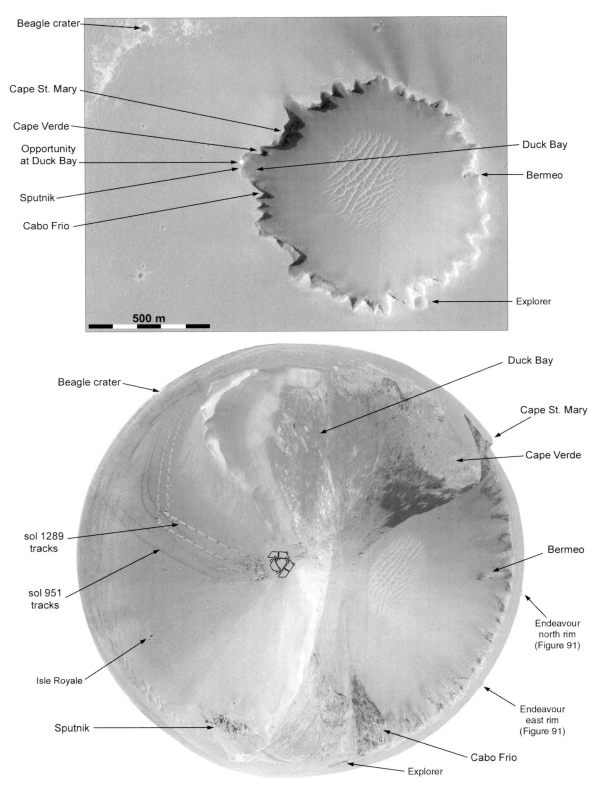

Figure 76. Part of HiRISE image PSP_001414_1780 (top) and a reprojected panorama from sol 1637 with exaggerated horizon relief (bottom), showing the locations of distant features. The eastern horizon of the circular panorama has been replaced with a Pancam panorama taken on sols 952–953.

coating it, and a track left by a rolling boulder inside Victoria was imaged on that sol.

On sol 1681 Pancam made super-resolution and multispectral images of Cape Victory. Then on sol 1682 Opportunity drove to a position for long-baseline stereoscopic imaging of the cape, took one set of images, and on the next sol moved 3 m to take the second set of stereo images before driving south. Targets called Dauphin and Iceland were also observed on Cape Agulhas. The team decided not to visit Explorer crater (Figures 68 and 76), so these were the last views of Victoria. On sol 1686 (MY 29, sol 309) the IAU formally named the next target crater Endeavour just as Opportunity set out for it, driving through a group of cobbles which was not interesting enough to stop for.

Endeavour was a long-distance goal which might never be reached, but many interim targets would be examined along a path chosen using the superb new HiRISE images. Matt Golombek and Tim Parker (JPL) chose a path they called the Yellow Brick Road which avoided larger drifts and crossed many outcrop areas, on one of which the rover would spend the next conjunction examining a good cobble if one was found. This area was so smooth that Opportunity drove 216 m in five hours on sol 1691, the second-longest drive of the mission. Larger drifts on the rocky surface of the Etched Terrain south of the annulus necessitated shorter drives beginning on sol 1695. The flash memory was nearly full now and needed to be cleared for the upcoming conjunction observations, which might include a long MB integration, argon measurements and possibly a seismic experiment. This used accelerometers in the rover's Inertial Measurement Unit as a seismometer, a procedure tested by Spirit on its sol 1805. Opportunity would do this in several locations, but it never detected any seismic events. The rover used its deep sleep capability to save power nearly every night in this area.

Sol 1700 was MY 29, sol 323, or 4 November 2008. On sol 1704 Opportunity imaged a "cracked ripple," where a linear feature seemed to pass through a drift, possibly related to a crack in the underlying bedrock. On the next sol argon was measured again and Mini-TES was given another shake to try to clean its mirror. A test on sol 1708 showed no improvement. On sol 1710 the rover reached a prominent outcrop called Crete, as its shape somewhat resembled that island, and many other

features on the southward drive also took their names from Greek islands. A specific target viewed on sol 1715 was named Crete, but here the name is applied to the whole outcrop as well. A cobble called Santorini was approached on sol 1713 (MY 29, sol 336), for study during superior conjunction as Mars passed behind the Sun on sol 1729. Santorini turned out to be another meteorite, the fourth found by the rover. Like Barberton and Santa Catarina, it was a mesosiderite, a mixture of rock and metal (Schröder *et al.*, 2010).

The rover team were still learning how to use the arm with its failed joint motor, but they placed it on Santorini on sol 1714 and left it there until sol 1743, collecting data on most sols around conjunction. The Pancam viewed targets here called Crete, Corfu and Sicily on sol 1715, and the team decided to drop the seismic experiment here during conjunction. Communications were interrupted by conjunction over sols 1723–1740 and argon was measured twice during the break. By the end of this period, flash memory was full and needed to be cleared again.

MI and APXS were used on two locations on Santorini over sols 1745–1747, and the Pancams observed targets Crete and Andros, and watched a Phobos transit of the Sun. MI viewed Crete on sol 1751 and MB collected data there for three sols as southern spring began on Mars on sol 1752. APXS inspected two targets here, Crete Rock Candia and Crete Soil Minos, and MI also viewed the soil target as the Pancams imaged Gavdos and the rover deck. Sol 1755, spent at this location, was 1 January 2009 on Earth. Then on sol 1759 Candia was imaged and a RAT grind scan was attempted on it for a brushing, but the arm stalled. Pancam viewed Samos and Kythira and APXS took another argon measurement. RAT diagnostics showed that a worn cable on the IDD, which had caused two of the RAT encoders to fail, had now deteriorated enough for it to lose its third and last encoder. The RAT would still work but would be harder to operate.

The planners decided to leave without brushing, as it would take time to develop new methods for using the RAT. The test rover at JPL was not immediately available to help with this. The RAT had to be "caged" or held securely for driving but an attempt to do so failed on sol 1764. Mini-TES was shaken again to clear dust. The last two shakes had been for 3 and 6 s, but now it was shaken

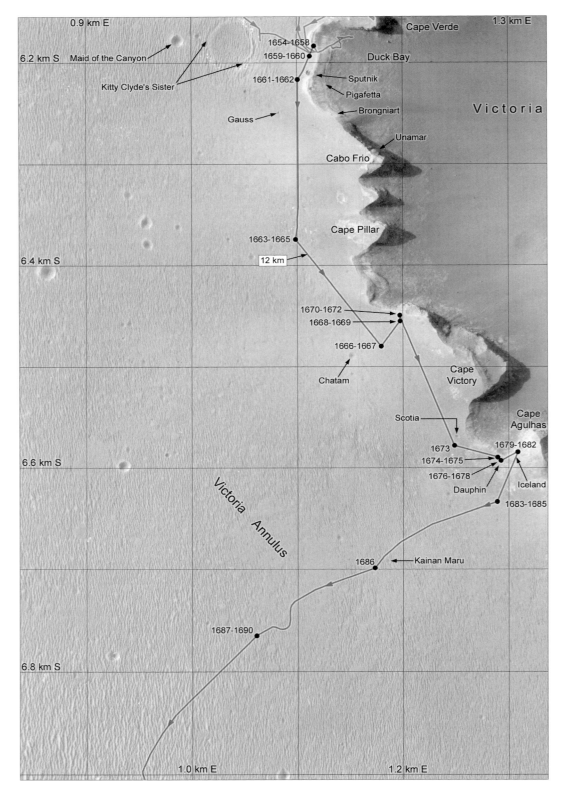

Figure 77. Opportunity route map, section 9. This map follows Figure 72 and is continued in Figure 78. Additional details are shown in Figure 79.

for 21 s, again without result. On sol 1768 the RAT was caged successfully and on sol 1770 the drive south resumed with a 104 m blind drive. When that finished D* was supposed to take over, but it could not find a path and backed up slightly before stopping. A target called Thassos 2 was imaged here before the next drive, and the next brief stop was at a 15 m diameter eroded impact crater called Ranger, reached on sol 1776. After some multispectral imaging of Ranger, Opportunity continued driving past the small crater Surveyor, largely filled with dust, stopping to image it during the drive on sol 1782.

The path chosen from here to Endeavour, the "Golden Path," meandered to drive on rocky outcrops where possible and to avoid the larger drifts. A target called Dusty was imaged on sol 1785, and then a planned 160 m drive was cut short at 80 m on sol 1786, but downlink was lost due to snow at the Madrid DSN station. When communications were recovered, a PMA fault had occurred, whose cause ultimately proved to be a cosmic ray strike in the computer. A Navcam image was taken on sol 1788 to recover pointing information, and more communication tests with MRO followed. Half Moon crater, off to the west of the route, was imaged on sols 1791 and 1792, and on sol 1794 new software was tested to permit drives on successive sols without intervention from Earth. After a drive on sol 1795, Pancam images were taken on sol 1797 to look for the rim hills of Endeavour crater. They had been visible from the western rim of Victoria (Figure 76) but were below the local horizon here. Another drive on sol 1798 was canceled because of new concerns about elevated right front wheel motor currents.

After a diagnostic drive on sol 1800 (MY 29, sol 423, or 15 February 2009), 11 m backwards and 16 m forwards, the team decided to spend more time driving backwards in future. Another multi-sol drive test was done on sol 1801 as on sol 1794. Then on sol 1803 the rover drove 5 m to get onto bedrock, as the wheels experienced more drag on the soil, crossed it, turned when back on the soil, as turning there was easier, and then drove backwards on rock again, all in an effort to redistribute lubricant. This did not yet make a difference and a week of rest was the next strategy. The rover was prepared for a flight software update on sol 1809 and a short marsquake seismic experiment was attempted, just

for 10 min. The flight software was updated on sol 1811 and driving resumed on sol 1813.

During the following week, Opportunity would rely on MRO as a communication relay while Mars Odyssey rebooted its computer system to clear a memory fault and test its backup system, so communications with MRO were tested during the drive on sol 1816. A concern arose now that a cable tie on the rear left wheel might be coming undone and could tangle with the wheel, though this turned out to be unfounded. On sol 1820 the rover was supposed to drive up to Resolution, one of a cluster of small craters in this area, but an erroneous command caused it to drive away from the crater instead (Figure 79D). The Endeavour rim hills were imaged again on sol 1821. The western rim was 16 km away, the northern rim 20 km away and the eastern rim 33 km away, and only the tops of the hills were visible above the local horizon. They were monitored periodically during the long drive.

On sols 1823 and 1824 Opportunity drove up to the 5 m diameter fresh crater Resolution, stopping slightly short of it due to slip, and made a final approach on sol 1825. Despite these successful drives, the cause of the sol 1820 error was not yet clear, and the rover would not drive south again until it was understood. A regional dust storm far to the west was darkening the sky noticeably now. New procedures for using the RAT were now ready for use, and an outcrop area called Cook Islands was the target for ISO on sols 1826–1829. The Pancams took a panoramic view of Resolution and imaged other targets including Surtsey, Lost Cobble and other cobbles, and Penrhyn, an area of cracked rock. Only the right side of the Pancam frames could be used for the Resolution panorama because of dust on the camera, increasing the number of images needed.

A RAT grind scan of Penrhyn on sol 1832 was followed by brushing and MI stereoscopic imaging on the next sol, and MB was used on an unbrushed area nearby called Takutea over sols 1833–1835 and 1838. APXS was used on Penrhyn on sol 1836 before a RAT grind scan there on sol 1837, a grind on sol 1838 and MI of the hole. The sky was still darkened by high-level dust from the storms to the west. Then on sol 1840 MI stereo images were taken of two cracks, Crack 1 and Crack 2, to investigate a hypothesis that cracks were caused by sulfate dehydration. ISO of Penrhyn followed until sol

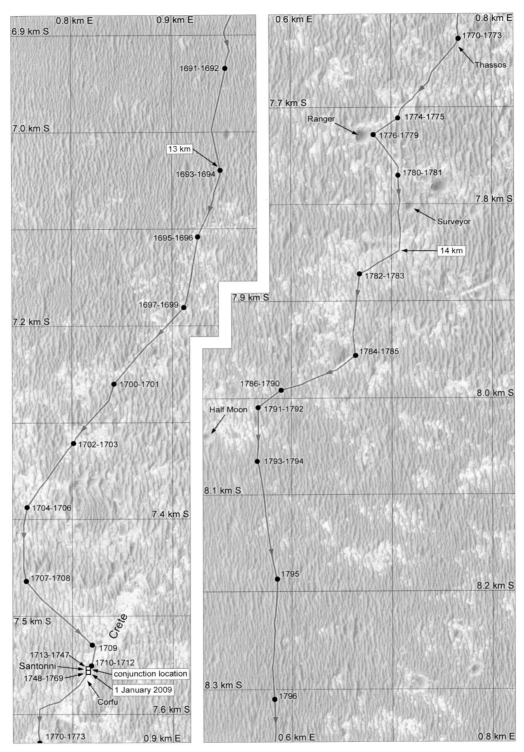

Figure 78. Opportunity route map, section 10. This map follows Figure 77 and is continued in Figure 80. Additional details are shown in Figure 79. White squares are main science stops.

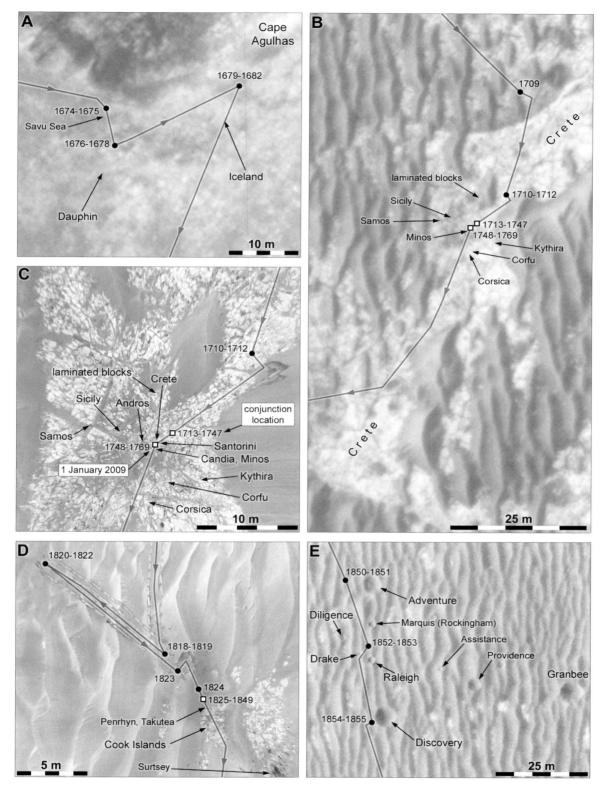

Figure 79. Activities south of Victoria. **A:** Cape Agulhas, sols 1674–1682. **B:** Crete, sols 1709–1769. **C:** Santorini, sols 1710–1769. **D:** Resolution crater area, sols 1818–1849. **E:** Resolution crater cluster, sols 1850–1855.

1849, while resting the troubled wheel. The sky was brightening again and a wind gust cleaned the solar panels, so power was more plentiful.

Driving was permitted again now and the rover moved south on sol 1850, driving backwards for 62 m and stopping near a crater named Adventure. The combination of rest and continued changes in drive direction resolved the wheel problem for the time being. Craters in this area (Figure 79E), known collectively as the Resolution crater cluster (Figure 80), were named after Captain James Cook's various ships, and were imaged as Opportunity drove past them. The craters were superimposed on the surrounding drifts and so were younger than the most recent episode of drift movement. Analysis suggested that the drifts in this area had been static for about 100 000 years (Golombek *et al.*, 2010b). Most craters seen along the traverse had been older than the drifts.

Golombek (2012) estimated some crater ages and interpreted the geological history of the area. The larger old craters such as Erebus, buried by the Burns Formation (the material forming the plains at Meridiani) and now being exhumed by erosion, were late Noachian in age. All or most of the Hesperian-age craters that would have formed on the plains had now been eroded away. The current surface was about ten million years old, and craters such as Aquarius and Apollo 7 (Figure 93) were about that old. Voskhod and Salyut craters were about seven million years old. The San Antonio pair of craters formed about five million years ago, Santa Maria about one to three million years ago, and Concepcion was only about 1000 years old.

Opportunity drove past the last of the Resolution cluster craters, Discovery, on sol 1856, and on sol 1857 it stopped near an old crater called Pembroke. There were no cobbles here, so the rover moved on, turning west to avoid larger drifts to the south and skirting a very old and almost buried crater west of the route. Targets called Phuket and Galveston were imaged on sols 1860 and 1864, respectively, and Angristi and Poros were viewed on sol 1865. Later that sol a drive ended early as the wheels sank in a soft drift. The rover backed partly out on sol 1866 and fully out on sol 1867, imaging the drift and tracks on sol 1868 (Figure 83A) before skirting the difficult area on its western side on sol 1870. A group of small craters southeast of here had been

a possible future target, but the rover planners decided to avoid that area of somewhat larger drifts and continue south.

The right front wheel current problem reappeared on sol 1873 after several daily drives. The wheel was rested for two sols, allowing another marsquake seismic experiment on sol 1875. Then on sol 1877 Opportunity drove to a nearby cluster of small cobbles to take a longer rest. MI stereo imaging of cobbles called Tilos and Kos, very close together, and the nearby Rhodes, was followed by APXS over sols 1879 and 1880. On sol 1881 MI was placed under the rover to try to image its left front wheel. The MI was not designed for this and its images were severely out of focus, but the ability to look under the rover might be helpful for Spirit, which was now stuck fast at Troy, and possibly for Opportunity in future. Indeed, this procedure was used repeatedly by Spirit to check the surface under the rover body.

APXS was used on Tilos and Rhodes on sol 1883 as Pancam imaged the targets Leros, Symi and Chalki. A 6 m drive on sol 1884 put the cobble Kasos in the work area and ISO followed over sols 1886–1889. It may have been another meteorite (Schröder *et al.*, 2010). More distant Pancam targets Adelphoi, Agathonissi, Arki and Astypalia were imaged here before another set of MI views of the rover's middle wheels were taken on sol 1890 to assist Spirit planning. Opportunity drove south again on sol 1891, imaging Halki on sol 1892 as it passed 16 km of driving, and the rover planners decided to use generally shorter drives of about 50–70 m at a time, to help the wheels. Argon was measured fairly frequently here, sometimes overnight, and targets Rho, Saria and Patmos were imaged as the rover moved south. On sol 1899 MI imaged the sky for calibration, and Pancam imaged the magnets and targets Karpathos and Kalymnos.

Sol 1900 was MY 29, sol 523, or 29 May 2009. Opportunity drove south, its wheel currents slightly elevated but not enough to cause a halt. Peaks on the rim of Endeavour had been visible from Victoria, but now a group of hills on the rim of Iazu (Figure 43A) could be seen when the atmosphere was clear enough (Figure 91). A drive on sol 1905 was shorter than usual, only 30 m, because of a limited time between uplink and downlink on that sol. MRO went into a safe mode on sol 1908, as Opportunity changed its driving direction. It had been

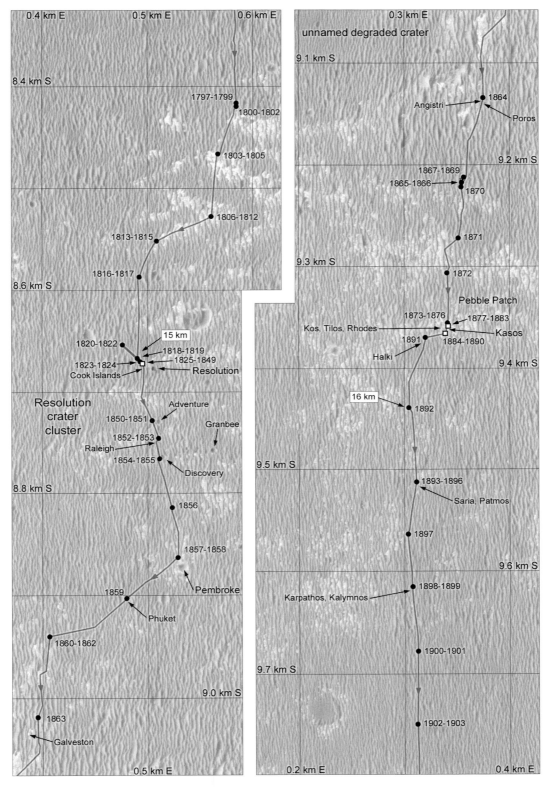

Figure 80. Opportunity route map, section 11. This map follows Figure 78 and is continued in Figure 81. Additional details are shown in Figure 79.

driving forwards and now switched to backwards driving to provide engineering information. On sol 1910 three wheels were on a ripple crest at the end of the drive, all drawing higher current because of excessive slip in this situation. On sol 1911 the rover moved south 1 m as tests were made. The rover rested again to help the wheels, and MI and Navcam took "sky flat" images to help subtract the effects of dust on the cameras. A new effort to recover Mini-TES was made now, leaving the instrument's cover (shroud) open for hours at a time in the hope that a wind gust might clean its mirror. This was repeated for over 500 sols, often overnight, but no improvement was ever seen.

While resting here Opportunity made another marsquake study on sol 1915 and observed two cobbles conveniently within reach. Pancam imaged Hydra on sol 1916 before ISO on Ios on sols 1918 and 1919. Then the rover moved south and imaged targets Delos, Donousa, Dryma and Naxos on sol 1923 before moving a short distance to Tinos, a patch of clean bedrock. Here a target called Mykonos was imaged on sol 1924 and ISO followed on Tinos over sols 1925–1926. Opportunity reached the next outcrop on sol 1927 (MY 29, sol 550), part of a broad rocky pavement surrounding an ancient eroded depression named Nimrod. Opportunity had now driven 2.5 km south and almost 1 km further west than Victoria on the "Golden Path" in order to avoid areas of large drifts, but still more westward travel would be required to avoid the last extensive area of "purgatoids." The rover took advantage of the current outcrop to turn west again, skirting Nimrod on a new route called the Pink Path from its color on maps prepared by Tim Parker. Paths to the east of Nimrod (the Golden and Blue Paths) offered shorter routes to Endeavour but crossed treacherous fields of drifts.

The outcrop here was called Brigantine, and a target named Absecon was investigated to characterize it. Brigantine was imaged on sol 1929 and a short move brought Opportunity to Absecon on sol 1930. Daily heating of the right front wheel was begun here, as it had been shown to be useful earlier, allowing lubrication to spread more easily through the bearing. After a marsquake study on sol 1931, ISO on Absecon occupied sols 1931–1941. MI viewed another target called Little Beach on sols 1932–1933 and the Pancams imaged Humbaba, Gilgamesh, Enkidu and Reeds Bay. The science team wanted

to find basaltic soil, soil free of blueberries, but none was found here.

On sol 1938 a target called Weakfish Thorofare was imaged and a RAT grind scan of Absecon was attempted, but failed. It succeeded on the next sol, permitting a RAT brushing on sol 1941, but that failed, and the outcrop was deemed not worth further delay. The last MI and APXS work was done later on sol 1941 and Opportunity drove west again on sol 1942. The wheel situation was very good now, helped by the heating. Targets called Sri Lanka, South Georgia and Hilton Head were imaged on sol 1945 and a rock called Block Island on sol 1946 before driving again. Another marsquake observation was made on sol 1948, and on sol 1949 the Navcams and Pancams took images in a coordinated observation with MRO as it flew over.

On sol 1950 the rover approached two craters first called Alvin and Dolphin, but as these names had already been used the names Kaiko and Nereus were substituted. Now the Block Island images from a few sols earlier were received on Earth, and it looked interesting. It was 0.7 m across, but a miscalculation at first suggested it was larger. Opportunity drove back to it starting on sol 1952, covering 110 m on this easy terrain. Mini-TES was still being exposed in the hope it would be cleaned, but warily now as a dust storm appeared northwest of Hellas, moving west across Noachis, but it broke up before threatening the rover.

The rover imaged Keros on sol 1956 and approached Block Island on sol 1959. Block Island was named after part of the state of Rhode Island and targets on the rock were named after its features. Its general appearance and ISO at targets called New Shoreham, Clayhead Swamp, Springhouse Icepond and Middle Pond over sols 1963–1972 suggested it was another iron meteorite similar to Heatshield Rock (Ashley *et al.*, 2011).

Other targets nearby called Bushy Rockrose, Corn Neck Road, Night Heron, Arrowwood, Little Bluestem and Switchgrass were imaged here, as well as a feature named Block Island Ripple Wake, and then a small move on sol 1973 put a soil target called Vail Beach in range for ISO on sol 1974. Opportunity imaged a new target on Block Island called Siah's Swamp and moved to reach it on sol 1975. Then ISO followed at two places on that target over sols 1976–1992. A very small move on sol 1981 gave better access to the second location. MI also

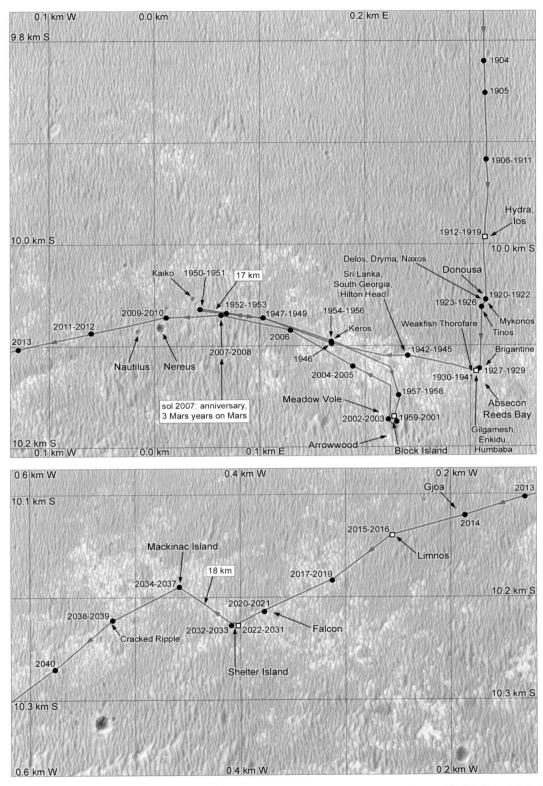

Figure 81. Opportunity route map, section 12. This map follows Figure 80 and is continued in Figure 82. Additional details are shown in Figure 83.

viewed Veteran's Park and Fresh Pond, and some other targets of interest here were called James Ashley and Great Salt Pond. Sol 1988 was a runout sol as the command uplink was lost during a spacecraft emergency on LCROSS, the lunar impact mission launched with Lunar Reconnaissance Orbiter. MRO was in safe mode on the following sol, and more time was lost on sols 1993 and 1994 as nearby brush fires forced JPL to close.

On sol 1995 a substantial wind gust cleaned the solar panels just 40 min before the Mini-TES cover was opened in the hopes it would be cleaned by just such an event. Opportunity now spent several sols driving around Block Island, imaging it from all sides on sols 1998, 2000 (MY 29, sol 623, or 9 September 2009) and 2001–2004. The rover drove away on sol 2004, heating the wheel periodically, and by sol 2010 it was back at Nereus imaging its ejecta. A crater west of Nereus, called Nautilus, was viewed on the next sol as the rover passed it, and another called Gjoa on sols 2014 and 2015. On the next sol a target called Limnos was examined with Pancam, MI and APXS, and MI also viewed tape near the IDD shoulder joint to see if saltating sand grains had eroded it.

Opportunity ran another marsquake survey on sol 2018, and several sols later approached another rock called Shelter Island, imaging it on sols 2022, 2024 and 2027. A target called Dering Harbor was selected and two sols of ISO began on sol 2029. Then the rover moved around the rock, imaging it from different sides before moving on to a nearby rock called Mackinac on sol 2034. When first seen, this was referred to as Khalid's Rock after JPL engineer Khalid Ali. These rocks, apparently all meteorites (Ashley *et al.*, 2011), were being named after islands as the craters had been named after ships. No ISO were scheduled at Mackinac, as multispectral images suggested it was very similar to Block Island, but the rock was viewed from several positions before Opportunity drove away on sol 2038.

Cracked Ripple, another linear feature crossing a big drift, was imaged on sol 2040, and a pair of cobbles called Kea were seen on sol 2042 as the rover moved to the southwest. The path turned south at a gravel patch adjacent to a large drift on sol 2043 with imaging of the tracks in the gravel. The gravel would have been trenched to estimate its properties if it could have been done without unduly stressing the wheels. The Mini-TES

cover was still being opened virtually every day without showing any results. Sol 2046 was the first sol of Mars year 30, and Opportunity now moved rapidly south on an area of outcrop and small ripples, stopping to view the small crater Trinidad on sol 2048 and the cracked ripple Fourni on sol 2049. Driving directions alternated every few sols to help the wheels.

A new target appeared now, a rock protruding vertically from the surface and nicknamed Sore Thumb before being named Marquette Island. On sol 2055 a drive was cut short on a soft drift, and at about the same time the right front wheel began drawing more current again. Harder surfaces generally seemed to help the wheel avoid that problem. Marquette was reached on sol 2056 and images showed it was not a meteorite or a piece of local bedrock. Opportunity stopped here to rest its wheels and investigate this unusual rock. Several short drives with imaging brought the rover to its first ISO position on sol 2063, and the target, Peck Bay, was examined over sols 2065–2071 with a grind scan on sol 2068 and a RAT brush on sol 2070. Meanwhile Pancam was imaging many nearby targets, including Wilderness Bay, Voight Bay and Hessel Bay, Echo Island, Birch Island and Gravelly Island, names taken from the region around Marquette Island, Michigan. On sol 2072 ISO began on the nearby target Islington Bay, with pre-RAT observations, a grind scan on sol 2074, brushing on the next sol and a long MB integration over sols 2076–2085. Mars Odyssey was in a safe mode now, slowing work here briefly, and when it was restored to health, MRO went into safe mode, illustrating the dependence of surface missions on orbital relays.

On sol 2086 the rover moved around the rock to view its less dusty side, and over sols 2089–2092 it studied the Loon Lake target with MI, APXS and MB. This target was not suitable for brushing, so Opportunity drove back to the Peck Bay area on sol 2093, moved slightly to position the IDD on sol 2095 and performed a RAT scan on sol 2097. It ground at a target called Peck Bay 2 on sol 2100 (MY 30, sol 55, or 20 December 2009), its first grind since sol 1838. The RAT cut about 0.8 mm deep into this very hard rock, but more depth was needed, so a second grind on sol 2103 cut a further 0.7 mm into the surface. Further work was precluded by a RAT contact switch fault until sol 2109 and the RAT bit was imaged on sol 2110 (1 January 2010 on Earth) to assess wear. MI viewed the hole on the same sol and MB collected data nearby at Cube Point for two sols. Then ISO continued at Peck Bay 2 over sols 2112–2117 and at a slightly offset position over sols 2118–2120. Debris was brushed from the grind hole on sol 2116, and the RAT brushes were still working very well despite having been bent when run backwards on sol 1348.

Marquette Island turned out to be an ejecta block from an unknown distant crater, unlike any other Mars rock yet seen (Arvidson *et al.*, 2011). This coarse-grained igneous rock contained olivine and pyroxene, quite different from the local bedrock but possibly related to the "protolith" or original material from which the Meridiani sandstones were derived. This was quite different from Bounce Rock, seen on sol 63, which was similar to SNC meteorites (meteorites named after the three representative members of the group, Shergotty, Nakhla and Chassigny, thought to originate on Mars). The rocky pavement around Marquette Island was named Lake Huron to reflect its setting in terrestrial geography.

New software called AEGIS was uploaded on sol 2112. It would analyze Navcam images on designated sols to identify a potential imaging target and command Pancam images of it without any involvement from the ground. This would avoid waiting at least a sol for human analysis and instructions from Earth and should increase the science output of the long traverse to Endeavour. Opportunity drove away from Marquette Island on sol 2122 and continued south towards a new target, a very fresh crater called Concepcion. The AEGIS software was tested on sol 2130 and the rover approached Concepcion on sol 2136. Dark rays around the crater were visible on the nearby drifts in HiRISE images, suggesting that this was likely to be the freshest crater Opportunity would ever see. The crater name was taken from one of Magellan's ships, and target names here were taken from the island of Bohol in the Philippines where the *Concepcion* was abandoned and burned in 1521.

The Pancams imaged targets Anda and Ubay on sols 2136 and 2138 (MY 30, sol 93) and tested AEGIS again at the rim of Concepcion on sol 2138. The crater itself and many blocks around it were imaged over several sols, including rocks called Loboc River and Chocolate Hills, both of which showed dark rinds possibly deposited by water in bedrock cracks and now exposed by the impact. Other targets called Mahoney Island,

Cebu Strait and Canuba Beach were imaged, and Opportunity moved closer to Chocolate Hills on sols 2147 and 2149. While here the Pancams imaged many more features of this target-rich location, including Santa Fe Beach, Camotes Sea and Chanigao Channel, during ISO of several parts of the rock and its rinds.

MI viewed a dark area called Aloya on sol 2150 and took more images on sol 2151 with better lighting, and collected some MI and MB data. After a very small move on sol 2152, ISO began on a lighter part of the rock called Arogo on sol 2154. Then on sols 2157–2158 another layer called Tears was analyzed, followed by ISO on a darker part of it called Dano over sols 2158–2160. Then Opportunity tried to find a path to another interesting rock called Napaling, but it was too difficult to reach. This was in one of the rays where rubble was superimposed on the local drifts. Many other targets were imaged as the rover drove around the ray on sols 2165–2169. This rocky area was good for AEGIS testing, so the new software was used on sol 2172 to identify a cobble and take multispectral images of it. None of the rocks were easy to reach, so after taking images of targets called Bugnao Beach on sol 2174 and Gakang Island on sol 2177, the rover left Concepcion and returned to its Pink Path around the area of large drifts blocking the direct path to Endeavour. Pancam images of the horizon from Concepcion on sol 2179 showed the distant rim of Bopolu crater (Figure 43A) 65 km away to the southwest, and images around this time also showed two small peaks on the rim of Miyamoto crater (Figure 43A).

The next waypoint was a double impact crater called San Antonio, 600 m further south. Power was quite low so roughly every third sol was devoted just to recharging the battery. Pancam images of the soil were taken often to look for regional variations in blueberry concentration. None of the little concretions were visible near Concepcion. The science team had decided on sol 2186 that the next significant cobble would be named Oileán Ruaidh after an Irish island, though in fact that name was not used until sol 2367. A less significant cobble called Syros was imaged on sol 2187 as well as a banded ripple, a drift that showed layers of different albedo. These had been seen many times along the path. Driving passed the 20 km mark on sol 2191, and to celebrate it Tim Parker suggested making a full 360° Pancam panorama of the

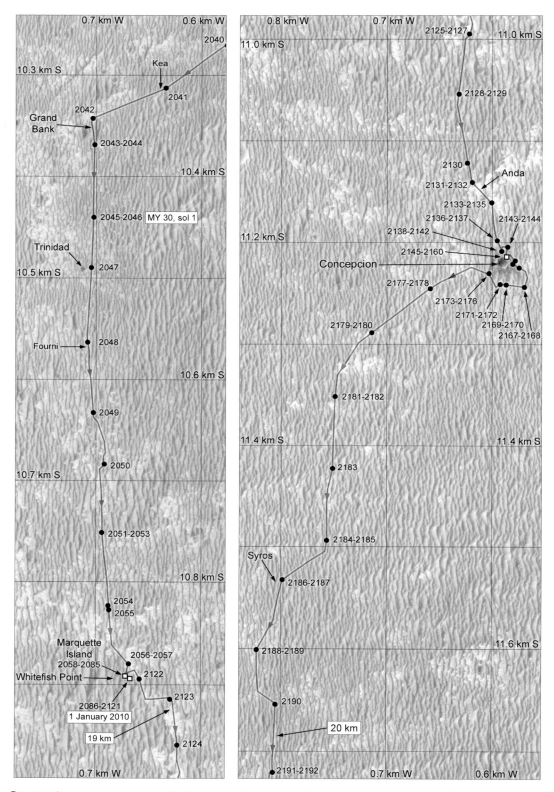

Figure 82. Opportunity route map, section 13. This map follows Figure 81 and is continued in Figure 85. Additional details are shown in Figures 83 and 84. White squares are main science sites.

horizon. Only part of it was actually made, covering the Endeavour and Bopolu hills.

Opportunity arrived on the northern rim of the San Antonio crater pair on sol 2194 (MY 30, sol 149). The *San Antonio* was another of Magellan's ships. The rover imaged the old, partly drift-filled craters, called San Antonio East and San Antonio West, on sols 2196–2198 and departed on sol 2199. Sol 2200 was MY 30, sol 155, or 2 April 2010, and Opportunity's planners were preparing to turn to the east for the remaining long drive to Endeavour, stopping only briefly to place APXS on the soil at roughly 1000 m intervals to examine regional variations. A steering test on sol 2201 was cut short by an elevated wheel current, and as the rover drove away from that location on sol 2204 the wheels slipped more than expected in the soil loosened by the test and that drive was cut short.

Driving resumed on sol 2206 and Opportunity moved south again, still testing AEGIS occasionally and taking cloud movies on the way. Power was getting low and temperatures were becoming colder. The rover had been sleeping more to conserve power, but now it was kept awake more often to allow the electronics to generate heat. An AEGIS test on sol 2221 was not successful because there were very few features to look at among the drifts. A soil target called Ocean Watch was examined on sols 2222 and 2224, the first of the regular ISO stops on soil along the drive. A rock called Elfin Cove was imaged here, and another science investigation in this region was a soil properties study involving images of ripple crests crossed by the wheels. The rover was now 38 percent of the way from Victoria to Endeavour and the expected date of arrival was sol 3108.

The rover imaged a ripple called Port Townsend on sol 2228 and a rock called Newfoundland on sol 2229, and also took an image called Moment in Time, to be used as part of a project by the *New York Times* to collect many images all taken at the same time. By now the power was becoming so low that Opportunity needed to find north-facing slopes to give it more power, as Spirit had done so often. In this flat landscape the only suitable "lily pads" were the north ends of drifts, so on sol 2233 images were taken to find a good one, and the rover drove towards it on the next sol. It helped a bit but was not really steep enough. The flash memory was nearly empty, so as many images as were needed could

be taken so downlink time was not wasted. Power was too limited to permit a planned drive on sol 2238, but on sol 2240, the winter solstice, Opportunity was driven 25 m to reach a better lily pad. Meanwhile the hills around Endeavour crater were becoming clearer, and on sol 2239 a set of super-resolution images (multiple almost identical images that could be combined to increase the effective resolution of the set) were taken. Figure 101 shows the place names given to the hills at this time, taken from the voyages of Captain James Cook. The destination point was called Cape York.

While on this lily pad Opportunity imaged ripples called Charleston on sol 2241 and conducted a mars-quake study on the next sol. A drive on sol 2245 was programmed to stop if a good northern tilt was encountered. None was found, so the drive did not stop, but a fortuitous gust of wind cleaned the solar panels and increased power, so lily pads were not needed any more. Unfortunately, although the Mini-TES cover was open at the time, it experienced no cleaning. Also on this sol (MY 30, sol 200, or 20 May 2010) Opportunity exceeded Viking Lander 1's operational lifetime to become the longest-operating robotic entity on the surface of Mars. Spirit had exceeded that record 20 sols earlier but its status was unknown and it never recovered from its enforced hibernation. Also at this location Opportunity finally passed the last of the larger drifts it had been avoiding, and turned east on sol 2252 (MY 30, sol 207) to drive directly towards Endeavour.

At the 41st Lunar and Planetary Science Conference, held at The Woodlands near Houston, Texas, in March 2010, Parker *et al.* (2010) showed a plan for the remaining route. The destination point at the time was the northern tip of a detached hill (Cape York) forming part of the rim of Endeavour (Figure 101). Clay minerals detected from orbit in the rim of Endeavour would make a scientific target very different from the rocks seen previously. In 2011 the target shifted to the south end of Cape York where clays seemed more abundant. Images from the Mars Reconnaissance Orbiter in 2009 showed that some large dark dunes in the East Dune Field inside Endeavour (Figure 91) had moved or dissipated since they had been imaged by Mars Global Surveyor in 2001, a kind of change rarely seen on Mars and suggesting strong local winds (Chojnacki *et al.*, 2011).

Table 22. *Opportunity Activities, Victoria to Marquette Island, Sols 1635–2070*

Sol	Activities
1635–1641	Imaging, drive to Ripple Dune (1637, 1639), move to Bagnold and Bright Patch With Salute (1641)
1642–1647	ISO Post Victoria Ripple Soil and perform IDD tests (1642–1647), imaging
1648–1668	Drive to Staging Area target (1648–1654), imaging, drive along rim to Cape Victory (1659–1668)
1669–1682	Image Cape Pillar (1669–1670), drive to Cape Agulhas (1670–1679), stereo imaging (1680–1682)
1683–1713	Drive south towards crater Endeavour, imaging (1683–1708), drive towards Santorini (1709–1713)
1714–1746	ISO Santorini cobble and imaging (1714–1746), conjunction (1724–1740)
1747–1750	ISO Santorini and Santorini2 (1747), drive to Crete (1748) and imaging (1749–1750)
1751–1762	ISO Candia (1751–1754), Minos (1755–1758), RAT problem (1759) and tests (1762)
1763–1776	Imaging (1763–1769), drive south (1770–1775), multispectral imaging of crater Ranger (1776)
1777–1819	Drive south towards Endeavour (1777–1797), right front wheel problem reappears (1797), drive on
1820–1832	Move to Resolution (1820–1825), study Cook Islands (1826–1830) and Penrhyn (1828–1832)
1833–1836	Brush and ISO Penrhyn (1833, 1836), ISO Takutea (1833–1836)
1837–1849	RAT (1837–1838) and ISO Penrhyn (1838–1849), and ISO Crack1 and Crack2 (1840)
1850–1857	Drive, image Adventure (1850), Rayleigh (1852), Discovery (1854) and Pembroke (1856–1857)
1858–1864	Imaging, continue driving south towards Endeavour crater, pass subdued 300 m crater (1864)
1865–1877	Delayed by large drift (1865), back up (1867), bypass drift (1870), approach Pebble Patch (1877)
1879–1883	ISO Tilos and Kos targets (1879–1880), Tilos (1881), MI wheel (1881), ISO Rhodes (1882–1883)
1884–1890	Drive to Kasos target (1884), ISO Kasos (1886–1890) and MI wheels (1890)
1891–1899	Drive south towards Endeavour and imaging (1891–1899), MI sky images (1899)
1900–1913	Drive south towards Endeavour and imaging (1900–1913), MI sky images (1913)
1914–1924	Imaging, ISO Ios target (1918–1919), drive south (1920), move to Tinos target (1923)
1925–1930	ISO Tinos (1925–1926), drive south (1927), move to Absecon target (1930), imaging
1931–1940	ISO Absecon (1931–1937) and Little Beach (1932–1933), RAT scan Absecon (1939)
1941–1951	Failed brush and ISO on Absecon (1941), drive west around Nimrod (1942–1950), imaging
1952–1964	Drive to Block Island (1952–1962), imaging, ISO New Shoreham target (1963–1964)
1965–1967	ISO Clayhead Swamp (1965), Springhouse Icepond (1966), Middle Pond (1967)
1968–1972	ISO New Shoreham (1968–1969) and Clayhead Swamp (1970–1972) targets
1973–1975	Move to soil pebbles (1973), ISO Vail Beach (1974), move to Siah's Swamp (1975)
1976–1980	ISO Siah's Swamp (1976–1979) and Veterans Park (1979), imaging
1981–1986	Move to (1981) and ISO (1982–1985) Siah's Swamp2, ISO Fresh Pond (1986)
1987–1997	ISO Siah's Swamp2 (1987, 1989–1992), move to Block Island positions 1 and 2 (1997)
1998–2003	Imaging, move to position 3 (2000), positions 4 and 5 (2001), position 6 (2002), imaging
2004–2016	Drive towards Endeavour, and imaging (2004–2015), ISO Limnos and study arm joint (2016)
2017–2028	Drive south towards Endeavour (2017–2023), move to Shelter Island (2024, 2027), imaging
2029–2033	ISO Dering Harbor (2029–2030), drive around Shelter Island and imaging (2032)
2034–2054	Drive to (2034) and around (2038) Mackinac, imaging, drive south to Endeavour (2040–2054)
2055–2067	Drive to Marquette Island (2055–2063), imaging, ISO Peck Bay target (2065–2067)
2068–2070	RAT scan Peck Bay (2068), imaging, brush and ISO Peck Bay (2070)

AEGIS was used on sol 2247 and detected an outcrop target. Pancam imaged a target called Anafi on sol 2249 and a cloud movie was made on the next sol, and then on sol 2252 driving resumed. The eastward drive was interrupted after an unexpected problem on sol 2256 with a camera azimuth pointing actuator on the PMA. Several sols of diagnostic tests found no problem with the PMA, which would have been very serious, but the fault was tracked to Mini-TES, which, though not collecting data now, was being tested at intervals. Now it would not be used at all until the problem was understood, though the cover was still opened on nearly every sol in the hope the mirror might be cleaned.

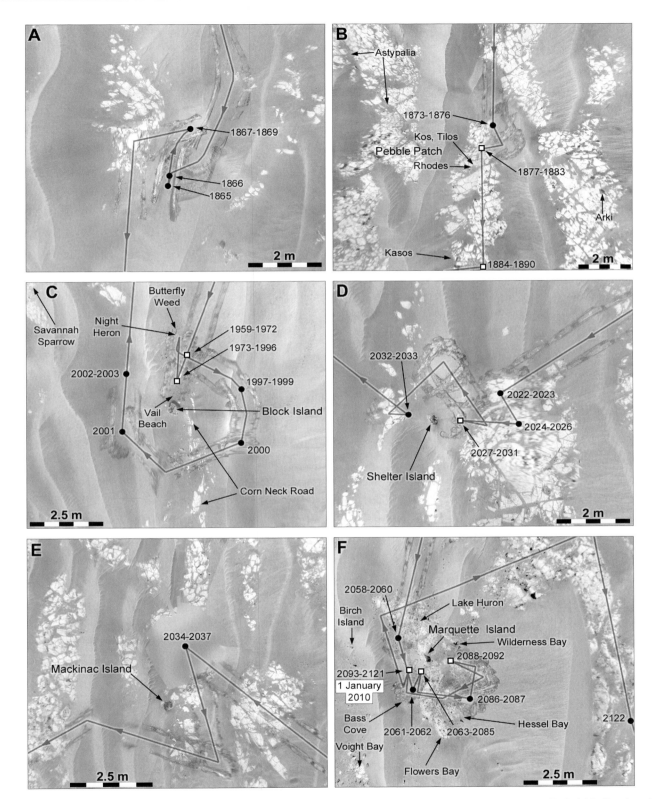

Figure 83. Opportunity activities between Victoria and Endeavour craters. **A:** Sol 1870 area, sols 1865–1870. **B:** Pebble Patch, sols 1873–1891. **C:** Block Island, sols 1959–2003. **D:** Shelter Island, sols 2022–2033. **E:** Mackinac Island, sols 2034–2037. **F:** Marquette Island, sols 2058–2122.

Opportunity drove again on sol 2270, now moving steadily eastward. A cobble was imaged on sol 2272 and AEGIS found a dark rock for imaging on sol 2278. Another apparently fractured ripple was imaged on the next sol, and a gravel bank along the path was viewed on sol 2281. The route here was slightly south of the planned Pink Path, and ISO would be commanded if interesting targets appeared to characterize interesting geology along the way. The sol 2288 drive stopped near another of the gravel piles which were seen occasionally, and it was imaged on sol 2291. On sol 2290 AEGIS found a high-albedo target for imaging. It could be directed to look for specific types of target, and on sol 2292 it was asked to find a low-albedo target, but did not find one. The pause here was over the US 4 July holiday.

A drive on sol 2295 was intended to stop on an outcrop but ended straddling a ripple crest, and the usual turn to improve communications resulted in the wheels cutting a trench in the ripple. On sol 2297 the rover examined the trench, called Juneau Road Cut, with MI and APXS. Super-resolution images of the Endeavour rim hills were taken on sol 2298 and another wind gust cleaned the solar panels. With more power and wheels behaving well, driving continued on every sol now on a path just south of the little crater Pond Inlet, with arrival at Cape York now expected around sol 3097. Sol 2300 was MY 30, sol 255, or 14 July 2010.

Activities were temporarily suspended when Mars Odyssey entered safe mode after a solar array fault on Opportunity's sol 2301, and pictures taken on that sol but only transmitted after Odyssey was returned to service showed a dust devil east of the rover. This was the first ever seen by Opportunity, though a few dust devil tracks were visible in orbital images of the plains around Endurance and the depression between Victoria and Endeavour. Tracks were less visible in the Etched Terrain. Small amounts of data were transmitted directly to Earth as imaging continued, including pictures of the magnets on sol 2303 and an argon measurement and an AEGIS search for dark rocks on sol 2304. Then on sol 2308 a large MRO data relay allowed the rover team to return to more normal operations, and on the next sol Opportunity imaged a target called Alert Bay and made an argon measurement.

When the traverse resumed on sol 2311 the first short drive, just 30 m long, was a test of a new strategy for driving further each sol. Previously the route had been chosen to avoid hazards seen in the previous sol's Pancam imaging of the drive direction. If the rover had enough power to drive beyond the range of reliable Pancam coverage, it could use its own autonomous Navcam-based hazard-avoidance system, but this had only been possible in the past while driving forwards. The rover's high-gain antenna blocked part of the view to the rear and interfered with hazard avoidance during backward driving. A strategy to overcome that problem by driving in roughly 1 m increments, each followed by a quick turn to allow unobstructed imaging, was devised at JPL and tested successfully with this drive. The success of the method was shown on sol 2353 with a drive of 111 m, at that time the longest backwards drive ever made by either rover, but that record was broken several times in the following months, the longest drive in this part of the journey being 166 m on sol 2616.

Opportunity used AEGIS again on sols 2312 and 2313, and observed a target called Port Hardy and looked for dust devils on sol 2314. The Mini-TES cover was still being opened almost every day to try to clear it of dust, without any effect. After a drive on sol 2315 the rover imaged an outcrop called Valparaiso on sol 2317, followed by ISO, as the team considered a name for a small crater which could be seen in Pancam images of the east rim of Endeavour, high up on the sloping wall. One suggestion was Nome, but in the end the crater was not named. The Endeavour rim hills were imaged again on sol 2318, followed by a dust devil search, and two sols later the rover moved on, imaging Valparaiso again as it departed because the previous images had been in shadow. This area contained several gravel patches or mounds and one, Portland, was imaged during the sol 2320 drive. The backward hazard-avoidance method was applied for the first time in a regular drive on sol 2322, giving a distance of 77 m, during which the total drive distance exceeded 22 km. Monterey Bay, another gravel deposit, was viewed on sol 2324.

Opportunity continued to use AEGIS software from time to time, including on sol 2325 when it failed to find a dark soil target for imaging. The rover drove past targets Boston and Halifax over the next few sols, and on sol 2329 it imaged Gravel Bank, a crater called Lightning and the outcrop Cayenne. Nowitna, viewed on the next sol, was another patch of gravel, and AEGIS

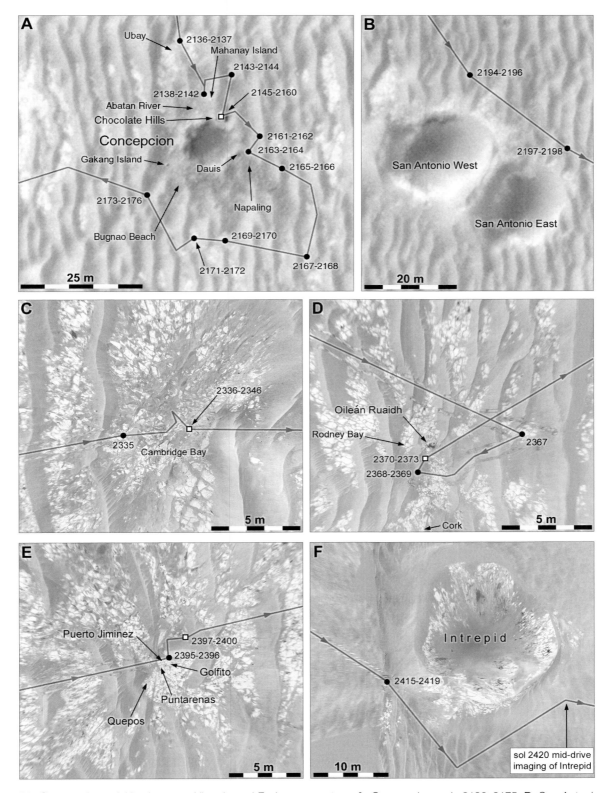

Figure 84. Opportunity activities between Victoria and Endeavour craters. **A:** Concepcion, sols 2136–2175. **B:** San Antonio, sols 2193–2197. **C:** Cambridge Bay, sols 2335–2346. **D:** Oileán Ruaidh, sols 2367–2373. **E:** Puntarenas, sols 2395–2400. **F:** Intrepid, sols 2415–2419. The object under the sol 2145 symbol in **A** is Opportunity, imaged on sol 2153 in HiRISE image ESP_016644_1780.

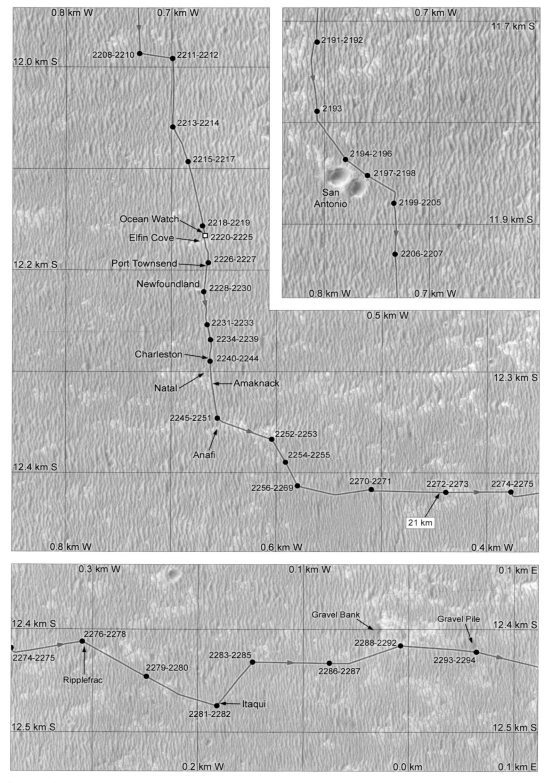

Figure 85. Opportunity route map, section 14. This map follows Figure 82 and is continued in Figure 86. Additional details are shown in Figure 84.

looked at another one on sol 2332. The rapid progress continued until a light tan outcrop caught the team's attention and Opportunity approached it on sol 2337. This was Cambridge Bay, and when seen closely it was divided into two parts, a somewhat red layer called Clarin Beach and a more buff-colored layer called Duero Beach, separated by Laya Beach, a transition zone with a slightly purple color.

Cambridge Bay was chosen for intensive ISO to explore the stratigraphy of the Burns Formation, the rocks of Meridiani Planum, with MI and APXS on Clarin Beach on sol 2340 and on Duero Beach on the next sol and on Laya Beach on sol 2342. MB was placed on Laya Beach at a target called Cervera Shoal for four sols after that, and two cloud movies were also made, though the flash memory was filling up again. Opportunity drove on again on sol 2347, but a possible problem with MB operation was noted now.

On sol 2349 (MY 30, sol 304), 11 km and almost exactly one Mars year after leaving Victoria, Opportunity arrived at the edge of the largest expanse of outcropping rock it had yet seen, making driving still easier. Argon was measured on sol 2350, as it was every one or two weeks over this period, and the record 111 m drive was made three sols later, marking the mid-point of the drive from Victoria to Endeavour. The MB instrument was subjected to tests at different times of day, a warmer time on sol 2355 and a cooler time on sol 2358, showing that it worked best when warmer. By sol 2360 the flash memory was full and data had to be transmitted and deleted, and as the rover drove on the MB diagnostic tests continued on sol 2363. A possible meteorite was noticed ahead and named Devon Island, but that name was changed to Oileán Ruaidh as had been suggested on sol 2183. As at Mackinac and previous rock targets, blocks not related to the plains material were named after islands.

On sol 2367 Opportunity drove around a crater called Gabriel (Figure 87) and approached Oileán Ruaidh. It was imaged thoroughly, as was an outcrop named Cork behind the rock. An ISO target called Mulroy on Oileán Ruaidh was examined by MI and APXS over sols 2371–2373, showing it to be an iron meteorite, and a color boundary called Rodney Bay in the outcrop nearby was imaged on sol 2372. A test of communications to try to contact Spirit was made on sol 2371, and

now the team decided that it would not stop for any additional meteorites for at least the next 3000 m, to speed up arrival at Cape York. Opportunity drove away on sol 2374, imaging another probable meteorite called Ireland from three directions as it passed.

Driving continued roughly every other day for the next week, with imaging of several targets, including Los Angeles and San Diego on sol 2376, a laminated outcrop called Choc Bay on sol 2382 and a rind or coating called Port Stanley on sol 2384. Other observations here included argon on sol 2376, backward autonav driving and Spirit communication tests on sols 2379 and 2386, a dust devil search on sol 2382 and a front left Hazcam image of sand in the adjacent wheel on sol 2383. There was a large data delete on sol 2382 but the flash memory filled up quickly, and after sol 2384 the need for further deletes and several communication faults held up driving until sol 2393. Opportunity did not have enough energy to wake up at night to relay data to an orbiter, but energy would increase with the changing seasons.

On sol 2389 (MY 30, sol 344) MRO's MARCI instrument observed a large dust storm 600 km south of Opportunity and moving north. A few sols later it appeared less directly threatening, but dust levels in the atmosphere increased, reducing energy again. Opportunity looked for dust devils on sols 2394 and 2400 (MY 30, sol 355, or 25 October 2010) and viewed its tracks on sol 2395. It had turned slightly north after sol 2381 to stay on the outcrop and avoid some large drifts, and this brought it to a possible geological contact between two rock types at Puntarenas on sol 2395. A small move put it in position for ISO on sol 2397. The analysis occurred at Puerto Jimenez with polygonal cracks adjacent to the apparent contact, and other targets nearby were Quepos and Golfito, the latter marked by festoons or undulating layers. When the rover drove away on sol 2410 it tried to image Golfito but missed its target.

The intention from now on was to stop rarely during the long drive to Endeavour, though interesting targets might appear unexpectedly at any time. On sol 2403, after imaging a target called Zancudo, Opportunity reached the end of the extensive rocky pavement forming the Etched Terrain (Figure 47), first encountered around sol 580 at Erebus Highway, 18 km and more than 1800 sols earlier. Now the rover moved onto an expanse of fairly smooth soil and small ripples which extended to

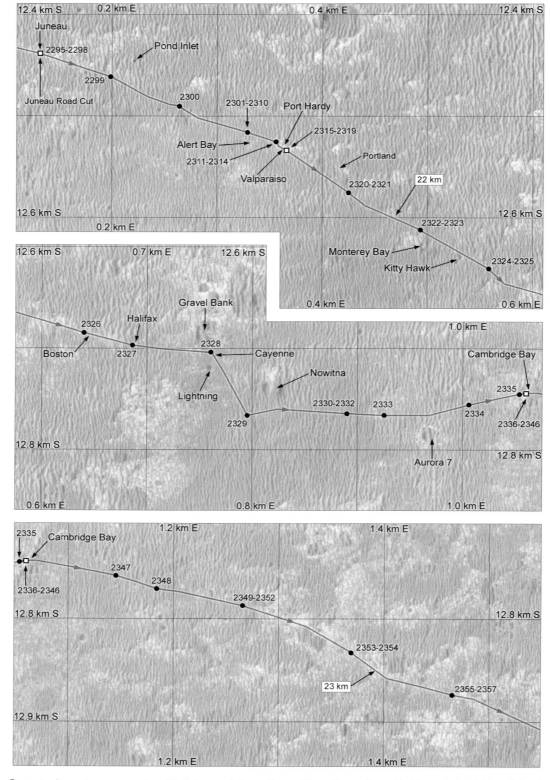

Figure 86. Opportunity route map, section 15. This map follows Figure 85 and is continued in Figure 87. Additional details are shown in Figure 84.

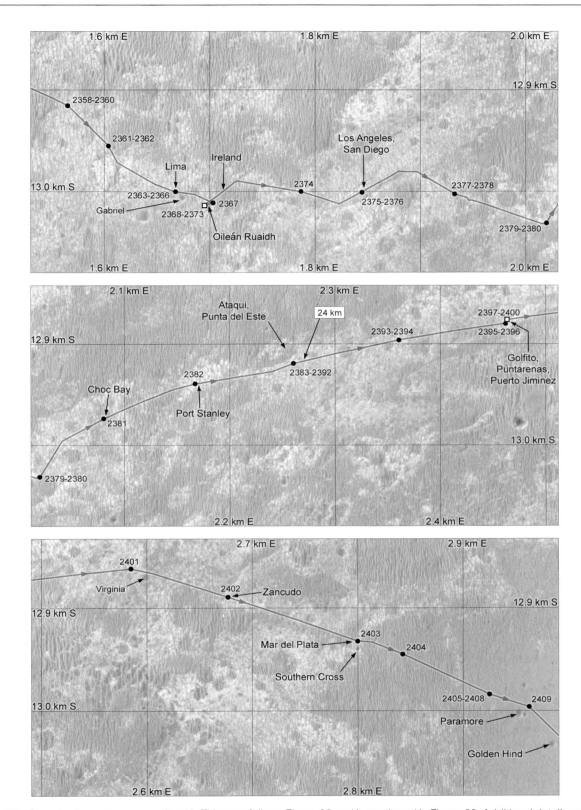

Figure 87. Opportunity route map, section 16. This map follows Figure 86 and is continued in Figure 88. Additional details are shown in Figure 84. White squares are main science stops.

Endeavour, broken only by a few intervening patches of outcrop. A crater, Southern Cross, and the layered rock Mar del Plata were imaged at the boundary on sol 2404. Driving was easy here and traverses of 120 m were fairly common.

The Endeavour rim hills were imaged again and AEGIS was used on sol 2407, probably unsuccessfully, as no images were produced. A double crater, Paramore, was imaged on sol 2409 and a Phobos transit and the single craters Golden Hind and Yankee Clipper were imaged on the next sol. The sol after that included a dust devil search and images of the setting Sun and Phobos. Another Phobos transit was seen on sol 2415 as the rover parked near Intrepid crater (Figure 84F), which was similar to Eagle in size but had more outcropping rock. Intrepid itself was imaged in color on sol 2417, and during a pause in the drive on sol 2420. Some nearby cobbles were also examined between the crater imaging sols. The craters Intrepid and Yankee Clipper commemorated the spacecraft of the Apollo 12 mission, whose 41st anniversary coincided with the time spent here.

The next goal was Santa Maria, a 100 m diameter crater resembling a smaller version of Endurance. Orbital remote sensing with MRO's CRISM instrument suggested that the composition of surface materials began to change in the vicinity of Santa Maria before becoming strikingly different at Cape York. Hydrated sulfate minerals were inferred from the CRISM data, and the first small exposure of them was on the southeast rim of Santa Maria, a location named Yuma. The eastward drive since sol 2250 had taken the rover down a gentle slope towards Endeavour, and Santa Maria itself lay on another east-facing slope just beyond a local high point, the rim of a subdued depression, so it was not seen until sol 2428 when Opportunity was 1100 m from it. The plains here were marked by numerous small craters and Anatolia-style linear depressions.

On sol 2424 Opportunity made super-resolution images of the Endeavour hills, and it tried using AEGIS again on sol 2428, finding several small dark cobbles. The rim of Santa Maria was now visible as a white line of rocks in the distance. The sol 2428 drive crossed a small linear depression or "ditch," requiring a short detour to pass through a suitable area. Then two craters, Hecla and Fury, were imaged along with a small rock called

Cornish Hen, and on sol 2441 the craters Vanguard (a cluster of three pits) and Voskhod were seen. The flash memory was too full to allow autonav driving, which saved many images for analysis, but the blind drives commanded from Earth still exceeded 100 m on many sols in this easy terrain.

Another crater called Salyut was imaged during the drive on sol 2442, and again at the start of sol 2443 because the first try missed the crater. Early on sol 2445 Opportunity had enough power for its first chance in many months to downlink data during an overnight Odyssey pass, and the flash memory problem began to ease. The recent US Thanksgiving holiday, 10 sols earlier, had also helped to clear the flash memory temporarily but this offered a longer-term solution. On the approach the rover imaged several large ejecta blocks, including Juan de la Cosa, Sancho Ruiz and Maestro Alonso on sols 2451 and 2452. The *Santa Maria* was one of the ships in Columbus's first voyage across the Atlantic, and rocks here took names from the crew. Sol 2451 was the last sol on which the Mini-TES cover was opened to try to clear it of dust. No change had been seen over 540 sols with this strategy and efforts to restore the instrument were now abandoned.

The rim of Santa Maria was reached on sol 2452, 1500 sols after the arrival at Victoria. Unlike the rims of Endurance and Victoria, this was covered with large ejecta blocks, indicating a relatively young age. One dark rock was initially thought to be part of the impactor that formed the crater, but was soon accepted as typical ejecta. The first crater viewpoint planned during the approach was too rocky to reach, but a second just 13 m south was acceptable. Opportunity took long-baseline stereoscopic images of the crater from two locations at this high point called Palos between sols 2452 and 2461, and obtained more views from a lower viewpoint called Wanahani on sols 2464 to 2467. Then the rover set out for the eastern side of the crater, a ledge called Yuma where orbital data suggested the presence of hydrated sulfate minerals. The small crater La Gallega was imaged on sol 2471. Several strings of boulders marked ejecta rays and three of them were observed as Opportunity drove around the crater's southern flank (Figure 90A). MRO's HiRISE imaged Opportunity at Wanahani on sol 2466 (MY 30, sol 419, or 31 December 2010) and at Yuma on sol 2524 (MY

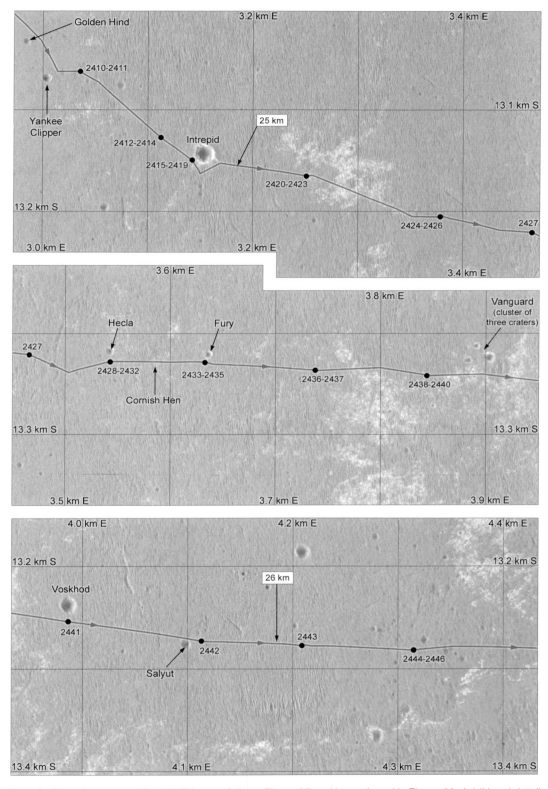

Figure 88. Opportunity route map, section 17. This map follows Figure 87 and is continued in Figure 89. Additional details are shown in Figure 84.

Table 23. *Opportunity Activities, Marquette Island to Santa Maria, Sols 2071–2452*

Sol	Activities
2071–2074	Imaging, ISO Peck Bay (2071) and Islington Bay (2072–2073), scan Islington Bay (2074)
2075–2085	Imaging, brush Islington Bay (2075), ISO (2075–2085) Islington Bay
2086–2088	Imaging, drive to (2086) and approach (2088) Marquette Island Unseen Side
2089–2094	Imaging, ISO Loon Lake (2089–2092), move around Marquette Island (2093)
2095–2109	Imaging, move to Peck Bay (2095), scan Peck Bay2 (2097), RAT Peck Bay2 (2100, 2103)
2110–2115	Imaging, ISO Peck Bay2 (2110, 2112–2114), Cube Point (2111), scan Peck Bay2 (2115)
2116–2117	Imaging, brush (2116) and ISO (2117) of Peck Bay2 target
2118–2121	Imaging, ISO Peck Bay2 offset1 (2118–2119) and Peck Bay2 offset2 (2119–2120)
2122–2142	Imaging, drive south towards Endeavour crater, approach Concepcion crater (2138)
2143–2149	Imaging, drive around Concepcion (2143, 2145), move to Chocolate Hills rock (2147, 2149)
2150–2153	Imaging, ISO Aloya target (2150–2151), move around Chocolate Hills (2152)
2154–2160	Imaging, ISO targets Arogo (2154–2155), Tears (2157–2158) and Dano (2158–2160)
2161–2178	Imaging, drive around Concepcion (2161–2174), leave Concepcion and drive south (2177)
2179–2197	Imaging, drive south (2179–2193), approach (2194) and move around (2197) San Antonio craters
2198–2224	Imaging, drive south towards Endeavour (2199–2223), ISO Ocean Watch (2224)
2225–2244	Imaging, drive towards Endeavour, atmospheric argon data (2230, 2236), marsquake test (2242)
2245	Break Viking Lander 1 record for longest operation on Mars (MY 30, sol 200)
2246–2256	Imaging, drive south (2246–2251), turn east (2252), drive east towards Endeavour crater
2257–2265	PMA fault (2257), diagnostics (2259–2261, 2265), battery recharge (2258, 2264), argon data (2263)
2266–2276	Imaging resumes (2266), recover attitude (2267), drive east towards Endeavour and imaging
2277–2296	Attitude data, argon data, imaging (2277), drive east and imaging (2278–2296)
2297–2301	ISO Juneau Road Cut trench (2297), imaging, drive east (2298–2300), image dust devil (2301)
2302–2310	Imaging (2302–2309), Odyssey safe-mode delay (2302–2308), argon data (2309), recharge (2310)
2311–2316	Hazard-avoidance tests (2311), imaging (2312–2316), approach Valparaiso outcrop target (2315)
2317–2339	ISO Valparaiso (2317), recharge (2319), imaging and continue drive east (2318–2339)
2340–2341	Imaging, ISO Clarin Beach (2340), Laya Beach and Duero Beach (2341) targets
2342–2346	Imaging, ISO Laya Beach2 (2342) and Cervera Shoal (2342–2346) targets on Cambridge Bay
2347–2370	Drive east towards Endeavour, imaging, Mössbauer diagnostics (2355, 2363)
2371–2398	Imaging, ISO Mulroy (2371–2373), drive east, problems prevent science (2385, 2388)
2399–2437	Imaging, ISO Puerto Jimenez (2399–2400), drive east towards Endeavour
2438–2452	Imaging, drive towards Santa Maria crater, reach crater rim (2452)

30, sol 478, or 1 March 2011). On Earth, the year 2011 began on sol 2467.

Opportunity arrived at Yuma on sol 2476 and positioned itself at Haomate, the high southern rim of Yuma, beside a flat rock slab called Luis de Torres for imaging and extended compositional data collection over conjunction, which was on sol 2498. The MB instrument would be essential here so a last set of diagnostic tests were undertaken on sol 2475. MB produced better data if it began taking data during the warmer part of the day. On sol 2477 the rover moved up to Luis de Torres and on the next sol it turned to put a suitable target in range of the arm. Other rocks were imaged on sol 2479, and then ISO began on Luis de Torres on sol 2481 with initial MI

and APXS data. A target was brushed and imaged by MI on sol 2485, probed by APXS on sol 2486, and by MB between sols 2487 and 2510, reflecting the very long integration times it now required. It started each day in the mid-afternoon, and collected a total of 96 hours of data. Images in this region showed that the ubiquitous blueberries were larger than previously seen.

The conjunction period lasted from sol 2499 to sol 2507 (MY 30, sols 454–462, or 27 January to 11 February 2011), and Opportunity experienced a solar panel cleaning event, a strong gust of wind, just before sol 2500 (MY 30, sol 455, or 4 February 2011). After conjunction the accumulated data were transmitted and the engineers checked the rover's condition, and then on

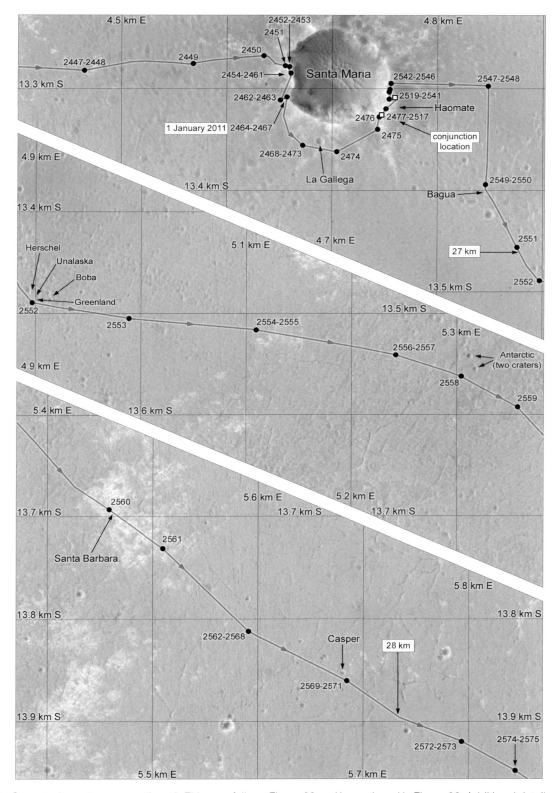

Figure 89. Opportunity route map, section 18. This map follows Figure 88 and is continued in Figure 92. Additional details are shown in Figure 90.

sol 2513 the RAT ground into Luis de Torres for MI and two sols of APXS data. Opportunity moved again on sol 2518, approaching another rock called Ruiz de Garcia. After imaging a target called Yarabi on sol 2519, the rover approached the new rock, making a final move on sol 2520. On the next sol it tried to begin ISO with MI and APXS, but the target was at the outer limit of the arm's reach and the instruments did not make contact, stopping the process. Three days of pre-programmed commands were carried out anyway, including cloud movies and images of other local targets, but the APXS collected three sols of atmospheric argon data. On sol 2524 the process was tried again and failed again, so on sol 2525 Opportunity moved closer to the rock and finally succeeded in placing APXS on the target for use over sols 2527–2530.

Next, on sol 2531, which coincided with perihelion for Mars, Opportunity moved northwards along the rim of Santa Maria to complete its stereoscopic imaging of the crater interior. A dust devil search was conducted on sol 2532, followed by a move to the desired stereo imaging location on 2534, but this drive stopped short and another move was needed on sol 2538. The first image set was taken over sols 2539–2541, with images of the RAT bit on the latter sol. On the next sol the rover moved to the second stereo station and took more images over sols 2543–2545. On sol 2547 Opportunity examined Balicasag Island and finished the stereo imaging, and then left Santa Maria to head for Endeavour.

The remaining drive began with an argon measurement and a test of AEGIS software on sols 2549 and 2550, and some final diagnostics with Mini-TES on sols 2550 and 2557. During the drive on sol 2551 Opportunity paused to view a large ejecta block called Bagua with an unusual texture southeast of Santa Maria. Several more rocks, including Greenland and Boba, were viewed on sol 2552. The backward drives were getting longer now, almost 120 m on sol 2554 and 131 m on sol 2556 as the rover dropped into a shallow bowl, a degraded crater, whose eastern rim hid the Endeavour hills for a while. Opportunity crossed a patch of outcrop called Santa Barbara on sol 2660, and on the next drive it crossed a linear trough or crack, using autonav in case it encountered trouble. The rover was moving rapidly and by sol 2568 the Endeavour hills could be seen again. A small crater called Casper was passed on the next sol, and here

the total drive distance of 27.8 km exceeded that of the Apollo 15 rover. One goal of the camera team around sol 2575 was to take images of a sunrise over the Endeavour hills, but power was a limiting factor. A significant crack in the rocky outcrop was imaged in the middle of the sol 2576 drive.

Opportunity drove whenever it could, to reach Cape York on Endeavour's rim as soon as possible. On sol 2578 it made another Pancam tau mosaic (images of the Sun in different parts of the frame to map dust coverage on the instrument). Wheel currents were seen to be slightly elevated on sol 2579, but long drives continued and distance limits were dropped to allow drives of up to 160 m per sol in safe areas, culminating in a 166 m drive on sol 2616. Another dust devil search was made on sol 2580.

The plains here were very smooth, textured by small dust ripples and cut by linear troughs, similar to but a little rougher than the area around Anatolia which had been seen on sol 70, but some patches had a very different appearance in HiRISE images and reprojected surface panoramas (Figure 92). Here the surface consisted of broad depressions only a few tens of centimeters below the surrounding smoother plains, but crossed by irregular linear ridges and isolated patches at the same level as the plains. One patch of this "reticulate terrain" had been crossed around sol 2556, and another was crossed on sol 2581, presenting no obstacle to driving and not appearing as distinctive at the surface as they did from orbit. The linear ridges resembled the nearby patterns of linear troughs, and they were explained as places where the regolith had been hardened by water rising up fractures. Subsequent wind erosion removed the unaffected regolith to form the shallow depressions between ridges.

Another common feature around here was a circular patch of smooth or finely rippled surface material level with or slightly elevated above the surrounding ripples (Figures 92 and 93). Some may have been remnants of older craters filled with dust and slightly protected from erosion. They had not been seen elsewhere along the traverse. The team considered examining one, but they were also concerned about running into one during a long blind drive.

A small crater cluster was encountered on sol 2585 just before the 50th anniversary of the first US space flight in

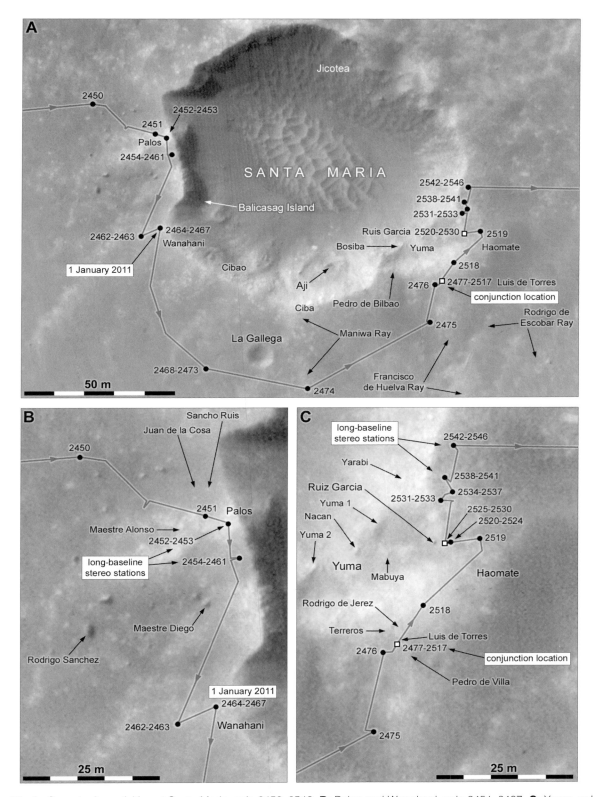

Figure 90. A: Opportunity activities at Santa Maria, sols 2450–2546. **B:** Palos and Wanahani, sols 2451–2467. **C:** Yuma, sols 2475–2546.

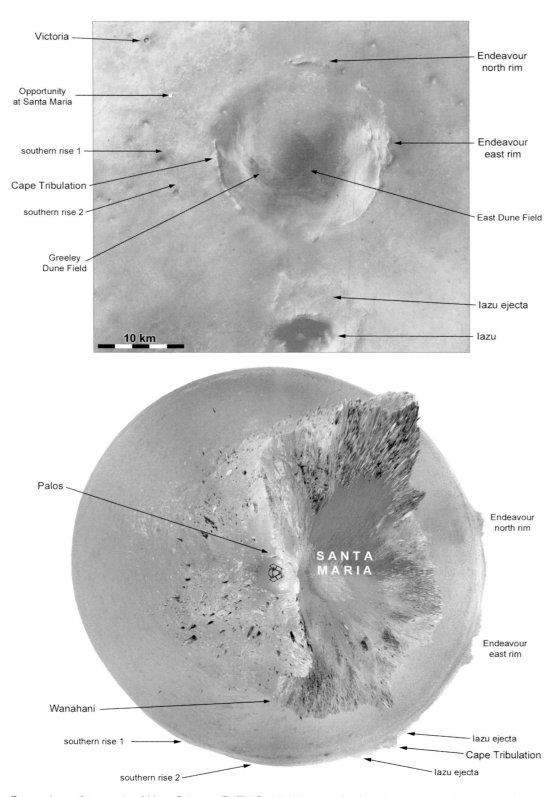

Figure 91. Comparison of a mosaic of Mars Odyssey THEMIS visible images (top) and a reprojected panorama taken on the rim of Santa Maria crater on sol 2452 (bottom). The two southern rises are the slightly elevated rims of two small subdued craters.

the Mercury program, Alan Shepard's suborbital flight in Freedom 7 on 5 May 1961. The largest crater was named Freedom 7 and several nearby craters were named for the other Mercury spacecraft. On sol 2589 Opportunity took images of the crater Molly Brown, another reference to the Mercury program. Liberty Bell 7, Virgil Grissom's capsule for the second suborbital Mercury flight, sank after splashing down in the Atlantic Ocean on 21 July 1961. When Grissom flew again on Gemini 3, he jocularly named that capsule Molly Brown after the stage musical *The Unsinkable Molly Brown*, which concerned the *Titanic* survivor Margaret Brown.

AEGIS was used on sol 2590 but found nothing. The blocky crater Skylab, an outcrop called San Francisco and the fracture Puerto Williams (Figure 93) were encountered next, and on sol 2597 (15 May 2011) the rover took a sunset image sequence to examine dust and haze in the upper atmosphere. Another dust devil search was made on the next sol, though no more dust devils were seen until much later in the mission.

On sol 2600 (MY 30, sol 555, or 18 May 2011) Opportunity viewed a 10 m diameter crater called Aquarius containing a smaller crater which might have excavated deeper rocks under Aquarius. The sol 2601 drive was cut short by a cosmic ray strike, flipping a bit in the computer memory on sol 2600. This did no lasting damage and had happened occasionally to both rovers, including on Opportunity's sol 1786, but it was a reminder that future human missions would have to contend with cosmic rays during Martian exploration. The thin atmosphere afforded little protection. The rover drove again on sol 2603 and tested a tank turn, running its left and right wheels in opposite directions, in case more of the steering actuators failed. It was around this time that the MER project abandoned attempts to contact Spirit.

Opportunity parked on an outcrop to examine a prominent crack or trough named Valdivia and a target called Puerto Montt adjacent to the small crater Gumdrop. A brief ISO campaign began on sol 2611 during a holiday weekend in the United States. MI saw vugs here, long narrow cavities marking places where a mineral crystal had been eroded or dissolved out of the rock, or possibly an ice crystal had formed and later melted or sublimed away. Vugs had been observed as early as sol 27 in Eagle crater. After the first MI observations the

IDD alternated between MB and MI placement to check that the MI poker, a surface sensor, was working. APXS was used on Valdivia on sol 2612, and two sols later the rover moved on, imaging a crater called Gemini 4 and passing the 30 km mark.

Sol 2626 set the record for a backward drive of 166 m. Three sols later a dust devil search was made and the young blocky crater Gemini 5 was observed, with additional images from three different positions on sol 2621. AEGIS was used on sol 2625 but no target was found. Dust cleaning events occurred on sols 2627 and 2628, and Opportunity drove past a crater cluster called the Gemini Cluster, commemorating the second phase of US human space flight. The right front wheel was behaving very well despite these long drives, helped by being heated before every drive. The target on the rim of Endeavour was now the south end of Cape York rather than the north end because orbital data suggested the presence of clay there. This area was named Spirit Point to commemorate Opportunity's fellow rover, now abandoned in Gusev crater.

DSN problems caused a delay over sols 2640 and 2641, and then on sol 2643 a rock called Bingag Caves was seen ahead. It may have been a meteorite. The next long drive passed it, and a second possible meteorite called Dia Island was noted, but the team decided not to stop at either rock. A drive on sol 2648 had to be canceled as the flash memory was too full, but Opportunity drove on the next sol, testing a new capability to drive autonomously on successive sols without ground intervention. This would speed up the drive still further. The rover computer's clock had slowly drifted off the correct time, so on sol 2651 the rover stayed awake late to keep the electronics warm in the hope this might reverse the drift, but it had no effect. Ultimately, 1000 sols later, the clock was reset a small amount at a time to correct the problem.

A two-toned rock called Milos and a blocky crater called Chrissi were imaged on sol 2651, and a filled crack called Paxoi on sol 2652. Then, just before another multi-sol autonomous driving test on sol 2654, a rind called Duljo and the crater Cabilao were seen, showing that even on these bland plains there were many interesting targets. An eroded bedrock target called Batungkay Cave was observed on sol 2656 and the Kalipayan outcrop and the dark rock Panglao Island two sols later.

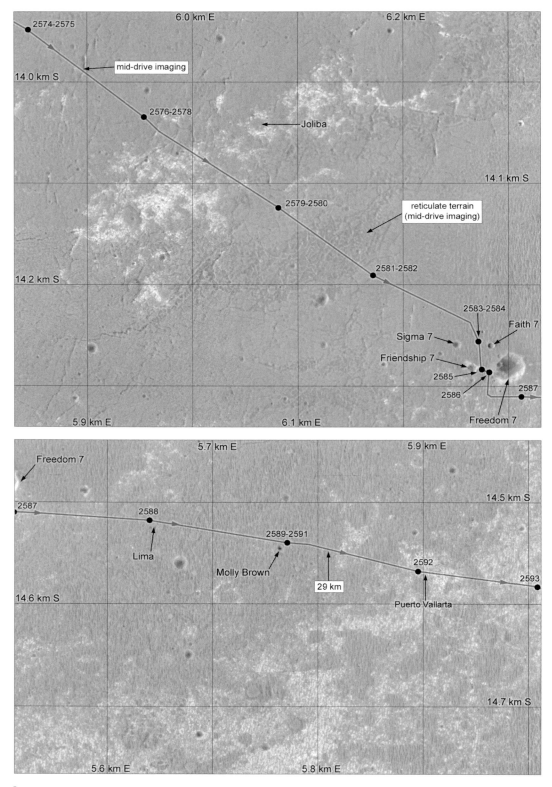

Figure 92. Opportunity route map, section 19. This map follows Figure 89 and is continued in Figure 93.

Table 24. *Opportunity Activities, Santa Maria to Endeavour Rim, Sols 2453–2680*

Sol	Activities
2453–2464	Imaging, long-baseline stereoscopic imaging (2452–2456), drive around Santa Maria (2462, 2464)
2465–2478	Imaging, drive around Santa Maria, Mössbauer diagnostics (2466), reach conjunction site (2478)
2479–2484	Imaging, ISO Luis de Torres target (2481, 2483), RAT scan Luis de Torres (2484)
2485–2492	Brush (2485) and ISO (2485–2491) Luis de Torres target, imaging
2493–2511	Conjunction, imaging, ISO Luis de Torres target
2512–2514	RAT scan Luis de Torres target (2512), imaging, RAT and ISO Luis de Torres (2513)
2515–2519	ISO Luis de Torres (2515–2517), drive around Santa Maria (2518, 2519)
2520–2530	Approach Ruiz Garcia (2520, 2525), imaging, ISO Ruiz Garcia (2527–2530)
2531–2540	Drive to LB2 (2531, 2534), imaging, move to LB2 Left (2538), take panorama (2539–2540)
2541–2549	Imaging, move to LB2 Right (2542), take panorama (2543, 2545, 2547), drive (2547, 2549)
2550–2582	Mini-TES diagnostics (2550, 2557), AEGIS testing (2550), imaging, drive to Endeavour
2583–2588	Drive (2583), approach Friendship 7 (2585) and Freedom 7 (2586), imaging, drive east
2589–2602	Drive past Molly Brown (2589), imaging, drive past Skylab (2594), drive east
2603–2610	Drive southeast, tank and turn test (2603), imaging, approach Valdivia (2606, 2608)
2611–2621	ISO Valdivia (2611, 2612), imaging, drive southeast (2614, 2616), pass Gemini 5 crater (2621)
2622–2644	Drive (2622, 2623, 2626–2630, 2633–2635, 2637), imaging, Mini-TES diagnostics (2625)
2645–2669	Alternate driving and imaging (2645–2668), ISO targets on Gibraltar and MI poker tests (2669)
2670–2680	Alternate driving and imaging (2670–2677), ISO Shipwrights (2678), drive southeast (2679–2680)

Most targets like these along the traverse were covered by multispectral Pancam images. AEGIS looked for meteorites on sol 2659 but found none. Two targets were imaged on sol 2660, an outcrop called Pangangan Island and the crater Pungtud Island, names that did not follow the pattern of using "island" names for isolated rocks.

Finally, the almost level surface of the plains began to slope downwards to the east, descending into Endeavour crater. The vista of the opposite side of the crater grew and the low shelf of Cape York came into view for the first time on sol 2662. Another crater cluster appeared now and was named after the Mariner series of planetary probes. Mariner 9 and Mariner 10 craters were imaged on sol 2663, but that sol's drive was cut short by a sensor indication that the arm had moved. Analysis suggested a sensor error. On sol 2664 Opportunity imaged an outcrop called Lisbon, and on the next sol it viewed the Endeavour rim hills and measured argon again. The surface was becoming rockier, with less soil and fewer drifts. Targets called Gibraltar, Madeira Island and Great Detached Reef were examined on sols 2669 and 2670 with ISO on Gibraltar, the last look at the Meridiani plains rocks, on the first sol. A very subdued crater named Pathfinder with a large gravel mound on its rim was imaged on sol 2671.

As the rover closed to within a few kilometers of Cape York late in 2010, the science team was planning activities in that area. Fraeman *et al.* (2011) illustrated possible routes to reach the older clay-bearing rocks detected by CRISM at the south end of Cape York (Figure 98). One route drove out onto the top of the cape before descending the steeper eastern slope into the target area, but at the time engineers considered it too dangerous. The rover could reach the area but might not be able to escape it later. The other route skirted the south end of Cape York to reach the bottom of the rugged target area. The safer southern route was chosen for the final approach.

AEGIS was used on sol 2673 but found nothing. The next imaging targets were a trough called Montevideo and a shallow depression named Inaccessible Island. Opportunity imaged the northern drive area noted by Fraeman *et al.* (2011) on sol 2677 and tested its ability to "touch and go," to combine ISO and a drive the day after arriving at a target, with MI views of Shipwright on sol 2678. A new type of feature appeared close to the edge of Cape York, bright veins in the rocky pavement, which resembled cracks filled with some light-toned material. There was no time to examine them here so Opportunity drove past them. The rover arrived on the

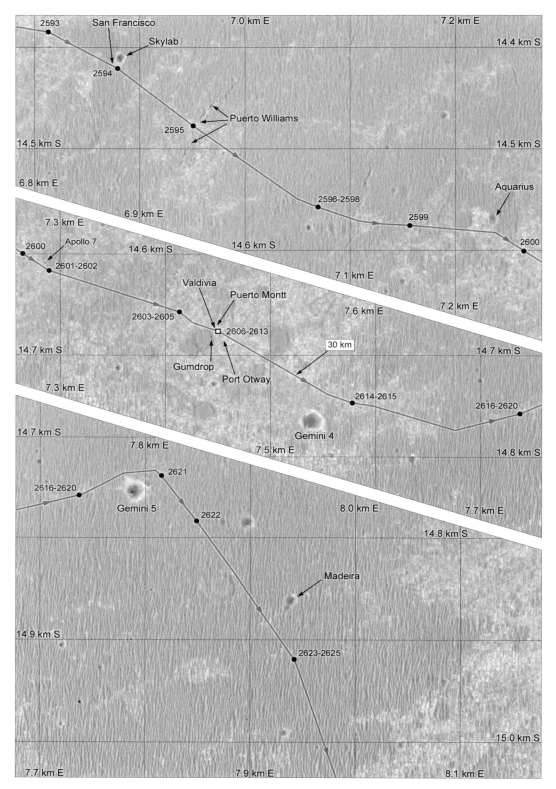

Figure 93. Opportunity route map, section 20. This map follows Figure 92 and is continued in Figure 94. White squares are main science stops.

rocky pavement surrounding Cape York (Figure 98) on sol 2681 (MY 30, sol 636). The cape was not rough and craggy, as HiRISE images might appear to suggest, but topographically subdued, its western side almost horizontal and its eastern slope falling more steeply down towards Endeavour crater.

The rover approached Odyssey, a relatively young crater with rugged ejecta blocks, including one called Ridout that was as large as the rover itself. Target names here would be taken from Earth's Archaean greenstone belts to reflect the age of these Noachian rocks. The first target, Tisdale 2, a flat-topped slab of breccia (a rock composed of many fragments) was approached on sol 2692. It was assumed to be a sample of Endeavour ejecta excavated by the Odyssey impact. Tisdale 2 was examined with multispectral images to select targets for ISO. Its flat top had a light coating, but it seemed too rough to brush successfully. After imaging many surrounding targets the ISO were done at three targets named Shaw 1, 2 and 3 on sols 2699–2702. On sol 2700 (MY 30, sol 655, or 29 August 2011) Mini-TES was tested one last time, as it would be so useful here, but to no avail. Tisdale 2 had a composition unlike any rock yet seen by Opportunity, an ancient basalt modified by the Endeavour impact and by underground water, but it was not the desired phyllosilicate the team was looking for. Tisdale 2 resembled rocks like Clovis and Wishstone seen by Spirit, but with more zinc than seen in Gusev crater.

On sol 2703 Opportunity began driving east and then north around Odyssey, climbing up onto Cape York to look for clay outcrops. The first target, Chester Lake, was one of a group of flat slabs eroded down to the level of the cape's sloping flank. It was reached on sol 2710 and a target on it called Salisbury was examined for 21 sols, allowing time to image many surrounding rocks. Sol 2712 was very quiet as NASA's GRAIL gravity mapping mission was launched to the Moon. Launches always carried the possibility that problems would require DSN resources normally used by other spacecraft.

Mars year 31 began on sol 2715. Two sols earlier on sol 2713, which was 11 September 2011 on Earth, the tenth anniversary of the World Trade Center attacks, Opportunity imaged its RAT, built by Honeybee Robotics of Manhattan, New York. Some metal from the fallen buildings, 1 km from the Honeybee facility, was incorporated into the RAT then being built for MER. ISO

began with MI and APXS on Salisbury later that sol, followed by a RAT brushing and more analysis on sol 2717 and a RAT grind on sol 2719 followed by APXS on sols 2722–2723. A two-sol delay followed as both MRO and Dawn, the asteroid mission on its way to 4 Vesta, went into safe mode. On sol 2726 the drill hole was brushed to clear it of tailings and analyzed again with APXS and then on sols 2729–2733 with MB.

The German physicist Rudolf Mössbauer, whose work led to the development of the MB instrument, died on 14 September 2011 (sol 2716), and the team considered naming a feature after him. In the end this was not done. Many other targets were imaged here, including Sutherland Point, a small rim hill south of Cape York. The last ISO on this rock was of Geluk, a clast embedded in the rock, on sol 2734. Chester Lake was a slab of modified impact breccia with coarse fragments (clasts) in a fine matrix, chemically similar to the materials analyzed by Spirit at Gusev, a sign that the ancient cratered terrain was indeed being observed here.

The winter season was now approaching and Opportunity was more threatened than in any of the four previous winters because of greater dust cover on its solar panels. Spirit had always had to find north-facing "havens" during southern hemisphere winters, and although the situation was not usually as bad for Opportunity at its site 10° further north, now it needed to find a north-facing refuge for the first time. The search began with a drive to Kirkland Lake on sol 2735, and imaging of it on the next sol, but this area sloped south and could not support a long visit. On sol 2737 the rover drove up onto Shoemaker Ridge, a name given to the central ridge of Cape York, especially here at the south end but extended to the north end occasionally. It was another commemoration of pioneering planetary scientist Gene Shoemaker.

From this ridge more images were taken to the south to examine Sutherland Point and the nearby Nobby's Head. A good north-facing slope was needed within about two months and the southern part of the ridge, called Shoemaker A, offered no suitable slope, but the next rise, Shoemaker B, might. As the rover moved north looking for "lily pads," it was also scouting for potential science targets after the winter. A rectilinear set of several low ridges, apparently thick veins, could be seen in HiRISE images east of the path (Figure 98), but the rover

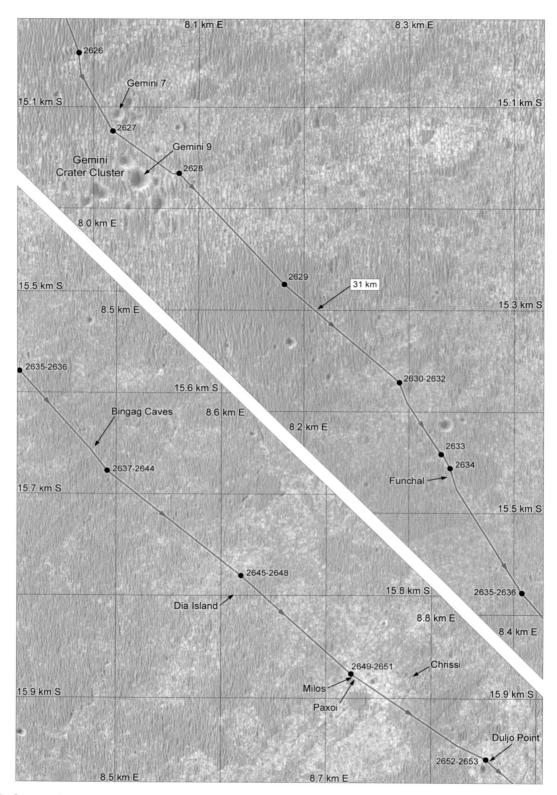

Figure 94. Opportunity route map, section 21. This map follows Figure 93 and is continued in Figure 95.

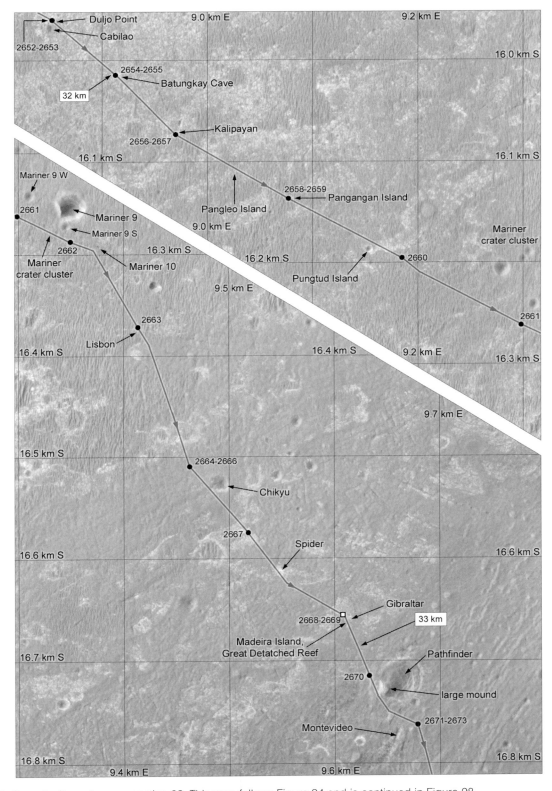

Figure 95. Opportunity route map, section 22. This map follows Figure 94 and is continued in Figure 98.

could not visit them as the east-facing slope would be bad for power. Outcrops called Onverwacht and Hooggenoeg were among potential future targets around sol 2750, and on that sol Opportunity took Navcam images of the Sun to support MSL operations. The northern tip of Cape York offered interesting targets and suitable slopes, so the rover was directed there with the intention of turning south again in the following spring.

On sol 2759 a super-resolution sequence of images of the large dark dunes in Endeavour crater serendipitously revealed a dust cloud raised by a wind gust on Endeavour's southern wall. These dunes had been observed to change in orbital images taken over several years (Chojnacki *et al.*, 2011), and they could now be examined from the surface. Further images for this study were taken on sols 2788, 2814 and 2946, but no changes were seen. The sol 2814 images had been accidentally pointed at the sky rather than the dunes. On the next sol Opportunity reached the level bench surrounding Cape York on its western side, which offered speedy driving to the northern slopes of the ridge. Here the veins seen on the approach to Cape York were seen again, and a prominent one named Homestake after a mine in the Black Hills of South Dakota was deemed so important that a brief stop could be made. Homestake was approached on sol 2763 and examined at two sites with MI and APXS over sols 2764–2767. It was 2 cm wide and about 40 cm long, elevated above the surrounding surface as if more resistant to erosion, and was found to consist of gypsum (hydrated calcium sulfate). This was the strongest evidence yet that water had soaked the ground here, depositing minerals in cracks in the pre-existing rocks (Squyres *et al.*, 2012). The surrounding material, an apron of debris eroded off Shoemaker Ridge, was called Deadwood, and it was targeted for ISO after small moves on sol 2769 and 2770. Another vein called Ross was seen nearby.

On sol 2772 Opportunity recharged its batteries while MRO was in a safe mode. Then on sol 2773 it was driven over Homestake to test its hardness, and then northeast to obtain images for a topographic map of northern Cape York, from which its winter resting place might be found. The slope here was 9° to the north and power was already noticeably improved. After skirting and imaging some dust ripples, the rover climbed to a possible winter site later called Turkey Haven where the US

Thanksgiving holiday (sol 2784, MY 31, sol 70, or 24 November 2011) could be spent on 10–15° slopes. A feature initially suspected of being a shatter cone, a diagnostic feature of impact-shocked rocks, was imaged on sol 2778. Turkey Haven was imaged on sols 2779 and 2780 before a final move to a 12° slope.

At Turkey Haven Opportunity imaged many targets, including Endeavour's rim on sol 2785 and its dune field on sol 2788, and performed ISO on the target Transvaal on sol 2787. A better site only 10 m northeast of Turkey Haven was initially called North Haven, but was renamed Greeley Haven to commemorate Ronald Greeley, the veteran planetary geologist from Arizona State University, who had died unexpectedly on 27 October 2010. Opportunity reached it on sol 2795, achieving a satisfactory tilt of 16° on the Saddleback outcrop, where it could spend a productive winter in close proximity to many interesting science targets.

The winter campaign on Cape York involved four major tasks: acquiring the Greeley panorama, a large image of the dramatic scenery; studying the rock outcrop on which the rover was perched; a long MB integration on the outcrop matrix; and a long radio tracking study, which replaced the plan to use Spirit in this way if it had survived its last winter at Troy. This study, using direct-to-Earth transmissions, would reveal variations in the planet's rotation, especially the precession and nutation of the Martian axis, from which information on the interior and core could be derived. Data from previous missions as far back as the Viking Landers could be incorporated to extend the study over several decades. Opportunity successfully completed the radio science campaign, which involved precise Doppler shift measurements at intervals during the winter without any movement of the rover (Le Maistre *et al.*, 2013). Radio science was the most important, and the MB study was least important, as it might no longer be usable. Its effectiveness would be evaluated after 40 hours of integration.

A dust devil search was undertaken on sol 2799, looking to the southwest over the plains. Several dust devils were seen later inside Endeavour crater, but none were seen from here looking back at the plains. The first ISO at Greeley Haven, on sol 2800 (MY 31, sol 86, or 10 December 2011), was on a target called Boesmanskop at Amboy on Saddleback, and the same target was brushed and analyzed again on the next sol, and imaged on sol 2803. The current tilt was a little more than was really needed at present, so the team briefly considered returning to Turkey Haven; its lesser tilt was still acceptable, but staying on the Saddleback breccia to analyze multiple clasts seemed more useful.

Opportunity turned to reach a clast called Komati on sol 2803, and imaged the magnets and a target called Oshowek on the next sol. It was not clear if Komati was fixed in place or loose and possibly transported from elsewhere, so the team considered nudging it with the RAT, but did not. ISO of Komati followed on sol 2805 and 2806. On sol 2808 a very small move to reach a filled fracture in the rock caused the right front wheel current to rise, but tests over sols 2810–2812 showed it was fine. Opportunity took a rover deck mosaic on sol 2813, the first since sol 1379, and tried to take super-resolution images of the dune field on the next sol, but the images were accidentally pointed at the sky.

The long MB observation began on sols 2819–2820 with ISO of the target point on Amboy. The terrestrial calendar marked the beginning of 2012 on sol 2821, and on the next sol MB was placed on Amboy to collect data for several hours per sol, gradually accumulating the required data with its very depleted radiation source. The Greeley panorama imaging began on the same sol, with a few images per sol as power permitted. On sol 2827 a hill on the ridge south of the haven was imaged and named Morris Hill, commemorating the former MER mission manager Richard Morris, who had died on 18 October 2011. Meanwhile the radio science study was also collecting data. On sol 2831 a very small move repositioned MB on a cleaner location, and APXS examined that area on sols 2833–2836. Opportunity also tested MB at different times of day and temperatures on sols 2835, 2838 and 2840 to try to improve the data. Imaging continued, including low-illumination-angle Pancam images on sol 2840 to improve viewing of local features. Power had to be monitored very carefully now as the winter progressed.

Opportunity passed eight Earth years on Mars on sol 2844 and took more low-lighting images three sols later as it gathered more APXS data at the revised ISO location, and on sol 2852 MI imaged the area and MB was placed back on it to gather more data with optimized instrument settings. Some other work was possible at times, including an argon measurement on sol 2858 and a set of MI images of the sky for calibration on sol 2845, unfortunately spoiled by glare. By sol 2861 the Greeley panorama was nearly finished, but it would be checked for gaps, and some parts would be retaken using all available camera filters. Specific targets could be imaged more often now, including Marble Bar on sol 2861 and Gorge Creek on sol 2863.

MB eventually accumulated nearly 100 hours of measurements on Amboy, but the data did not look good by sol 2875, and the effort was abandoned after sol 2893. It was now clear that the instrument's radiation source was too weak to provide any more useful results. This made power available for more imaging, as did wind gusts cleaning the solar panels on sols 2878–2879. Pancam imaged targets including Coonterunah on sol 2874 and Strelley Pond on sol 2880, and MI began to compile a very large transect across Amboy on sol 2879. The planning team considered leaving Greeley Haven early because they had increased power from the cleaner solar panels, and wondered if Opportunity could reach the summit of Morris Hill. The radio science experiment continued, but two sols of direct-to-Earth transmission were spoiled briefly by interference from MRO on sols 2885 and 2886.

Several cloud searches were made in this period, and argon was measured on sol 2888. This was the time of minimum sunlight, but the cleaning events meant it was not the time of least power, and the wind gusts raised the possibility that surface changes might be seen, so Pancam imaged an area near the right front wheel on sol 2890 to monitor any changes. Another wind gust occurred on sol 2891, raising power again, and a high Sun panorama was taken on sol 2895 to help with surface change detection. Then as MI continued its work on a mosaic of Amboy, an IDD joint stalled on sol 2899, possibly because MB unexpectedly struck the uneven surface, and images on the next sol showed a very small rover motion, as if its left front wheel had slipped off a small rock. The IDD was tested on sols 2901 and 2904, and the Pancam imaged the area where MB might have touched the surface. On sol 2906 the rover team tried unsuccessfully to reproduce the stall to help understand it.

Sol 2900 had been MY 31, sol 186, or 21 March 2012, and sol 2909 was the winter solstice, slightly after the minimum sunlight date because of the planet's elliptical orbit. More cleaning events occurred around this

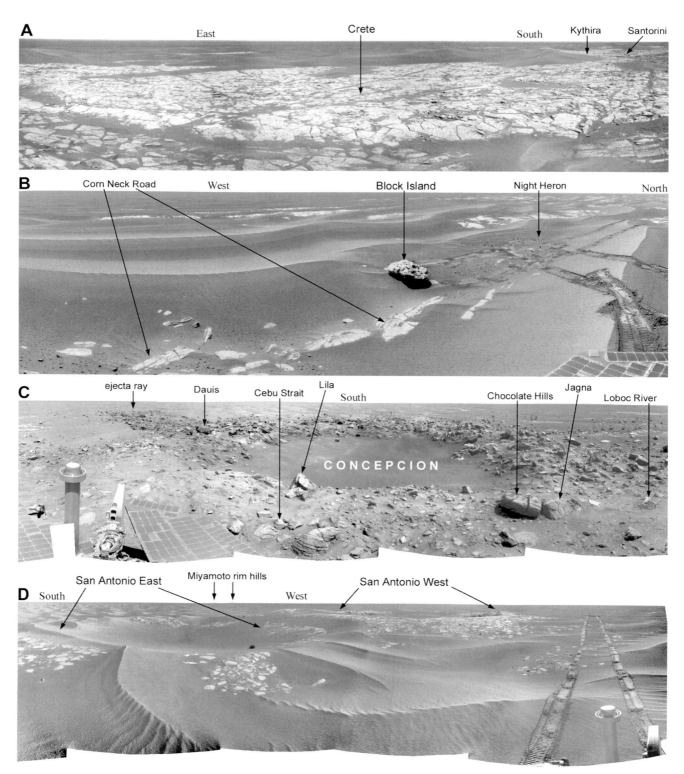

Figure 96 (both pages). Panoramas between Victoria and Endeavour craters. **A:** Santorini (Crete), sol 1709. **B:** Block Island, sol 2000. **C:** Concepcion crater, sol 2145. **D:** San Antonio craters, sol 2197. **E:** Santa Maria, sol 2464. **F:** Santa Maria, sol 2543. **G:** Freedom 7, sol 2586. **H:** Cape Tribulation and Botany Bay, sol 2671.

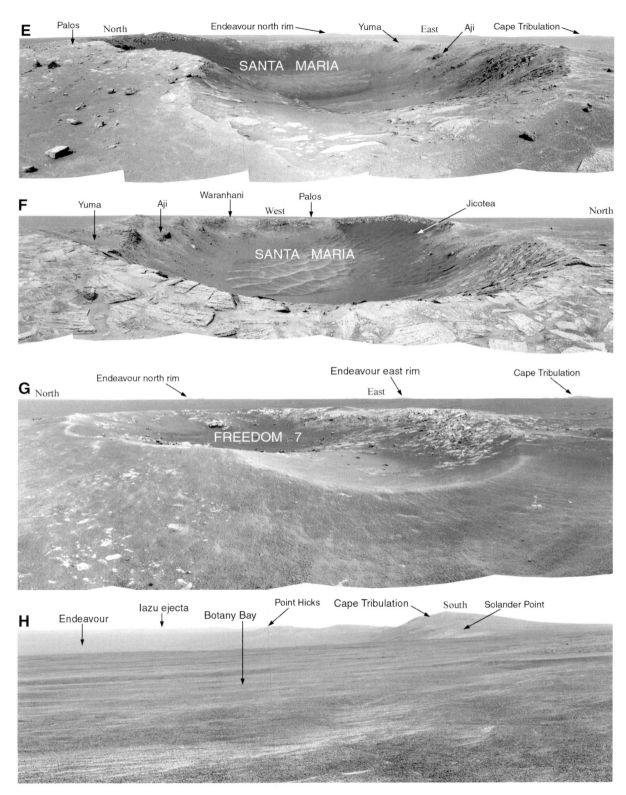

Figure 96 (continued)

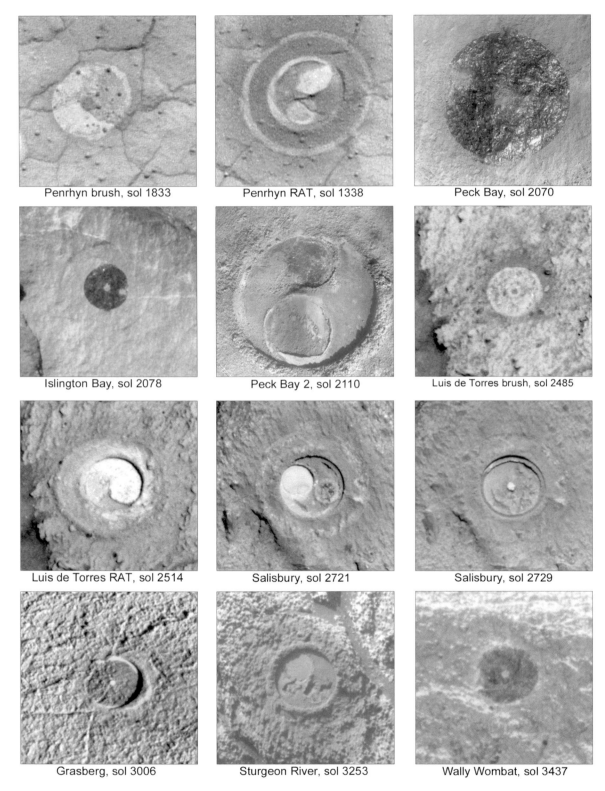

Figure 97. Selected RAT brush and grind marks from Victoria crater to Solander Point.

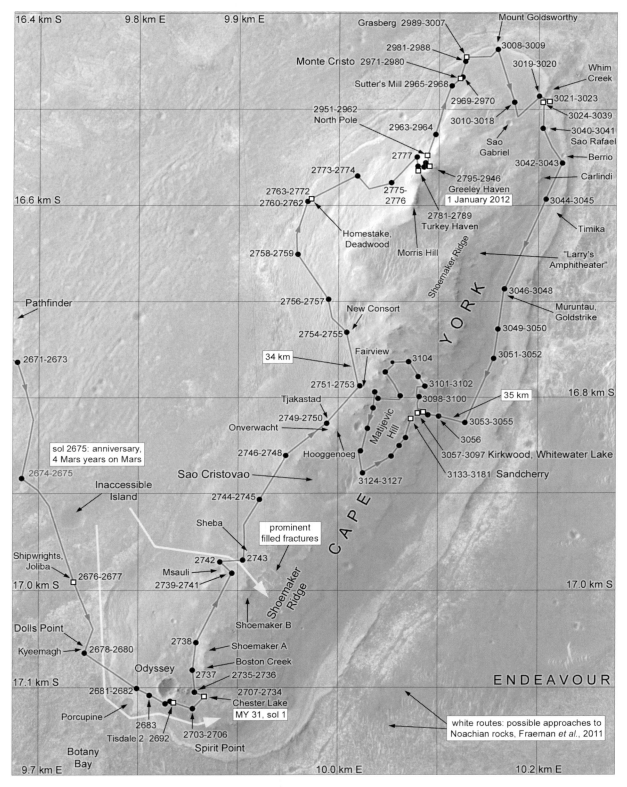

16.4 km S   9.8 km E   9.9 km E

Grasberg 2989-3007   Mount Goldsworthy

2981-2988   3008-3009

Monte Cristo 2971-2980   3019-3020   Whim Creek

Sutter's Mill 2965-2968   3021-3023

2969-2970   3024-3039

2951-2962 North Pole   3010-3018   3040-3041 Sao Rafael

2963-2964   Sao Gabriel   Berrio

2777   3042-3043

2773-2774   2775-2776   Carlindi

2763-2772   Greeley Haven   3044-3045

16.6 km S   2760-2762   1 January 2012

2781-2789 Turkey Haven   Timika

Homestake, Deadwood

2758-2759   Morris Hill   "Larry's Amphitheater"

Shoemaker Ridge   3046-3048

2756-2757   New Consort   Muruntau, Goldstrike

Pathfinder   2754-2755   3049-3050

Y O R K

2671-2673   34 km   Fairview   3104   3051-3052   16.8 km S

2751-2753   3101-3102

Tjakastad   3098-3100   35 km

2749-2750   Matijevic Hill   3053-3055

Onverwacht   3056

sol 2675: anniversary, 4 Mars years on Mars   2746-2748   Hooggenoeg   3057-3097 Kirkwood, Whitewater Lake

2674-2675   Sao Cristovao   3124-3127   3133-3181 Sandcherry

Inaccessible Island

2744-2745

C A P E

Sheba   prominent filled fractures

Shipwrights, Joliba   2742   2743

2676-2677   Msauli   Shoemaker Ridge

17.0 km S   2739-2741   17.0 km S

Shoemaker B

Dolls Point   2738   Shoemaker A

Kyeemagh   2678-2680   Boston Creek

Odyssey   2737   2735-2736

17.1 km S   2681-2682   2707-2734   E N D E A V O U R

Porcupine   Chester Lake   MY 31, sol 1

2683   white routes: possible approaches to Noachian rocks, Fraeman *et al.*, 2011

Tisdale 2 2692   2703-2706

Spirit Point

Botany Bay

9.7 km E   10.0 km E   10.2 km E

Figure 98. Opportunity route map, section 23. This map follows Figure 95 and is continued in Figure 105. Additional details are shown in Figures 99, 100, 102, 103 and 104.

Table 25. *Opportunity Activities, Cape York to Greeley Haven, Sols 2681–2956*

Sol	Activities
2681–2693	Drive onto Cape York (2681), imaging, alternate drives to Tisdale2 and imaging (2685–2693)
2694–2700	ISO Timmins targets (2694–2696), move to Shaw1 (2697), ISO Shaw1 (2699), Mini-TES tests (2700)
2701–2709	ISO Shaw2 (2701), Shaw3 (2702), drive towards Chester Lake (2703, 2707), imaging
2710–2716	Approach Salisbury (2710), ISO Salisbury1, Salisbury2 (2713), scan Salisbury1 (2715), imaging
2717–2725	Brush and ISO Salisbury1 (2717), RAT (2719) and ISO (2719, 2722, 2723) Salisbury1, imaging
2726–2734	Brush (2726, 2729) and ISO (2726, 2727, 2729–2733) Salisbury1, MI test and ISO Geluk (2734)
2735–2743	Drive towards Kirkland Lake (2735) and Shoemaker Ridge (2737, 2738, 2742), imaging
2744–2763	Alternate driving on Cape York and imaging, seeking a north-facing winter haven
2764–2770	ISO Homestake (2764, 2765), Homestake2 (2766, 2767), imaging, move to Deadwood (2769–2770)
2771–2780	ISO Deadwood (2771), alternate imaging and driving to top of ridge (2772–2780)
2781–2790	Approach Turkey Haven (2781, 2783), imaging, ISO Transvaal (2787), imaging, drive north (2790)
2791–2801	Imaging, drive to (2792, 2795), ISO (2798–2800), brush and ISO (2801) Boesmanskop target
2802–2810	Imaging, drive to (2803) and ISO (2805, 2806) Komati target, drive to Amboy (2808) target
2811–2822	Rover deck images (2811–2814), approach (2816) and ISO (2819–2822) Amboy target
2823–2835	Start Greeley panorama (2823), ISO Amboy (2823–2834), MB temperature tests (2833, 2835)
2836–2867	ISO Amboy (2836–2839, 2847, 2852–2867), MB temperature tests (2838, 2840), imaging
2868–2894	Radio science (2868), ISO Amboy (2869–2882, 2224–2894), Greeley panorama (2882–2891)
2895–2906	ISO Amboy3 (2895, 2897, 2898), Amboy (2899), imaging, IDD diagnostics (2901, 2904, 2906)
2907–2920	Imaging, stability check (2914), MI front wheel (2916), ISO 2nd AmboyL8, Amboy4 (2919, 2920)
2921–2925	ISO Amboy4 (2921), Amboy5 (2921, 2922, 2925), R12 and Amboy6 (2924), imaging
2926–2932	ISO R13 (2926), Amboy7R13 (2926–2929), Amboy8R14 (2929, 2930), Amboy9R15 (2931–2932)
2933–2941	Imaging, ISO Amboy10R16 (2935–2936), Amboy11 (2937–2939), Amboy12 (2940–2941)
2942–2956	ISO R10 (2942), R8, R9 (2944), Amboy12 (2946), drive to North Pole (2947–2955), imaging

time, so power was increased again. On sol 2414 another high Sun panorama was taken to look for changes, and the left front wheel was spun to try to settle it on the surface. On the next sol the rover was moved back a very small amount to try to set the wheel firmly on the ground, and on sol 2916 MI looked under the rover to see if the left front wheel was touching the ground properly. The images were out of focus but appeared to show a good contact, so the IDD could be used again. After taking super-resolution images of an area called Whim Creek, north of Greeley Haven (Figures 98 and 99E) on sol 2917, the rover made ISO on a clast on Amboy over sols 2919–2920. The MI images were out of focus but were retaken with proper focus on sol 2921.

The rover's Inertial Measurement Unit (IMU), which measured the rover's position and motion and was used in seismic experiments occasionally (sols 1809 and 1875), was tested on sols 2922 and 2938, as it was not as stable as it should be. MI and APXS observations were taken on many different locations on Amboy over the next few sols, and wind gusts continued to make

small improvements in power. More change detection images were taken on sol 2926, and the flash memory was being cleared around this time as it had become too full. Another large imaging project at this time was a multispectral mosaic of the foreground, and super-resolution image sets were taken of Mt. Goldsworthy at the northern tip of Cape York on sol 2935 and of the Endeavour dunes on sol 2946, finally ending work at Greeley Haven. The team had decided that Amboy was too rough to use the RAT successfully.

The first movement of the new season was on sol 2947 (MY 31, sol 233) as Opportunity backed off Amboy for multispectral imaging and then turned north towards a patch of bright dust called North Pole. The first drive was only 4 m long but immediately reduced the northward tilt to just 8°. The last radio science data should have been before that move, but two final tests were done on sols 2948 and 2950 to see if movements of a few meters could be accommodated. Then Opportunity was driven north and downhill to North Pole, leaving tracks in it on sol 2951

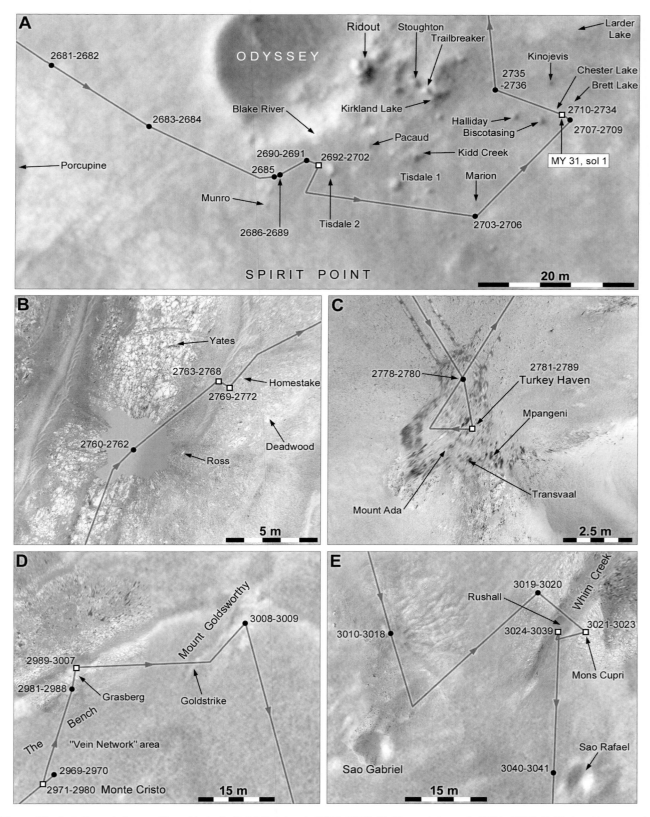

Figure 99. Activities at sites on Cape York. **A:** Spirit Point, sols 2681–2736. **B:** Homestake, sols 2761–2772. **C:** Turkey Haven, sols 2777–2789. **D:** Northern tip, sols 2969–3009. **E:** Whim Creek, sols 3010–3041.

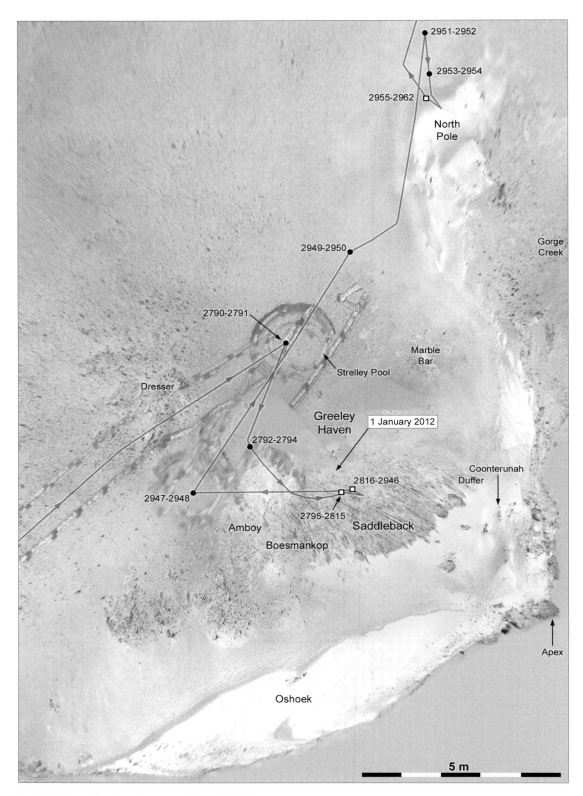

Figure 100. Activities at Greeley Haven, sols 2790–2962.

(Figure 100). After moving back to those tracks on sols 2953 and 2955, the rover performed ISO on disturbed and undisturbed areas over sols 2957–2961. This replaced the lost opportunity to study bright dust at Duck Bay on sol 1668. Another IMU test was done on sol 2956.

On sol 2963 the rover began descending the gentle slope towards the northern tip of Cape York, seeking a vein larger than Homestake which could be analyzed more thoroughly. Opportunity still had to be careful of the tilt, but had some flexibility because of the helpful wind gusts. The wide slope around the ridge was called The Bench, and veins proved to be so common here that this part of it was referred to as Vein Network. Veins were named after gold mines, and one called Sutter's Mill was imaged on sol 2969. A better example called Monte Cristo was investigated between sols 2971 and 2980 but was still not large enough for safe use of the RAT brush. ISO on the widest part, called Ootsark Canyon, were done over sols 2974–2977, but work was slowed by a problem with Mars Odyssey's attitude control system. Some data were transmitted through Mars Express on sol 2984, and via MRO, and a little was sent directly to Earth, but progress was slow.

Nevertheless, a nearby vein called Mt. Edgar was imaged, as well as Mill Creek, a part of Monte Cristo broken by the wheels as Opportunity drove over it. The rover was driven north on sol 2981 to approach the northern edge of Cape York, and on sol 2989 it reached the contact between the cape and the Meridiani plains material. A target called Grasberg was analyzed at two spots over sols 2990–2994, and when Odyssey was back in operation on sol 2995 the target was brushed and analyzed again for three sols. Then on sol 2998 the RAT ground into Grasberg, followed by RAT bit imaging on sol 3000 (MY 31, sol 286, or 2 July 2012). The grind tailings were brushed out of the hole on sol 3001 and the hole was analyzed over sols 3001–3004. Meanwhile the rocks Cortez and Veladero were imaged on sol 3002. Another target nearby was analyzed on sols 3006–3007, leaving the flash memory very full, given the recent orbiter problems. Grasberg was the lowest layer of the Burns Formation, the rocks seen everywhere on the plains since Eagle crater, but it had a different composition enriched with calcium and sulfur. Back on Earth, JPL was now training extra rover operators to be

sure it would have enough when Curiosity landed later in 2012.

On sol 3008 Opportunity was driven eastwards to Mt. Goldsworthy, a shallow notch cut into the smooth platform surrounding Cape York, passing a large vein called Goldstrike during the drive. The route turned to the south on sol 3010, along the inside edge of Cape York, but activities were again interrupted by an Odyssey safe mode caused by loss of a reaction wheel, part of its attitude control system. This was only a few weeks before the Mars Science Laboratory (Curiosity) landing, in which Odyssey would play a crucial role, and the ongoing problems were troubling. Opportunity undertook only limited atmospheric work supported by direct data transmission until sol 3019 when Odyssey was working again. MARCI, the color imager on MRO, had detected a local dust storm near Endeavour, which raised dust levels briefly, and ice clouds were also seen as ice condensed onto dust grains.

The sol 3019 drive included imaging of Sao Gabriel crater (one of three here named after Vasco da Gama's ships) before stopping at the edge of Whim Creek, a deep notch cut into the rocky pavement surrounding Cape York, a larger version of Mt. Goldsworthy. Opportunity needed a science target to study while Curiosity landed on 5 August 2012 (MY 31, sol 319, Opportunity's sol 3033), and Whim Creek might offer a cross-section of the edge of the pavement. On sol 3022 a target called Mons Cupri just outside Whim Creek on its southeast flank was examined with MI and APXS. On sol 3024 the rover repositioned itself inside the narrow depression of Whim Creek and examined a target called Rushall with MI and APXS starting on sol 3027.

On the next sol a test of direct radio transmission from Opportunity (not via an orbiter) was performed to see if Curiosity could be detected by the Parkes Radio Observatory in Australia during its final approach to Mars. The test was successful, so Parkes would listen for Curiosity's signals. The fact that the two rovers shared a common transmission frequency contributed to the selection of Curiosity's landing site at Gale crater on the other side of the planet, minimizing conflicts between the two missions.

APXS analysis of Rushall continued until sol 3037 and Opportunity also undertook multispectral imaging of the rocks of Whim Creek, including Cistern

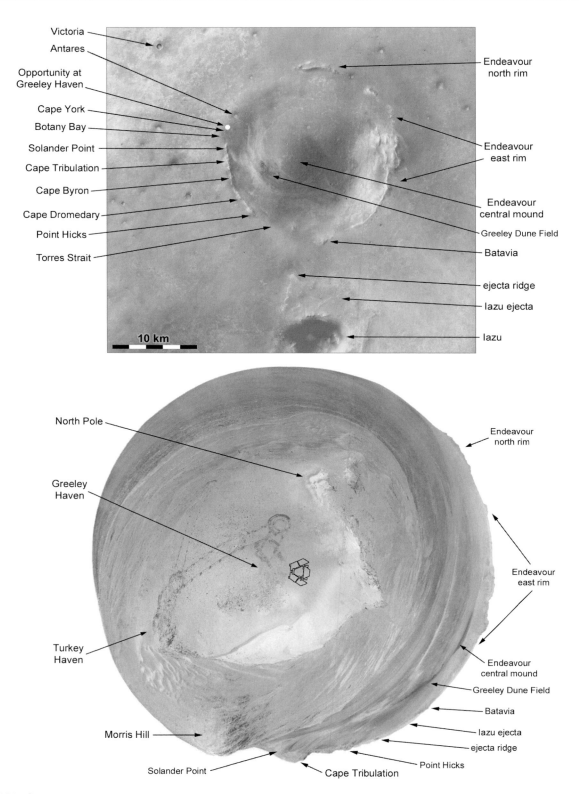

Figure 101. Comparison of a mosaic of THEMIS visible images of Endeavour crater (top) and a reprojected panorama from Greeley Haven on Cape York with exaggerated horizon relief (bottom), showing locations of horizon features and regional feature names.

on sol 3038. Then on sol 3040 the rover drove south past a small impact crater called Sao Rafael and imaged it and a second crater called Berrio on sol 3042. Many craters had been seen on the plains, and their degrees of degradation indicated their relative ages (Golombek *et al.*, 2010b), so the small craters here on a different geological material offered the potential for a new study. The air cleared to give spectacular views across Endeavour around this time.

As Opportunity drove south from Berrio it began examining the rocky outcrops of Cape York with multispectral images on any sol where interesting targets

Table 26. *Opportunity Activities, North Pole to Matijevic Hill, Sols 2957–3315*

Sol	Activities
2957–2962	ISO North Pole 1 (2957), North Pole 2 (2957–2959, 2961), North Pole 3 (2960), imaging
2963–2970	Drive to Vein Network area (2963, 2965), approach Monte Cristo (2969), imaging
2971–2977	Approach (2971) and ISO (2974–2975) Ootsark Canyon, ISO Ootsark Canyon Offset (2976–2977)
2978–2991	Imaging, approach Mt. Edgar (2981), approach Grasberg (2989), ISO Grasberg (2990–2991)
2992–2997	ISO Grasberg2 (2992–2994), brush (2995) and ISO (2995–2997) Grasberg1, imaging
2998–3005	RAT and ISO Grasberg1 (2998), brush (3001) and ISO (3001–3004) Grasberg1, imaging
3006–3018	ISO Grasberg3 (3006–3007), MI Grasberg1 (3006), drive east (3008) and south (3010), imaging
3019–3022	Drive past Sao Gabriel to Whim Creek (3019), cross Whim Creek (3021), ISO Mons Cupri (3022)
3023–3039	Imaging, approach Rushall (3024), ISO Rushall1 (3027–3037), Rushall2 (3027)
3040–3055	Drive to Sao Rafael (3040), imaging, pass Berrio (3042), drive south and image hills (3044–3055)
3056–3066	Drive west onto hill (3056), approach Kirkwood (3057, 3060, 3063), imaging
3067–3072	Brush (3067) and ISO (3067–3068) Kirkwood, imaging, drive to Whitewater Lake (3070, 3071)
3073–3076	ISO Azilda1 (3073–3074), imaging, Phobos transit images (3075), brush and ISO Azilda1 (3076)
3077–3079	ISO Azilda1 (3077), brush and ISO Azilda2 and image satellite transits (3078), imaging
3080–3084	Brush (3080) and ISO (3080–3082) Azilda3, RAT Azilda2 (3083), imaging
3085–3089	RAT and brush (3085) and ISO (3085–3086) Azilda2, brush (3087) and ISO (3087–3089) Azilda2
3090–3093	Move to Chelmsford (3090), imaging, move to Chelmsford2 and ISO Chelmsford1 (3092), imaging
3094–3097	Brush (3094) and ISO (3094–3095) Chelmsford2, brush (3096) and ISO (3096–3097) Chelmsford2
3098–3132	Begin Matijevic Hill survey, drive around hill and imaging (3098–3132), MI calibration (3126)
3133–3143	Drive to Sandcherry (3133, 3135), imaging, ISO Sandcherry (3137–3138), ISO joint stall (3139)
3144–3146	Brush (3144) and ISO (3144–3145) Sandcherry, brush, RAT and ISO Sandcherry (3246), imaging
3147–3154	Imaging, ISO Sandcherry (3148–3150), move to Dowling (3151), to Copper Cliff (3153), imaging
3155–3165	Move to (3155) and ISO Onaping (3158–3160), ISO Onaping2 (3162), move to Vermilion (3165)
3166–3176	Imaging, ISO Vermilion Cliffs1 (3168–3173), ISO Vermilion Lake 1 (3174–3177)
3177–3187	ISO Vermilion Lake 2 (3177–3180), move to Whitewater Lake (3182), to Ortiz (3185, 3187)
3188–3199	ISO Ortiz (3189), Ortiz1 (3189–3190), Ortiz2 (3192–3193), Ortiz3 (3194–3195), imaging
3200–3205	ISO Ortiz1 (3200), Ortiz2b (3200–3202), drive to Flack Lake (3203), image RAT bit (3205), imaging
3206–3209	Drive to Flack Lake (3206), ISO Fullerton1 (3207), Fullerton2 (3208), brush Fullerton3 (3209)
3210–3214	ISO Fullerton3 (3209–3211), drive to Fecunis Lake (3212–3213), brush and ISO Fecunis Lake (3214)
3215–3224	ISO Fecunis Lake (3215), drive to Stobie (3216), to Maley (3219), brush and ISO Maley (3224)
3225–3230	ISO Maley (3225–3226), drive to Big Nickel (3227), approach and attempt scuff on Lihir (3230)
3231–3239	Imaging, move to Lihir (3233), problems delay science (3235–3236), ISO Lihir1 (3239)
3240–3248	Drive to Kirkwood (3240, 3246), imaging, ISO Sturgeon River (=SR) 1 (3247), SR 2 (3248)
3249–3250	ISO SR 1a (3249–3250), MI Lake Laurentian and South Range (3250), imaging
3251–3254	RAT (3251) and ISO (3251–3252) SR 3, RAT (3253) and ISO (3253–3254) SR 3a, imaging
3255–3263	Drive to Esperance (3255, 3257, 3260), imaging, ISO Esperance (3262–3263)
3264–3272	MI Esperance (3264), ISO Esperance 2 (3264–3266), ISO Esperance 3 (3267–3272), imaging
3273–3295	Conjunction (3273–3293), ISO Esperance 3 (3273–3283), anomalies prevent science (3287–3295)
3296–3302	Move to (3296) and ISO (3298–3300) Esperance 4, RAT (3301) and ISO (3301–3302) Esperance 5
3303–3315	RAT and brush and ISO Esperance 5 (3305), ISO Esperance 6 (3305–3306), drive south (3308–3315)

appeared. The CRISM instrument on MRO had identified phyllosilicate clay minerals in this area, and finding them was now perhaps the most important remaining goal for the venerable rover, so much so that MER team leader Steve Squyres referred to the ridge as "our Mt. Sharp," referring to Curiosity's main science target. An outcrop called Carlindi on the hillside, still called Shoemaker Ridge, was viewed on sol 3044. Other targets on the bench below the hill included the outcrop Timika on sol 3046, a vein called Muruntau and a rock called Goldstrike on sol 3049. A recess in the hillside was nicknamed Larry's Amphitheater after team member Larry Crumpler.

On sol 3053 the route turned slightly west to improve the view of an interesting area, and on sol 3055 the RAT was imaged by the Pancams to confirm it could still be used for grinding any new targets. After viewing a sinuous vein early on the next sol, Opportunity drove directly west towards an unusual dark outcrop higher on the slope, as the rover's total distance exceeded 35 km. The drivers moved the rover closer to the outcrop on sols 3057 and 3060 so multispectral images of the outcrop could be taken to choose targets, and part of a dark protruding fin-like ridge called Kirkwood was selected. Images of several mound-like outcrops to the north were taken on sol 3059 for later planning. The names here were taken from Canadian mines.

Opportunity moved closer to Kirkwood on sol 3063 for ISO on sols 3064 and 3065. Kirkwood was studded with small spherical concretions. They were superficially similar to the blueberries on the Meridiani plains but were not made of hematite, and might have been impact spherules formed of solidifying droplets of impact melt, so the nickname "newberry" was applied to them. On the next sol a communication fault occurred as Earth moved below the deck of the tilted rover while the high-gain antenna was tracking it. The tilt had been calculated incorrectly, but the necessary instructions had been received before the error. Kirkwood was imaged on sol 3066, brushed on sol 3067 and then studied with MI and APXS. The communication error was repeated on sol 3069 but avoided after that, and on the next sol the high-albedo rocks on the uphill side of Kirkwood were approached to search for the anticipated clay minerals.

Multispectral images had hinted at hydrated minerals here, and the rocks were partly covered with a distinctive eroded layer or rind unlike anything seen previously. This outcrop was called Whitewater Lake. Opportunity viewed a target called Azilda 1 with MI on sol 3073, and APXS measured its composition for several sols. Azilda 1 was brushed on sol 3076, followed by ISO, but this target was too rough for the brush to be effective. Another nearby target called Azilda 2 was brushed and observed with MI and APXS starting on sol 3078, and a third target called Azilda 3 received the same treatment starting on sol 3080. A Deimos transit on sol 3078 was missed by images, but a partial Phobos transit later that sol was seen.

The RAT was used sparingly now as it had only a limited life remaining, but it was deployed on Azilda 2 on sol 3083 for a shallow grind. This cut in only 0.8 mm so it was followed by another grind on sol 3085, leaving a hole 3.6 mm deep, which was brushed clean later that sol and again two sols later and analyzed for two sols. The next target was a nearby dark rind called Chelmsford, reached by a small move on sol 3090. MI viewed Chelmsford 1 on sol 3092 and then the rover turned to reach a better location called Chelmsford 2. ISO here began on sol 3094 but the APXS was not placed exactly where MI had imaged the surface, so it was repeated on sol 3095, and a third spot called Chelmsford 3 was brushed and analyzed on sol 3096. Other targets nearby including Lindsay were imaged during the analysis.

Now that the unusual outcrops at Whitewater Lake and Kirkwood had been characterized, Opportunity was sent on a survey of the top of Matijevic Hill, part of Shoemaker Ridge, which was named after Jake Matijevic, a rover engineer, who had died on 20 August 2014. It would "walk the outcrop" to explore the extent of these units, determine their place in the stratigraphy of the hill, and help planners select new targets for study. Figure 102 shows the survey route. At each stop full panoramas were taken for mapping, with multispectral images of many interesting features. Back on Earth the rover team produced a geological map of the area (Crumpler *et al.*, 2013a) to assist in target selection and analysis. The drive commenced on sol 3098, moving north, west, south, east and north again to close the loop, roughly circumnavigating one of the regions CRISM had identified as containing clays. Each stop was given a numbered station designation and many outcrops were named.

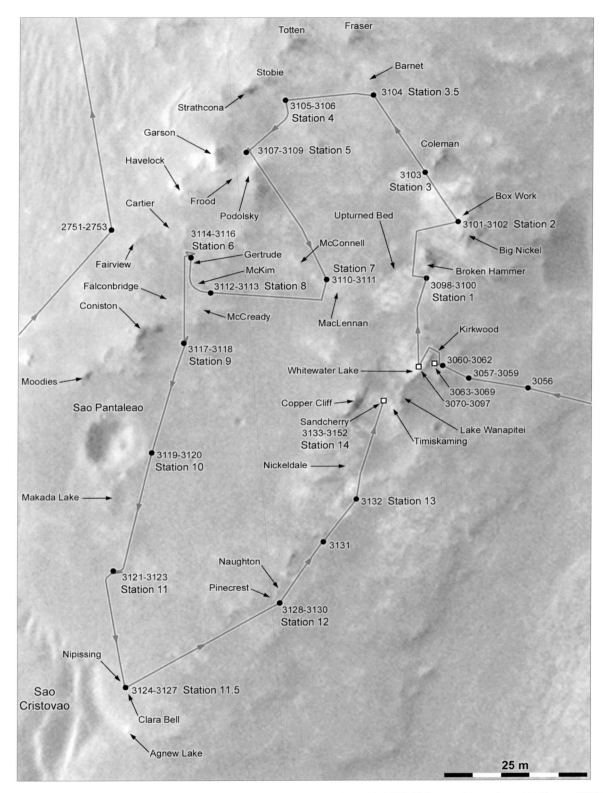

Figure 102. Opportunity reconnaissance of Matijevic Hill, Cape York, sols 3056–3152. This map is continued in Figure 103B.

The circuit, nicknamed Larry's Loop after Larry Crumpler, began on sol 3098 with a drive to Broken Hammer, followed by imaging of the target Upturned Bed on sol 3101. Sol 3100 was MY 31, sol 386, or 13 October 2012. Big Nickel was reached later on sol 3101, Coleman on sol 3103, Barnet on the next sol and Stobie on sol 3105. This was near the top of the ridge, and Opportunity now turned south to approach Garson on sol 3107. This was Station 5 (Figure 102), and after imaging several targets, including a boulder called Frood, the rover moved on to Station 7 at MacLennan, closer to the middle of the survey area. The path to Station 6 was not readily visible from Station 5 for route planning, so the stations were encountered out of order. By sol 3112 Opportunity was at Station 8, McKim, and Station 6, Gertrude, was reached on sol 3114. This was only about 15 m from the location on sol 2751, seen about one Earth year earlier during the northward drive to Greeley Haven.

On sol 3117 Opportunity reached Coniston, a possible remnant of an eroded crater, imaging it and several other targets, including the crater Sao Pantaleo on sol 3119. The next stop, later that sol, was beside the crater where a rocky debris ray was imaged on sol 3121. Sao Cristovao, a large crater filled with sediment, was visited over sols 3124–3127 at Station 11.5 (Figure 102), a station added to the original survey plan on an extension of Larry's Loop jokingly referred to as Ray's Loop after Ray Arvidson. Mars Odyssey was switching from one computer system to its backup at this time, so MRO took over the communication relay work for Opportunity until sol 3129. After imaging several targets, including Agnew Lake, Nippissing and Clara Bell at Sao Cristovao, the rover was driven back towards Whitewater Lake, stopping at Pinecrest on sol 3128, Nickeldale on sol 3132 and finally Sandcherry on sol 3133. This was only 10 m from Whitewater Lake and the survey was complete. MI imaged the sky for calibration purposes on sol 3126 and argon measurements were made by APXS during the circuit of the hill on sols 3099, 3109, 3122 and 3129 to measure atmospheric pressure.

The distribution of rock types seen during this drive suggested that the Whitewater Lake material was the unit containing the clay. Sandcherry was another exposure of the same material, and on sol 3135 Opportunity approached its new target as the skies darkened with dust. MRO's MARCI instrument had observed a regional dust storm approaching the rover on sol 3133, which passed by to the south but greatly increased dust in the air and reduced power at Endeavour. The rover would spend the US Thanksgiving holiday period here.

On sol 3137 (19 November 2012, MY 31, sol 423), Opportunity began studying Sandcherry with ISO on an unbrushed target. A RAT brushing on sol 3139 was prevented by an arm joint stall, and an APXS observation on the next sol became an atmospheric argon measurement instead because it was not placed on the surface. Testing on sol 3144 showed no problems, and the rock was brushed and viewed by MI later that sol, followed by APXS use overnight. Meanwhile a large color panorama called the Matijevic panorama was being compiled. One science question here was whether the Kirkwood material formed a layer between the Whitewater unit and the impact breccia, now called the Shoemaker Formation. On sol 3146 the RAT ground into Sandcherry, taking care to cut through any weathered layer but remain in the Sandcherry layer or rind, followed by ISO. The flash memory was very full now and had to be cleared.

A small move on sol 3151 was intended to put a vein target called Dowling in reach of the IDD, but Opportunity was not quite in the right place. Dowling was dropped and two sols later the rover moved on to Copper Cliff, 3 m west of Sandcherry, an outcrop with two distinct components. Opportunity arrived at the first target, Onaping, on sol 3155 and began ISO, slowed by two IDD stalls on sols 3160 and 3163. Images of a group of raised veins called Ministic were taken on sol 3164. Then Opportunity moved to the second target at Copper Cliff, called Vermilion Lake, on sol 3165. The rover carried out ISO here over sols 3168–3173, and on a nearby target called Vermilion Lake 1 on sols 3174–3176 during the Christmas holiday period, and measured atmospheric opacity and argon. A dust cleaning event occurred on sol 3175, and Opportunity collected more ISO data at Vermilion Lake 2 on sols 3177–3180.

Copper Cliff was only 10 m from Whitewater Lake, and on sol 3182 Opportunity drove back to that intriguing site, closing the loop around Matijevic Hill, to examine small light-toned veins seen earlier but not studied in detail. Images looking south towards Cape Tribulation on that sol revealed a dust devil in Botany

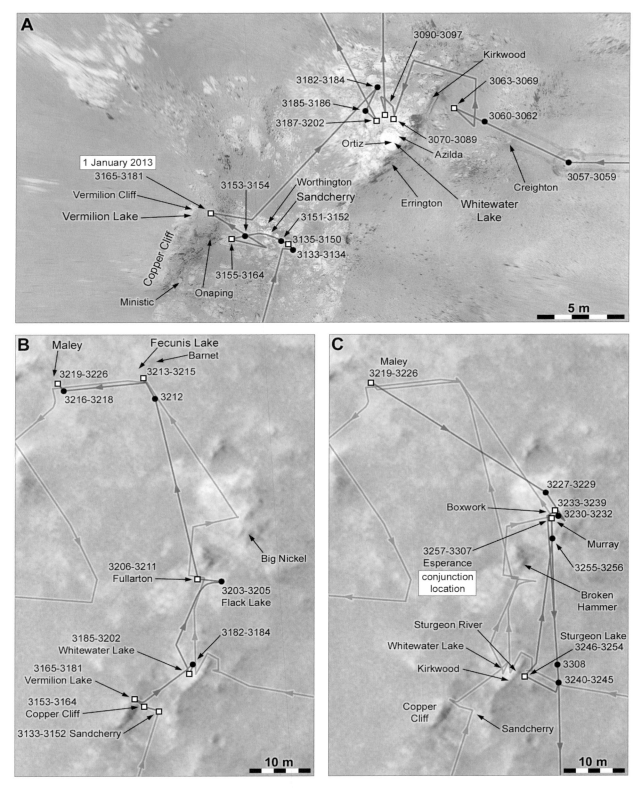

Figure 103. Opportunity exploration of Matijevic Hill, Cape York. **A:** Whitewater Lake, sols 3057–3097 and 3133–3202.
B: Sandcherry to Maley, sols 3133–3226. **C:** Maley to Kirkwood, sols 3219–3308.

Bay, only the second ever seen by Opportunity. Around this time Opportunity started exhibiting memory problems similar to events seen with Spirit around its sol 2065. Data stored in the flash memory would be lost occasionally if not transmitted before the rover shut down overnight. One of these events occurred on sol 3183. Spirit's memory had been reformatted on its sol 2083 to correct the problem, so a solution was available if needed. To help diagnose the problem, the rover began to store some information in an unaffected memory area on sol 3189. The rover moved again on sol 3185 and made its final approach to the vein target Ortiz on sol 3187. A large MI mosaic was made on sol 3189 and APXS was positioned on Ortiz 1 for analysis. The process of MI mapping and APXS analysis was repeated at the nearby target Ortiz 2 on sol 3192 and at Ortiz 3 on sol 3194. The small veins only filled a small part of the APXS data area, so the strategy was to analyze three targets with 0 percent, 10 percent and 20 percent vein coverage, to be able to separate the vein signal from the background.

A dust devil had been observed by chance on sol 3182, so targeted dust devil searches were made occasionally after this, the first on sol 3196, using a strategy designed to save flash memory. Several identical images would be taken and compared. The first would be saved for transmission to Earth and the others only saved if a change was noted. The RAT was supposed to brush Ortiz 2 on sol 3197 for an APXS observation, but another arm fault interrupted the plan. When it recovered, the brushing was dropped, leaving ISO in the plan for sols 3200 (MY 31, sol 486, or 21 January 2013) to 3202.

Work here was finished on sol 3203 and Opportunity drove north to seek more newberry-rich sites, beginning with Flack Lake, 20 m further north. Two sols later it imaged the RAT bit to assess its remaining capability. The small spherules called newberries were the target here, and the rover approached its new target, Fullarton 1, on sol 3206. On the next sol an MI mosaic was made and an APXS measurement started, the regular pattern of work at most of these outcrops. This was repeated on sol 3208 at Fullerton 2, and on the next sol the nearby target Fullerton 3 was brushed before repeating the MI and APXS work. The Fullerton 1 site was berry rich, Fullerton 2 had few berries, and Fullerton 3 also had few berries but was closer to the rover to avoid a berry that

might interfere with brushing. An embedded newberry called Teardrop was the target of multispectral imaging on sol 3208.

The season turned windy now. An unsuccessful dust devil search was made on sol 3209, but a dust devil appeared by chance on sol 3219 in the southern part of the floor of Endeavour, and later gusts stirred up clouds of dust on the crater floor on sols 3194 and 3230.

One remaining goal was to find the contact between the Whitewater Lake material and the overlying Shoemaker Formation material, impact ejecta that formed the top of the hill. Opportunity approached the next target, part of the Barnet outcrop called Fecunis Lake, on sol 3212 after a 35 m drive, brushed it on sol 3214 and followed that with ISO. Two sols later the rover moved back a little to image the brushed area, turned to the west and drove 16 m to a target called Maley in the Stobie area. A short drive on sol 3219 positioned the rover next to Maley, but its work was interrupted on sol 3221 when a Deep Space Network problem delayed transmission of the sol's commands. Opportunity was supposed to turn to reach its new target, but instead of waiting to do the turn the team decided to work on a target already within reach. This spot was brushed on sol 3224 before performing the usual ISO on it. The contact between rock types should have been observed by now, but it was becoming apparent that it was gradational, not a sharp dividing line. The newberries embedded in the rock became less numerous as the rover climbed the hill.

The next winter haven was planned to be on Solander Point, 2 km south of Maley, and Opportunity needed to be on its northern slopes in six months, so it was time to leave Cape York. On sol 3227 the rover was driven southeast to Big Nickel to examine an important target, a "boxwork" feature called Lihir made of intersecting veins exposed by erosion. This would be studied quickly, and then the rover would have to leave the area. On sol 3230 it approached Lihir, intending to scuff it with its wheels, but it was deflected by the uneven terrain and missed the target. A final small move on sol 3233 placed the target Lihir in reach of the IDD, but work was delayed by another flash memory fault that occurred late on sol 3234, the first since sol 3183. Work resumed on sol 3239 with ISO on Lihir, and on the next sol, after imaging Spanish River and Broken Hammer, Opportunity was driven south to Kirkwood for a final experiment.

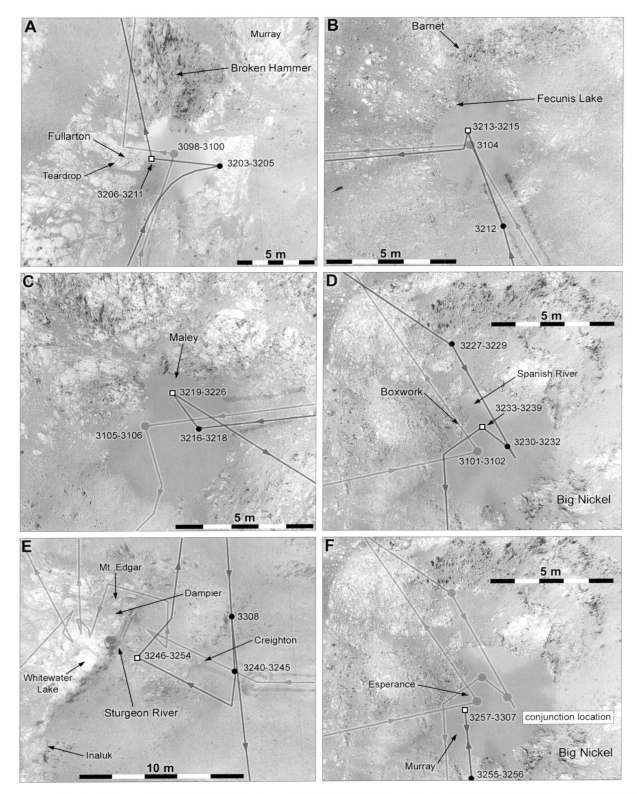

Figure 104. Final activities on Matijevic Hill. **A:** Fullarton, sols 3203–3211. **B:** Fecunis Lake, 3212–3215. **C:** Maley, sols 3216–3226. **D:** Boxwork, 3227–3239. **E:** Sturgeon River (Kirkwood), sols 3240–3254. **F:** Esperance, 3255–3308.

This last task was a further study of the "newberries," whose composition was still unclear. A geologist's streak test, made on Earth by scraping a mineral across a white ceramic plate to examine the color of its powdered form, was duplicated by lightly grinding the nodules at Kirkwood. Opportunity reached that area on sol 3240, stopping close to its sol 3057 location. Images of Creighton were taken on sol 3242, and of its tracks from 200 sols earlier on sol 3244 to look for changes. Another flash memory problem, not serious but different from the earlier events, occurred on sol 3244, and as each such event occurred, the engineers had a better idea of the locations of bad areas of memory.

Two sols later Opportunity moved 9 m back up the hill to a point near its sol 3063 location at Kirkwood for its last analysis of the newberries. On sol 3247 a target called Sturgeon River 1 with many newberries was analyzed. On the next sol the rover took an MI image of an area called Sturgeon River 2 with fewer newberries and placed APXS on it for an overnight analysis. This was repeated for a target called Sturgeon River 1a on sol 3249, and on sol 3250 it made MI mosaics of the nearby Lake Laurentian and South Range targets before returning APXS to Sturgeon River 1a. The much anticipated streak test was performed by grinding a new target, Sturgeon River 3, on sol 3251, but it did not generate enough powder for effective analysis. The experiment was repeated two sols later at Sturgeon River 3a. Each RAT grind was followed by an MI mosaic, an APXS analysis and multispectral imaging of the powder, which detected iron oxides.

Despite the need to leave, one more compelling target was added to the schedule. The analysis at Kirkwood concluded on sol 3254, and wind blew dust off the solar panels again. On the next sol the RAT bit was imaged and then Opportunity drove 25 m north to the Lihir area where superior conjunction would be spent on a final analysis. The hurried study on sol 3239 was not enough, and now a vein target would be brushed if one could be found without newberries in it which might jam the RAT. Another cleaning event occurred on sol 3256, and the cleaner panels helped by postponing the need to be on a north-facing slope.

Opportunity climbed onto the Big Nickel outcrop on sol 3257 and rotated on sol 3260 to position its instruments on a new target called Esperance. Multispectral images were taken on sol 3262, followed by ISO. This spot had too many newberries for the RAT to be used safely, but the MI mosaic was extended on sols 3264 and 3267 to find a good location. APXS collected data over sols 3264–3266 at Esperance 2, but it also had too many berries. With time running out, the APXS was placed on Esperance 3, which had few berries but was too far from the rover for the RAT to be used on it. The nearby tracks were imaged on sol 3269 to look for changes after conjunction, and then APXS was set to collect data over the conjunction period, which lasted from sol 3272 to 3293 (MY 31, sols 558–579). The actual conjunction date was sol 3281. Other tasks during this time were deep sleeps to conserve power, and imaging, though some of this was canceled before conjunction, including more views of the tracks and a panorama across Botany Bay, called the Rubicon panorama, since the rover had to cross this area soon.

When operations resumed after conjunction, the rover team discovered that Opportunity had entered a standby mode on sol 3286, possibly during a routine Pancam sequence of images of the Sun taken to estimate dust levels in the atmosphere. They tried to recover the rover on sol 3294 and succeeded with a second attempt on the next sol, resuming operations with a very small clockwise move on sol 3296 to reposition the IDD for work at a better site on Esperance. Argon was measured on sol 3297 and images showed the solar panels were becoming dustier again, as dust in the atmosphere was also increasing, limiting power again. ISO were done on a suitable location called Esperance 4 over sols 3298–3300 (MY 31, sol 586, or 6 May 2013), and a nearby site called Esperance 5 was ground into and brushed on sol 3301 with ISO there over two sols. A light-toned outcrop called Murray was imaged on sol 3302, and then the rover recharged for two sols. A final grind, brush and ISO sequence followed on sols 3305 and 3306 at Esperance 6, another small offset from the previous site. Esperance appeared to be a montmorillonite clay altered by neutral water, not the acidic water suggested by rocks in the plains. This might once have been a habitable environment.

The geology of this region was becoming clear now (Arvidson *et al.*, 2014a). A pre-existing surface consisting of the ejecta of Miyamoto crater (Figure 43A) was covered by ejecta from Endeavour crater. These two

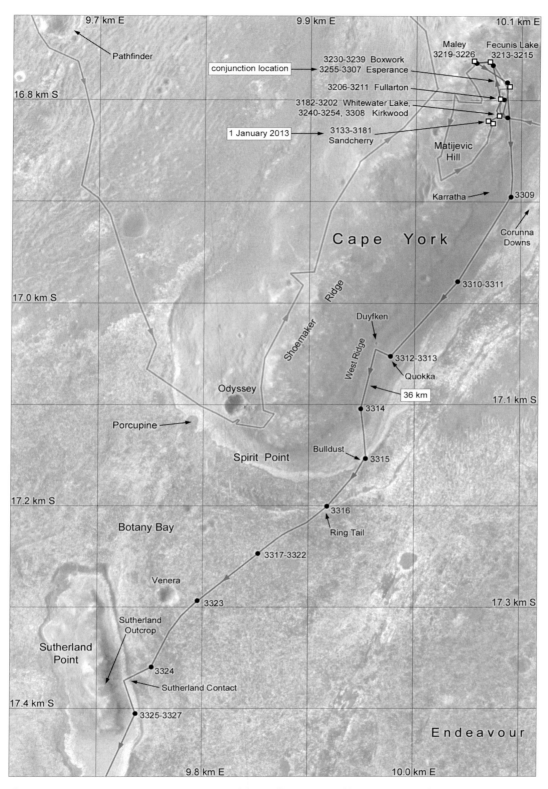

Figure 105. Opportunity route map, section 24. This map follows Figure 98 and is continued in Figure 106. Additional details are shown in Figure 104. White squares are main science stops.

units were the Whitewater and Shoemaker Formations, respectively, and Whitewater had been soaked in water to produce veins such as Esperance. The Shoemaker Formation formed the hilly rim of Endurance, and had been eroded to produce the gently sloping aprons surrounding the hilltop, now cut by gypsum veins like Homestake. The sulfate-cemented sandstones of the Burns Formation that covered Meridiani Planum had been deposited over and around those hills and on the floor of Endeavour, and the lowest stratum of the plains material, with a slightly different composition, was the Grasberg Formation.

At last, on sol 3308 (14 May 2013, MY 31, sol 594), the drive to Solander Point commenced. This northern spur of Cape Tribulation offered extensive outcrops on a broad north-facing hillside and a productive winter season for Opportunity. Two paths were considered, a direct route called "rapid transit" across Botany Bay with few or no stops, and a route that zig-zagged between interesting targets, including Nobby's Head and several craters and outcrops of gypsum. This route was designed by Tim Parker and Matt Golombek (JPL) and discussed by the rover team in early March 2013, but it was rejected in favor of the shorter route. In fact, however, several science sites were fitted into the traverse.

Opportunity drove from Esperance to Kirkwood on sol 3308 and then descended from Matijevic Hill to move rapidly to the southwest on the broad smooth debris apron (Figure 105). On the next sol an 80 m drive brought the rover's total driving distance to 35.76 km, breaking the Apollo 17 LRV distance record for any previous NASA rover. Several targets were imaged along the path, and on sol 3314 Opportunity made a short detour to view the small crater Duyfken as its traverse distance exceeded 36 km. Finally, on sol 3315 Opportunity reached Cape York's southern edge, tilting enough for its software to halt the drive, and on the next sol it imaged the Bulldust outcrop and then drove onto the rocky slope of Botany Bay nearly a full Mars year after reaching Endeavour on sol 2681. The air was clearing after the dust storm and wind gusts cleaned the solar arrays. A faint dust devil had been seen on the floor of Endeavour on sol 3308, and a strong gust raised a cloud of dust on the inner slope of Cape Tribulation on sol 3317, attesting to the windy environment here.

Ring Tail, a contact between two rock types here, was imaged on sol 3317 before moving out into the sloping plains, and the RAT bit was imaged on sol 3312. The rover imaged Venera crater on sol 3323 and approached Sutherland Point on sol 3324. This and the neighboring Nobby's Head formed a miniature version of Cape York surrounded by its own rocky ledge, so Nobby's Head had been referred to as Cape York B by the Science Operations Working Group before it received its own name. The contact between Botany Bay and the Sutherland Point bench, and the rubble-strewn slopes of the Point itself, were viewed on arrival, and Opportunity mounted the bench and drove south on sol 3325. A cluster of rocks called Walleroo was imaged during the drive on sol 3330, and a gravelly soil called Gnarleroo was an ISO target on sol 3332.

Opportunity undertook a quick survey on sol 3334 to identify potential winter havens at Nobby's Head if Solander Point could not be reached, keeping in mind what happened to Spirit as it approached McCool Hill around its sol 800. Slopes of up to 15° were found here, but the hill was less enticing than the much larger Solander Point. The drive to Solander Point was interrupted on sol 3336 by another flash memory problem causing a computer reset, but it was soon restored to full operations and drove 75 m on sol 3339. The drives on sols 3339 and 3342 tested new autonomous driving software, which would permit drives of 100 m per sol or more as the rover hurried towards Solander Point. Opportunity passed its fifth anniversary, in Mars years, near the south end of Nobby's Head (Figures 106 and 107A) on sol 3344 (MY 31, sol 630), and the science team looked for a target for ISO in this area. A rock called Chameleon was imaged for this purpose on sol 3344 but another called Gibber Earless, about 30 m further east, was preferred. These names were taken from plants and animals of Australia to reflect the name Botany Bay and the general Australian theme for names at Endeavour.

When Opportunity drove to Gibber Earless on sol 3344 it overshot its goal and there was no time to try to reach it, so the ISO were dropped. The rover departed on sol 3346, imaged its RAT bit on the next sol and exceeded 37 km of driving on sol 3348. This distance had long been reported as the length of Lunokhod 2's traverse on the Moon in 1973, and there was considerable interest in whether or exactly when Opportunity

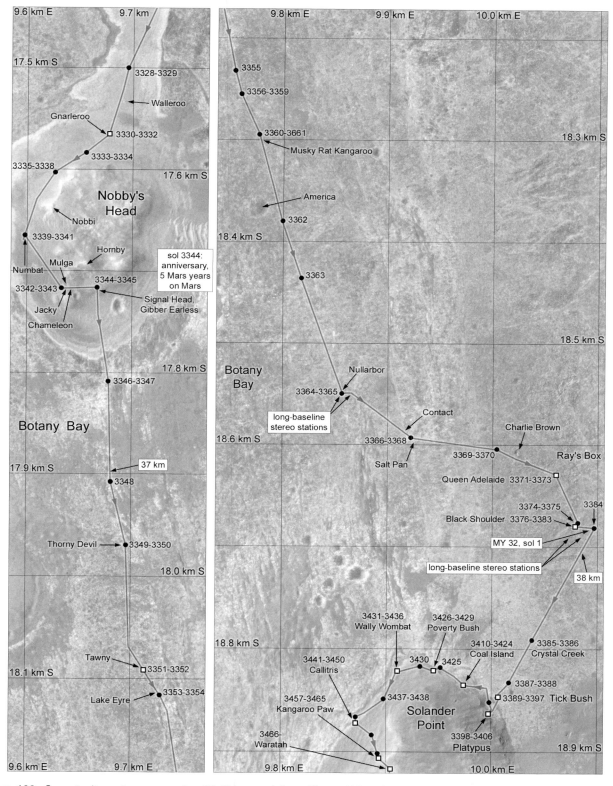

Figure 106. Opportunity route map, section 25. This map follows Figure 105 and is continued in Figure 112. Additional details are shown in Figures 107, 108 and 109.

might exceed that record drive. Lunar Reconnaissance Orbiter images of the Lunokhod tracks suggested that the old Soviet rover might have driven more than 37 km, so no claim of a record was made while Russian and US researchers looked carefully at the question.

A drive on sol 3349 was interrupted after 63 m by troubling signals from a sensor on the arm, which was carried in an unstowed position during driving, as it had always been since sol 1612 in Victoria crater. The arm was not supposed to move during a drive but the sensor suggested it did. Images showed it had not moved, but the sensor reported another movement on sol 3350 almost as soon as the drive commenced. Testing on sol 3351 suggested an error in the sensor, so the operating software was directed to ignore its readings and Opportunity continued driving. The rover now had about two sols to spare in its schedule to reach Solander Point. Images of the surface of Botany Bay revealed good hard ground with many blueberries, very easy to drive on. The drive on sol 3351 was 123 m long, and ended with a layered rock target called Tawny within reach of the arm, so it was analyzed on sol 3352.

Opportunity imaged a jagged rock called Lake Eyre on sol 3355 before a 118 m drive. On sol 3357 APXS measured atmospheric argon, and early in the morning of sol 3360 the rover observed Phobos eclipse Deimos. Two sols later the broken rock Musky Rat Kangaroo was observed. Solander Point was imaged after every drive, including that of sol 3364, but the two Pancams alone did not yet provide a sufficient stereoscopic view for detailed topographic mapping at a range of 300 m. After ripples at Nullarbor were imaged early on sol 3366, Opportunity turned east and drove downhill, stopping after 8 m to view Solander Point again for long-baseline stereoscopic mapping of the Solander area and to begin the search for winter haven sites.

The drive continued later that sol, descending further to reach an area showing a distinctive hummocky appearance and composition in orbital data, which were interpreted to indicate hydrated gypsum. This region was nicknamed Ray's Box after Raymond Arvidson, Deputy Principal Investigator for MER and a member of the MRO CRISM science team, who identified the area of interest. Opportunity was ahead of its driving schedule, and new analyses of power generation pushed back the date when north-facing slopes were needed, so the rover

could afford time for a look at this area. Observations here included imaging wavy bedding in the rock Salt Pan on sol 3369 and a small crater called Charlie Brown on sol 3371, and ISO on a soil area called Queen Adelaide on sol 3373.

Opportunity drove another 50 m on sol 3374, viewed a Deimos solar transit on sol 3376, missed another on sol 3377, and investigated three ISO targets on a slab of coarse-grained layered rock called Black Shoulder on sols 3378, 3380 and 3381 (Figure 107B). This area was 200 m further east than the sol 3365 area and offered the best view ever likely to be obtained of the inward slope of Solander Point, so images were collected for a long-baseline stereo view of the ridge early on sol 3381. On the next sol the RAT was used to grind into Black Shoulder and brush it for 20 min to get it as clean as possible, followed by more ISO. On sol 3384 the rover drove 10 m further east, as far east as it had ever been, for the second part of the long-baseline stereo observation, as the Mars calendar moved into Mars year 32. This stereo imaging was done early on sol 3385, just before Opportunity drove 115 m towards the base of Solander Point, exceeding 38 km of total distance in the process.

A plan for Opportunity's initial exploration of Solander Point was illustrated by Larry Crumpler (Crumpler *et al.*, 2013b) in August 2013. The rover would work its way around the northern tip of Solander from east to west, climbing onto the base of the hill by October 2013 and reaching a good north-facing slope by December. This strategic path, with science targets designated as numbered stations, was followed closely at first and is partly illustrated in Figure 108 after the rover deviated from the plan. To assist planning, Ray Arvidson used new CRISM data to locate possible clay targets on Solander for later study at an area referred to first as Winter Wonderland and later named Cook Haven. The clays might have been related to a fracture called Hinge Line extending in from the plains to the west (Figure 110). Opportunity was expected to spend much of the winter at Cook Haven.

A layered rock slab called Crystal Creek was imaged at the start of the drive on sol 3387. It was considered as an ISO target but power was now too limited. The rover approached the foot of Solander Point on sol 3389 (Figure 107C) for ISO of a rock slab called Red Hot Poker on the surface below the hill on sol 3390. The next target was a pitted rock called Tick Bush, 5 m further east. Opportunity compiled an MI mosaic of the rock on sol 3392 and placed APXS on it for three sols, while also taking multispectral images of other features nearby, including the base of the hill and the rock Cheese Tree. APXS was moved to an adjacent target called Tick Bush 2 on sol 3396, and the rover also observed solar transits by Phobos and Deimos. Another Phobos transit and rocks including Quandong and Peach Tree were observed on the next sol, and a large block called Mulla Mulla was chosen over Cheese Tree as the next target. Opportunity was driven 23 m to the base of the hill on sol 3398, but stopped uncomfortably close to Mulla Mulla, which was large enough to damage the solar panels if they struck it (Figure 107D). Here APXS measured argon in the atmosphere again.

A 40 cm move away from Mulla Mulla on sol 3400 (MY 32, sol 18, or 17 August 2013) put a new target called Platypus within reach of the arm for brushing on sol 3403 and two sols of ISO. Opportunity backed away from Platypus on sol 3405 for multispectral imaging and then drove around the rocky area, imaging its tracks near Mulla Mulla to check how close it had been to the rock. Three rocks here were considered for the next ISO, Quandong, Coolabah (a cluster of four rocks) and Banksia, a small rock near the front wheels after the 3405 drive, but all were hard to reach with the arm, so the chosen target was Coal Island at the geological contact between the hill and the plains. This boundary was covered with debris in many places but seemed to be exposed better here. The rover drove further north on sol 3407, measured argon again on the next sol, and turned west to approach the contact on sol 3410.

Opportunity approached Coal Island, driving backwards, on sol 3412, turned to face the contact and moved forwards before taking multispectral images to plan IDD work (Figure 107E). On sol 3415 MI made a mosaic of the target Dibbler and APXS collected data for several sols as the multispectral imaging continued. That work ended on sol 3418 and Opportunity moved 3 m along the scarp to a dark spot called Monjon on sol 3419. The rock had two parts, gray and slightly purple in color. On sol 3420 Opportunity measured argon again and imaged the nearby target Woylie, and two sols later the MI made a mosaic of Monjon. APXS was placed on the purple and gray areas of Monjon on sols 3422 and 3423–3424,

respectively, and on sol 3425 the rover moved northwest, further along the contact.

The next target, Poverty Bush, was reached on sol 3426, and on sol 3427 the RAT bit was imaged again to assess its remaining effective life. Then MI was calibrated by imaging the sky and then made a mosaic of Poverty Bush, and finally APXS was placed on Poverty Bush for two sols to collect data. Two drives totaling 25 m on sols 3430 and 3431 brought Opportunity to the next target area, now fully on Solander Point rather than on the plains at its foot and stopping each time at a "lily pad" with a northerly tilt.

Operations here began with multispectral imaging of several targets, including Long Nosed Potoroo and Little Red Kaluta on sol 3432. A 2 m drive placed a target called Wally Wombat in range of the IDD, and on sol 3433 the RAT brushed it in preparation for ISO on sols 3434 and 3435. Wally Wombat was not ground into to preserve the RAT bit for higher-priority targets. Pancam images of several features including the brushed target and a vein called Agile Antechinus were taken on sols 3436 and 3437, and then Opportunity made a 32 m drive along the west side of the hill, followed by another argon measurement.

On sol 3439 a 31 m drive brought Opportunity to the next target, a rock called Callitris on the western edge of Solander Point, where it surveyed the area and took multispectral images of a dust drift. Callitris was imaged on sol 3441 before a small move to put the rock within range of the IDD, and Comet C/2012 S1 (ISON) – the International Scientific Optical Network (ISON) is an international project with telescopes in many countries – was imaged at night on sols 3441 and 3443 but could not be seen in the Pancam images. Curiosity also attempted to image the comet but failed.

Opportunity brushed Callitris on sol 3444, beginning two sols of ISO. Next, several sols were taken up with color stereoscopic imaging of the ridge forming the backbone of Solander Point, which had now been named Murray Ridge after former JPL Director and veteran planetary scientist Bruce Murray, who had died on 29 August 2013. More APXS data were collected at Callitris on sol 3448, and the rover left this area at the base of the ridge for a more north-facing "lily pad" on sol 3451. Most of Opportunity's stops over the next few months were on these north-tilting areas.

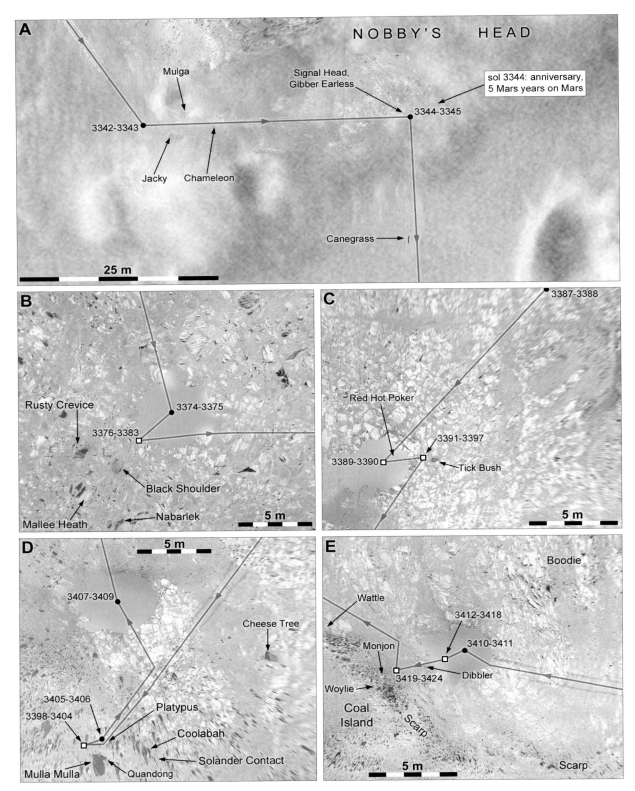

Figure 107. Opportunity activities in Botany Bay. **A:** Nobby's Head, sols 3342–3345. **B:** Black Shoulder, sols 3374–3383. **C:** Tick Bush, sols 3387–3397. **D:** Platypus, sols 3398–3409. **E:** Coal Island, sols 3410–3424.

Table 27. *Opportunity Activities, Botany Bay and Solander Point, Sols 3316–3700*

Sol	Activities
3316–3324	Drive south (3316–3317), RAT bit imaging, MI calibration (3321), drive south (3323–3324)
3325–3332	Reach Sutherland Point (3325), imaging, drive to Nobby's Head (3328–3331), ISO Gnarlaroo (3332)
3333–3345	Drive around Nobby's Head (3333–3345), anomalies prevent science (3336–3338), imaging
3346–3373	Drive south (3346–3351), ISO Tawny (3352), drive south (3353–3372), ISO Queen Adelaide (3373)
3374–3384	Drive south (3374–3376), ISO Black Shoulder (3378–3383), Solander long-baseline stereo (3384)
3385–3397	Drive south (3385–3387), ISO Red Poker (3390), move to (3391) and ISO (3392–3397) Tick Bush
3398–3404	Reach Solander Point (3398), move to (3400), brush and ISO (3403–3404) Platypus, imaging
3405–3421	Drive around Solander (3405–3414), imaging, ISO Dibbler (3415–3416), drive west (3419)
3422–3426	ISO Monjon, Monjon Purple (3422), ISO Monjon Gray (3423–3424), drive west (3425–3426)
3427–3436	ISO Poverty Bush (3427–3428), drive, imaging, brush (3434) and ISO (3434–3435) Wally Wombat
3427–3450	Drive, imaging, brush (3444) and ISO (3444–3445, 3448) Callitris, ridge panorama (3446–3450)
3451–3465	Drive to Kangaroo Paw, brush (3461) and ISO (3462–3464) Spinifex, ISO Paper Bark (3462)
3466–3476	Drive to Waratah, ISO Baobab (3468–3469), drive south, ISO Yellow Bellied Glider (3475–3476)
3477–3500	Drive around drifts (3478), drive to Moreton Island, imaging, ISO Tangalooma (3498)
3501–3510	Imaging, ISO Mt. Tempest (3502), drive to Great Keppel Island (3505), imaging, drive south
3511–3539	Imaging, move to tilt north (3512), ISO Cape Darby (3519–3532), Cape Darby 2 (3535–3536)
3540–3547	Move to (3540) and ISO (3541–3544) Cape Elizabeth, ISO Pinnacle Island (3546), imaging
3548–3559	ISO Pinnacle Island 2 (3548), Pinnacle Island 3 (3551–3554), small moves (3555, 3557), imaging
3560–3568	ISO Pinnacle Island 4 (3560–3562), Pinnacle Island 5 (3562–3564), move to new target (3566–3567)
3569–3574	Brush and ISO Green Island (3569), drive (3571), ISO Stuart Island 1 (3573), Stuart Island 2 (3574)
3575–3581	ISO Stuart Island 3 (3575), Stuart Island 4 (3576–3577), drive (3578), ISO Anchor Point (3581)
3582–3585	ISO Anchor Point 2 (3582–3583), drive to Sledge Island and crush it with wheels (3585)
3586–3590	ISO Sledge Island 1, Sledge Island 2 (3587), drive to Stuart Island (3589), imaging
3591–3598	Drive to Cross Sound (3591), imaging, drive (3594, 3596), brush and ISO Turnagain Arm (3598)
3599–3600	ISO Turnagain Arm (3599), drive to Augustine, imaging (3600)
3601–3608	Move to Augustine (3601–3602) and ISO (3603–3605), imaging, drive to Sugarloaf (3607)
3609–3617	Drive to Point Bede (3609–3615), imaging, deck mosaic (3611, 3613), ISO Point Bede (3616–3617)
3618–3632	Drive south along ridge (3618–3628), reach summit area of Murray Ridge (3630), imaging
3633–3637	Endeavour panorama (3633–3634), drive to highest point (3635), imaging, drive downhill (3637)
3638–3651	Imaging, drive south (3639–3644), computer reboot fault (3645–3647), drive south (3649–3650)
3652–3658	ISO Mulgrave Hills (3652), drive (3653, 3655), ISO Ash Meadows (3657, 3658), imaging
3659–3666	Drive (3659), move to Bristol Well (3662), ISO Bristol Well 1 (3664–3665), Bristol Well 2 (3666)
3667–3673	ISO Bristol Well 3 (3667–3668), move (3669), brush (3671) and ISO (3671–3673) Sarcobatus Flat
3674–3678	ISO Sarcobatus Clast (3674–3675), Sarcobatus Clast2 (3676), drive to Landshut (3676–3678)
3679–3685	ISO Landshut (3679, 3682), faults preclude science (3680–3681, 3683), drive south (3684), imaging
3686–3690	Computer reset precludes science (3686–3688), drive south (3689), imaging
3691–3697	Drive south (3691, 3696), image Besboro Island (3693, 3696), sky images for night opacity (3697)
3698–3700	Image Cape Lisburne, Norton Sound, Phobos (3698), drive south (3698), ISO Mayfield (3700)

Opportunity now drove uphill, generally southwards, imaging features, including a vein called Lilli Pilli on sol 3454 and a small rock called Desert Pea on sol 3456, and measuring argon on sol 3453. The Jupiter-bound spacecraft Juno was in a safe mode on Opportunity's sol 3456, limiting DSN availability briefly. The rover reached an outcrop called Kangaroo Paw on sol 3457, and short drives on sols 3458 and 3459 positioned the arm for ISO on the next sol, in an area with a northerly tilt of 16°. After recharging the battery on sol 3460 a target called Spinifex on Kangaroo Paw was brushed on the next sol, followed by several sols of ISO. The next drive up the ridge on sol 3466 was followed by an argon measurement, and on the next sol a short move was made towards

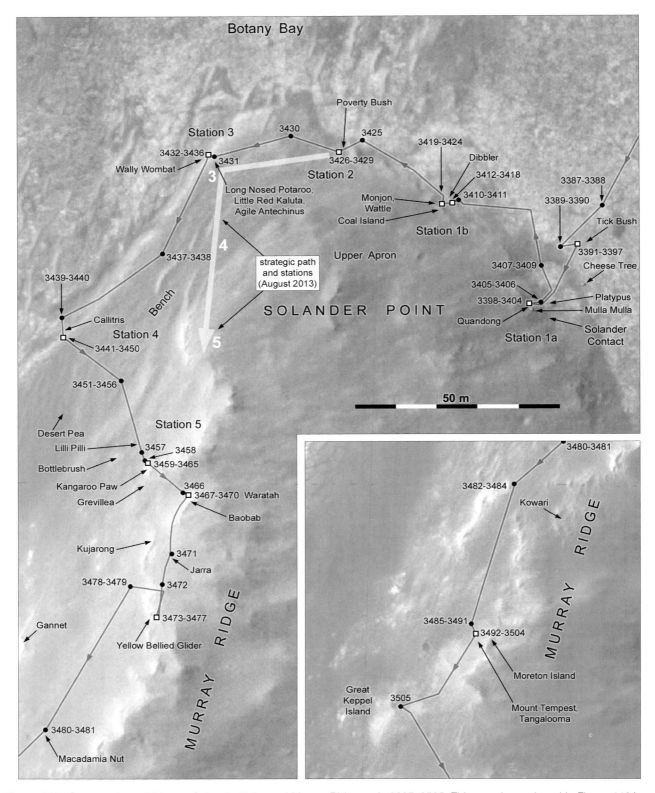

Figure 108. Opportunity activities on Solander Point and Murray Ridge, sols 3387–3505. This map is continued in Figure 110A.

Waratah, the next outcrop. The target here was called Baobab, and an MI mosaic was compiled on sol 3468, followed by a multi-sol integration by APXS. These outcrops on the ridge all consisted of impact ejecta.

Opportunity was driven 30 m uphill to the next lily pad on sol 3471, imaged its surroundings, including a cobble cluster called Jarra, and drove again on sol 3472, heading generally south along the crest of the ridge. These drives were often designed to end when the rover's tilt was in a suitable range rather than after a specified time or distance. Murray Ridge sloped steeply to the east into Endeavour and more gently to the west out towards the plains of Meridiani Planum. A thin layer of dust was almost ubiquitous on Mars, but thicker deposits of pure dust were very rare, previously seen only on the rim of Duck Bay at Victoria crater, where it was impossible to reach, and at North Pole on Cape York, but now a patch of pure dust was found on the ridge crest. By measuring its composition here the effects of dust contamination could be subtracted from other APXS data, improving the interpretation of those targets. This site, called Yellow Bellied Glider, was reached on sol 3473 and investigated with MI and APXS on sol 3475–3476.

On sol 3478 Opportunity drove around the dusty ripples to avoid an area of poor traction on the slope, and moved south again. Gannet, a group of boulders on the slope to the southwest, was imaged on sol 3480, Macadamia Nut on sol 3482 and the rubble pile Kowari on sol 3485, and then a bright outcrop called Moreton Island was examined on sol 3487. Opportunity imaged its tracks here to assess slip and view the disturbed soil, and also viewed a crater called Matthew on sol 3488. Then the rover moved to the outcrop with several small moves, arriving on sol 3496. Moreton Island was very heterogeneous, and the area had a very acceptable slope of 18°. The first target was Tangalooma, investigated on sol 3498. Sol 3500 was MY 32, sol 118, or 28 November 2013, and two sols later Opportunity moved its IDD to a new ISO target called Mt. Tempest with somewhat blue-toned clasts for several sols of data collection.

The next drives southwards and uphill occurred on sols 3505 and 3506, but on the second sol the right front wheel began to draw more current, as seen from time to time in the past. All recent long drives had been made backwards, which generally improved the wheel's

behavior, so on sol 3507 after moving 20 m to a better lily pad the rover drove forwards and backwards as a test, and the wheel now seemed to do better driving forwards. The sol 3505 drive brought Opportunity to Great Keppel Island, but the rocks looked very much like Moreton Island and the tilt was only 8°, so the rover moved on. Odyssey was in a safe mode on sol 3510, slowing work for a few sols as Opportunity tried to reach its winter area. This had been nicknamed Winter Wonderland but it was now given the name Cook Haven.

The slope here was about 10° to the north, increasing power to some extent, but the winter nights were becoming colder and now the battery heaters began to turn on in the coldest periods, so more power would be needed very soon. The rover imaged Cook Haven and measured argon on sol 3510 and then recharged for a sol as its ground team looked for better lily pads in the images. Opportunity drove to its next target, Cape Douglas, on sol 3512, during a period of very limited communication caused by the safe mode on Mars Odyssey. On sol 3514 a dust devil was seen by chance on the floor of Endeavour to the northeast, and the team tested revised procedures for IDD operations to reduce the risk of future faults. Images of the Cape Douglas outcrop on sols 3512 and 3516 looked interesting, and the northward tilt here was 17°, so the rover would stay here over the upcoming holiday period.

MI and APXS took data at Cape Darby, a target on Cape Douglas, on sol 3519, but the arm was placed too high. The measurements were repeated successfully on sol 3521, with APXS continuing intermittently for several sols until sol 3532 and then taking data at Cape Darby 2 on sols 3535 and 3536. MI data showed loose pebbles which could jam the RAT, so the target was not safe to brush. Pancam imaged a possible geological contact between rock types at Trinity Island on sol 3522, and other observations included mapping of Cook Haven, views across Endeavour on sol 3526 and cloud sequences on sol 3534. A DSN problem on sol 3530 prevented commands for two sols from being received, but the lost observations were recovered over the next two sols.

Topographic mapping now suggested that much of Cook Haven would be inaccessible because north-facing slopes were too shallow, but this was about to change. Unexpected wind gusts cleaned the solar panels on sols

3533 and 3534 (1 January 2014), and the cleaning was documented with a rover deck Pancam mosaic taken on sols 3536–3538. Already this was enough to make Icy Cape, a target just north of Cape Douglas, seem accessible for the first time, and cleaning continued in many small increments. Icy Cape appeared to differ from the rocks at Cape Douglas.

On sol 3539 Opportunity imaged its recent tracks to commemorate the tenth anniversary, in Earth years, of Spirit's landing, and on the next sol the rover moved a short distance to reach Cape Elizabeth, a light-toned outcrop on the western edge of Cook Haven. Images taken just after the move revealed an unusual small rock named Pinnacle Island near the rover, which had not been present only a few sols earlier. It may have been moved by the rover wheels, but even ejecta from a new impact was considered as a possibility, and MRO's HiRISE camera took an image on Opportunity's sol 3577 to see if any new crater was visible nearby. The RAT scanned and brushed Cape Elizabeth on sol 3541 and began ISO, also finding time for MI work on the jagged, bright upper surface of Pinnacle Island.

APXS work on the outcrop continued until sol 3544 and the rover magnets were imaged on sol 3545, but a DSN fault prevented data transmission to Earth. The lost transmission was resent quickly and APXS was deployed at two sites on Pinnacle Island on sols 3548 and 3551–3544, with some additional MI observations of the rock. Other work included images of the rock Green Island and some disturbed soil and rocks. The cleaning events had continued and Opportunity now needed only to stay on slopes steeper than about 5°.

The ISO campaign on Pinnacle Island continued at many separate, overlapping, locations until sol 3562, providing extensive MI coverage and compositional data from areas of different color and revealing a unique composition rich in sulfur, magnesium and manganese. To reach different targets with the arm's restricted movement, the rover was nudged slightly on sol 3555, but the turn was insufficient, so on sol 3557 a "tank turn" was made by spinning wheels in opposite directions on the left and right sides of the rover. Sol 3556 was the tenth anniversary, on Earth, of Opportunity's landing, and sol 3558 was the sol with minimum illumination, the least sunlight of the winter.

The recent dust-cleaning events suggested winds were increasing, so Opportunity looked for dust devils across Endeavour on sols 3559 and 3566, separated by ISO at two targets on Pinnacle Island. Later on sol 3566 Opportunity drove about 2 m downslope (north) and looked back at the area previously hidden under the rover to see if a source area for Pinnacle Island could be identified in the area disturbed by its wheels on sol 3540. This was variously referred to as "Socket" or "Pinnacle Divot," and when the source was found to be a rock similar in color to Pinnacle Island it was nicknamed Pork Chop and later more formally called Stuart Island.

Opportunity undertook multispectral imaging, a RAT brush and ISO on Green Island, an outcrop of the lowest layer of rock in this area, over sols 3568–3571, and then drove back up to Stuart Island later on sol 3571 to analyze the new target. It took Pancam multispectral images on sol 3571 to identify areas with different colors, and then performed ISO at four locations, obtaining full MI coverage and deploying APXS on areas of different color over sols 3573–3577. Sol 3577 (14 February 2014 or MY 32, sol 192) was the winter solstice, so sunlight would soon become stronger, and although several recent wind gusts had cleared dust from the solar panels, Opportunity still needed to find north-facing slopes. One of these, a rocky slope just south of Cook Haven, was now named McClure–Beverlin Escarpment to commemorate the two men, Bill McClure and Jack Beverlin, whose quick actions saved the Mariner 6 launch vehicle from a potentially catastrophic failure on the launch pad on 14 February 1969.

Two more of Opportunity's flash memory faults occurred on sols 3576 and 3577, but they were soon overcome, and on sol 3578 the rover moved a short distance backwards and then forwards, imaging the rock Two-Headed Island during the drive. A rock called Sledge Island was under a wheel at the end of the drive, and the science team hoped it would be crushed to help confirm that Pinnacle Island could have been broken off another rock by the wheels, but Sledge Island was not broken this time. APXS measured argon on the next sol, and the left Navcam was used to measure atmospheric opacity, a job usually done on almost every sol by the Pancams. The future Insight lander would only carry a Navcam and Hazcams, and this test and another on sol 3585 would help plan activities for that mission.

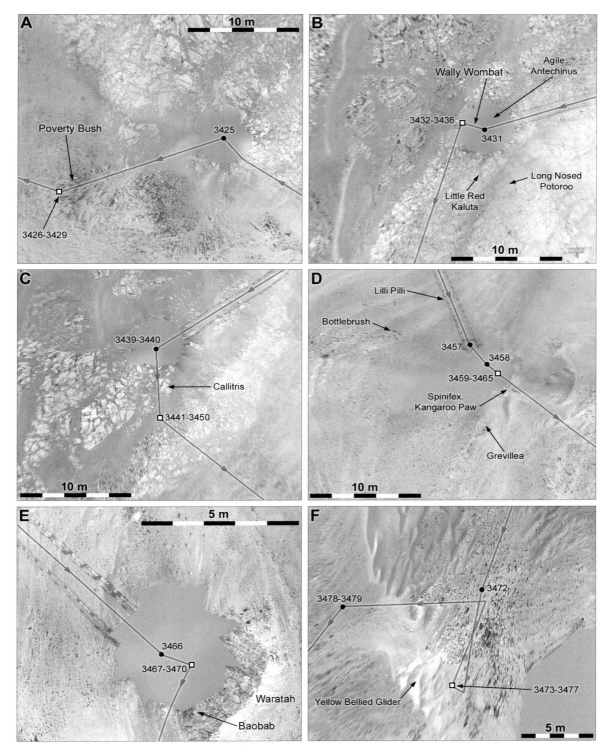

Figure 109. Opportunity activities on Solander Point. **A:** Poverty Bush, sols 3425–3429. **B:** Wally Wombat, sols 3431–3436. **C:** Callitris, sols 3439–3450. **D:** Kangaroo Paw, sols 3457–3465. **E:** Waratah, sols 3466–3470. **F:** Yellow Bellied Glider, sols 3473–3477.

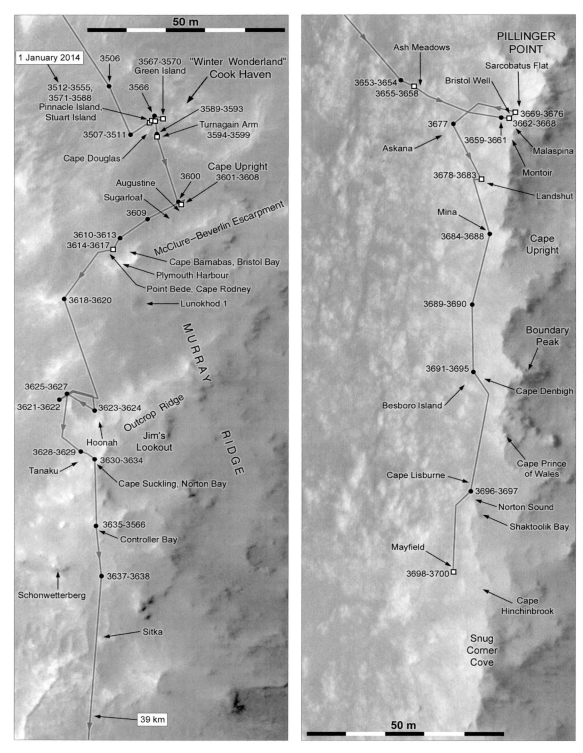

Figure 110. Opportunity activities on Murray Ridge. **A:** Sols 3506–3638, following on from Figure 108. **B:** Sols 3653–3700. The two maps are not contiguous.

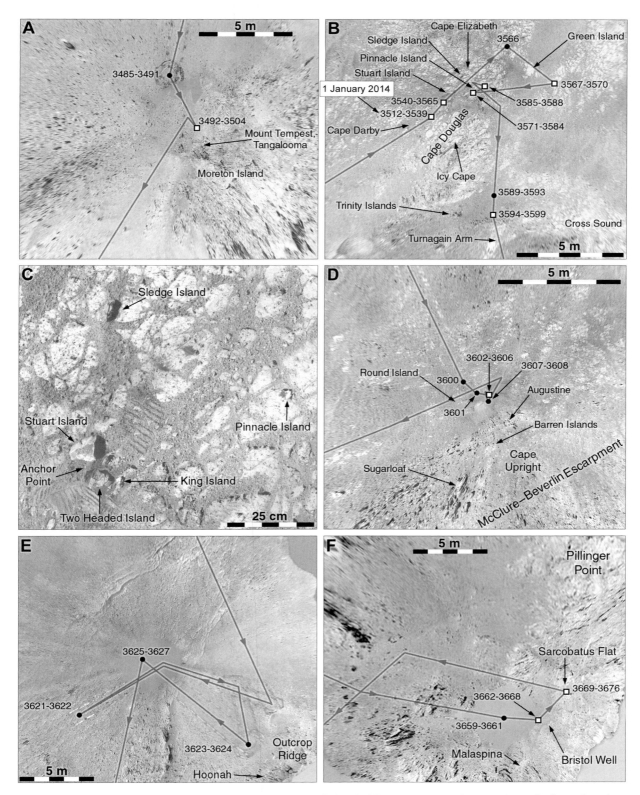

Figure 111. Opportunity activities on Murray Ridge and Pillinger Point. **A:** Moreton Island, sols 3485–3504. **B:** Cape Douglas, sols 3512–3599. **C:** Pinnacle Island targets. **D:** Cape Upright, sols 3600–3608. **E:** Outcrop Ridge, sols 3621–3627. **F:** Bristol Well, sols 3662–3676.

The new location offered access to Anchor Point, a patch of red soil near these rocks. Two places at Anchor Point were examined with ISO on sols 3581–3584, and on sol 3585 the right front wheel was rotated on Sledge Island, crushing it as the other wheels were locked to hold the rover steady. A cloud movie was made on sol 3586, followed by ISO of the broken rock on sol 3587 after Opportunity moved back. When work around Cape Douglas was completed on sol 3589, Opportunity was driven briefly forwards to crack Stuart Island with one of its wheels. Stuart Island had a crack in it and the wheel activities broke the rock along that fracture, exposing a fresh face, which was then imaged.

On sol 3591 the rover moved 4 m south, up the ridge towards an area called Cross Sound, also referred to as Station D. Two other drives on sols 3594 and 3596 brought Opportunity to an outcrop called Turnagain Arm in the Cross Sound area, for a RAT brush and ISO on sol 3598. Work at Cook Haven was now wrapping up, and on sol 3600 (MY 32, sol 218, or 11 March 2014) the rover moved 18 m south to a part of the McClure–Beverlin Escarpment called Cape Upright (Station E) where the surface was littered with dark sharp-edged rocks. Opportunity imaged Sugarloaf, an unusual light-toned rock with a rough surface, on sol 3601 and the dark rock Augustine on sol 3602, driving on each sol to reach Augustine despite slipping on the steep slopes. Three sols of ISO at Augustine followed as the cameras viewed the surroundings.

Sugarloaf would have made another good target, but an attempt to reach it on sol 3607 was halted by the steep rocky slope and elevated wheel currents, and several drives might have been needed to reach it. Ray Arvidson pointed out that Opportunity "is not a mountain goat," so the rock was imaged on sol 3608 and then abandoned. Two drives along the base of the slope on sols 3609 and 3610 brought Opportunity to another part of the escarpment called Plymouth Harbour. A gust of wind substantially cleaned the dusty rover on sol 3607, so Opportunity imaged its solar panels on sol 3611 and 3613 to monitor the cleaning, but an even more dramatic cleaning event occurred on sol 3612. A new ISO target was noted here and named Point Bede, a small rock with possible layering on one face. Short drives on sols 3614 and 3615 put Point Bede within reach of the arm for ISO on the next sol, and imaging targets nearby included Cape Barnabus and a large rock called Bristol Bay.

The science team was looking for a good vantage point from which to image the interior of Endeavour, so now the rover drove south on sols 3618 and 3621 to reach the next escarpment south of McClure–Beverlin. This was initially referred to as Jim's Lookout after team member Jim Rice of Arizona State University, and later named Outcrop Ridge. Pancam images of the rocks here were taken near the end of the last drive, before the rover turned west to stop at the end of the ridge, but on sol 3623 it was directed back to Outcrop Ridge again to inspect the rocky bluff at Hoonah. White parts of the outcrop might have resembled Sugarloaf and could be easier to reach, but they were not sufficiently enticing, so Opportunity was driven south again. One target near here was a crater called Lunokhod 1. The team was very aware that Opportunity was approaching the distance record set on the Moon by Lunokhod 2 in 1973, and worked with Russian planetary scientists, including Irina Karachevtseva of the Moscow State University of Geodesy and Cartography, to determine the exact distance traveled by the old rover.

Opportunity mounted the ridge at Cape Suckling on sol 3630, and for the first time since the approach to Solander Point the top of Cape Tribulation could be seen over the hilltop. A color panorama of Endeavour crater was made from this area before the southward drive resumed on sol 3635, and very clear cloud images were taken on sol 3640. The right front wheel was showing higher currents along here, so the team applied the usual procedures of warming the wheel and reversing driving directions, and also imaged tracks to see if they provided clues to the problem. They would also rest for a while at a suitable location.

Opportunity drove quickly south towards the shallow saddle that divided Murray Ridge into two separate hills, and in order to avoid south-pointing slopes and rougher terrain the rover descended from the ridge crest starting on sol 3641, passing several small craters including X15 and Lunokhod 2 (Figure 110). The latter crater was named as it was close to the point at which, on sol 3643, the rover distance record was finally broken. The latest analyses suggested that Lunokhod 2 drove 39.0 ± 0.1 km, so the record passed to Opportunity at the 39.1 km mark (Figure 112).

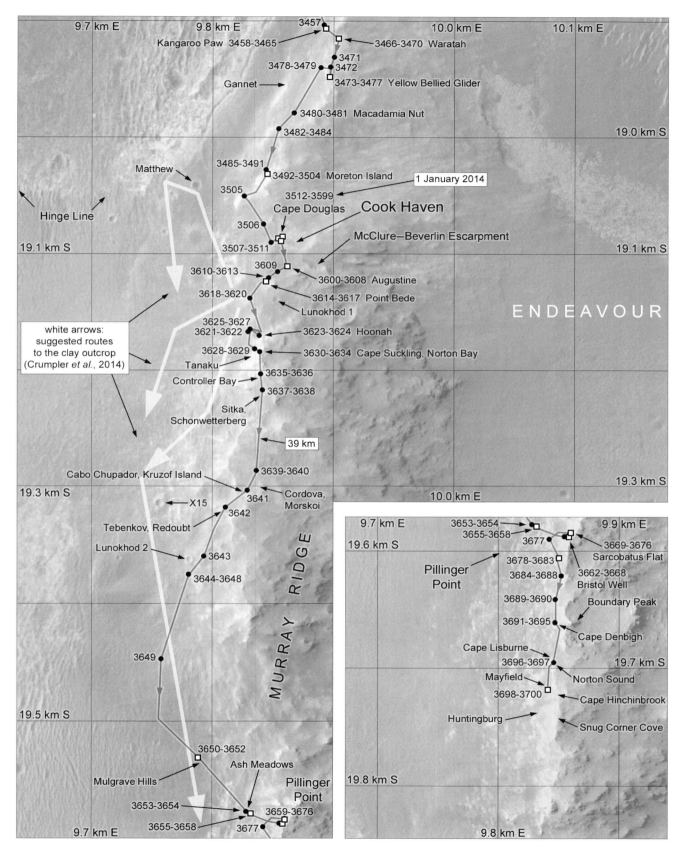

Figure 112. Opportunity route map, section 26. This map follows Figure 106 and concludes Opportunity mapping in this atlas. Additional details are shown in Figures 108 and 109.

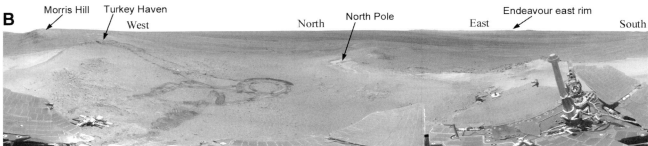

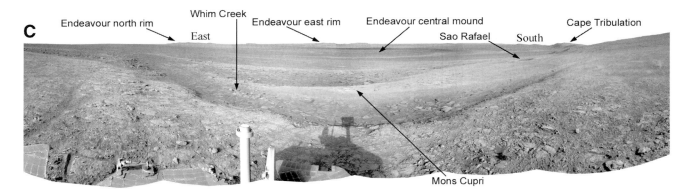

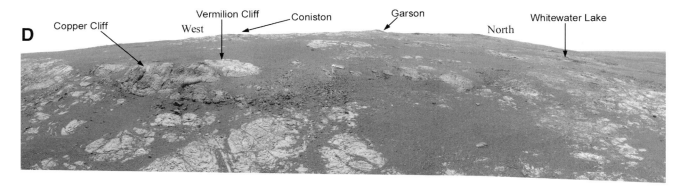

Figure 113. Panoramas from Cape York. **A:** Odyssey crater, sols 2681–2683. **B:** Greeley Haven, sols 2811–2947. **C:** Whim Creek, sol 3020. **D:** Matijevic Hill, sols 3137–3150.

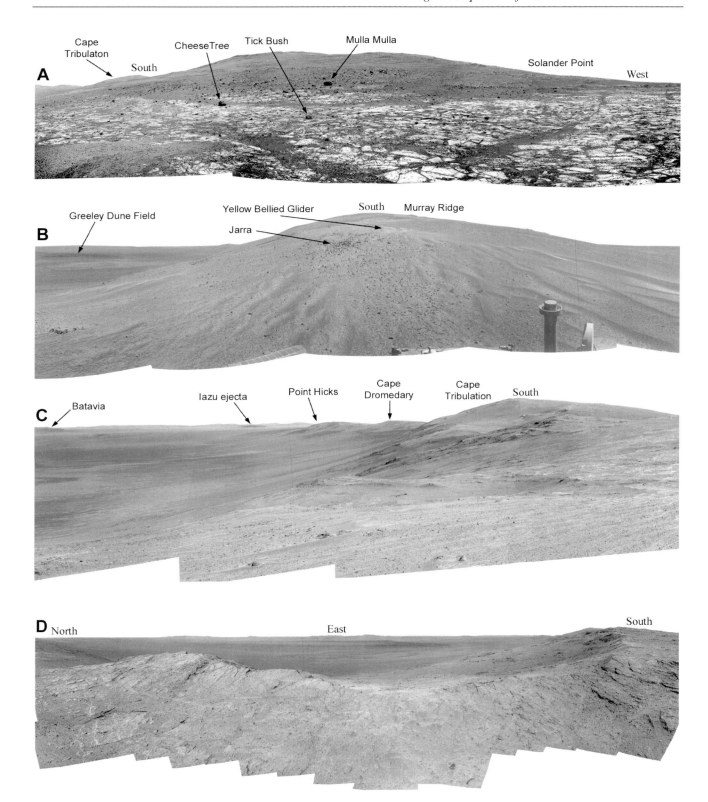

Figure 114. Panoramas from Solander Point. **A:** Solander Point, sol 3389. **B:** Solander Point, sol 3471. **C:** Murray Ridge, sol 3637. **D:** Pillinger Point, sol 3663.

A computer reset on sol 3645 delayed the rover for a few sols and prevented a planned APXS analysis of a ripple, but then Opportunity drove out onto the smooth western flank of Murray Ridge, turned south to bypass the saddle and approached the next target. This was a large outcrop on the western slope of Murray Ridge's southern section in which MRO's CRISM instrument had detected hydrated aluminum-rich clays. The rover stopped briefly to analyze the local dust ripples at Mulgrave Hills on sol 3652, and on the next sol it climbed up onto the broad rocky outcrop. Opportunity now began a year-long process of correcting its clock, which had drifted slowly away from the proper time since landing. Too great a discrepancy could affect operations, so on sol 3655 a change of a few seconds was made, and over a year many small changes would solve the problem.

APXS measured argon on sol 3656 and was used with MI on a rock called Ash Meadows on the next sol, and then Opportunity climbed almost to the summit of the ridge on sol 3659. This area was given the name Pillinger Point to commemorate Colin Pillinger, the British planetary scientist and Principal Investigator for Beagle 2, who had died on 7 May 2014. The rover conducted ISO at three places on a light-toned vein called Bristol Well over sols 3664–3668, and the Pancams observed winds blowing dust on the plains north of Cape York on sol 3666 while making a large panorama of the dramatic scenery. During this period as Opportunity drove along the ridge it made regular measurements of atmospheric opacity with its Navcams to support planning for the future Insight mission.

The next target, Sarcobatus Flat, was reached after a small move on sol 3669 and brushed two sols later for several sols of ISO. This was the matrix of the impact breccia that formed all the outcrops along the ridge. A second target, a clast embedded in the rock matrix, was analyzed on sols 3674–3676, interrupted by a computer fault and a small arm move to position the instrument better. Then, after imaging several targets, including Askana and Montoir, Opportunity moved another 30 m south in two drives on sols 3677 and 3678 to reach a new location for ISO, a soil target called Landshut in the rover tracks. The wheels were now behaving properly again. The analysis of Landshut was done on sol 3679, but two more computer resets on sols 3680 and 3683 delayed the next drive until sol 3684. Part

of the ridge here was named Cape Upright despite the name having been used 60 sols earlier.

A regional dust storm was a concern on sol 3688. It was seen from orbit about 1000 km from the rover, but it had dissipated a few sols later. The rover drove past Boundary Peak (Figure 108) to Cape Prince of Wales, reaching it on sol 3691. Several more outcrops were passed and imaged before Opportunity arrived at Mayfield on sol 3698, and imaged Phobos late that evening. From this location Cape Tribulation was visible again, only 1000 m further south. The most significant exposures of clay detected from orbit were located in a valley extending roughly east to west across the north–south oriented cape. At first that area was nicknamed Smectite Valley from the type of clay detected, but later it took the name Marathon Valley (Figure 47) and became Opportunity's next major target. The full driving distance from Eagle crater to the valley would be about 42 km, close to the official length of a marathon race. On sol 3700 (MY 32, sol 318, or 23 June 2014) the rover began two sols of ISO on Mayfield. This concludes Opportunity coverage in this volume.

2004: Vision for Space Exploration

Following the loss of the Space Shuttle Columbia on 1 February 2003, NASA's goals in space were reviewed under President George W. Bush. On 14 January 2004 Bush announced new goals for NASA, including a return to lunar exploration, with the eventual goal of sending astronauts to Mars at an unspecified date generally expected to be in the 2030s. This was termed the Vision for Space Exploration (NASA, 2004). NASA's Office of Exploration Systems solicited suggestions under a Concept Exploration and Refinement Broad Agency Announcement (CE&R BAA), and among the responses to this were several presentations by aerospace contractor Lockheed Martin (2004, 2005). They were mainly directed at the Moon, but also described a possible human exploration concept for Mars.

The Lockheed Martin concept involved sending three robotic long-range rovers to diverse destinations on Mars, including high-latitude sites, Tharsis, Ares Vallis, the northern plains and southern highlands, and three previous landing sites. These rovers would collect

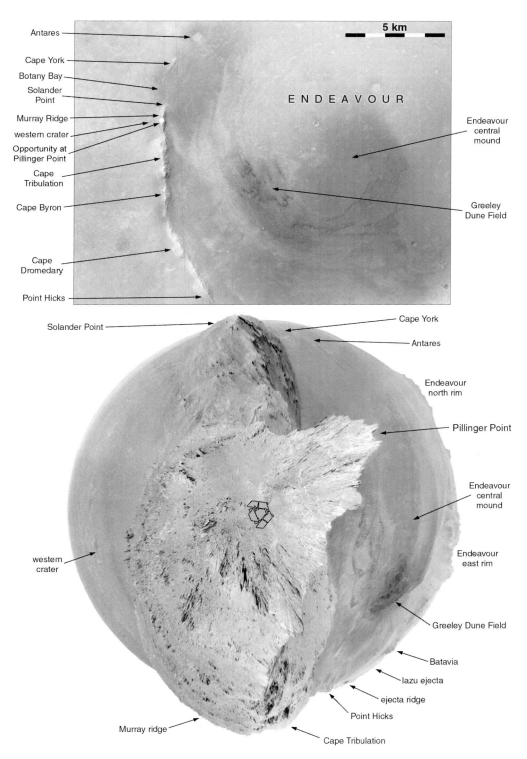

Figure 115. Comparison of CTX image B02_010341_1778_XI_02S005W (top) and a reprojected composite panorama from sols 3659, 3664 and 3665 with exaggerated horizon relief (bottom), showing locations of horizon features and regional feature names. Features on the east rim of Endeavour are located in Figure 101.

samples during traverses of several thousand kilometers before converging on the human landing site in Chryse Planitia, where the crew would collect the samples for analysis and return to Earth. Figure 116A illustrates this exploration concept. The rover landing sites were shown at $65°$ N, $275°$ E (1200 km east of the Phoenix landing site), $14°$ N, $261°$ E and $60°$ S, $350°$ E, and the human landing site was illustrated at $30°$ N, $325°$ E in Chryse Planitia.

After several years of development, NASA's plan to implement the Vision for Space Exploration, called Constellation, was re-examined by an expert panel headed by former Lockheed Martin executive Norman Augustine (Augustine *et al.*, 2009) at the direction of President Barack Obama. The Constellation program had never been adequately funded and the Augustine panel found it to be unsustainable. The main Constellation goal of lunar exploration was replaced with a technology development program directed at sending humans to near-Earth asteroids in the 2020s and to Mars in the 2030s. The Human Exploration of Mars Science Analysis Group (HEM-SAG) study in 2008 prepared more detailed plans for future Mars missions (Figures 123 to 131).

2005: Balloon mission concept

Jones *et al.* (2005) described a balloon mission that could be deployed over high northern latitudes during a period of constant summer sunlight. It would drift around the pole for a month or more, blown by the circumpolar winds, and might make several circuits of the polar cap. The balloon would be deployed during a parachute descent following atmospheric entry and would fill with atmospheric gases until properly inflated, using solar heating for buoyancy. It could deploy payloads at several locations, land for soil or ice sampling and analysis, or undertake prolonged flights for remote sensing or meteorological studies. Figure 116B, derived from an illustration in that paper, shows a deployment at about $78°$ N, $70°$ E, predictions of the balloon's path around the pole based on global climate models of wind patterns, and the midsummer terminator latitude. North of that latitude the balloon would experience continuous sunlight at midsummer. A Montgolfiere balloon like this

might also be able to assist landing in place of a parachute and rockets (Blamont and Jones, 2002).

2005: Mars sample return concept

O'Neil and Cazaux (2000) and Squyres (2005) described a sample return concept involving the May 2003 launch of a NASA lander with a rover and Mars Ascent Vehicle (MAV). The lander would touch down in December 2003 between $5°$ S and $15°$ N, limits determined by solar power considerations. Here the rover would collect a sample during a 90 day surface mission and deliver it to the MAV. A contingency sampling system on the lander itself, provided by the Italian Space Agency, would also collect some material and place it in the MAV, which would launch into Mars orbit in March 2004.

A French mission would be launched in August 2005 to deliver a second lander, rover and MAV and a sample return orbiter to Mars in July 2006. The landing zone for this mission would be from $5°$ N to $25°$ N. The landing mission would collect samples from a second site and place them in orbit. The orbiter would then retrieve both samples and return them to Earth, departing Mars in June 2007 for a gravity-assist Earth flyby in April 2008 and atmospheric entry and landing in October 2008. The total sample weight would be about 1000 g between the two sites and its temperature would be controlled so as not to exceed $50°$ C. The rovers designed for this mission were forerunners of the MER rovers, differing mainly in having a rock coring mechanism mounted under the rover body and a Raman spectrometer on the IDD in addition to the MER instruments. The French mission in 2005 would also deliver four Netlander probes, described below.

10 August 2005: Mars Reconnaissance Orbiter

Mars Reconnaissance Orbiter (MRO) was designed to characterize the planet's climate, analyze the surface composition from orbit at high resolution, and use the highest-resolution camera ever operated at Mars to identify future landing sites. MRO had a mass of 1030 kg (2180 kg when fueled), and consisted of a body containing several instruments and other components, with two

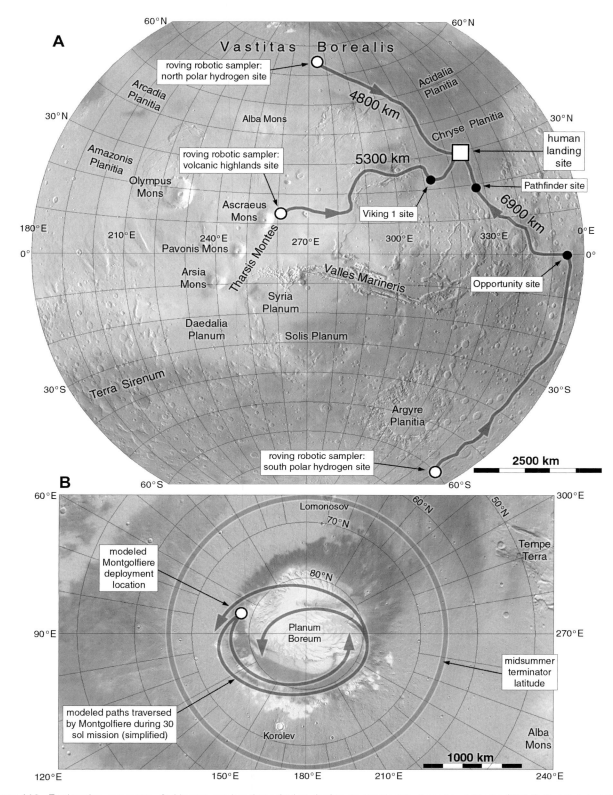

Figure 116. Exploration concepts. **A:** Human exploration mission design suggested by Lockheed Martin (2004). **B:** Modeled 30 sol balloon flightpaths around the north pole, simplified from Jones *et al.* (2005).

large solar arrays and a 3 m diameter high-gain antenna to transmit a high volume of data to Earth. Its instruments were a very high-resolution camera called the High Resolution Imaging Science Experiment (HiRISE), a separate wide-angle Context Camera (CTX), a Mars Color Imager (MARCI) to monitor clouds and dust storms, a visible/near-infrared spectrometer to study surface composition (the Compact Reconnaissance Imaging Spectrometer for Mars, CRISM), an infrared radiometer for atmospheric studies (the Mars Climate Sounder, MCS) and an Italian Space Agency sounding radar (SHARAD) to search for underground layering or other structures and buried water. There were also three engineering instruments: a relay for use with Mars landers and rovers, an optical navigation camera, which took images of Phobos and Deimos from a distance to triangulate a highly accurate spacecraft position for possible use on future missions, and a communication system test package. Mission operations are summarized by Zurek and Smrekar (2007), Graf *et al.* (2007) and Johnston *et al.* (2007, 2011), and the mission summary presented here is compiled from those descriptions and the contemporary MRO mission website.

MRO was launched from Kennedy Space Center at 11:43 UT on 10 August 2005 and made two trajectory corrections on 27 August and 18 November, so accurately that planned corrections later in the flight were not required. MARCI observed Earth and the Moon at very low resolution three days after launch, and HiRISE and CTX imaged the Moon and a star cluster on 7 September to check focus. The spacecraft was ten million kilometers from the Moon at the time. Further star images were taken in December 2005 to monitor the camera focus during flight. Optical navigation images were taken between 28 and 3 days before arrival, 103 images of Phobos and 479 of Deimos. MRO entered Mars orbit on 10 March 2006 (MY 28, sol 46) with a 27 min engine burn beginning at 21:24 UT.

The initial 425 by 35 000 km orbit was polar with a 35.5 hour period. Aerobraking was used to modify the orbit between 23 March and 30 August 2006 when a thruster burn lifted periapsis out of the atmosphere. MCS imaged the spacecraft at low resolution on 25 March during aerobraking. The orbit was adjusted again on 11 September, and then the SHARAD radar antenna was deployed and the CRISM protective cover was removed.

The mapping orbit was 255 by 320 km with a period of just over 2 hours and periapsis over the south pole, and it was Sun synchronous so the surface was always seen at 15:00 local time. Mars Odyssey was in a similar orbit, crossing the equator at 17:00 local time initially and 15:45 after a 30 September 2008 adjustment, but the two spacecraft moved in opposite directions (the orbit planes were nearly 180° apart).

Science operations started on 7 November 2006 after solar conjunction. Test images were taken on 23 March 2006, but otherwise no observations were made during aerobraking. A highly successful orbital mission followed, in which thousands of exquisitely detailed images were obtained. Past and future landing sites were seen in unprecedented detail, and HiRISE images made it possible to identify the Viking 2 landing site for the first time and to refine the Mars Pathfinder location. HiRISE imaged the Earth and the Moon, even the Moon appearing as a distinct crescent, on 3 October 2007. Despite the distance, excellent HiRISE images of Phobos and Deimos were obtained on 23 March 2008 and 21 February 2009, respectively, better in places for both satellites than many of the images taken by Viking, Mars Global Surveyor and Mars Express during their close approaches. Figure 117 shows examples of MRO images of Mars, and images of Phobos and Deimos are presented in Figure 204.

The primary science phase of MRO's mission ran from November 2006 to December 2008, after which MRO entered its relay and extended science phase, including responsibility for relaying data from landers and rovers. This was interrupted in 2009 with faults on 23 February, 4 June and 6 August, putting the spacecraft into a safe mode, from which it recovered each time in just a few days. After another fault on 26 August, the spacecraft was left in safe mode until the underlying problem could be corrected. It was put back in service on 8 December 2009. The first extended science phase lasted until September 2010 and was followed by several additional extended missions.

All instruments performed superbly. On 27 August 2011, the electronics associated with one of 14 charge-coupled device (CCD) detectors at the edge of the HiRISE array failed, placing the spacecraft in safe mode. After a second attempt to use that CCD on 6 September caused another safe mode, its use was discontinued. This had no effect on data acquisition because, although they were narrower, the images could be lengthened to cover the same area. MRO directly supported surface missions by certifying and monitoring landing sites, providing high-resolution images for rover traverse planning, imaging Phoenix and MSL during their parachute descents, imaging hardware and rover tracks on the surface of the planet, and relaying data from surface missions. By late 2014 HiRISE had taken more than 36 500 high-resolution images and CTX had obtained almost 70 000 images covering over 93 percent of the planet. CRISM had observed about 83 percent of the planet in a 72-channel survey mode at 200 m/pixel, and had taken over 26 000 hyperspectral images at much higher spatial and spectral resolutions.

MRO had two identical inertial measurement units to provide data for attitude control. When a gyroscope in one unit was approaching the end of its life, the spacecraft switched to the other system on 12 August 2013. The first unit was still functional for the time being, so switching now preserved it for emergency backup use in future.

Asteroid 2007 WD5, discovered on 20 November 2007, was thought initially to have a small chance of striking Mars somewhere in its northern hemisphere on 30 January 2008. Later analyses showed this would not occur, but the possibility of observing an impact immediately after it occurred provoked considerable interest and some concern. Another potential Mars impactor was Comet C/2013 A1 (Siding Spring), discovered by Australian comet observer Rob McNaught at the Siding Spring Observatory, New South Wales, on 3 January 2013. It passed only 138 000 km from Mars on 19 October 2014 (MY 32, sol 433), but the earliest predictions suggested a shorter distance and gave it a very small chance of impacting Mars.

As a precursor which would allow evaluation of comet imaging methods a year in advance, a relatively close flyby of Comet ISON was observed in 2013. ISON (International Scientific Optical Network) is an international project with telescopes in many countries, but the comet was discovered by Vitaly Nevski and Artyom Novichonok using the ISON telescope in Kislovodsk, Russia. Images and other data were obtained by MRO, Mars Express, Opportunity and Curiosity, though the rover images did not show the faint comet. MRO's ISON

images were taken on 20 August, 28 and 29 September, and 1 and 2 October 2013. The closest approach to Mars was 10.8 million kilometers on 1 October 2013 (MY 32, sol 60), and the first successful HiRISE image was taken on 29 September. ISON fragmented just after its perihelion on 28 November 2013, having passed only 1.2 million kilometers above the solar photosphere.

Siding Spring was thought to pose a threat to orbiters as well as an appealing target. Its tail would envelop the planet briefly after closest approach, and dust particles in the tail would impact at velocities around 56 km/s because the comet's orbit was retrograde relative to the planets. Plans were made to position NASA's active Mars orbiters behind Mars during the most vulnerable period, or turn them to face in the least vulnerable orientation. MRO was moved in its orbit on 2 July and 25 September 2014 to place it behind Mars during the encounter with the tail.

HiRISE and CTX imaged the comet, though the CTX images showed little detail. Over 50 HiRISE images were taken, revealing the nucleus to be about 400 m across, smaller than had been anticipated, with an 8 hour rotation period. The 130 m resolution and the opacity of the coma prevented HiRISE from imaging any surface features. SHARAD radar data were blurred by effects of the tail-induced ionosphere, and CRISM detected the comet but did not provide useful compositional data.

2005: Netlander

Several Mars geophysical and meteorological network missions had been described in the 1990s, including Marsnet, MESUR and InterMarsnet (Figures 142–148 in Stooke, 2012). Network missions were so desirable that plans continued to be developed in the next decade. Netlander was a proposed surface network mission studied by the French space agency CNES (Centre National d'Études Spatiales) with possible contributions by NASA. It was intended for an August 2005 launch and July 2006 arrival, accompanied by an orbiter. An earlier plan for this network assumed a launch in 2003, not associated with the sample return mission described above. A later version called Mars Premier was intended to launch in 2007 as a precursor of a Mars sample return mission, testing automated rendezvous hardware and

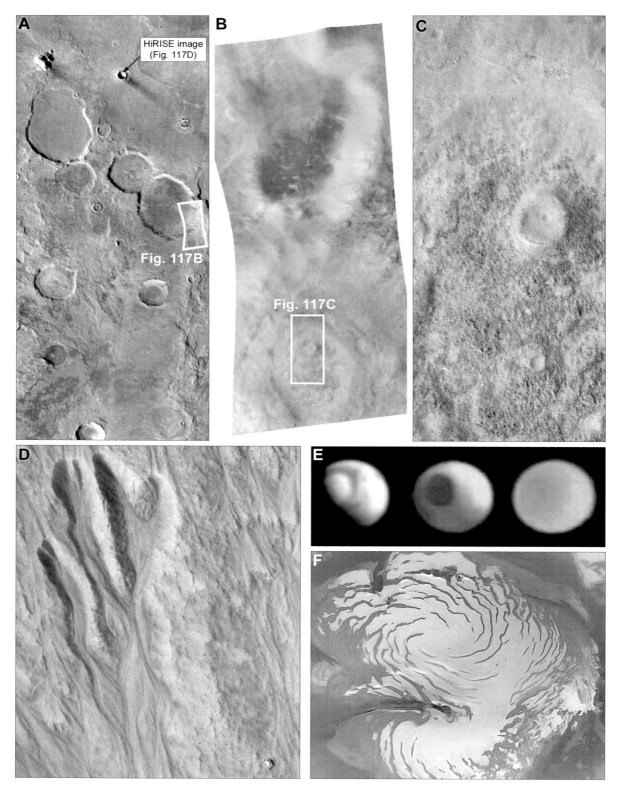

Figure 117. Mars Reconnaissance Orbiter images of Mars. **A:** Figure 188A from Stooke (2012) for context. **B:** CRISM observation 0000A42D, visible and near-infrared composite. **C:** Detail of CTX image G23_027168_2269_XN_46N050W. **D:** Detail of HiRISE image ESP_016553_2310 showing gullies in the small crater indicated in **A**, image width 250 m. **E:** MCS aerobraking phase images of the northern hemisphere of Mars, visible wavelengths (left) and infrared (middle and right). **F:** MARCI mosaic of the north polar cap, courtesy NASA/JPL/MSSS.

Table 28. *Netlander Landing Sites*

Source	Name	InterMarsnet site	Location
Marsal *et al.* (1999),	Lycus Sulci	near 59	27.5° N, 230° E
"reference network"	Memnonia	near 148	12.5° S, 200° E
	Tempe Terra South	near 18	35° N, 290° E
	Hellas East	near 277	32.5° S, 85° E
Barriot *et al.* (2001),	Isidis Planitia	near 100	16.0° N, 84.9° E
antipodal site not specified	Tyrrhena Terra	near 211	12.1° S, 94.8° E
	Utopia Planitia	near 31	35.1° N, 118.3° E
Lebleu and Schipper (2003)	East Hellas	near 277	32° S, 85° E
	Isidis	near 29	30° N, 90° E
	Amazonis North	near 45	30° N, 205° E
	Amazonis South	near 137	10° S, 205° E

procedures in Mars orbit. That plan assumed that the sample return mission described above in association with Netlander had not occurred. The Russian firm NPO Lavochkin also studied a 30 kg Phobos lander to be carried on Mars Premier. Netlander was not funded for flight and the concept was later taken over by a Finnish design called MetNet.

Netlander would require four landers equipped with cameras, seismometers, ground-penetrating radar, atmospheric sensors and magnetometers. The landers would be released from the orbiter individually during the approach to the planet, after which the orbiter would divert into a flyby trajectory from which it would brake into orbit. The nominal deployment dates would be 28, 24, 20 and 16 days before arrival, and the backup dates 20, 16, 12 and 8 days before arrival. The landers would use parachutes and airbags to reach the surface. The orbiter would provide communications support to the landers, and also to possible future landers such as Mars Scout missions, from a 1000 km circular orbit inclined 95° to the equator (Matousek, 2001).

Landing sites for Netlander are listed in Table 28 and plotted in Figure 118A. Marsal *et al.* (1999) described what they called a reference network, which was useful for mission planning but was subject to future confirmation. Sites had to lie below +2 km elevation and between 40° N and 40° S. Limits of 35° N and 30° or 35° S, and a maximum of +1 km elevation, were also discussed. The landing ellipse size was 500 by 100 km. In this list the first three sites form a triangle around Tharsis and the fourth is on the edge of Hellas, roughly antipodal to

Tharsis, to favor seismic studies of the planet's core. The individual locations are similar to InterMarsnet sites (Table 56 in Stooke, 2012) and the array somewhat resembles the InterMarsnet baseline array. Barriot *et al.* (2001) assumed a different hypothetical configuration, effectively reversed on the planet, for simulations of radio tracking as part of a study of the planet's rotation and ionosphere. A different description of Netlander site selection was presented by Lebleu and Schipper (2003) at an International Planetary Probe Workshop held in Lisbon, Portugal, in October 2003. Their latitude limits were 35° N to 35° S and their four sites included one in Hellas, but the sites are arranged in two pairs, not a triangle and a lone antipodal site. The locations are taken from an illustration and are only approximate.

2000s: Mars Scout proposals

Mars Scout was a program similar to Discovery (Table 60 in Stooke, 2012), instituted in 2001 and intended specifically for Mars, which was then excluded from further Discovery competitions. The intention was to have numerous small spacecraft flying at frequent intervals rather than a few large infrequent missions (Blaney and Wilson, 2000). This continued with two cycles of mission proposal and selection (Phoenix and MAVEN, launched in 2007 and 2013, respectively) until 2010, when the program terminated and Discovery was opened up to Mars proposals again.

Table 29. *Mars Scout 2007 Proposals Selected for Further Study, June 2001*

Mission	Description
SCIM	Sample Collection for Investigation of Mars. A flyby passing through the atmosphere to collect samples of dust and gas in aerogel and return them to Earth
Kitty Hawk	Three or four gliders to study the composition and layering of the Valles Marineris walls; 120 km routes beginning outside the canyons, lasting 10 min, taking 20 images each, one probably targeted over the Candor Chasma hematite deposit
Urey	Mars 2001 lander and Marie Curie rover with instruments to determine the absolute ages of surface materials in the Cerberus region or Lunae Planum
MACO	Mars Atmospheric Constellation Observatory. A network of microsatellites orbiting Mars to characterize the three-dimensional structure of the atmosphere
Artemis	An orbiter would deploy three or four small landers, each landing with airbags and carrying a robotic arm and nanorover; landing sites include polar regions and equatorial layered deposits, for climate studies and to seek water and organics; polar lander would obtain much of the planned Mars Polar Lander data
Mars Environmental Observer	Orbiter to explore the role of water, dust, ice and other materials in the atmosphere to understand the hydrologic cycle
Pascal	A network of 18 or 24 hard-landing surface weather stations (Figure 118)
Mars Scout Radar	Synthetic aperture radar imaging to map the surface and shallow subsurface
The Naiades	Two or four hard landers within 20 km of each other to seek subsurface liquid water using low-frequency electromagnetic sounding, possibly near Dao Vallis
CryoScout	Lander using heat to melt polar ice and descend at least 100 m through a polar cap, measuring composition and stratigraphy (Figure 120B)

A workshop was held from 21 to 24 May 2001 in Pasadena to allow investigators to present their proposals for a mission in 2007. Some 43 missions were described (Matousek, 2001), but in general the proposals are considered proprietary and many have not been publicized. Table 29 lists the ten Mars Scout missions selected by NASA for further study in June 2001, and some of the rejected candidates are identified in Table 30. Several of the missions are described briefly below, particularly if landing sites were identified.

Urey was a lander and rover mission designed to measure the absolute age of an igneous rock unit on Mars. It would use the Mars Surveyor 2001 lander, which eventually became Phoenix, and the miniature rover from that mission, deployed from its stowed position on the lander deck to the surface by a robotic arm. The rover, Marie Curie, would collect a bedrock sample with an ultrasonic drill and return it to the lander for potassium–argon dating. An imaging spectrometer on the lander would identify suitable targets for drilling. Analysis would also identify minerals and any small traces of organic compounds. Cameras would image the landing site, which was intended to be in the very young volcanic rocks of the Cerberus region (Figure 120A), not far from the Athabasca site considered for MER (Figure 3B). An alternative location was Lunae Planum, an extensive region of Hesperian lava flows and wrinkle ridges, but these older lavas might be more altered by weathering, possibly complicating the analysis (J. Plescia, personal communication, 3 April 2013).

After landing, the solar panels and rover would be deployed and all systems checked and prepared for operation. A panorama of the site would be transmitted to help identify targets and routes for the rover. Target distances would be on the order of 10–20 m from the lander, requiring at least a sol of driving to reach. On the sol after arrival, and making use of images from the rover to choose a drilling spot, the sample would be collected. The rover would return to the lander over the next sol or two and deliver its sample by ejecting it into the robotic arm scoop. Analysis and data transmission would occupy another sol as the rover battery was recharged by its solar panel. This cycle would be repeated up to 20 times with

Table 30. *Mars Scout 2007: Partial List of Rejected Candidates*

Mission	Description
VERBAL	Very Low Frequency Radio Beacon on an Aerobotic Laboratory
MAGE	Mars Airborne Geophysical Explorer. An aircraft to explore Valles Marineris; modified from a 1998 Discovery proposal (Figure 150 in Stooke, 2012)
Piccard	A helium balloon-borne geoscience mission
Polar	Solar-heated Montgolfiere balloon for polar regions
SSTAMP	Subsurface Science from Targeted Mars Penetrators
BEES	Bio-inspired Engineering of Exploration Systems (Figure 193 in Stooke, 2012)
AMEBA	Autonomous Mars Environmental and Biological Assay. Reusing the 2001 Mars Lander, then in storage but used instead for Phoenix; a mini-greenhouse on top of the lander would hold an *Arabidopsis thaliana* (thale cress) plant
Mother Goose	A wing-shaped balloon would descend through the atmosphere, land and release a rover; that rover would drive to a target, especially an opening such as a lava tube cave, and deploy a group of small mobile robots to survey the target
MarsLab	A cluster of instrument packages to look for life on or below the surface of Mars
Long Day's Drive	Rover operating during a period of prolonged sunlight near a pole
AME	Airplane for Mars Exploration. Repeat of a Discovery proposal

Table 31. *Mars Scout 2007 Finalists, 2003*

Mission	Description
Phoenix	Lander to seek ice in surface materials at high northern latitudes
MARVEL	Mars Volcanic Emission and Life. A polar orbiter using solar occultations at orbital sunrise and sunset to characterize atmospheric particles
SCIM	Flyby, atmospheric gas and dust sample return
ARES	Aerial Regional-scale Environmental Survey. An aircraft designed to map magnetism in the southern highlands, and atmospheric composition

different targets, limited by the number of sample receptacles in the analytical instrument. The nominal mission would last for 90 sols.

By 2003 these missions and several new candidates had been reduced to a shortlist of four (Table 31). MARVEL would place an orbiter in a near-polar orbit, with an infrared spectrometer to examine light from the Sun to probe the atmosphere at sunrise and sunset. The polar orbit provided a sunrise and sunset on every orbit and also over time would give data over the whole surface of the planet. It would be especially sensitive to methane, a possible signature of life or recent volcanic or hydrothermal activity. A camera on MARVEL would show cloud cover at the time of the atmospheric readings. The mission would last a full Mars year, and after it

the spacecraft would be moved into a higher orbit to prevent contamination of the planet.

SCIM was designed to sample the atmosphere of Mars, including atmospheric dust, during a low pass through the atmosphere during a flyby. The mission would make two high-speed passes over the same near-equatorial area of Mars one year apart. On the first it would scan the area to characterize the atmosphere. On its second close pass one year later, it would fly low enough to take dust and air samples. This pass would be only about 40 km above the surface. SCIM would collect a liter of atmospheric gas and at least 1000 dust particles 10 µm or more in diameter, plus millions of smaller particles. Aerogel collectors would trap the dust. This technology was used successfully in the Stardust mission

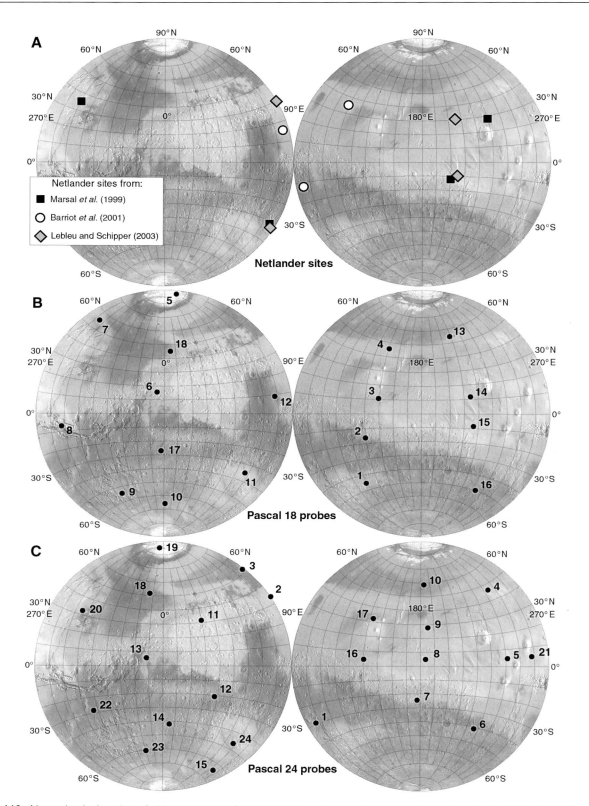

Figure 118. Network mission sites. **A:** Netlander sites from Table 28. **B:** Pascal 18-probe network from Table 32. **C:** Pascal 24-probe network from Table 33.

to return comet dust to Earth in 2006. About 10 months after the second Mars flyby, SCIM would use its thrusters to target the return trajectory to Earth. Two years later it would fly past Earth and eject the capsule holding the samples. Leshin (2003) described sample collection in April 2007 at 14° S, 290° E (70° longitude was stated and is assumed to be west longitude) and an altitude within 2.5 km of 37.2 km. This is on the southern edge of Valles Marineris near Melas Chasma (Figure 120C).

Pascal would deploy a network of 18 or 24 miniature landers on Mars to monitor the climate for 10 Martian years. Table 32 and Figure 118B identify possible landing sites for the 18-lander option if launched in 2007 (Miller *et al.*, 2003; Haberle, 2003). Table 33 and Figure 118C show a possible network of 24 stations for the same date (Haberle *et al.*, 2000, 2003). A larger number of landers could achieve better results, and a minimum mission might launch only 12 landers, with 8 surviving the landing, and still fulfil the basic goals. The 9 kg probes would land using parachutes and airbags, taking descent images to establish their locations.

On the surface, pressure, atmospheric opacity, temperature, wind speed and water vapor data would be collected as often as every 15 min for a detailed global study of the climate and weather. A panoramic camera would transmit images every 30 sols to observe any changes in the landing site. The deployment sequence was designed to release six probes at once towards landing sites arranged in a circle near the limb of the planet's disk as seen by the approaching spacecraft (Miller *et al.*, 2003). As each group of six probes was released, the rotation of the planet between deployments would distribute the probes in the desired global network. The deployments (sets in Tables 32 and 33) are identified by number to differentiate them, but the numbers may not match the order of deployment.

The ARES (Aerial Regional-scale Environmental Survey) mission was intended to fly a powered aircraft on a 500 km trajectory at an altitude between 1000 and 2000 m, collecting data on the atmosphere, surface and magnetic field. The mission had to be conducted in daylight and in the absence of dust storms, constraining the local time and season of arrival, and the magnetic

Table 32. *Pascal 18-Site Network for 2007 Launch (Haberle, 2003)*

Site	Set	Location	Site	Set	Location	Site	Set	Location
1	1	47° S, 130° E	7	2	55° N, 285° E	13	3	54° N, 210° E
2	1	17° S, 140° E	8	2	8° S, 287° E	14	3	12° N, 215° E
3	1	11° N, 150° E	9	2	55° S, 315° E	15	3	8° S, 218° E
4	1	45° N, 150° E	10	2	67° S, 1° E	16	3	50° S, 233° E
5	1	85° N, 70° E	11	2	36° S, 67° E	17	3	27° S, 358° E
6	1	15° N, 355° E	12	2	7° N, 77° E	18	3	45° N, 6° E

Table 33. *Pascal 24-Site Network for 2007 Launch (Haberle* et al., *2000)*

Site	Set	Location	Site	Set	Location	Site	Set	Location
1	1	29° S, 98° E	9	2	27° N, 186° E	17	3	32° N, 143° E
2	1	33° N, 90° E	10	2	57° N, 183° E	18	3	52° N, 348° E
3	1	52° N, 88° E	11	2	32° N, 30° E	19	4	84° N, 335° E
4	1	47° N, 244° E	12	2	21° S, 38° E	20	4	32° N, 295° E
5	1	4° N, 242° E	13	3	4° N, 349° E	21	4	4° N, 258° E
6	1	42° S, 226° E	14	3	43° S, 5° E	22	4	29° S, 307° E
7	2	25° S, 177° E	15	3	65° S, 77° E	23	4	61° S, 341° E
8	2	4° N, 183° E	16	3	4° N, 140° E	24	4	47° S, 69° E

field goals dictated a site among the planet's largest magnetic anomalies (Figure 154A in Stooke, 2012).

ARES would launch in September 2007 and arrive a year later at solar longitude $L_s = 121°$ and local time about 14:30. Modified trajectories could reach different latitudes within the desired longitude region, with more southerly targets accessible later in the season and earlier in local time. The most important constraints were to avoid conjunction around $L_s = 169°$ and the start of the dust storm season at $L_s = 180°$, within range of Mars Reconnaissance Orbiter for data relay. On arrival at Mars, it would enter the atmosphere and eject a capsule containing a folded aircraft. A parachute would slow the capsule, which would then release the aircraft. The ARES aircraft would unfold automatically and fire its small rocket engine.

Typically, remote geophysical sensors can resolve features at a scale comparable with their altitudes, so an orbiting magnetometer resolves magnetic anomalies at a scale of about 100–200 km. ARES would operate at a height of only 1000–1500 m, resolving the local structure of the magnetic anomalies at kilometer scale. ARES would fly at a speed of about 500 km/h, making three parallel flights over the selected magnetic anomaly. Two flights over the same terrain in opposite directions are necessary to enable the magnetic field of the vehicle to be separated from the surrounding field. It would then observe other targets before running out of fuel and crashing. The flight would last only an hour and all data transmission would have to be accomplished during that time, as the crash would probably not be survived.

The ARES instruments were a pair of magnetometers providing 400 times the spatial resolution of MGS, a mass spectrometer to analyze the atmosphere during the flight, and a down-facing spectrometer to map the composition and mineralogy of the surface. A forward-looking camera would image areas studied by the spectrometer, and a video camera mounted in the aircraft tail would provide a view of the flying craft and its route. Data would be relayed through the carrier spacecraft (bus) that supported ARES during its cruise to Mars, which would fly past Mars after deploying the entry capsule. Existing orbiters would serve as a relay backup.

Figure 119 shows two possible ARES routes within the region of interest coinciding with the largest magnetic anomalies, extending from 20° S to 60° S and

150° E to 210° E (Figure 119A). Kenney and Croom (2003) and Braun *et al.* (2004, 2006) illustrated a route starting with atmospheric entry at about 35° S, 189° E. The aircraft would fly northwards to 28° S with a loop around 31° S to 32° S to provide three passes over the magnetic gradient in this area, crossing from positive in the south to near zero values in the north (Figure 119B). Kuhl (2008, 2009) showed a different route further south, entering the atmosphere at about 51° S, 180° E and flying southwards with a loop around 52° S to 53° S (Figure 119C). This crossed a very steep magnetic gradient between negative and positive anomalies (Connerney *et al.*, 2012).

CryoScout was a proposal that would use the Mars Surveyor 2001 lander, like its competitor Phoenix, which eventually won this competition. It would land near the north pole, high on the ice cap in Planum Borealis. Its main goal was to deploy a small instrumented probe, which would melt its way down as much as 200 m into the ice cap, analyzing layering in the polar cap to study millions of years of climate change (Zimmerman *et al.*, 2002). The lander would also provide surface images and meteorology data. Launch was planned for mid-August to early September 2007 and arrival at Mars between 18 May and 18 June 2008, according to an unpublished JPL presentation made on 3 May 2002 by Stephen Saunders, the mission's Principal Investigator. The mission would operate on the surface for 90 sols, centured around solar longitude $L_s = 93°$, just three sols after the summer solstice. The landing zone was between 83° N and 88° N.

The description of CryoScout by Zimmerman *et al.* (2002) differed slightly from that by Stephen Saunders. Its launch date was 4 September 2007 and the landings were to occur later. Two possible landing sites were considered, one at 85.2° N, 40° E with a landing on 18 July 2008, and the second at 87.2° N (longitude unspecified) with a landing on 18 August 2008. The July landing was preferred, as it allowed a longer period of operations. Figure 120B shows this region and an ellipse illustrated by Zimmerman *et al.* (2002), which is centered at 83.8° N, 35° E. The preferred site for CryoScout was the middle of the ridge connecting the two main domes of the northern ice cap, where a large flat area was almost completely free of rocky material. Other areas of the polar cap contained dark lanes of exposed layered rock, forming a prominent spiral pattern in the ice cap,

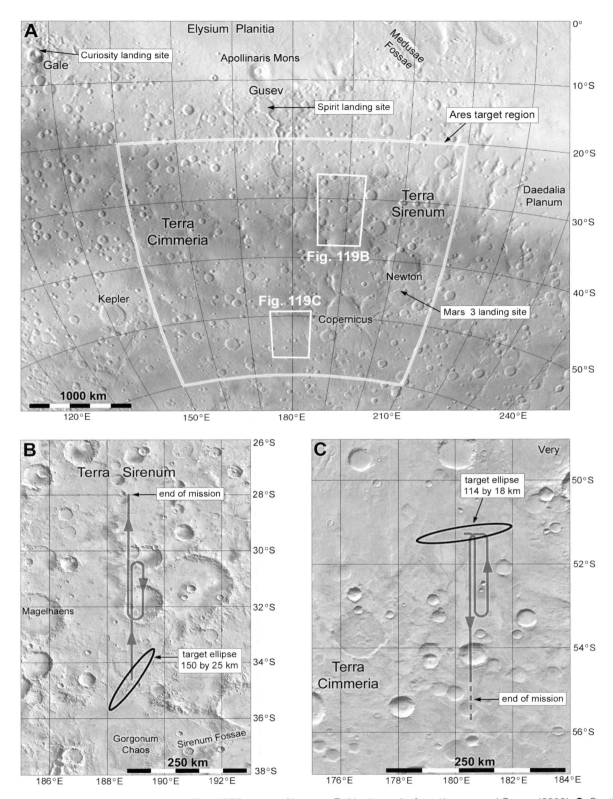

Figure 119. ARES airplane flightpaths. **A:** The ARES region of interest. **B:** Northern site from Kenney and Croom (2003). **C:** Southern site from Kuhl (2008).

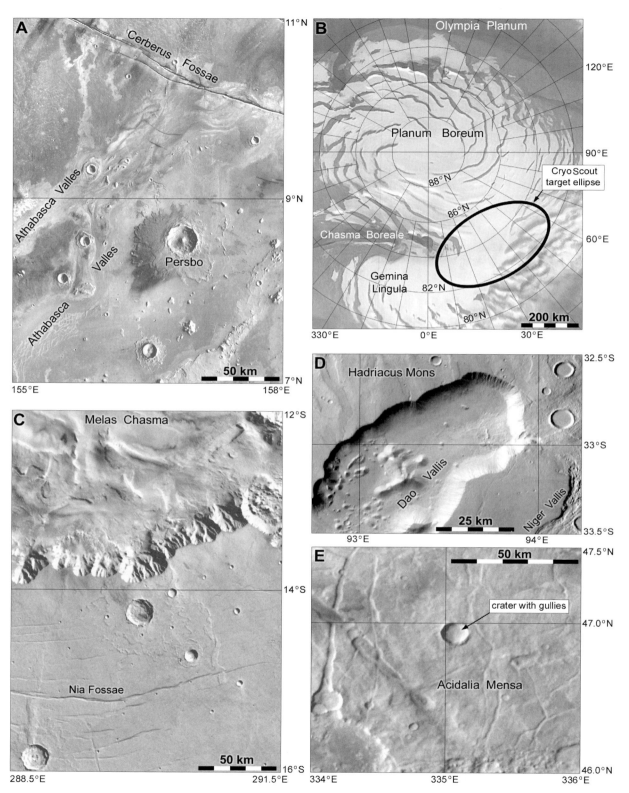

Figure 120. Mars Scout candidate sites. **A:** One Urey target: lava flows in the Cerberus region (THEMIS infrared mosaic, inverted shading). **B:** CryoScout target ellipse near the north pole (MARCI mosaic, courtesy NASA/JPL/MSSS, modified by P. Stooke). **C:** SCIM closest-approach region during its sample collection flyby (Viking MDIM v2.1 mosaic). **D:** Naiades landing region near Dao Vallis. **E:** Final Naiades target in Acidalia Planitia. **D** and **E** are THEMIS infrared mosaics.

increasing the possibility of landing in ice-free material. The more northerly landing site was less satisfactory for that reason, as well as for its reduced mission length.

Another Mars Scout polar mission called Atromos was described at the Sixth Planetary Probe Workshop held in Atlanta, Georgia, on 23–27 June 2008 (Iskander *et al.*, 2008). This concept, developed at NASA Ames Research Center, involved a novel "mechanical airbag" landing system and an ultrasonic drill (Murbach *et al.*, 2007). Two small landers would be launched as secondary payloads traveling with a European Space Agency (ESA) Mars mission such as ExoMars, and ejected from the carrier during the approach to Mars. After a parachute descent, the final landing would be softened by an array of spring-loaded spokes, the "mechanical airbag," which would not require inflation and was simpler than the MER airbags.

One probe would land at the north pole, the second several hundred kilometers away to measure latitudinal changes in the atmosphere. The meteorology data would extend the Phoenix results across different latitudes and also by operating through at least part of the polar winter and possibly an entire Mars year. It might also be feasible to target the south pole instead. The drill would obtain ice samples for analysis. No specific sites were identified, assuming that the north polar site was only intended to be in the vicinity of the true pole. A later version of the Atromos mission, really a very different mission with the same name (Murbach *et al.*, 2010, 2012), would be targeted at Hellas.

The Naiades were a set of four hard landers, designed to cope with a rough landing to simplify the hardware requirements and reduce mass and cost. Versions of the mission with only two landers were also considered. The landers would search for subsurface water using electromagnetic sounding. In early planning a possible landing site was at the head of Dao Vallis, where heat from magma associated with the nearby volcano Hadriacus Mons may have melted ice to trigger the vast floods that carved Dao and other nearby valleys. Four landers would be released as a carrier spacecraft approached Mars, and would land in a square pattern roughly 16 km across (Figure 120D). By the time the proposal was submitted, the proposed landing site had been moved to 47° N, 335° E in Acidalia Planitia (Figure 120E) where shallow ground water was suspected to be feeding contemporary

flows in crater wall gullies. The target was a 6 km diameter crater with numerous gullies seen in Mars Orbiter Camera (MOC) images E1001090 and R1303765, and the landers would set down around it to probe the aquifer suspected of feeding the gullies (R. Grimm, personal communication, 4 April 2013).

The winner of this Scout competition was Phoenix, described below.

2000s: Low-cost planetary missions

In September 2003 a meeting on low-cost planetary missions was held at ESA's European Space Research and Technology Centre (ESTEC) in Noordwijk, South Holland. Papers presented at that meeting were published in 2006, and a few are summarized here to illustrate a range of low-cost mission concepts suitable for periods of limited planetary exploration budgets, or for increasing the number of fundable missions that could fly in a given period.

Kerstein *et al.* (2006) proposed an orbiter with a very small lander weighing only 15 kg. The orbiter would obtain high-resolution images of Mars, Phobos and Deimos from an elliptical orbit and map magnetic and gravitational anomalies in the crust. The atmosphere, ionosphere and radiation environment would also be studied from orbit for a full Martian year. The lander, deployed from orbit, would image the surface and measure environmental conditions, including temperature, pressure, wind, illumination and radiation at the surface during a mission lasting at least seven sols. The landing site was not specified but would be between 25° N and 25° S.

Walker *et al.* (2006) described three mission concepts. One was a set of four small orbiters using occultations for global atmospheric monitoring from polar orbits. The second, Mars Phobos and Deimos Survey, was a Mars orbiter that would make flybys of Phobos and Deimos to study their compositions and origins. The Phobos–Deimos mission would enter Mars orbit and take four months to adjust its orbit to match Deimos before spending a month near that outer satellite. Then it would take two further months to adjust its orbit to match that of Phobos before spending a month studying the inner satellite. An option for placing a small lander on Phobos, based on the Philae lander from ESA's Rosetta mission to a comet, was also considered. The third, the Mars Sub-Surface Ice mission, was a larger orbiter carrying four penetrators designed to measure the composition of sub-surface ice. The penetrators would be deployed from low orbit, not while approaching Mars, and would carry small cameras to survey the landing sites while the subsurface portions of the penetrators analyzed ice and other materials. Two would land near the equator and two at high latitudes, but specific locations were not identified.

Ellery *et al.* (2006) described a mission called Vanguard which would place a small rover in Gusev crater. A modified Mars Express spacecraft would carry a small lander to Mars. The lander would release a Sojourner-sized rover carrying three ground-penetrating "moles" or digging probes similar to one carried on Beagle 2. Mole deployment sites would be chosen based on results from a ground-penetrating radar carried on the rover. The moles would analyze surface materials to depths of 5 m, communicating via tethers with the rover. The rover in turn would transmit data to the lander, staying within 1000 m of it during the mission, and the lander would transmit to the orbiter when it passed over the landing site. The results, expected to reveal some evidence of water, as the site was considered the probable site of a paleolake (Cabrol *et al.*, 2003), would be relevant to astrobiology and *in situ* resource utilization (ISRU) for future human exploration.

Griebel (2006) proposed a mission called ARCHIME-DES (Aerial Robot Carrying High-resolution Imaging, a Magnetometric Experiment and Direct Environmental Sensing instruments) to deploy a 14 m diameter balloon in the atmosphere of Mars. It would float at about 2000 m altitude carrying a high-resolution camera, a magnetometer and a meteorology instrument in a gondola suspended beneath the balloon. Although no specific area was mentioned for deployment, the author hoped that it might cross the putative northern ocean shoreline (Parker *et al.*, 1993) several times, implying deployment in the mid-northern latitudes. Conversely, the magnetometer was supposed to examine the magnetic anomalies found by Mars Global Surveyor (Connerney *et al.*, 2012), but the best examples are in mid-southern latitudes. Northern latitudes were preferred so the availability of magnetic anomalies would be limited. The 2000 m altitude would be with reference to the Mars Orbiter Laser Altimeter (MOLA) datum, not the surface, and the balloon's height

above some low-elevation regions could be as much as 7000 m. ARCHIMEDES might be carried to Mars with AMSAT P5-A, a dedicated Mars communications satellite then contemplated for a 2007 or 2009 launch. After that spacecraft entered Mars orbit, ARCHIMEDES would be deployed into the atmosphere.

25 February 2007: Rosetta Mars flyby

The European Space Agency's comet orbiter and lander mission, Rosetta, was launched from Kourou, French Guiana, on 2 March 2004 at 07:17 UT, on a mission to rendezvous with and land on the nucleus of Comet 67P/Churyumov–Gerasimenko in 2014. Its complex trajectory took it past Earth three times, on 4 March 2005, 13 November 2007 and 13 November 2009. Also, at 01:54 UT on 25 February 2007 (MY 28, sol 389) it made a close flyby of Mars at an altitude of 250 km. Rosetta passed over the planet's northern hemisphere, approaching the planet over Tharsis, with closest approach over Tempe Terra at 43.5° N, 298.2° E and departure over Syrtis Major Planum. Rosetta had a wide variety of instruments, only a few of which were used at Mars, and it carried a comet lander named Philae, which also observed Mars with one of its cameras.

Rosetta's optical, spectroscopic and infrared cameras (OSIRIS) observed Mars (Pajola *et al.*, 2012a) and Phobos (Figures 121 and 204) for 12 hours beginning 8 hours before closest approach. The VIRTIS-M imaging spectrometer made infrared observations of day and night atmospheric conditions and ozone concentrations (Coradini *et al.*, 2010). Philae's CIVA infrared and visible camera imaged Mars partly obscured by one of the main spacecraft's solar panels (Figure 121D), and its ROMAP magnetometer and plasma instrument also made measurements during the flyby from 12 hours before to 12 hours after the flyby. The Mars observations helped calibrate the instruments and also provided useful Mars data, particularly on the atmosphere.

2007: Hydrothermal sites

Schulze-Makuch *et al.* (2007) extended earlier work by Dohm *et al.* (2000) (Figure 179 in Stooke, 2012) which

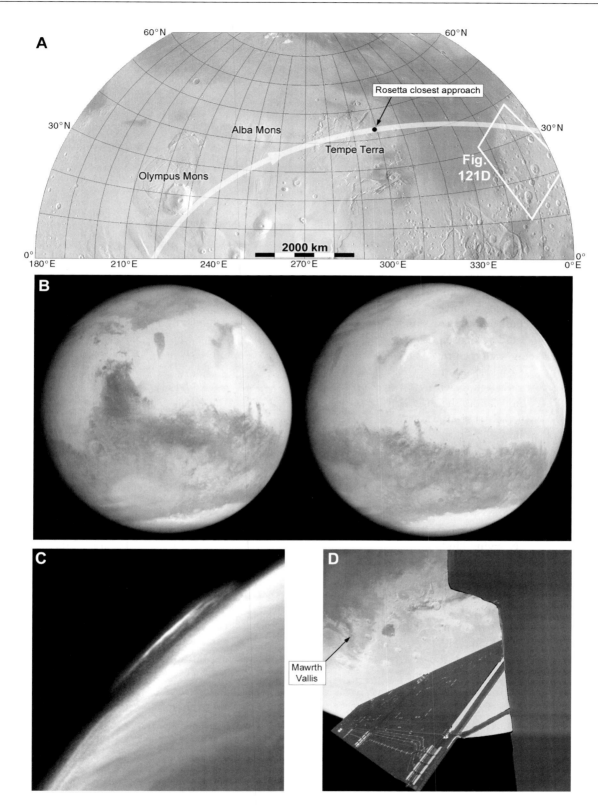

Figure 121. Rosetta Mars flyby. **A:** Closest-approach point and ground track. **B:** OSIRIS approach images showing Syrtis Major (left) and Elysium (right). **C:** OSIRIS image showing hazes or clouds on the limb. **D:** CIVA image from the Philae lander showing the Mawrth Vallis region. OSIRIS images: ESA © 2007 MPS for OSIRIS Team MPS/UPD/LAM/IAA/RSSD/INTA/UPM/DASP/IDA. CIVA image: ESA © 2007 ESA/Rosetta/Philae/CIVA.

had concentrated on hydrothermal environments in the Tharsis region. These locations showed evidence for both water and thermal energy sources and therefore were important sites for biological investigations. The new study incorporated new data to identify globally distributed sites of hydrothermal activity with strong astrobiological potential and implications for future mission planning. The 42 sites described in this study are listed in Table 34 and illustrated in Figure 122. The locations in Figure 122 are taken from an illustration in Schulze-Makuch et al. (2007) and are only approximate. Many are still close to the largest volcanic center at Tharsis, but others are found at high latitudes and near Elysium, Hellas and other locations. Several impact craters were also identified for their hydrothermal potential, where impact-induced heating and water might come together. These included Gusev and Gale craters.

Other studies involving these and other researchers added to the identification of biologically significant sites. Dohm et al. (2004) described the Northwest Slope Valleys region at the southwestern edge of Tharsis (site 5 in Table 34a), a group of enormous outflow channels whose sources are buried under Tharsis lavas. That location was suggested for MSL (Figures 161 and 165, Table 43).

Fairén et al. (2005) described a model of geological history for Mars involving two distinct periods with different climate and geological conditions. The first, in the Noachian, could have included an internal dynamo and magnetic field, plate tectonics, and a northern lowland ocean covering up to a third of the surface of Mars. Later the interior cooled, producing a planet with sporadic volcanism and flooding and a generally cold dry surface as seen today. These have very different implications for biology, so the authors suggested three prime candidate sites for astrobiological exploration, corresponding to different water-related periods: Meridiani Planum (Noachian/Early Hesperian), Mangala (Late Hesperian/Early Amazonian) and Orcus Patera (Amazonian). The Meridiani site illustrated by these authors is very close to the Opportunity traverse region (Figure 47). The Mangala site is beside Labou Vallis, part of the Mangala Valles system of channels. At Orcus Patera the waters of Marte Vallis may have entered the southern end of the elongated basin and deposited relatively young sediments. These sites are illustrated in Figure 122.

2007: ASI Telesat (Marconi)

At the beginning of the new millennium, the Italian Space Agency (Agenzia Spaziale Italiana, ASI) considered flying a telecommunications/navigation orbiter called Marconi as a contribution to the global effort to explore Mars (Matousek, 2001; Noreen et al., 2002). This mission, a collaboration between ASI and NASA, would launch in 2007 and arrive in mid-2008 to provide a communication relay for landers, including Netlander and the Mars Scouts. It would have a circular orbit at an altitude of 4200–4450 km, inclined 125–130° to the equator (retrograde) to maximize its visibility from landers on the surface. A larger version of the orbiter might have flown in 2009 with other science instruments, including a synthetic aperture radar. The project was canceled late in 2002.

2007: Human Exploration of Mars (HEM-SAG)

In March 2007 the Mars Exploration Program Analysis Group (MEPAG) established a Human Exploration of Mars Science Analysis Group (HEM-SAG) to study science goals for the human exploration of Mars. This would be used to define the strategy for Mars exploration as part of the Constellation program and the Vision for Space Exploration outlined by President George W. Bush in January 2004. The study was based on a set of three missions in consecutive opportunities. This followed numerous past studies, including Drake (1998), which had examined the scientific goals and hardware needed to accomplish human exploration of Mars, but now the work was backed up with an official mission in mind.

HEM-SAG considered both long-stay missions (500 days at Mars) and short stays (40 days at Mars), and single-site or multiple-site options. The single-site option brought all three crews to the same site to build up infrastructure, as the Group for Lunar Exploration Planning had considered for Apollo in June 1968 (Stooke, 2007b). The short-stay, single-site option (three brief visits to one place) was deemed scientifically inadequate for the cost and effort. The science goals were most readily satisfied by long stays at multiple sites with mobility of tens of kilometers provided by long-range rovers. The study included a list of possible landing sites,

Table 34. *Hydrothermal Targets From Schulze-Makuch* et al. *(2007)*

34a. Preferred sites

Site	Name	Description
1	Apollinaris Patera	Long-lived volcano, water-rich region, valleys, chaotic terrain
2	Elysium plateau	Younger volcanic complex, source of complex outflow channels
3	Nili Fossae	Isidis basin rim graben, younger basalts, extensive evidence for clays
4	Hadriaca/Tyrrhena area	Low shield volcanoes and lava plains with large outflow channels
5	Northwest Slope Valleys	Channels, lava flows, chlorine enrichment, magnetic anomalies
6	West Candor Chasma	Layered sediments, hydrated sulfates, volcanic materials
7	Melas Chasma	Plume-driven uplift, giant rift system, canyon floor deposits
8	Cerberus/Marte Vallis	Plains and channels, geologically recent water and lava flows
9	Tharsis: Warrego Rise	Thaumasia highlands, magnetic anomalies, channels, fractures
10	Malea Planum	Volcanoes, lava flows, channels, possible glaciation

34b. Additional sites

Site	Name	Description
11	Chasma Boreale	Valley in polar cap, possible catastrophic floods, volcanic hills
12	West Olympia Undae	Volcanic hills and hydrothermal minerals
13	Tharsis: Tempe Terra	Fractured igneous plateau, volcanoes
14	Xanthe Terra	Highlands with outflow channels, release of ground water
15	East Valles Marineris	Chaotic terrain, source of outflow channels, possible magmatism
16	Northeast Valles Marineris	Juventae Chasma, source of Maja Valles, hills may be volcanic
17	Tharsis: Kasei	Near source of Kasei Valles, magmatic and tectonic features, channel
18	Tharsis: Alba Patera	Large low shield volcano, valley network on north flank
19	Tharsis: Arsia aureole	Base of large volcano, tectonic and water-related features
20	Tharsis: Pavonis aureole	Base of large volcano, tectonic and water-related features
21	Tharsis: Ascraeus aureole	Base of large volcano, tectonic and water-related features
22	Tharsis: Olympus aureole	Base of large volcano, tectonic and water-related features
23	Tharsis: Acheron Fossae	Hills, fractures, magmatism, mass wasting and water-related features
24	West Valles Marineris	Noctis Labyrinthus canyon network
25	Tharsis: Syria Planum	Fracturing, uplift, lava flows, volcanic shields, pit crater chains
26	Central Thaumasia	Rifts, fractures, channel networks, possible large volcano
27	Southwest Thaumasia	Rifts, fractures, channel networks, possible large volcano
28	Meridiani Planum	Sedimentary rocks, concretions, past sulfur-rich hydrothermal water
29	Tharsis: northwest	Lava flows, valleys, fractures
30	Ceraunius Tholus	Volcanic mountain, channels possibly from snow melt, delta in crater
31	Tharsis: southern Coprates	Valley network, faults, wrinkle ridges
32	Tharsis: west Thaumasia	Volcano, fractures, valley networks
33	Tharsis: Claritas rise	Elongated hill, fractures, magmatism, hydrothermal activity
34	White Rock	Possible lacustrine deposit in crater, possible evaporites

34c. Crater sites

Site	Name	Description
A	Gusev	Impact crater, possible impact-heating hydrothermal system
B	Gale	Impact crater, possible impact-heating hydrothermal system
C	Boeddicker	Impact crater, possible impact-heating hydrothermal system
D	Lowell	Impact crater, possible impact-heating hydrothermal system
E	Lampland	Impact crater, possible impact-heating hydrothermal system
F	Voeykov	Impact crater, possible impact-heating hydrothermal system
G	Hale	Impact crater, possible impact-heating hydrothermal system
H	Bond	Impact crater, possible impact-heating hydrothermal system

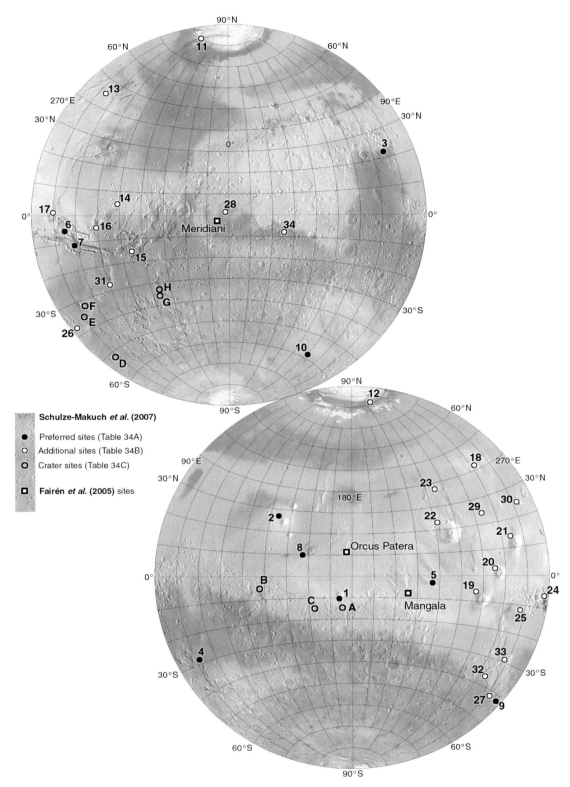

Figure 122. Hydrothermal sites from Schulze-Makuch *et al.* (2007), from Table 34, and astrobiological sites from Fairén *et al.* (2005). For scale, 10° of latitude spans 600 km.

Table 35. *Human Landing Sites From MEPAG HEM-SAG (2008)*

Name	Location	Description
1. Impact crater near Nili Fossae	18.4° N, 77.7° E	Crater lake with deltas on Isidis basin rim. Valleys, layering, old crater and basin deposits, possible volcano, phyllosilicates and olivine
2. Newton crater gullies	40.5° S, 202.1° E	Lineated valley fill, gullies, crater stratigraphy, possible hydrothermal system. Possible active water seeps with extant biology potential
3. Meridiani	2.0° S, 354.5° E	MER-B site, old water-modified deposits, hematite. Possible site for early life, but now dry and geologically inactive, present-day life less likely
4. Gusev crater	14.6° S, 175.4° E	MER-A site, impact and volcanic deposits in Columbia Hills. Channel or lake deposits, possible old hydrothermal systems, early biological potential
5. Chasma Boreale	82.6° N, 312.7° E	Polar layered deposits and basal unit. Origin of Chasma Boreale – if water-cut, possible biological potential. Climate history
6. South polar deposits	71.8° S, 292.7° E	Dorsa Argentea Formation, stratigraphy, possible ancient water and recent ice flow. Ancient permafrost with biological preservation potential
7. Gale crater	5.1° S, 137.5° E	Old crust, valley networks, volatile-rich layered central mound. Crater wall stratigraphy, possible hydrothermal systems, early biological potential
8. Valles Marineris	7.0° S, 287.3° E	Interior layered deposits, wall talus and stratigraphy, relationship to Chryse outflow channels and early biology. Depth may allow liquid water today
9. Holden and Eberswalde	24.0° S, 326.4° E	Valley network deltas, well-preserved and accessible sedimentary deposits. Ancient crust preserved along crater walls
10. Olympus Mons East	17.7° N, 231.8° E	Recent volcanic, tectonic and fluvial activity, possible hydrothermal systems with biological potential
11. Elysium Planitia	5.0° N, 150° E	Recent volcanic flows and outflow channels with high biological interest. Platy units suggested recent ice activity (not supported by radar data)
12. Olympus Mons West	19.6° N, 220.3° E	Olympus Mons scarp, lava flows, talus, and recent glacial deposits, access to aureole deposits
13. Eastern Hellas	38.7° S, 97.0° E	Basin massifs, ancient stratigraphy, recent ice-rich deposits. Gullies, including Centauri Montes "active gully" site (site 29)
14. Nili Fossae	24.2° N, 79.4° E	Lava plains, possible basin impact melt, clays and olivine. Possible micro-environments for ancient life, but probably not suitable for biology today
15. Lyot crater	50.3° N, 29.1° E	Large crater with central deposits. Stratigraphy, recent ice-rich mantle deposits, impact melt hydrothermal system. Good habitat for life
16. Coloe Fossae	41.3° N, 54.2° E	Dichotomy boundary stratigraphy. Recent lobate debris aprons and lineated valley fill. Role of water uncertain
17. Utopia Planitia	28.5° N, 134.4° E	Basin floor deposits, volcanic and fluvial materials, polygonally patterned ground. Dead today but possible fluvial episodes in past
18. Aram Chaos	2.6° N, 339° E	Outflow channel deposits, stratigraphy of mineral assemblages observed from orbit. Transient water availability, not suitable for life today
19. Arsia Mons	4.8° S, 233.7° E	Recent dyke-fed volcanism cutting recent glacial deposit on the northwest flank of Arsia Mons, but may not be active today. High-elevation climate study
20. Slope streaks	14.4° N, 241.8° E	Study slope streaks, possible subsurface water, springs, landslide deposits. Evidence of contemporary activity?
21. Atlantis Chaos	34.8° S, 182.6° E	Examine hills, assess proposed lacustrine origin. Explore regional magnetic anomalies, stratigraphy, fluvial processes, possible link to Gusev crater
22. Central Alba Patera	40.7° N, 250.4° E	Young volcanic activity, ice-rich mantle deposit, old valley networks. Possible hydrothermal systems with biological potential
23. Chryse Planitia	27.0° N, 319.0° E	Outflow channel, possible ocean deposits. Ridged plains, impact craters. Viking 1 and Pathfinder context. Transient water flow, not enough for life
24. Medusae Fossae	1.6° N, 186.8° E	Relationship between Tharsis lavas and Medusae Fossae Formation. Stratigraphy, role of water

Table 35. (*cont.*)

Name	Location	Description
25. Hellas basin floor	41.9° S, 69.6° E	Outflow channel and possible Hellas ocean deposits. Unique low-elevation climate, strong potential for biology. Transient liquid water today?
26. Arsia Mons northeast	7.4° S, 238.8° E	Cave skylight site with high biological interest if water is available. Access to Arsia Mons lava flows
27. Dao Vallis walls	33.7° S, 92.5° E	Well-developed gullies, possible melting snowpacks. High relief with local microclimates and related surface processes
28. Terra Sirenum	39.3° S, 198.3° E	Recent high-albedo deposit beneath gullies, possibly active today. Ancient highland terrain, younger with lava flows, some recent fluvial activity
29. Centauri Montes	38.7° S, 96.7° E	Recent high-albedo deposit beneath gullies, possibly active today. Extensive volatile-rich deposits, local and regional climate studies
30. Terra Cimmeria	70.0° S, 180.0° E	Region of preserved crustal magnetism and ground ice. Deep drilling in the region of 180° W between 60° and 80° S
31. Mawrth Vallis	25.3° N, 340.7° E	Fluvial geomorphology, heavy weathering, clays. Stratigraphic analysis, access to cratered terrain and northern lowlands
32. Olympia Planitia	75.0° N, 180.0° E	Sulfate-rich dunes around the north pole, very recent alteration products. Access to residual polar cap and large polar troughs, climate history
33. Valles Marineris	6.2° S, 290° E	Sulfate-rich deposits only accessible by human operations. Extensive stratigraphic analysis and access to landslides and talus
34. Arsia lobate deposit	7.4° S, 236.2° E	Recent glacial activity, interaction with volcanism, climate history, possible residual ice. Examine moraines, obtain ice cores, lava flow stratigraphy
35. North polar cap	86.0° N, 79.0° E	Very recent ice deposits. Drilling and stratigraphic analysis of polar troughs
36. South polar cap	88.0° S, 30.0° E	Very recent ice deposits. Drilling and stratigraphic analysis of polar troughs
37. Syrtis Major Planum	7.0° N, 69.0° E	Possible SNC meteorite ejection site, older lava flows and silicate-rich deposits in caldera. Interactions with Isidis and the northern lowlands
38. Mangala Valles	18.0° S, 210.6° E	Outflow channel floor, residual ice-rich deposits, dyke-related vents, volcano–water interactions. Study outflow chronology
39. Nilosyrtis Mensae	35.0° N, 71.0° E	Complex stratigraphy on the dichotomy boundary. Multi-stage glacial activity. Recent climate change chronology
40. Olympus Mons caldera	18.3° N, 227.0° E	Age of Tharsis volcanism, caldera wall stratigraphy, dust and ash deposits, landslide deposits. Seismic studies and search for possible gas venting
41. Milankovič crater	55° N, 213.5° E	Rare, large impact crater in northern lowlands. Ejecta stratigraphy and study of nearby high-latitude ice-rich mantling deposits
42. Kasei Valles	21.0° N, 286.2° E	Massive old fluvial deposits, channel wall stratigraphy. Evidence for fluvial and glacial activity and subsequent Tharsis lava flows
43. Vastitas Borealis	65.7° N, 20.2° E	Vastitas Borealis Formation, widespread northern plains, possible ocean sublimation residue. Polygons, mantling deposits, Phoenix context
44. T-shaped valley	37.6° N, 24.0° E	Complex glacial deposits along the dichotomy boundary. Recent climate history, possible recent glacial ice preserved
45. Isidis basin floor	12.0° N, 88.5° E	Possibly part of an ancient northern ocean, with volcanic materials from Syrtis Major
46. Utopia basin floor	43.8° N, 117.0° E	Patterned ground with near-surface volatiles, possible young preserved ice. Outer parts of the Elysium flow deposits, relationship to Viking 2 results
47. Hecates Tholus	32.0° N, 150.3° E	Young valley networks near extensive volcanic activity. Access to nearby northern lowlands
48. Magnetic anomalies	60.0° S, 175.0° E	Geophysical analysis of early magnetic field, ancient cratered terrain. Relationship of surface geology to subsurface magnetism

Table 35. (*cont.*)

Name	Location	Description
49. Hesperian calderas	59.4° S, 60.7° E	Volcanism in middle part of Martian history. Climate interaction between Hellas and south pole. Relationship between volcanism and valley networks
50. Hesperia Planum	23.3° S, 110.6° E	Potential comparison with lunar mare terrain. Structure of wrinkle ridges. Age of a classic unit of the Mars timeline
51. Huygens ridge	12.3° S, 66.3° E	Access to exhumed dyke or older ridged terrain. Geochemistry of intrusive volcanic rocks. Dyke may have been a feeder for ridged plains volcanism
52. Argyre floor deposits	51.5° S, 319.0° E	Large impact basin and melt sheet. Volatiles from south polar cap and recent small fluvial features. Possible eskers, evidence for shorelines
53. Warrego Valles	38.6° S, 270.6° E	Thaumasia valley networks. Post-emplacement modification, ice-rich crater fill. Geophysical study of thrust structures at edge of Tharsis rise
54. Syria Planum	7.7° S, 259.5° E	Structural evolution of Tharsis and western end of Valles Marineris. Highest point on Tharsis dome and key to its early volcanic evolution
55. Proctor crater	47.5° S, 30.2° E	Dune field on crater floor. Study of recent dune formation and migration, relation to climate change. Stratigraphy of ancient crust
56. White Rock	8.0° S, 25.2° E	High-albedo crater floor deposit and its aeolian modification. Study early mineralogy and resurfacing processes
57. Complex ridges	66.0° S, 140.0° E	Structural analysis of ancient tectonic features. Possible exposure of deep crust material by massive faulting
58. Mie crater	48.5° N, 139.7° E	Rare, large crater in northern plains. Stratigraphy, periglacial processes. Compare present with Viking 2 data. Study Viking 2 lander after 50 years
59. Crustal magnetism	47.2° S, 176.8° E	Radial magnetic field reversal location within 100 km of a crater with gullies on its walls
60. North polar dome	84.0° N, 7.0° E	Deep drilling for stratigraphy and climate studies on the permanent ice cap, followed by long drive south to a return vehicle

which is summarized in Table 35 and illustrated in Figures 123–131 (MEPAG HEM-SAG, 2008; Drake, 2009; Drake *et al.*, 2010; Levine *et al.*, 2010a, 2010b, 2010c, 2010d). The sites considered in this period for HEM-SAG, MSL and other missions are very different from the old Landing Site Catalog (Table 47 in Stooke, 2012) and other earlier site lists because of the greatly improved imaging and topographic data, and especially the surface compositional data from orbiting spectrometers. The name of site 32 (Olympia Planitia) is informal, taking its name from the nearby upland called Olympia Planum. In some cases, including sites 55 and 60, the location might be considered as the science target rather than the landing point.

The HEM-SAG study assumed a first flight between 2030 and 2040. It recommended crew mobility capabilities (rover ranges) of up to 200 km radially from the landing site, and the ability to drill several hundred meters deep at one site or less at multiple sites.

Instrument stations should be established around the landing area for long-term remote operation, like the lunar ALSEPs, and at least 250 kg of samples should be returned to Earth.

A group at Brown University led by James Head selected four of the landing sites in Table 35 for detailed planning purposes. Three of them were chosen to sample the three basic subdivisions of Martian geological time, Noachian (site 1, in Jezero crater, also considered for MSL), Hesperian (site 34, west of Arsia Mons) and Amazonian (site 38, in Mangala Valles). At each site, traverses capable of providing the necessary science data were laid out to help with equipment and logistics planning. Missions at the Centauri Montes crater gully location (site 29) and the north pole (site 35) were also described. These sites are illustrated in Figures 129–131. The first three sites constituted the reference mission sequence of three landings, which would form the first phase of the human exploration of Mars.

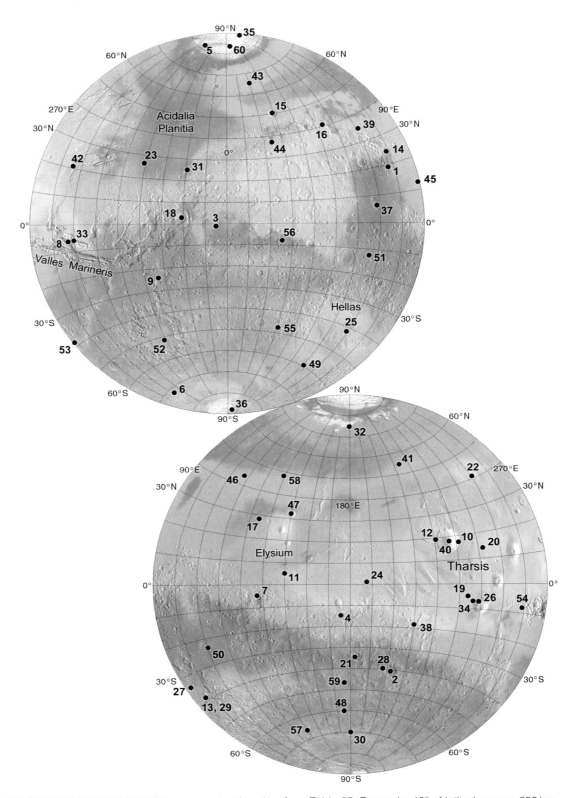

Figure 123. MEPAG HEM-SAG (2008) human exploration sites from Table 35. For scale, 10° of latitude spans 600 km.

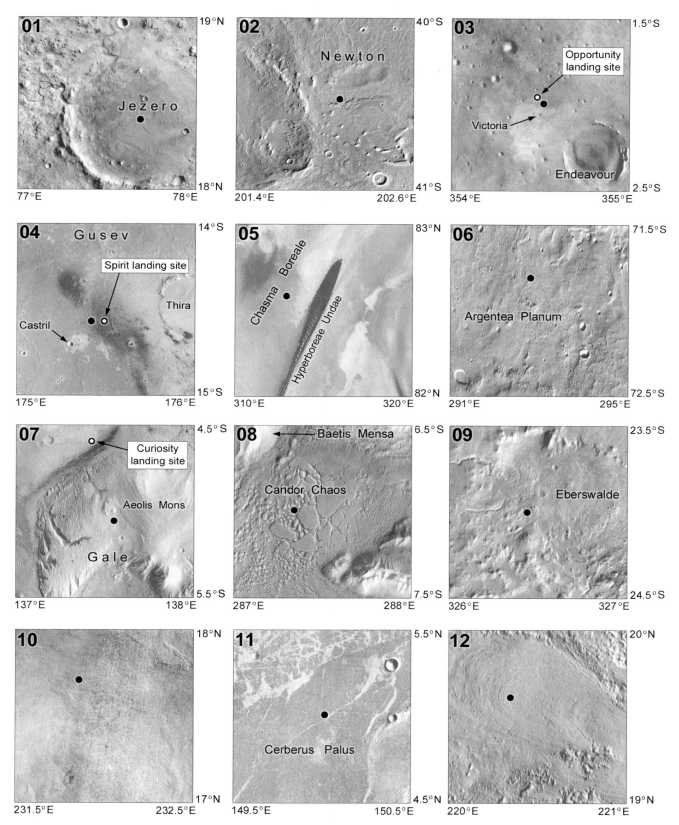

Figure 124. HEM-SAG sites 1 to 12. Each image is a THEMIS infrared mosaic with inverted shading, spanning 1° (60 km) north to south. The HEM-SAG sites are black dots; other sites are depicted as circles with white centers.

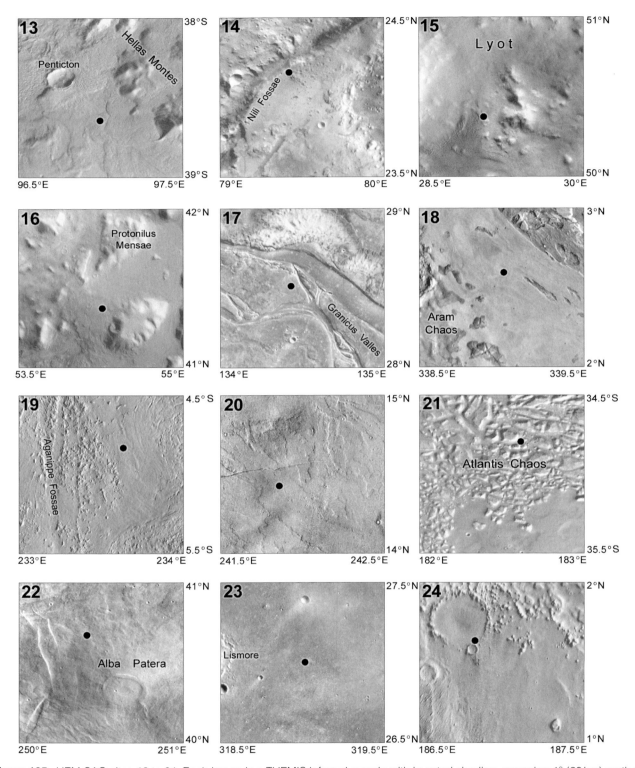

Figure 125. HEM-SAG sites 13 to 24. Each image is a THEMIS infrared mosaic with inverted shading, spanning 1° (60 km) north to south. The HEM-SAG sites are black dots.

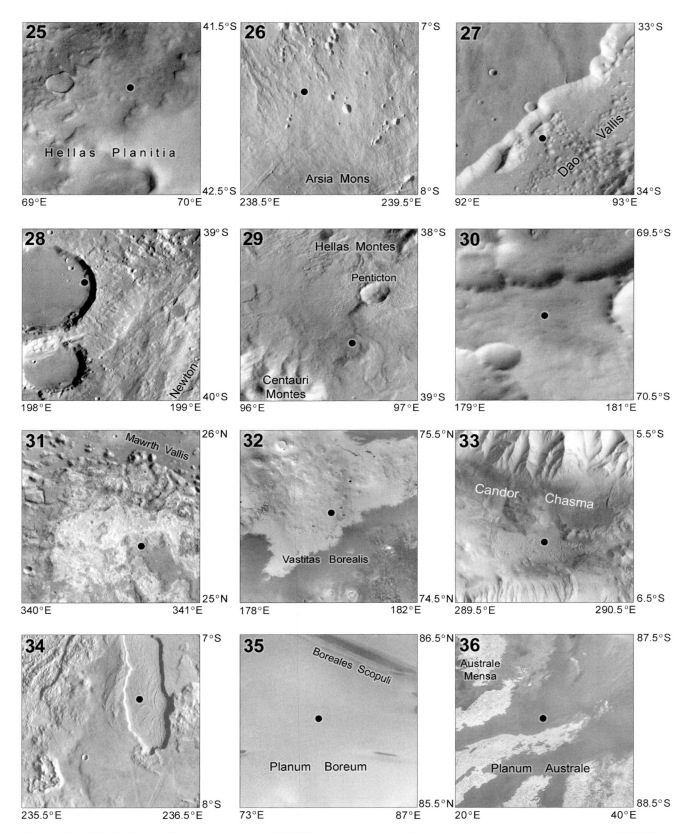

Figure 126. HEM-SAG sites 25 to 36. Images are THEMIS infrared mosaics with inverted shading, except sites 32 and 36 (CTX mosaics) and 35 (Viking mosaic, orbit 122B), spanning 1° (60 km) north to south. The HEM-SAG sites are black dots.

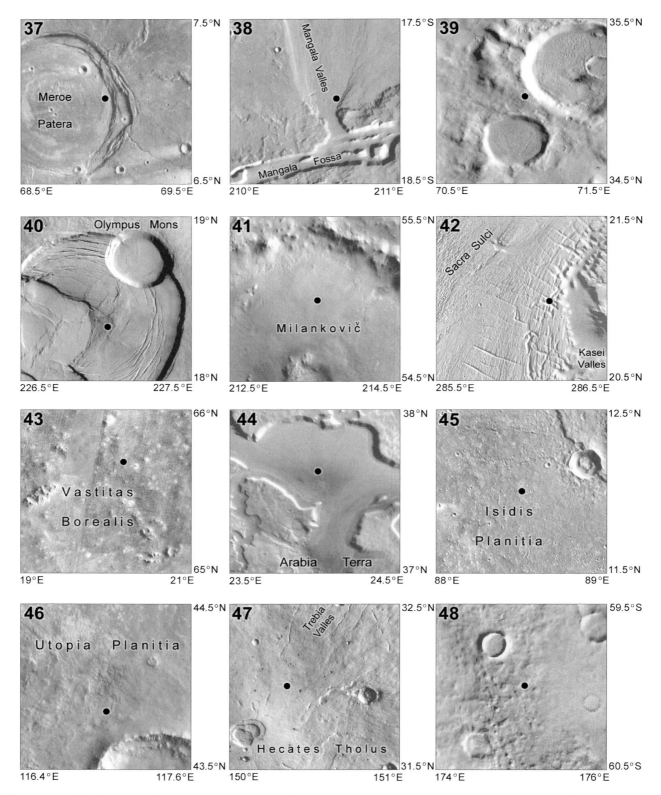

Figure 127. HEM-SAG sites 37 to 48. Images are THEMIS infrared mosaics with inverted shading, spanning 1° (60 km) north to south. The HEM-SAG sites are black dots.

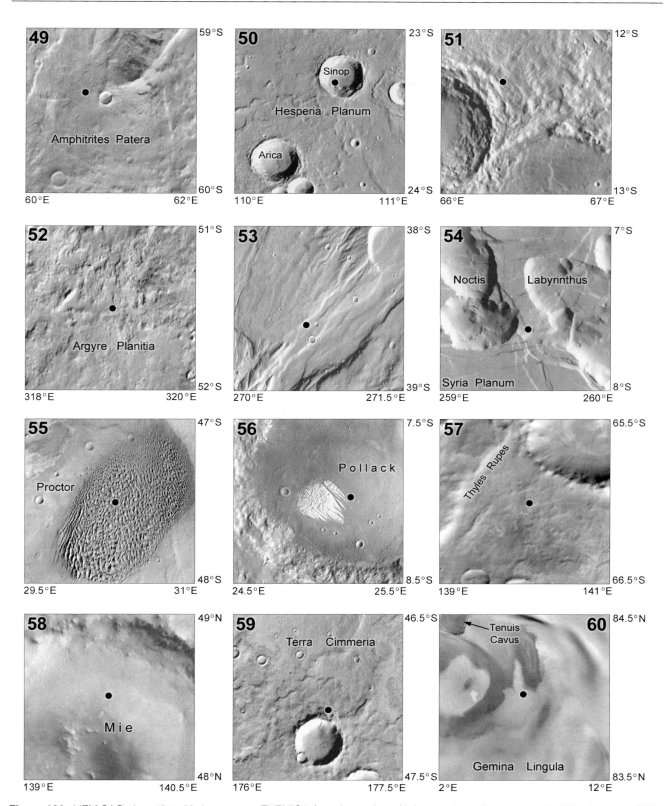

Figure 128. HEM-SAG sites 49 to 60. Images are THEMIS infrared mosaics with inverted shading, except site 60, which is a CTX mosaic, all spanning 1° (60 km) north to south. The HEM-SAG sites are black dots.

Jezero is a crater with a diameter of 45 km, in the Nili Fossae region north of Syrtis Major Planitia and on the northwest rim of the 1200 km diameter Isidis impact basin (Figure 129A). The site would provide ancient samples from the Isidis basin rim, extensive clay deposits from the subsequent modification of this ancient terrain, and younger volcanism. The crater rim has been broken in two places by channels extending from the Isidis rim massifs to the west. Another breach on the eastern rim of the crater allowed ponded water to flow east into the Isidis Planitia lowlands. The deposits of the inflowing channels created well-preserved deltas on the crater floor and include clay deposits. Other parts of the crater floor may have been covered by lavas, possibly associated with Syrtis Major Planum. The standing water implied by the deltas suggests biological potential at this site. The landing site would be at the center of the crater, and traverses would extend into the three associated channels, to a massif probably associated with the Isidis basin rim, and to the nearest exposures of Syrtis Major volcanism. Figure 129A shows the locations of central and satellite geophysical stations and areas of localized detailed study, discussed in more detail below for the Centauri Montes site.

Mangala Valles is an outflow channel that starts at a long fracture radial to Tharsis and extends northwards across a region of cratered terrain for 900 km. The region shows evidence for volcanic, tectonic, fluvial and glacial activity. The landing site would be in a smooth part of the channel floor. Traverses would extend to the graben at the head of the channel to examine glacial and volcanic features (Figure 129B), and along the channel floor and the surrounding scoured plains to study erosional and depositional processes. Some ice-rich deposits may be found in these areas. Samples of Tharsis volcanic materials may be found near the graben, intruded into the surrounding rock along the fracture. If life exists on Mars today, it is most likely to be found in the subsurface, and the Mangala site offers a chance to look at water erupted from a significant depth which might bring organic materials to the surface.

The Arsia Mons graben site (number 34 in Table 35) addresses several aspects of younger geology and climate events. All of the large Tharsis volcanoes display young glacial deposits on their northwestern flanks, best seen on Arsia Mons. The landing site (Figure 130A) is at the foot of the western slope of the vast volcanic shield. Traverses would cross the glacial deposits and volcanic flows from the shield. One traverse would visit a 5 km wide graben in the glacial deposit, and drilling at selected locations could reveal recent climate history. Figure 130A shows the locations of central, satellite and local geophysical stations.

A detailed study of activities at the Arsia site was presented in this report. Five months would be spent at the graben just north of the landing site and nearby smooth plains units, examining recent glacial deposits and landforms and climate history. A search would be made for samples carried downslope from higher elevations on the volcano, and for buried ice. The large graben would be entered from the north, to study wall stratigraphy and to collect samples for dating. Drilling, seismic experiments and ground-penetrating radar would provide subsurface data here and at other sites. Three months would be spent on the eastern flank of the volcano, examining lava flows and pyroclastic deposits, volcano–ice interactions, and looking for the highest elevation at which ice had accumulated on the slope. Seismometers would be deployed to study the internal structure of the volcano and to detect any current subsurface volcanic activity.

Another three-month period would be devoted to glacial and volcanic features just south of the landing site. These may include the youngest glacial deposits in the area. Some very young lava flows cover parts of the glacial deposit and must be very young. Samples would be collected for dating, evidence of volcano–ice interactions, including water flow, would be sought, and ejecta from small craters would be inspected for subsurface material. Any ice encountered would be cored to examine its composition and to search for trapped atmosphere samples for recent climate change studies. Four months would be spent at another graben 200 km west of the landing site, within the large glacial moraine deposit. Glacial processes and materials would be examined, and the graben would be explored for clues to its origin. If it had formed over a volcanic intrusion, volcano–ice interactions might be found there, with biological implications.

These three sites constituted the reference mission sequence in the HEM-SAG report. Not only do they provide diverse samples, but also they are sufficiently widely distributed to form a useful seismic network and

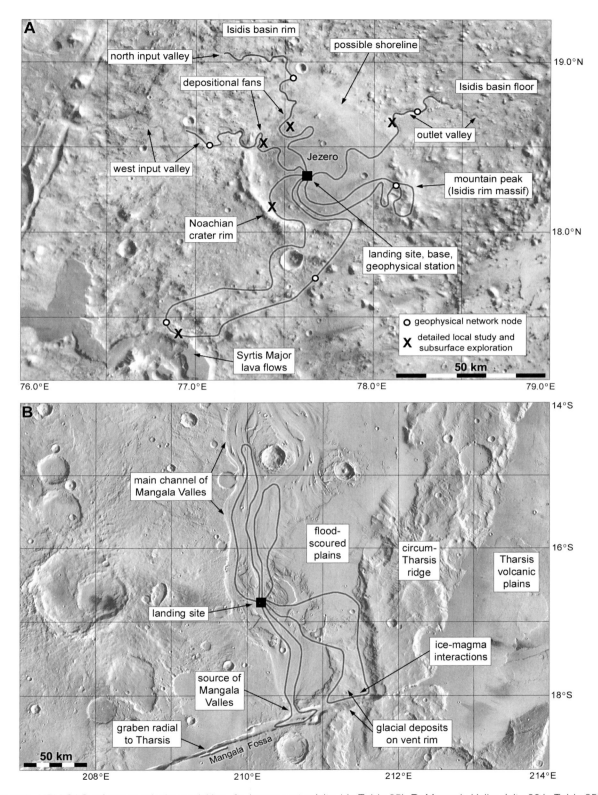

Figure 129. HEM-SAG reference mission activities. **A:** Jezero crater (site 1 in Table 35). **B:** Mangala Valles (site 38 in Table 35). The Mangala landing site is 80 km northwest of that shown in Figure 127. Dark lines show rover traverses.

to sample different crustal settings: Tharsis, the dichotomy boundary and the rim of Hellas. However, geophysical considerations suggested other possible site combinations. The report observed that a site in the region of strong magnetic anomalies, in the higher latitudes of the southern hemisphere near 180° longitude, would be desirable. This "crustal magnetism site" was not included in the published site list, though site 48 is similar, but it was described in the report and is included here as site 59. A favored location was at 47.2° S, 176.8° E (Figures 208 and 213), where a crater with gullies on its interior wall lies within 100 km of a radial magnetic field reversal seen in MGS data. A polar site would also be desirable, and a mission to one such site is described below. Table 36 lists various possible groupings of sites chosen for geophysical reasons.

Two reference missions to study climate and meteorology were described in the report. One was for general climate studies, which in fact would be expected during any human mission. Generally, landing site selection for a climate mission would be less critical than for geology or geophysics, but some constraints were mentioned. The fetch of the wind, the distance over which the wind blows unimpeded by obstacles, should be at least 1 km, so the wind data are properly representative of the region. Local or regional slopes and albedo variations would also influence climate and should be taken into account. The climate mission would set up a central meteorology station and satellite stations for regional characterization or three-dimensional sampling. Balloons could be used to collect high-altitude data, tethered for lower altitudes, and free floating for heights of up to 20 km. The climate mission would study atmospheric dynamics and the energy and mass budgets, and also atmospheric chemistry, including sources of trace gases such as methane.

The second climate mission was to a polar cap (referred to as a dome, from its topography) for deep drilling. The polar regions are especially challenging because of very low temperatures and seasonal illumination variations. Since the Sun is above the horizon at the pole for only half a Mars year at a time, a 500 day mission must either land or launch in darkness, or include a long traverse off the cap.

The north polar dome was suggested for the first human polar mission. This landing site was not on the published list, but is included in Table 35 as site 60, with details for Figure 130B taken from a poster at the 39th Lunar and Planetary Science Conference (Garvin *et al.*, 2008). After landing, a deep drilling program would be undertaken during polar daylight, followed by a long traverse to lower latitudes to avoid the polar night. A return vehicle would be waiting for the crew at the new location, a scenario somewhat reminiscent of Wernher von Braun's Mars Project of 1952 (Figure 7 in Stooke, 2012). The scientific goal would be to study ice deposition chronology and climate history. Biology would be important as well, as organisms or biological materials might be preserved between layers of ice.

Table 36. *HEM-SAG Suggested Site Combinations*

First site	Second site	Third site (antipodal)
19. Arsia Mons	23. Chryse Planitia	1. Impact crater near Nili Fossae
26. Arsia Mons northeast	24. Medusae Fossae Formation	14. Nili Fossae
34. Arsia lobate deposit	28. Terra Sirenum	25. Hellas basin floor
	38. Mangala Valles	27. Walls of Dao Vallis
		29. Centauri Montes
		37. Syrtis Major Planum
		45. Isidis basin floor
		46. Utopia basin floor

Notes: (1) Regions near Arsia Mons are preferred for a single-site mission scenario.

(2) Restricting the second site to Mangala Valles and the third to Nili Fossae, Syrtis Major Planum and Isidis basin floor may give better geophysical data (seismic, heat flow, magnetization).

(3) Sites 23, 25–29 and 46 may have higher probability of near-surface recent activity and thus of high priority for ground water studies.

(4) The Terra Sirenum site could be relocated near a magnetization reversal zone.

A deep core, to bedrock if possible, would be collected, as well as shorter cores to observe regional variations in layering. Geophysical sensors would measure heat flow and map layering remotely. Orbital topographic and sounding data would be used to locate a site.

The mission outlined by HEM-SAG (Figure 130B) would land within 2 km of a site chosen from orbital data and spend 40 sols setting up meteorological stations and making geophysical traverses to locate a drill site. The landing site was not specified in the report, but a site near 84° N, 7° E was shown by Garvin *et al.* (2008) in an otherwise unpublished conference poster. A total of 220 sols would be used for deep drilling, while some crew members undertook traverses of up to 10 km for local studies and short-core collection.

The deep drill would have to reach depths of about 2 km, the thickness of the polar ice as revealed by orbital radar sounding. It would be designed to avoid biological contamination to preserve the ice in pristine condition. The core might be studied in place with borehole instruments rather than being extracted, a much more challenging problem, though some samples might be extracted by melting and pumping out. Then, as the seasons progressed, the deep drill would be left in place and the crew would commence their "grand traverse," a 200 sol drive off the dome to the south to rendezvous with their pre-landed launch vehicle. On the way they would deploy seismic stations along a 500 km baseline to probe the deeper crust and interior of the planet. Sites of biological or geological interest would be visited along the roughly 1500 km route.

The Centauri Montes site was presented as an example of a geophysical exploration mission. The landing site and surface traverses are shown in Figure 131. This landing site was adjacent to Penticton crater, in which MOC images from August 1999 and September 2005 showed a surface change (Malin Space Science Systems, 2006; Malin *et al.*, 2006), interpreted as a small flow of water or mud down a gully on the crater wall some time between those dates. Exploration targets at this site included recent gully deposits possibly indicating contemporary liquid water flow, ancient Hellas basin rim massifs, and young debris aprons or glacier-like flow features suggesting recent climate changes. The traverses in Figure 131A extend out about 50 km from the landing site, but, if mobility was

increased, then additional targets could be reached (Figure 131B). Two of the shorter traverses include short walking segments across difficult terrain to reach steep massif slopes and the floor of Harmakhis Vallis. Supply caches or remote camps could be set up at convenient locations to support longer traverses, and if placed carefully they might contribute to two different traverses.

A central geophysics station would be set up adjacent to the landing site. During four long looping traverses, smaller satellite geophysics stations would be deployed at eight locations. Additional gravity, crustal sounding and active seismic experiments would be undertaken at various locations along the traverses to study subsurface structures, ice deposits and overlapping debris flows. Similar sounding along the rim of Penticton crater might reveal the presence or absence of an aquifer feeding the active gully (Figure 131C), helping to locate a drilling site to reach any volatiles. The gully study would then deploy a tethered human or robot explorer from above to descend to the gully site for direct sampling, and later traverse to a location below the gully to sample the deposits. A second location on the floor of the crater would be drilled if it was feasible.

Another subject considered in the HEM-SAG report was the search for life. Jakosky *et al.* (2007) suggested high-priority sites for this study. Contemporary living organisms might live where orbital data suggested present or recent water near the surface. These were the polar caps, springs or outflows of water, subsurface hydrothermal systems, and active gullies observed from orbit. Extinct organisms (fossils) might be preserved where ancient water was suggested, including source regions for the great catastrophic flood channels, ancient cratered highlands that formed when surface water might have been stable at the surface, and mineral deposits associated with water or hydrothermal systems. The Centauri Montes site was used as an illustration of this study (Figure 131).

The biology mission would land just south of a crater called Penticton and would begin with a 30 sol period of site preparation at a point on the crater's rim just above the gully described above (Figure 131C). For the next 200 sols drilling and local traverses would be undertaken, and crew members not needed for the drilling could prepare supply caches and undertake reconnaissance for the later traverses. The rest of the mission, another period of 200 days or more, would be taken up

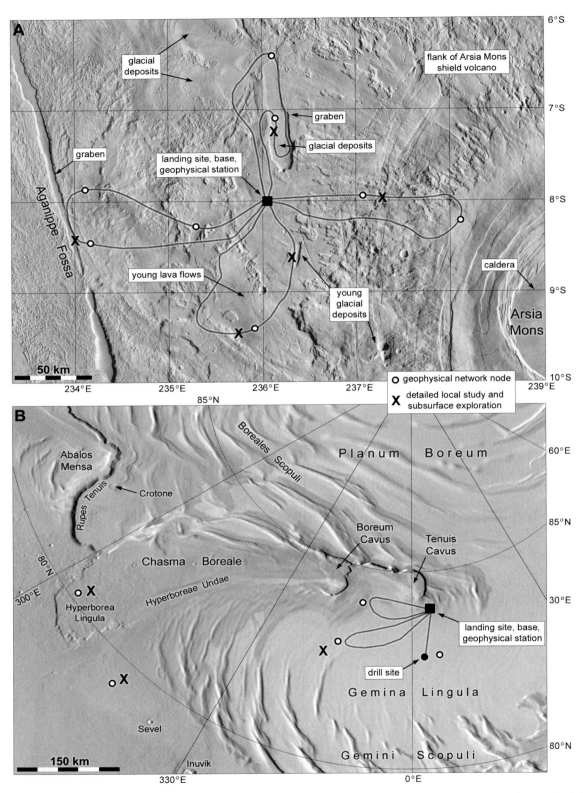

Figure 130. HEM-SAG reference mission activities. **A:** Arsia Mons (site 34 in Table 35). **B:** North polar dome (site 60 in Table 35). Dark lines show rover traverses. At the northern site a long traverse off the polar cap would follow the ice cap operations. **A** is drawn on a THEMIS infrared mosaic, **B** on a composite of MOLA and MARCI data.

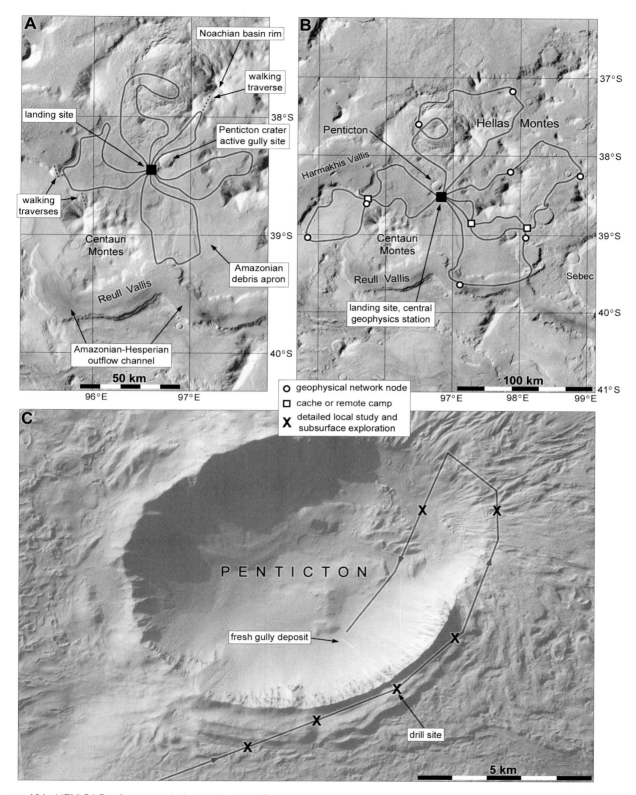

Figure 131. HEM-SAG reference mission activities at Centauri Montes (site 29 in Table 35). **A:** The 50 km radial traverse limit. **B:** The 100 km radial traverse limit. **C:** Penticton crater area showing gully exploration activities. Dark lines show traverses. **A** and **B:** THEMIS infrared mosaics. **C:** CTX image G19_025818_1413_XI_38S263W.

with the traverses shown in Figures 131A and 131B. The biology study suggested that a 50 km traverse radius was sufficient for their purposes.

Phoenix landing site selection

The goal of the Phoenix mission was to sample and analyze ice contained in or covered by regolith in areas suggested by Mars Odyssey orbital data. The initial Phoenix proposal illustrated a landing site at 65° N, 240° E (Figure 132), which lay at the centre of a large hydrogen concentration in the Mars Odyssey HEND data (Figure 188D in Stooke, 2012) and represented a starting point in the landing site selection process. Arvidson *et al.* (2005) also identified 70° N, 230° E as an early site. Engineering constraints on the landing site were that it lie between latitudes 65° and 72° N, and below an elevation of −3500 m, to meet the entry constraints and to provide adequate lighting for an expected five-month mission. In addition, it should have winds no faster than 20 m/s, local slopes no steeper than 16° and rock abundance no greater than at the Viking 2 landing site. Within the specified latitude zone, sites were sought in areas where Mars Odyssey data suggested the presence of ice under a thin layer of dust or regolith. The site selection process is summarized by Arvidson *et al.* (2008b), with additional details in Bonfiglio *et al.* (2011), Spencer *et al.* (2009) and Prince *et al.* (2008). Three candidate areas were initially selected from the latitude band to limit MOC imaging requirements (step 1 in Table 37, areas A, B and C in Figures 132 and 133). Two other areas (D and E) were also considered (Putzig *et al.*, 2006; Sizemore and Mellon, 2006). All these areas spanned 20° east to west and 7° north to south, or about 400 km in each direction. Area D contained the preliminary site mentioned in the mission proposal because it had the highest ice concentration in the latitude band, but it was rejected because the amount of ice might be too great. A mixture of ice and soil would provide more useful information. Area E showed little evidence for ice and was rejected at a very early stage.

Next, the limited MOC and THEMIS images available in 2005 were used to find the most satisfactory site in each of the candidate areas. Arvidson *et al.* (2005) illustrated two versions of these potential sites, referred to as boxes, in areas A to D. Step 2 in Table 37 identifies the larger boxes, and step 3a narrows them to rectangles 150 km east to west and 75 km north to south (Figure 133). The landing ellipse for Phoenix planning was 150 km long and 20 km wide, but its orientation depended critically on the date of launch, varying by 50° between the first and last days of the 20 day launch period in 2007. As the ellipse rotated about its center between the first and last day extremes, it swept out a shape called a butterfly (Figure 133), whose envelope was roughly the box defined in step 3a. Tamppari (2005) illustrated promising butterflies in area A and area B, and two butterflies in area C, though they did not coincide exactly with the nearby boxes (step 3b in Table 37). After analyzing the images contained within these boxes, the three areas were ranked, B having the best characteristics and C being least satisfactory.

Area B was now subjected to a more thorough analysis. Three boxes within this area (step 4 in Table 37, Figure 133) were selected for further study, including the acquisition of additional Mars Global Surveyor and Mars Odyssey images in 2005 and 2006. These boxes were chosen to avoid obvious hazards and to provide a range of landing latitudes. The southern box was geologically preferable, the northern box was best for communication via orbital relays, and the middle box was a compromise location. This work was interrupted when early HiRISE images from Mars Reconnaissance Orbiter at the end of 2006 showed unacceptably high rock concentrations in area B. With time now limited, the project was forced to re-examine the other areas.

Initially, area A showed some promising locations in MRO images. Additional sites were located using Mars Odyssey thermal inertia data, made by comparing THEMIS day and night infrared images to identify rock-free areas. Rocky surfaces warm more slowly during the day and cool more slowly at night in comparison with rock-free surfaces. These locations were then scrutinized by MRO. Three promising boxes in areas A and D were considered at a 22 January 2007 meeting (step 5 in Table 37, Figure 134A) and a preferred site was located within each box (step 6 in Table 37). These boxes were 2.5° by 9° across, roughly 150 km square.

The most promising site was found to be box 1, an almost rock-free area situated in a shallow 50 km wide valley between rougher upland areas in the Scandia Colles (Scandia Hills). This was the only site that could fit ellipses of different orientations for different launch dates. The other boxes were limited to one ellipse each in areas with few rocks, and therefore tied to very specific launch dates. Hazard maps prepared by Tim Parker (JPL) used green shading to show the safer areas, so the safe area in box 1 was informally named Green Valley. It was also referred to as Dogbone, from its shape, during the site selection process. Green Valley is also the name of a location near Tucson, Arizona, home of mission operations for Phoenix.

Box 1 was the final choice, and the best site in it, identified in March 2007, was at 68.25° N, 232.5° E.

Illustrations of this site before launch showed three ellipses suited for launches at different dates within the launch period (Figure 134B), since the approach azimuth varied significantly with the launch date. This was another butterfly, as described above, but the ellipses were moved to avoid obstacles and did not intersect at their centers. The specific target ellipse was determined by the actual launch date. During the cruise, new MRO HiRISE images suggested that the ellipse should be moved 13 km southeast to avoid rougher terrain at the west end of the ellipse. The final target, a 100 by 20 km ellipse, was centered at 68.25° N, 233.4° E.

Table 37. *Phoenix Landing Site Selection*

Selection phase	Site	Latitude	Longitude (east)
Step 1:	Area A	65° to 72° N	250° to 270° E
Select areas	Area B		120° to 140° E
(D – rejected)	Area C		65° to 85° E
(E – rejected)	Area D		230° to 250° E
	Area E		300° to 320° E
Step 2:	Area A box	67° to 70° N	256° to 264° E
Identify initial boxes in each area	Area B box	67° to 70° N	126° to 135° E
	Area C box	69° to 71° N	75° to 84° E
	Area D box	69° to 71° N	238° to 249° E
Step 3a:	Area A box	68° N	260° E
Smaller boxes,	Area B box	67.5° N	130° E
150 × 75 km	Area C box	70° N	80° E
(center coordinates)	Area D box	70° N	243° E
Step 3b:	Area A	69° N	259° E
Identify possible "butterflies"	Area B	67.5° N	130° E
(center coordinates)	Area C site A	70° N	77° E
	Area C site B	68° N	70° E
Step 4:	Area B, box 1	67.5° N	130° E
Seek additional boxes in the best area	Area B, box 2	66° N	136° E
	Area B, box 3	70.5° N	136° E
Step 5:	Box 1 (Area D)	67° to 69.5° N	229° to 238° E
Broaden search for safe boxes	Box 2 (Area D)	66° to 68.5° N	242.5° to 251.5° E
	Box 3 (Area A)	69.5° to 72° N	248.5° to 257.5° E
Step 6:	Box 1 site	68.35° N	233.0° E
Find safest site in each box	Box 2 site	66.75° N	247.6° E
	Box 3 site	71.20° N	253.0° E
Step 7:	Green Valley	68.25° N	232.5° E.
Safest site			
Step 8:	Green Valley	68.151° N	233.975° E
Adjusted site			
Step 9:	Green Valley	68.25° N	233.4° E
Final target			

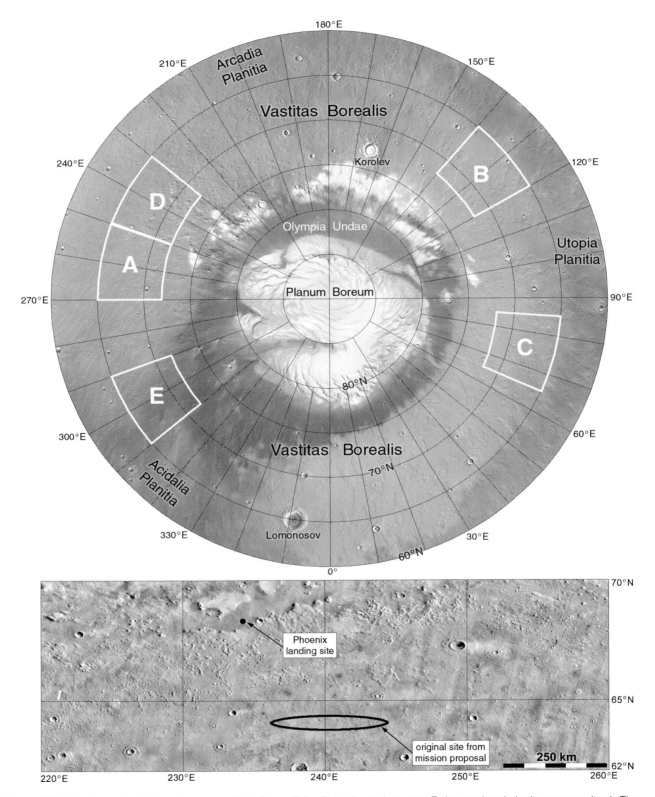

Figure 132. The four potential landing areas A, B, C and D for Phoenix, and an area E dropped early in the process (top). The Phoenix landing site illustrated in the mission proposal (bottom), superimposed on a THEMIS infrared mosaic with inverted shading. The final site is 300 km northwest of the original proposal site.

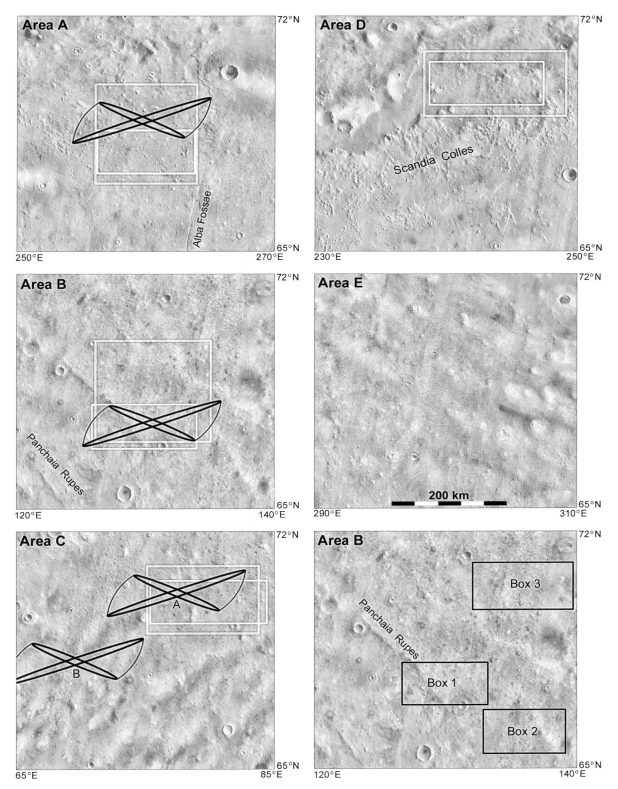

Figure 133. Phoenix landing areas and early candidate sites. White boxes are areas chosen for further imaging (Arvidson *et al.*, 2005). Dark cross shapes are "butterflies" illustrated by Tamppari (2005). At lower right the rectangles in area B were targeted for analysis and HiRISE imaging (Table 37, step 4). Images are THEMIS infrared mosaics with inverted shading, and the scale in area E applies to all areas.

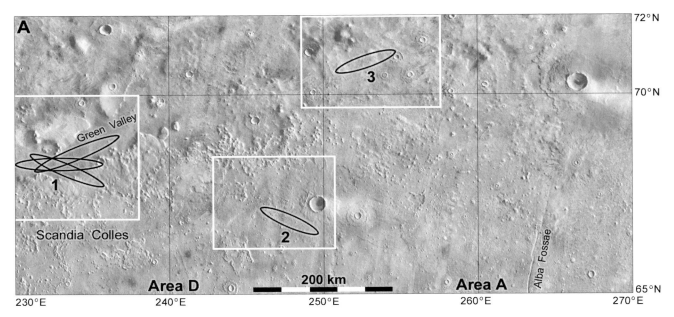

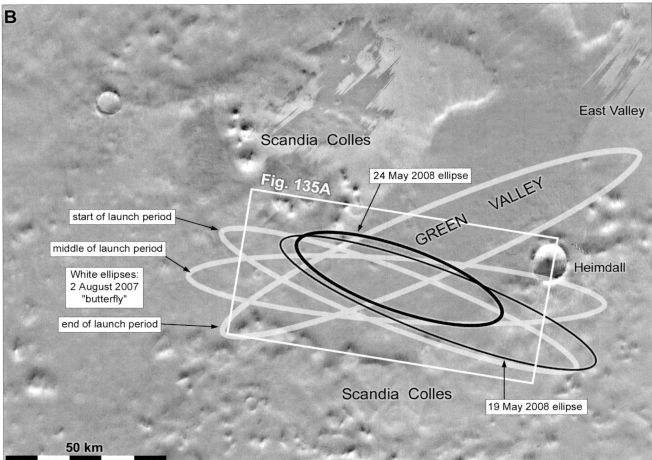

Figure 134. Phoenix landing site. **A:** Areas D and A combined, showing the safer sites chosen using THEMIS thermal inertia and HiRISE data, drawn on a THEMIS infrared mosaic with inverted shading. The white boxes and black ellipses are listed in Table 37. **B:** The Green Valley landing area in Scandia Colles, roughly corresponding to box 1 in **A** above. The image is a composite of Mars Express HRSC images and Mars Global Surveyor topography. Pre-launch ellipses are shown in white and pre-landing ellipses in black.

4 August 2007: Phoenix

Phoenix, the first Mars Scout mission, made use of the flight hardware built for Mars Surveyor 2001 before that mission was canceled following the Mars Polar Lander failure in 1999. The spacecraft had an octagonal frame with three legs and two octagonal solar panels spanning 5.5 m when open. The lander deck was 1.5 m across and 1.2 m above the surface, and the science instruments and a robotic arm were mounted on the deck. Phoenix communicated primarily through Mars Odyssey and Mars Reconnaissance Orbiter, with Mars Express as a backup and an X-band antenna to communicate directly with Earth if necessary. During its flight to Mars, the lander was enclosed within its heatshield and backshell and was controlled by a small cruise stage with two solar panels. The cruise stage would have partly burned up in the atmosphere, but parts may have struck the surface tens of kilometers northwest of the landing site.

The instruments were a Microscopy, Electrochemistry, and Conductivity Analyzer (MECA), a Thermal and Evolved Gas Analyzer (TEGA), with eight ovens to heat samples for analysis, a Canadian meteorological station (MET), and cameras on a mast (the Surface Stereo Imager, SSI) and on the robotic arm (the Robotic Arm Camera, RAC). RAC included red, green and blue LED lights to illuminate the scoop contents for three-color imaging with the monochrome camera. The arm also carried a Thermal and Electrical Conductivity Probe (TECP), which could be held up to measure atmospheric water vapor, or pressed into the soil to measure thermal and electrical properties of the surface material with its needle-like probes. MECA had four components, an Atomic Force Microscope (AFM), which mapped sample grains at resolutions down to 10 nm, an Optical Microscope (OM), with a resolution of 4 μm for imaging soil grains, a Wet Chemistry Lab (WCL), with four cells for sample analysis, and the TECP on the arm. A Mars Descent Imager (MARDI) was mounted under the lander to obtain images during landing, but it was not used until sol 146. Pre-flight testing revealed a problem with data handling on the spacecraft which could have threatened the landing, so MARDI was not used during the descent. On sol 146 it failed to turn on, so no data were ever received from it.

MET consisted of an upward-pointing lidar, a pressure sensor and three temperature sensors at different elevations on a mast rising 1 m above the deck. A small free-hanging "telltale" on the mast moved in the wind, so images of it provided estimates of wind speed and direction, and a tilted mirror under the telltale showed the direction of its motion. The Organic-Free Blank (OFB) was a carbon-free ceramic plate attached to the RAC base and prevented from contamination by a biobarrier wrapped around the Robotic Arm (RA) during spacecraft construction and flight. It could provide samples for TEGA to calibrate any potential detection of organic materials. The total science payload mass was 55 kg.

The goals of Phoenix were to study water, ice and organics in the regolith, to examine the ice–soil boundary and its biological potential, and to collect meteorological data at a location where ice was within reach of the sampling arm. The GRS instrument on Mars Odyssey had suggested that up to 50 percent of the top 50 cm of regolith at high latitudes consisted of water ice, putting it within easy reach of a lander's robotic arm. Phoenix would also characterize the geomorphology and current processes influencing the northern plains.

Phoenix was launched from Cape Canaveral Air Force Station at 09:27 UT on 4 August 2007, and arrived at Mars on 25 May 2008 (MY 29, sol 164) after mid-course trajectory corrections on 10 August 2007, 30 October 2007 and 10 April 2008. The October correction had been delayed for a week when a cosmic ray strike on a computer chip caused a "single event upset" fault, a rare problem but well understood. Spirit, Opportunity and Curiosity all experienced these faults while on the surface of Mars. The cruise stage solar panels generated more power than was needed when close to Earth's orbit, so they were oriented obliquely to the Sun to reduce power output. On 6 November 2007 the spacecraft was turned to provide full power as the distance from the Sun increased. The instruments were checked during cruise and TEGA was found to contain excess humidity, so it was heated to remove the water.

Phoenix used a heatshield and parachute to reduce its velocity before landing on thrusters, as Viking had done 32 years earlier. The spacecraft was too heavy and delicate to use the airbag style of landing adopted for Mars Pathfinder and the Mars Exploration Rovers. It landed just before the northern summer solstice and north of the

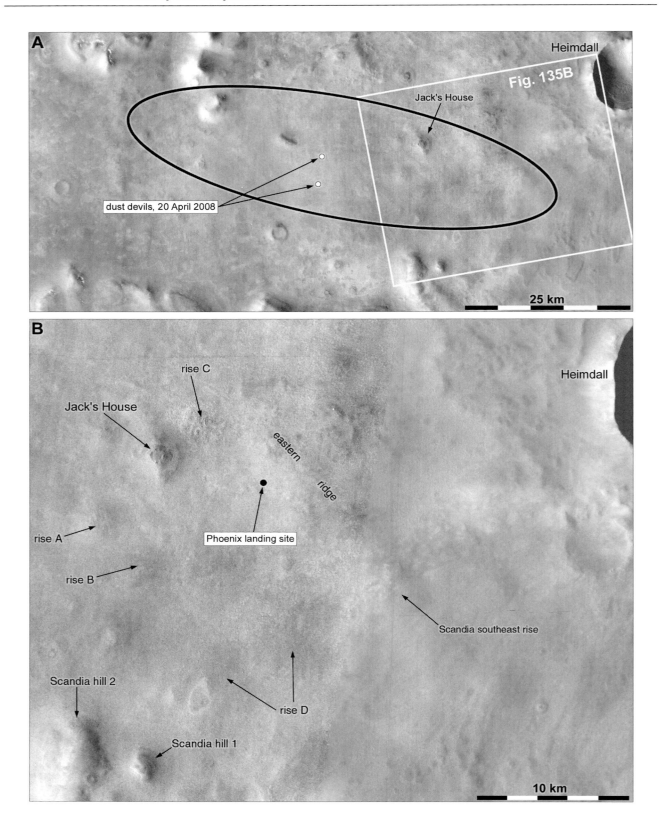

Figure 135. Phoenix landing site. **A:** The landing ellipse. **B:** The region around the landing site. The backgrounds are Mars Odyssey THEMIS visible image mosaics. **A** is oriented as shown in Figure 134B, **B** has north at the top. The rises in **B** are identified in Figure 138.

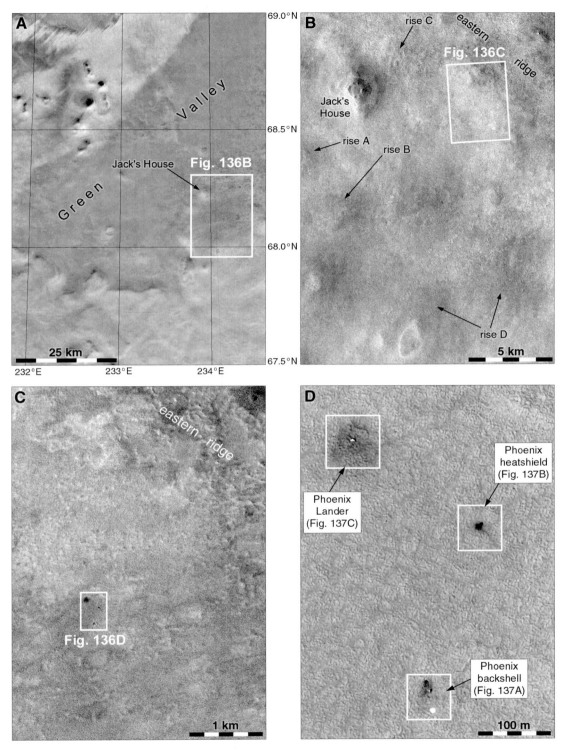

Figure 136. Phoenix landing site. **A** is part of Figure 134B. **B** is part of Figure 135B. **C** and **D** are parts of HiRISE image PSP_008591_2485 taken on Phoenix sol 1 (26 May 2008, MY 29, sol 165).

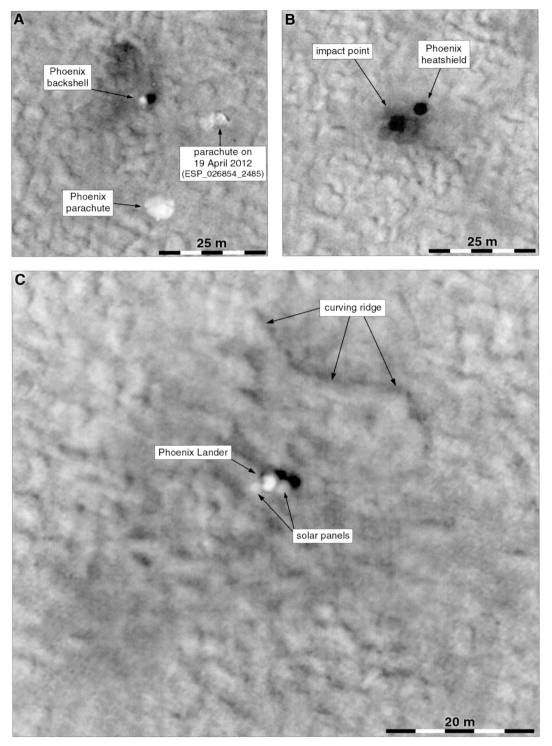

Figure 137. Phoenix components on the surface of Mars. The base is HiRISE image PSP_008591_2485 taken on 26 May 2008 (MY 29, sol 164). The parachute had moved by 19 April 2012 (MY 31, sol 213).

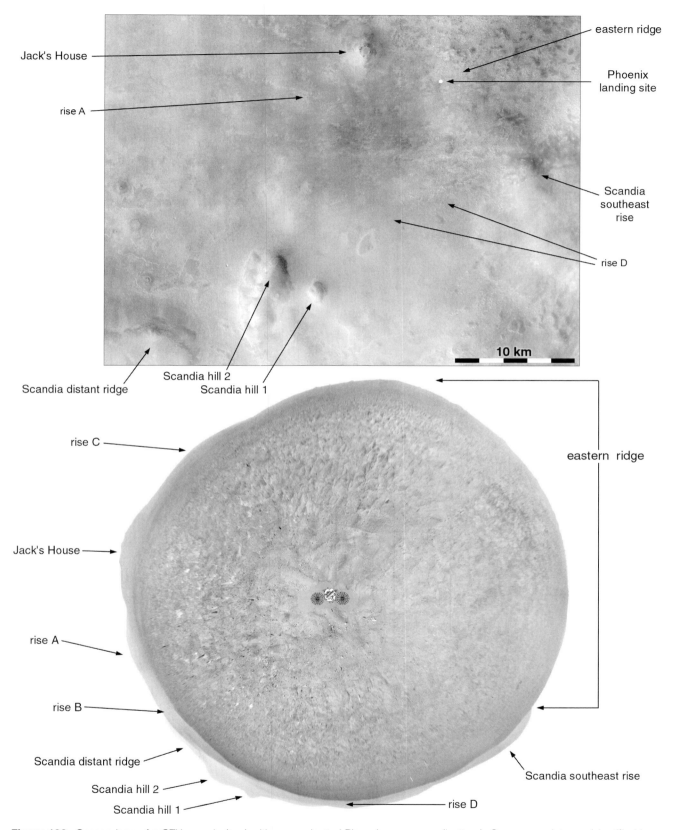

Figure 138. Comparison of a CTX mosaic (top) with a reprojected Phoenix panorama (bottom). Common points are identified here and in Figure 135.

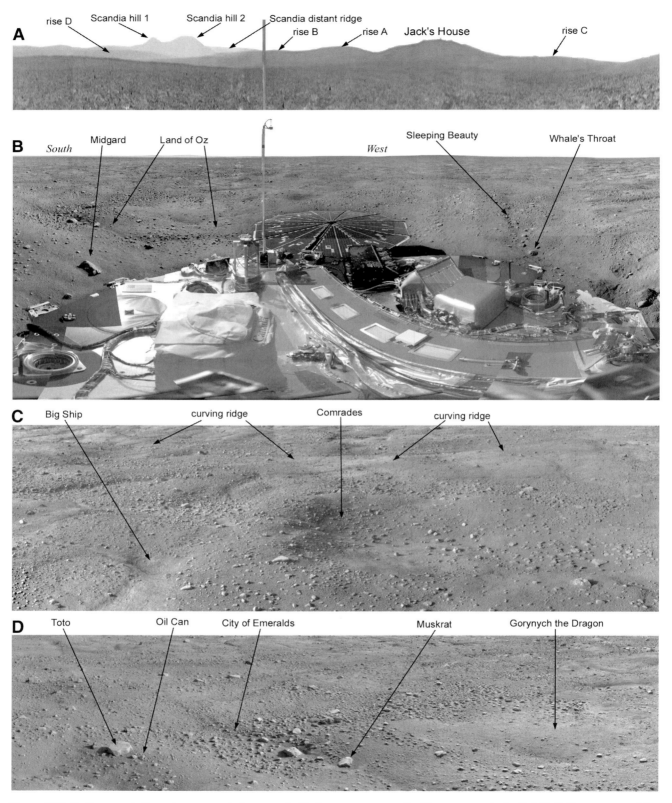

Figure 139 (both pages). Phoenix panoramic images. **A:** Horizon panorama with 10× exaggerated relief. **B:** Full panorama of the landing site. **C–F:** Enlarged details of the full panorama.

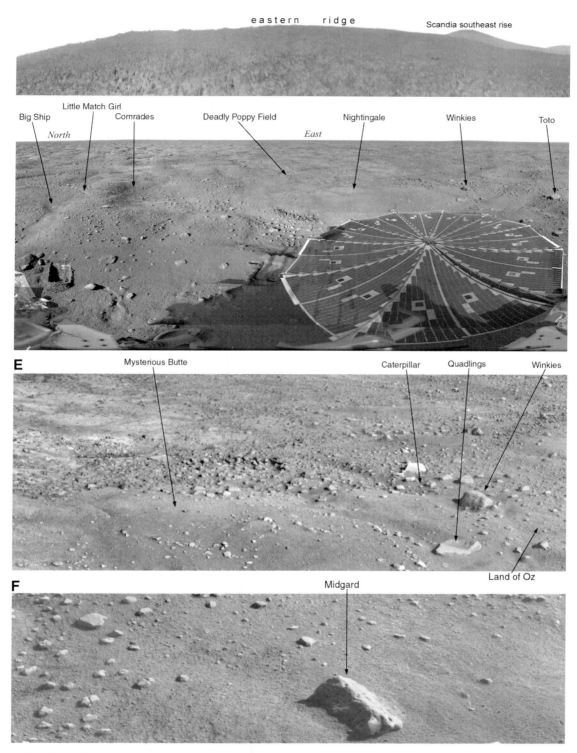

Figure 139 (continued)

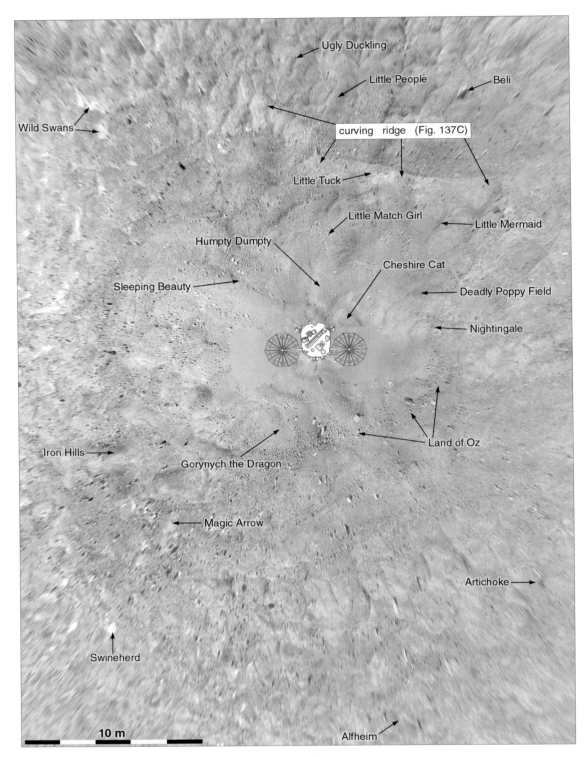

Figure 140. Names of surface features around the Phoenix lander. The background is a reprojected surface panorama.

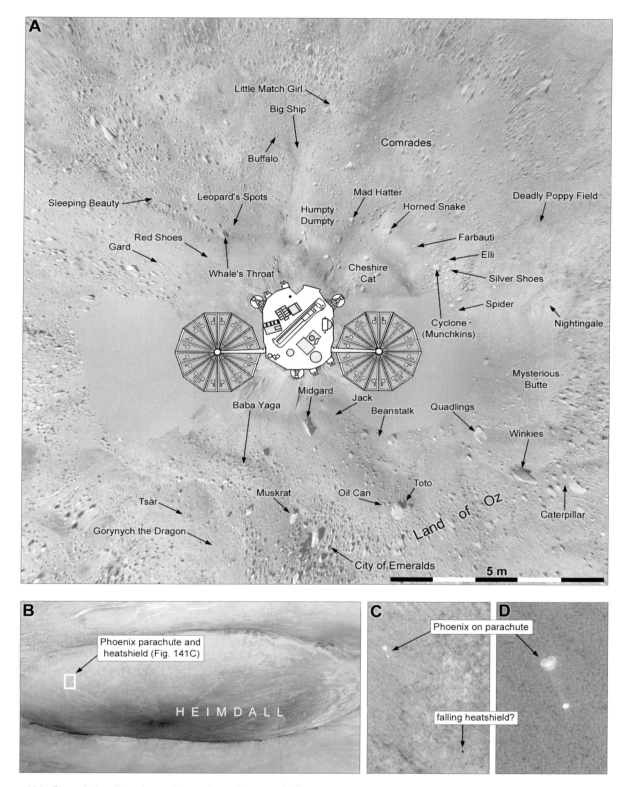

Figure 141. Phoenix landing site and parachute descent. **A:** Feature names near the lander. The base consists of surface images projected to fit a map in Mellon *et al.* (2009). **B:** Part of HiRISE image PSP_008579_9020 showing Phoenix descending on its parachute with Heimdall crater in the background. **C** and **D:** Details of **B**.

Table 38. *Phoenix Activities During the Primary Mission*

Sol	Activities
0	Land, image landing site, solar arrays, horizon, footpad, MET mast, confirm equipment deployment
1	Begin site panorama, image RA work area, check WCL and TECP, HiRISE image of landing site
2	MRO communication relay problem, color imaging of landing site
3	Begin RA deployment, complete first full panorama at reduced scale, check AFM and TEGA
4	Begin color panorama, check lidar, WCL, OM, note TEGA anomaly, unstow RA, Snow Queen seen
5	RAC images of third footpad viewed underneath the lander, Holy Cow ice exposure noted
6	RA touch test at Yeti, attempt RAC color imaging of Snow Queen (unsuccessful)
7	RA test dig at Dodo, color imaging of scoop contents, sample dumped, TEGA cover opened
8	RA positioning tests, additional images of Holy Cow, diagnose TEGA door failure to open completely
9	RA dig at Dodo, image and dump sample, first OM dustfall images, TEGA atmosphere analysis
10	Mars Odyssey safe-mode delays sampling, color panorama extended, image wind telltale and sky
11	TEGA atmosphere, Baby Bear sample dug at Goldilocks, RA moved to TEGA, image Lory, Mad Hatter
12	Baby Bear sample delivered to TEGA 4 but does not enter oven, OM dustfall images
13	Delay TEGA analysis, RA dig Dodo and image Snow Queen, locate biobarrier spring, extend panorama
14	Shake TEGA screen, image Baby Bear sample site, RA dig Mama Bear sample, extend panorama
15	Shake TEGA screen, test RA sprinkling method of sample delivery to instruments, extend panorama
16	TEGA atmosphere, shake Baby Bear into TEGA 4, color images of soil on lander, extend panorama
17	RA sprinkle Mama Bear sample to OM, OM operations, image Cheshire Cat area, extend panorama
18	TEGA 4 oven low heat, extend Dodo–Goldilocks trench, arm movie, image trench, extend panorama
19	Check AFM, extend and image trench, arm movie, OM image magnets, image Snow Queen, panorama
20	TEGA 4 oven medium heat, scrape trench, arm movie, infrared panorama for albedo studies
21	OM magnets, monitor trench, debris, Snow Queen for changes, TECP wind/humidity profile
22	TEGA 4 oven high heat, RA dig new trench, Snow White, in Wonderland area
23	Computer flash memory problem prevents science activities, causes loss of stored RAC images
24	Observe disappearing ice fragments in Dodo–Goldilocks, RA extend Snow White trench
25	Prepare OM for sample, TEGA 4 high heat, TEGA door 5 barely opens, RA dig Rosy Red, shake sample
26	Sprinkle Rosy Red on OM, image TEGA doors to diagnose problem, TEGA 4 high heat
27	Lander enters safe mode due to computer scheduling conflict, no science activity
28	Thaw WCL reagent tank, RA sample delivery to WCL fails, monitor Dodo–Goldilocks, check TEGA
29	OM images of Rosy Red, reposition RA over WCL, apply software patch, collect atmospheric data
30	Solstice, Rosy Red sample to WCL 0, test TEGA delivery, extend panorama, WCL analysis
31	Scrape to dirty ice at Snow White, image Holy Cow, fill panorama gaps (RePeter Pan), TECP profile
32	Extend Snow White to expose large ice area for sampling, collect atmospheric data
33	Multiple scrapes at Snow White to gather ice fragments, OM image magnets, fill gaps in panorama
34	WCL analysis, RA dig Sorceress sample from Snow White, practice delivery to TEGA and WCL
35	Prepare OM for sample, image spacecraft deck to complete panorama, hold sample in scoop over TEGA
36	Hold sample, monitor ice areas in both trenches, coordinated atmospheric studies with SSI and MET
37	Continue to hold Sorceress sample, make coordinated photometry observations with MRO
38	WCL thaw test, sprinkle small sample to OM after ice has sublimed out of it, OM image sample
39	Panoramic imaging and atmospheric observations during team break (4 July holiday)
40	Panoramic imaging and atmospheric observations
41	Sorceress sample delivered to WCL 1, WCL analysis, TECP touch test fails to touch surface at Vestri
42	More scraping at Snow White, image scoop for signs of icy material for TEGA, test AFM
43	WCL analysis, Wonderland scraping and scoop monitoring, TECP touch test, panorama gaps filled
44	AFM tests completed, site imaging, RAC imaging under lander, TEGA calibration, OM operations
45	Scraping continues at Snow White, more site imaging, AFM diagnostics
46	Image TECP in soil, monitor Dodo–Goldilocks, atmospheric data, night imaging, AFM tests
47	Image Sun as it nears horizon, retract and image TECP, RA accidental contact (Runaway), TEGA heating

Table 38. (*cont.*)

Sol	Activities
48	Diagnose RA contact, image arm and Midgard, night imaging, AFM calibration, atmospheric observations
49	Extend Snow White, RA tests, examine scoop rasp for future use, examine TECP, AFM diagnostics
50	RA load plate test, rasp Snow White, fail to collect ice, monitor solar panel, AFM calibration, WCL test
51	Scrape in Snow White stops after striking buried object, AFM test, OM images, TECP atmospheric data
52	Image arm over Snow White, site imaging, night and atmospheric observations (still not dark at night)
53	Open TEGA 0 doors, TEGA heat, test rasp, dig four holes to sample ice, little or no sample collected
54	Make animation of RA load plate test, deploy TECP at Vestri, image midnight Sun near the horizon
55	Coordinated atmospheric observations, SSI and lidar with MRO, TECP at Vestri, RAC frost search
56	TECP soil data, retract TECP, AFM tests and temperature data, RA rasp test, dig four holes, collect ice
57	Begin Happy panorama, scrape Snow White 80 times to clean ice surface, monitor changes as ice sublimes
58	TEGA bake, RA five scrapes, frequent images to monitor sublimation, RA test, OM pre-sample images
59	TEGA pump, site imaging and change monitoring in Dodo–Goldilocks and Snow White
60	RA scrape and 16 rasps in Snow White, Glass Slipper sample sticky, none to TEGA 0, RAC image TEGA
61	RA sprinkle test to MECA, study sample delivery fault, sample is unexpectedly cohesive, possibly moist
62	Repeat scrape and rasp, but Shoes of Fortune sample also sticky, none to TEGA 0, RAC image TEGA
63	Atmospheric observations while sample problem is considered, test TEGA pump, RAC dust devil search
64	RA scrape Wicked Witch sample from Snow White, deliver to TEGA 0, TEGA low heat, AFM calibration
65	Coordinated atmospheric observations with SSI, lidar, TECP as TEGA low heat analysis continues
66	RA collect Rosy Red 2 sample, deliver to WCL 0 and analyze, atmospheric data, TEGA 0 medium heat
67	Mother Goose sample to OM, RA Upper Cupboard trench, image Neverland, TEGA 0 organics analysis
68	TEGA 0 high heat, RA dig Neverland trench, dig Snow White headwall, image laser shot, AFM scans
69	TEGA 0 organics analysis, RA test sample delivery, dig Lower Cupboard, AFM scans, TECP at Vestri
70	Photometric studies, night observations, TECP at night, frost search, AFM scans, TEGA 0 high heat
71	Frost monitoring, image trenches, Snow Queen, Holy Cow, RAC calibration, TECP soil, RA Neverland
72	RA dig Rosy Red 3 sample, image in scoop, attempt delivery to TEGA 5, prepare OM for sample
73	RA extend Neverland, TEGA shake to deliver sample, OM and AFM data, RA image under lander
74	RA dig Stone Soup, collect Wicked Witch 2, vibrate Rosy Red 3 into TEGA 5, OM and AFM imaging
75	Frost search, Wicked Witch 2 to OM, RA dig Burn Alive, TEGA images, TEGA 5 low heat, OM analysis
76	Test TEGA 7 delivery, TEGA images, TEGA 5 high heat, AFM and OM diagnostics, RA dig Stone Soup
77	Frost search, RA scrape Burn Alive trench, trench imaging, TEGA 5 high heat
78	Frost search, add to Happy Pan, AFM scans, WCL thermal diagnostics
79	Open TEGA 7, TEGA pump test, RA extend Upper Cupboard, scrape Burn Alive, AFM tests
80	Frost search, AFM diagnostics, extend Happy Pan, image Caterpillar dump pile
81	RA fault prevents sampling at Burn Alive, atmospheric studies, OM and AFM images, WCL tank thaw
82	Frost detected, extend Happy Pan, AFM scan, image Dodo–Goldilocks and Burn Alive trenches
83	RA dig Burning Coals sample at Burn Alive, atmospheric observations, surface photometry, OM data
84	Frost search, coordinated atmospheric data, map workspace, image TEGA magnets, AFM scans
85	RA deliver Burning Coals to TEGA 7, dig Stone Soup, extend Happy Pan, AFM scans, atmospheric data
86	Frost search, TECP in Upper Cupboard, image Caterpillar, OM night dustfall sample collected
87	Frost search, scrape Cupboard and examine scoop, complete infrared panorama, WCL thermal test
88	Extend Stone Soup trench, image Caterpillar dump pile, test sample delivery to WCL 3, extend Happy Pan
89	RA test move to TEGA 0, sample to TEGA 7 and low heat, image Holy Cow, scrape Snow White, AFM
90	Image sunrise, Holy Cow, workspace, frost search, RA dig Golden Goose sample, TEGA medium heat

Notes: AFM, Atomic Force Microscope; OFB, Organic-Free Blank; OM, Optical Microscope; MCS, Mars Climate Sounder (on MRO); RA, Robotic Arm; RAC, Robotic Arm Camera; SSI, Surface Stereo Imager; TECP, Thermal and Electrical Conductivity Probe; TEGA, Thermal and Evolved Gas Analyzer; WCL, Wet Chemistry Lab.

Martian arctic circle (64.8° N). On 20 April 2008 (MY 29, sol 129), 35 days before Phoenix landed, the MRO Context Camera had seen afternoon dust devils inside the landing ellipse (Malin *et al.*, 2008).

Figures 134B and 135A show the landing ellipse and surrounding area, and Figure 135B shows the landing site in Green Valley at still larger scale. The section of Green Valley north of Heimdall crater was called Eastern (or East) Valley. The site was generally very flat, but several distant hills were visible southwest of the lander, and formed part of Scandia Colles, a hilly region extending from 60° N to 70° N between 200° E and 250° E. Figure 139A is a panorama of the horizon with greatly exaggerated relief, which helps reveal the topography of this very flat site. A panorama of the landing site without exaggeration is shown in Figure 139B. The surface northeast of the lander formed a gentle rise, the eastern ridge in Figure 138 which hid Heimdall crater, and west of the lander a small crater was situated on a local rise informally called Jack's House. Surface features around the lander were given names from children's stories and fairy tales, and named features are identified in Figures 135, 136 and 140 and subsequent maps and images.

The surface operations spanned the range of solar longitudes $L_s = 77°$ to $148°$. The Sun was above the horizon all day at the time of landing, and did not dip below the horizon until August 2008. Surface temperatures varied between about 190 and 245 K at the start of the mission, and between 178 and 235 K at the end (Davy *et al.*, 2010). Eventually, the solar power generated by the two panels was too limited to keep the spacecraft operating, and the mission ended on 2 November 2008 (MY 29, sol 317). Attempts to revive it in the following spring (early 2010) were unsuccessful, as expected.

Mellon *et al.* (2009) mapped the landing site using stereoscopic images from SSI. The surface in the landing area was covered with closely packed polygonal mounds roughly 2–4 m across, separated by narrow troughs and pits. On Earth, this kind of surface, called patterned ground, is caused by alternating freezing and thawing in high-latitude or high-altitude areas. In some cases the polygons lie over buried ice with sand-filled wedges extending downwards under the troughs, but the opposite arrangement can also occur with ice wedges under the

troughs. One goal of Phoenix was to sample both polygon and trough landforms to explore this relationship, and indeed excavation by Phoenix showed that these troughs at the Phoenix site were sand wedges, not ice wedges. In the Dodo–Goldilocks trench (Figure 144), massive ice (that is, masses of clean ice without significant soil contamination) was exposed by digging. In the Snow White trench (Figure 145), pore ice (that is, ice occupying the spaces between grains) was encountered, so there was considerable variability in ice distribution even within the very limited area accessible to Phoenix.

All the ice referred to above is frozen water, and some was visible from orbit. Cull *et al.* (2010) described CRISM data showing water ice on the north-facing (southern) inner wall of Heimdall crater (Figure 141B). During the winter the Phoenix landing site would be covered with 30–40 cm of carbon dioxide frost, enough to damage the spacecraft so that it was unlikely to be revived in the following spring. The geology of the landing region is described by Mellon *et al.* (2008), Arvidson *et al.* (2009) and Heet *et al.* (2009).

The lander touched down on the surface of Mars at 23:54 UT, when Mars was 275 million kilometers from Earth, with a one-way light time delay of about 15 min. The final target coordinates were 68.25° N, 233.4° E, and the first estimate of the landed location was 68.22° N, 234.3° E, downrange of the ellipse center. MRO obtained a remarkable image of the spacecraft descending on its parachute with Heimdall crater in the background (Figure 141B). The falling backshell might have been visible beneath it. Only a day after landing, Phoenix was imaged on the ground by MRO (Figure 137) and its exact location was determined to be 68.219° N, 234.251° E. The site was imaged frequently by HiRISE after the landing, and in images taken during and after April 2012 the parachute was seen to have been blown to a different location 30 m from its original resting place (Figure 137A). It was still at the new location early in 2014.

After landing, the early views of the surface showed that Phoenix was resting on top of one of the polygons. Two other polygons lay within reach of the robotic arm on the north side of the lander, Humpty Dumpty to the northwest and Cheshire Cat to the east, separated by a low area called Sleepy Hollow (Figure 143). A small trough whose deep western end was called Rabbit Hole

Table 39. *Phoenix Sample Collection and Delivery From Arvidson* et al. *(2009)*

Sample	Sol collected	Source area	Instrument	Sol delivered
Baby Bear	11	Goldilocks	TEGA oven 4	12 (failed), 16
Mama Bear	14	Goldilocks	MECA OM 2	17
Rosy Red-Sol 25	25	Rosy Red	MECA OM 1	26
Rosy Red-Sol 25	25	Rosy Red	MECA WCL 0	30
Sorceress	34	Snow White	MECA OM 10	38
Sorceress	34	Snow White	MECA WCL 1	41
Glass Slipper	60	Snow White	TEGA oven 0	60 (failed)
Shoes of Fortune	62	Snow White	TEGA oven 0	62 (failed)
Wicked Witch	64	Snow White	TEGA oven 0	64
Rosy Red-Sol 66 (Mother Goose)	66	Rosy Red 2	MECA OM 8	67
Rosy Red-Sol 72	72	Rosy Red 3	TEGA oven 5	72, 74
Wicked Witch 2	74	Snow White	MECA OM 7	75
Burning Coals	83	Burn Alive	TEGA oven 7	85
Golden Goose	90	Stone Soup	dumped	92
Golden Goose 2	95	Stone Soup	MECA WCL 3	96
Golden Key	99	Dodo–Goldilocks	MECA OM 6	99
Golden Goose 3	101	Stone Soup	MECA WCL 3	102, 147 (failed)
Sorceress 2	105	Snow White	MECA WCL 2	107
Golden Goose	110	Stone Soup	MECA OM 5	110
Sam McGee	113	Snow White	TEGA oven 1	113 (failed)
Sam McGee	120	Snow White	TEGA oven 1	120 (failed)
Blank	122	Organic-Free Blank	TEGA oven 2	122 (failed)
Galloping Hessian	125	under Headless rock	MECA OM 4	125 (failed)
Wicked Witch 2	126	Snow White	TEGA oven 1	126 (failed)
Galloping Hessian	128	under Headless rock	MECA OM 4	128
Rosy Red-Sol 130	130	Rosy Red N	TEGA oven 6	131 (failed)
Rosy Red-Sol 136	136	Rosy Red 2	TEGA oven 6	138

Table 40. *Phoenix TECP Activities, Modified From Zent* et al. *(2010)*

Sol	Location	Notes
43	Vestri	Touch test in Sleepy Hollow trough
46–47	Vestri	Sleepy Hollow trough
54–56	Vestri	Sleepy Hollow trough
69–71	Vestri	Sleepy Hollow trough
86	Upper Cupboard	Trench floor in Humpty Dumpty polygon
98	Gandalf	Sleepy Hollow trough
103	Sindr	Edge of Humpty Dumpty polygon
111	Rosy Red 2	Rim of sample pit on Cheshire Cat polygon
119	Upper Cupboard	Trench floor in Humpty Dumpty polygon
122–124	Vestri	Sleepy Hollow trough
149	Alviss	Humpty Dumpty polygon, end of mission

Note: TECP also made 233 measurements of temperature and humidity in the air.

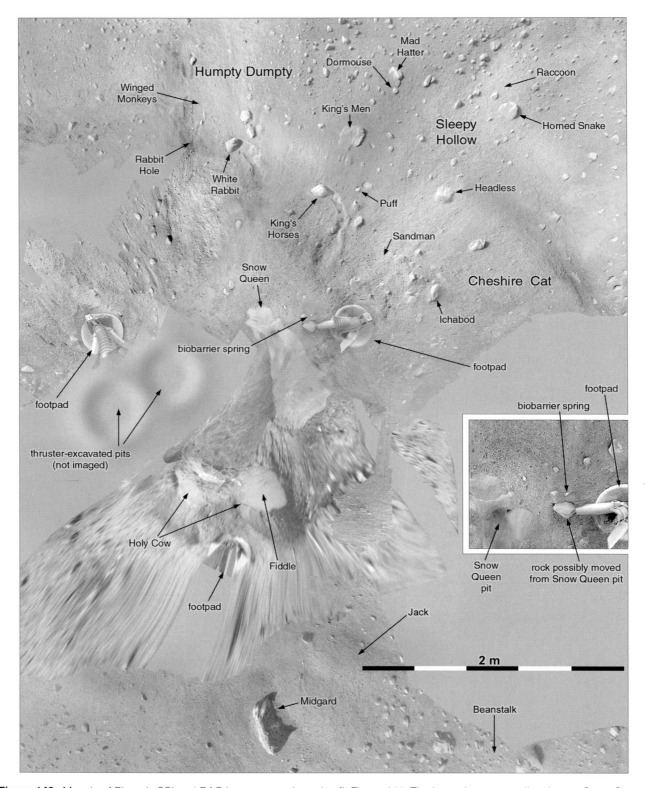

Figure 142. Mosaic of Phoenix SSI and RAC images reprojected to fit Figure 141. The inset shows a small rock near Snow Queen, which may have been moved from the surface above the pit on Snow Queen.

separated Humpty Dumpty from the polygon under the lander, and its northern slope, leading up to Humpty Dumpty, was called Wall. Two rocks, White Rabbit and King's Horses, originally perched on the south slope of the Rabbit Hole trough, were displaced by Phoenix during its landing and slid down into the depression, leaving distinctive tracks in the soil. The highest part of the Cheshire Cat polygon was called Wonderland, a name often applied to the whole polygon in post-mission literature. Its level surface, nearly free of rocks, became one of the main targets for analysis. Excavation in the polygons showed that Humpty Dumpty consisted of clean ice under a thin soil cover and Cheshire Cat contained dirty ice, as described above.

The polygon under the lander also appeared to contain clean ice, exposed after the thrusters blew soil away during landing. It is mapped in Figure 142, a composite of SSI and RAC images showing the maximum surface imaging coverage. The RAC images taken under the lander were viewed from a very low elevation, so topographic distortions in the center of the map are severe. The landing thrusters were mounted in groups inboard from each landing leg, and as the exhaust plumes impinged on the loose surface they blew fine material away to create pits, as also seen under Viking Landers 1 and 2 (Figures 76 and 106 in Stooke, 2012). The northeastern excavation exposed massive material, which might have been either rock or ice, but it degraded over time as if subliming slowly, confirming it was ice. It received the name Snow Queen. The southern excavation dug two pits which exposed ice on their floors. They were collectively named Holy Cow. The northwestern excavated area was not imaged, but radial scars made by ejected material were visible beyond the footpad.

The icy patch at Snow Queen contained a pit about the same size as a rock 40 cm away near the footpad (Figure 142 inset). Was this rock previously situated over the hollow and moved by the lander exhaust? Modeling suggested that the ice table should be depressed under the rock due to thermal conduction (Sizemore *et al.*, 2009).

A plan was devised for arm operations so the early testing of sampling procedures would not compromise later sampling, for instance, by mixing subsurface material with what was supposed to be a pristine surface

sample. Early tests of digging and sampling procedures would be conducted at the left (northwestern) end of the sampling area, while the rest was left untouched to avoid compromising data interpretation (Figure 143B). The untouched area was called "National Park," but it did not stay pristine for long. Soil would be dumped in several discrete piles beyond most of the digging, and the TECP would be used mainly in the outer parts of the sampling area.

The following description of the mission's surface activities is derived from Arvidson *et al.* (2009) and augmented by lists of day-by-day activities on the contemporary mission website, NASA's Planetary Data System (Phoenix Analyst's Notebook) and the Planetary Society website. The robotic arm activities are depicted in sol-by-sol images in Figures 144 to 157, with maps showing the state of the sample arm work area as it was on sols 0, 45, 90 and the end of the mission (Figures 143, 146, 152 and 158, respectively). The photomaps of the landing site (Figures 141, 142, 143, 146, 152 and 158) were constructed by reprojecting a polar projection of the highest-resolution surface panorama to match the geometry of the topographic map published by Mellon *et al.* (2009), and combining that with a mosaic of RAC images of areas hidden from the SSI. Sol-by-sol lander activities are listed in Tables 38 and 41, for the primary and extended missions, respectively, and sampling and TECP activities are summarized in Tables 39 and 40.

The day of landing, 25 May 2008 (MY 29, sol 164), was called sol 0. Following the landing and a short pause to vent helium to depressurize the thruster propellant system and to allow any raised dust to dissipate, the solar arrays and camera mast were deployed. The biobarrier that kept the RA sterile before use was partially removed from the arm, a task completed only on sol 4, and the first images were taken to confirm deployment of the solar panels and the nature of the landing site. These included a view of a footpad to show that it was firmly set on the ground, not perched on a rock, as successful arm operations would require a stable spacecraft.

The MET began operations on sol 0 and operated almost continuously throughout the mission, amassing the most detailed meteorological dataset acquired for Mars at that time. MET detected pressure changes suggestive of the passage of a dust devil on almost every sol of the mission, with a notable increase after sol 75

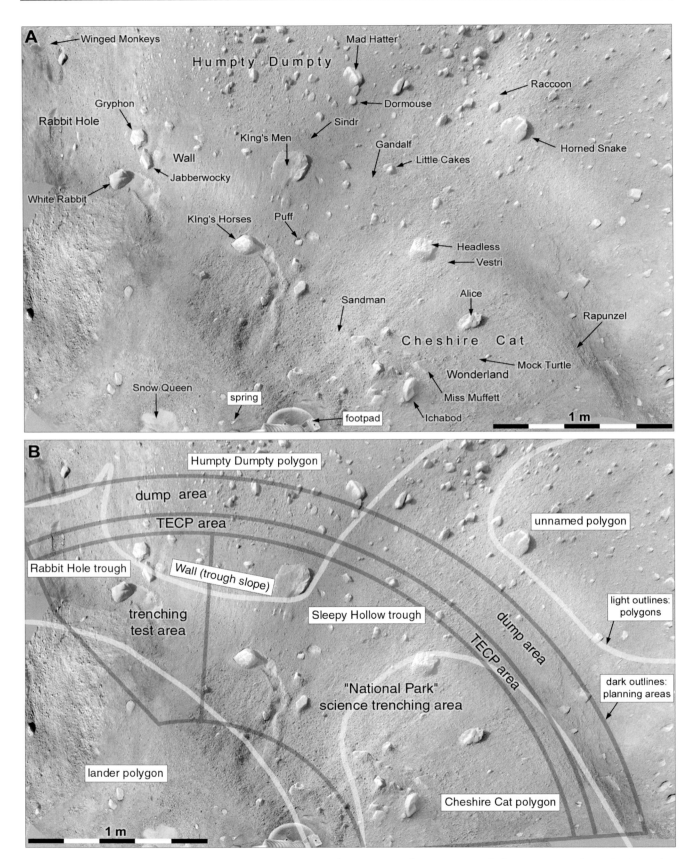

Figure 143. Phoenix sampling area. **A:** Mosaic of SSI and RAC images with feature names. **B:** Polygons, troughs and schematic outlines of areas used to plan Robotic Arm operations.

(Ellehoj *et al.*, 2010). Dust devils had already been seen from orbit before landing (Figure 135A; Malin *et al.*, 2008), but only showed up in surface images after sol 104.

On sol 1 the RA temperature sensor collected its first data, and the camera took preparatory images of the work area for initial planning of arm operations and pictures of a DVD on the spacecraft deck carrying the names of 250 000 people, Mars-related art and literature, and greetings from people including Carl Sagan and Arthur C. Clarke, all assembled by the Planetary Society, Pasadena, California. Two HiRISE images of the lander hardware on the surface were transmitted. This was the first mission for which a lander was found almost immediately in orbital images rather than by comparing features in its surface images with orbital data (Viking 2 was not identified until HiRISE images became available, but that was in 2006, some 30 years after the landing). On the next sol a color "postcard" (partial panorama) was transmitted, but other activities were delayed by a communications anomaly with MRO, the main link with Earth.

The arm was mostly unstowed on sol 3, and a full panorama at reduced resolution was transmitted to help characterize the landing site. The horizon was almost flat, but if viewed with substantial vertical exaggeration many distant features became visible (Figures 138 and 139). The hill called Jack's House, a rise in the direction of Heimdall's rim, and distant hills on the south side of Green Valley were all visible. The Canadian lidar meteorology instrument was used for the first time on sol 3 and tested on sol 4, a potential problem with the TEGA ion source was identified but did not prevent its operation, and the final steps in unstowing the RA were completed. RAC images of the footpad revealed a solid mass near it, which was named Snow Queen, a piece of ice previously buried but exposed by the landing thruster exhaust. A pit on its surface might have been the original location of a rock pushed by the thrusters towards the nearby footpad (Figure 142). The same image of the footpad showed a small object on the ground, first described as a screw and suspected of coming from the biobarrier release mechanism. A thin dust cloud passed over Phoenix on sol 4 and was seen by MARCI on the following day.

Images of the underside of the lander taken by RAC on sol 5 to document the third (rear) footpad's stability resulted in the discovery of another ice surface named Holy Cow exposed by thruster exhaust during landing. The existence of ice within a few centimeters of the surface had now been established without any arm operations at all. RAC also imaged the sample delivery locations for the OM, WCL and TEGA in preparation for later sample deliveries. The arm had now been tested and used for RAC imaging under the lander, but on sol 6 it made its first test surface contact, leaving a footprint-like mark, which was given the name Yeti (Figure 144). An attempt to obtain color images of Snow Queen using red, green and blue LED lights on the RAC failed because the lights were not intended to be used at that distance or in daylight.

Then on sol 7 the TEGA instrument cover was opened and the arm made its first test dig at Knave of Hearts, creating a trench called Dodo, which partly obscured Yeti. The soil was dumped at a place called Porridge at the western limit of the arm work area (Figure 144), after the RAC had taken color images of the contents of the scoop. TEGA experienced a potentially serious problem on sol 8 when its first door opening on oven 4 was only partially successful. This problem persisted throughout the mission, but useful analyses were still possible. Holy Cow was imaged again on that sol, and the arm tested its positioning at TEGA and the OM. On sol 9 the Dodo trench was expanded to test sample collection, completely obliterating Yeti and revealing the first glimpse of ice at a depth of 3 cm in the bottom of the trench. The excavated soil was dumped just to the west of Dodo at Lory after colour images had been taken of it in the scoop. The first OM images were also taken on sol 9.

The next day, sol 10 (MY 29, sol 174, or 4 June 2008) was a "runout" day when pre-planned contingency operations were performed as Mars Odyssey, the communication link for that day, entered a safe mode. The sol was used to take part of a large color panorama. On sol 11 a sample of surface material was picked up adjacent to Dodo in a new trench called Goldilocks. The scoop was left poised over TEGA at the end of the sol's operations, and only attempted to deliver its sample, called Baby Bear, to TEGA's oven 4 on the next sol. The soil was dumped over the partially open door but its cloddy nature prevented it from entering the analysis chamber.

On sol 13 the TEGA problem was considered further while the arm expanded Dodo towards the lander

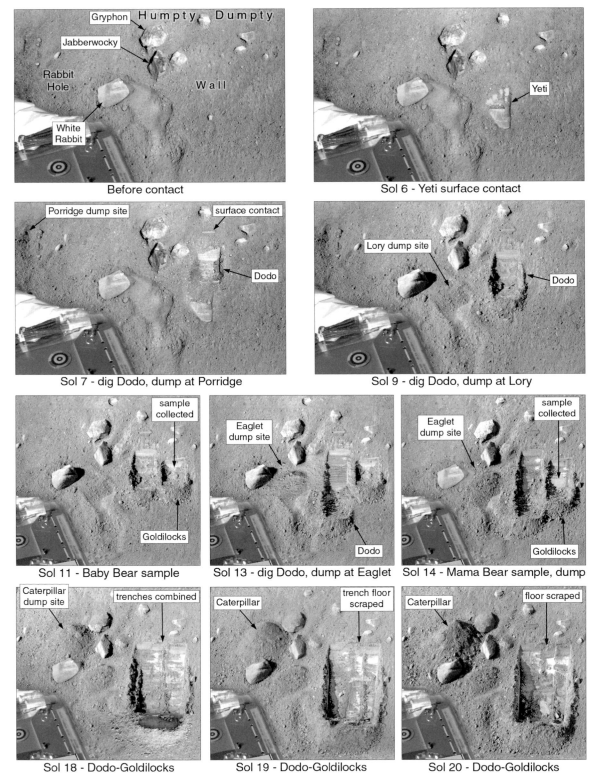

Figure 144. Phoenix surface activities at Dodo–Goldilocks, sols 0–20.

(Figure 144). The excavated material was dumped at Eaglet, just to the left of the trench. Also, more of the large color panorama was taken, RAC imaged Snow Queen again and SSI imaged the biobarrier area of the deck to help identify the object near the footpad (Figure 142). It was identified now as a spring from the biobarrier release mechanism. More of the panorama was taken on sol 14, as well as multispectral imaging of Goldilocks trench, which was made larger by the excavation of a sample called Mama Bear. An attempt was made to shake some of the soil sample into TEGA by vibrating its cover screen, which was supposed to stop large particles from blocking the entry to the small analysis chamber. This first attempt did not succeed.

On sol 15 more of the panorama was obtained and the arm tested a new way to deliver samples to TEGA by sprinkling some of the Mama Bear sample gently onto the MECA cover instead of dumping large amounts at once. The TEGA screen vibration finally allowed some soil from Baby Bear to enter the TEGA oven on sol 16, and the camera took more of the large panorama and observed a lump of soil that had fallen onto the DVD. Then on sol 17 some soil from Mama Bear was sprinkled onto the OM for the first time, RAC imaged the sample on a recessed area called the divot at the front of the scoop, and the cameras imaged more of the panorama and the large polygon called Cheshire Cat which would be the next trenching area. Sol 18 saw the first heating of the sample in TEGA oven 4 to 35 °C. The arm extended the two existing trenches, now given the combined name Dodo–Goldilocks. A movie sequence of this dig was obtained as well as multispectral imaging of the trench and soil dump areas, and more of the large color panorama.

Soil had been dumped into the shallow trench dug by White Rabbit, the rock displaced during landing, on sols 9, 13 and 14, but from sol 18 on it was deposited nearby to form a conical pile later named Caterpillar (Figure 144). On sol 19 Dodo–Goldilocks was expanded again. White ice patches were seen in the trench at a depth of about 3 cm, and a loose white fragment probably derived from the ice layer was seen in the trench. Imaging on sol 19 included more of the panorama, multispectral views of Caterpillar, an arm movie and continued monitoring of Snow Queen to look for changes as the ice responded to the surface environment. Many

images of the wind sensor, the little hanging "telltale," were taken almost every day, but from this time onwards they were made using very small subframe images to free space for transmission of other data.

The interesting white fragment in the trench was obliterated by pre-planned digging on sol 20 (15 June on Earth, or MY 29, sol 184), and an arm movie and an infrared panorama were obtained. Material dug out of Dodo–Goldilocks was dumped at Caterpillar. On sol 21 images were taken of Caterpillar, Dodo–Goldilocks and Snow Queen, and repeated often to monitor surface changes. The exposed ice in Dodo–Goldilocks gradually sublimed, leaving a small pit, which showed the ice was quite pure, not just occupying pore space in the soil. The surface became darker as the ice disappeared, and the dark material would be sampled on sol 99. This area of ice was, unfortunately, too far from the lander and too steeply sloping and irregular for easy sampling, so ice would be collected in a polygon named Wonderland at the other end of the arm work area instead. Also on sol 21, TECP was calibrated with atmospheric measurements, and HiRISE obtained a new image of Phoenix on the surface, while MARCI observed a dust cloud several hundred kilometres to the west.

A new trench called Snow White was begun on sol 22 in Wonderland (Figure 145). It was one scoop wide and 3–5 cm deep at this stage, and the excavated soil was dumped to start a new pile called Croquet Ground beyond the trench. Later that sol the spacecraft experienced a problem with its file-handling system, and no work was done on the next day while this was fixed. Further monitoring images of Dodo–Goldilocks on sol 24 showed that several small fragments seen in the trench on sol 21 had now disappeared, indicating that they were composed largely of ice. The arm dug again at Snow White on sol 24, widening the trench to two scoop widths, and added more material to Croquet Ground. The arm stopped digging when it encountered very hard material.

A small atmospheric event was observed on sol 25 (Ellehoj *et al.*, 2010). Earlier, on sol 19, MRO had observed a small dust storm in the largely frost-covered dune field of Olympia Undae (coordinates). Two days later the dust activity had moved further south, and by sol 22 most of the frost had cleared from those dunes. The expanding dust cloud reached the Phoenix area on sol 25,

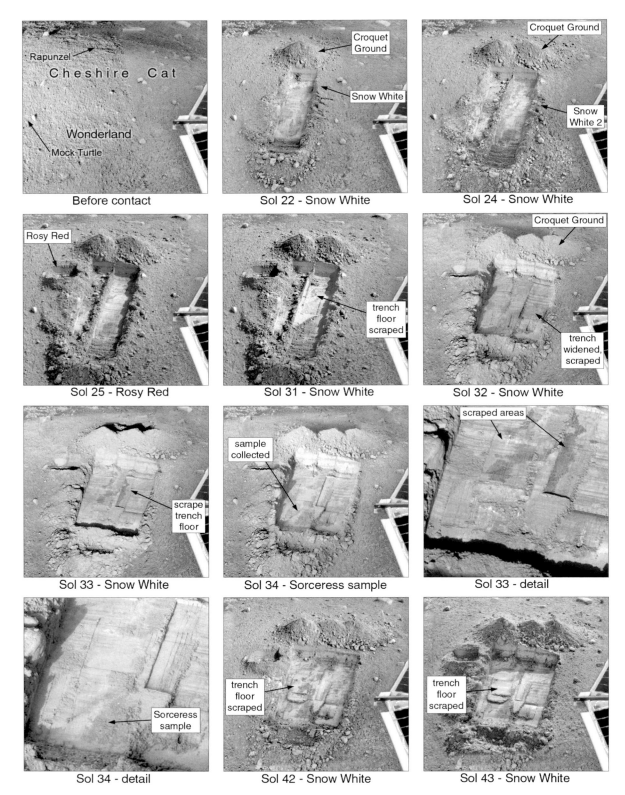

Figure 145. Phoenix surface activities at Wonderland, sols 21–43.

reducing the amount of sunlight and raising the atmospheric pressure slightly. TEGA door 5 was commanded to open on sol 25, but it opened only slightly. A sample of surface material referred to as Rosy Red-Sol 25 was taken from a small trench called Rosy Red, just to the left (north) of Snow White, followed by a test of the sample sprinkling procedure over Croquet Ground. As clods fell out of the scoop and hit the ground, they seemed to break up, suggesting that they were only weakly held together and should not clog the TEGA doors too much. Dodo–Goldilocks was imaged again, and areas of ice exposed in that trench were seen to darken gradually as the ice evaporated. Part of the Rosy Red-Sol 25 sample was sprinkled onto the OM on the next sol, the second sample for that instrument, RAC imaged the divot again, and pictures of the TEGA doors were taken to try to understand their opening problem.

No science was done on sol 27 because of a safe mode caused by a command sequence conflict. Operations resumed on sol 28 with an attempt to deliver material to the WCL, but the arm was positioned incorrectly. More imaging was done for the almost complete large color panorama whimsically named the Peter Pan (after Peter Smith, the mission Principal Investigator), to diagnose the TEGA door issue and to continue monitoring Dodo–Goldilocks. On sol 29 the OM was tested for future imaging (neither of its samples had been imaged yet) and the arm was placed correctly over the WCL 0 funnel. A computer code patch was successfully loaded on the lander to correct the cause of the sol 22 file system problem, and atmospheric observations continued.

Sol 30 was MY 29, sol 194, or 25 June 2008 on Earth. The remainder of the Rosy Red-Sol 25 surface material sample was finally delivered to WCL 0. The arm was moved to test delivery of a sample to TEGA oven 5 and then to document Burn Alive. The camera system obtained some additional panorama images, monitored Dodo–Goldilocks, and viewed its shadow to make a kind of self-portrait to mark the Martian northern summer solstice. More digging, 50 scrapes in a single column on the very hard surface, finally reached the ice interface in Snow White on sol 31. Whereas the ice in Dodo–Goldilocks was clean when exposed, here in Snow White it was dirty but very hard. Images of Holy Cow were taken in the middle of the day on the same sol, and some repeat imaging of parts of the main panorama was done

to fill gaps. This was humorously called the "RePeter Pan" (repeater), and when finished it marked one of the criteria for full mission success. TECP was held at different levels to obtain a vertical temperature profile 2 m high, and OM imaged the sample it received on sol 17.

The arm dug again at Snow White on sol 32 to try to reach cleaner ice, widening the trench to three scoop widths and grooming (scraping) the floor, with the excavated material dumped on Croquet Ground (Figure 145). More atmospheric observations were made, as on most days, showing cloud and dust movements. These horizon and zenith cloud movies were collected throughout the mission, beginning on sol 6 (Moores *et al.*, 2010). On the next sol another 50 scrapes were made in each of three columns in the central part of Snow White to try to build small piles of ice-rich soil. OM took more images, this time of the sol 26 sample, and panoramic imaging continued as well.

On sol 34 the arm scooped some of the collected soil sample, called Sorceress, from the lander end of Snow White and practiced delivering it to TEGA and WCL to test arm placement. The sample was held in the TEGA delivery position on the next sol while its temperature was monitored and images of the lander deck were taken. The sample continued to be held on sol 36 while Dodo–Goldilocks and Sorceress, the latest sampling site in the Snow White trench, were monitored. On the next sol, coordinated photometric observations were made with Odyssey. The sample in the scoop had been held long enough that it had probably lost any water it had originally held, so it was referred to as a sublimation lag. Then on sol 38 RAC imaged the sample on the divot, and a very small part of this sample was sprinkled onto the OM. Sols 39 and 40 (MY 29, sols 203–204) fell on the US holiday weekend, 4 and 5 July 2008. The spacecraft team took a break while pre-planned panoramic imaging and atmospheric observations continued on Mars.

When work resumed on sol 41, part of the Sorceress sample was delivered to WCL 1 and its analysis commenced. The TECP was moved to the ground for the first time, initially to try a gentle surface contact to ensure proper arm placement. TECP had been used in the air to measure humidity and temperature and to estimate wind speed on almost every sol since its first checkout on sol 1, and these atmospheric observations continued until sol 148 (Zent *et al.*, 2010), but it was placed in the soil on

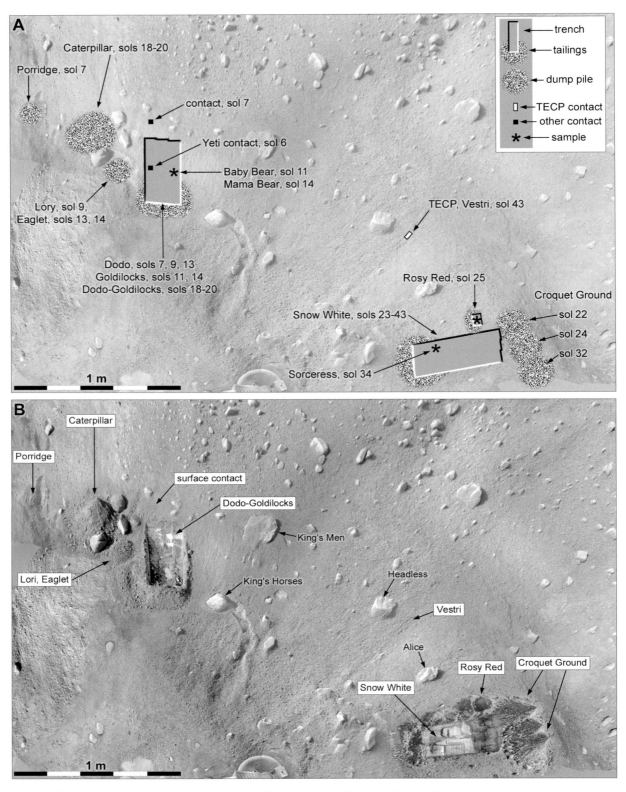

Figure 146. Phoenix activities between sols 0 and 44. This map is continued in Figure 152.

only 11 occasions (Table 40). This first contact was done very conservatively and TECP did not reach the ground this time. On the next sol the arm cleaned and scraped the bottom of Snow White, with 10 scrapes in each of two columns and 80 in a third column. Images were taken of the scoop to see if an icy sample had been acquired, in preparation for later procedures. The arm actuator heater was also tested, and the AFM was prepared for operation.

The WCL analysis continued on sol 43 while Snow White was scraped again (Figure 144) and once more the scoop was examined for evidence of an icy sample. Next, TECP performed a successful touch test at Vestri, contacting the surface but not collecting any data. This location and subsequent TECP placements at Vestri are shown in Figure 150. Later on sol 43 the last images for the mission success color panorama were obtained. The AFM checkout concluded on sol 44 and the cameras took more images, including views obtained by RAC underneath the lander. Arm temperatures were monitored and the delivery pose required to drop a sample into TEGA oven 0 was tested. OM took images of the sample it received on sol 38. Activities up to this point are mapped in Figure 146.

Sol 45 saw continued scraping on the bottom of Snow White trench (Figure 147), 20 scrapes in each of four columns, and imaging of interesting targets near the lander. On the next sol RAC obtained images of the north side of the TEGA instrument, and then TECP was successfully placed in the soil at Vestri for its first measurements and imaged in place (Figure 150). Dodo–Goldilocks monitoring continued, and atmospheric data were collected, including some night-time sky imaging to measure sky brightness. At this season the Sun was always above the horizon so there was no real night, but these images were taken when the Sun was at its lowest elevation. Then on sol 47 the "midnight Sun" (the solar disk at its lowest point near the horizon) was observed. TECP was removed from the soil and imaged, after which the arm was moved towards Snow White, but it accidentally struck the surface near the rock Alice, making a small trench called Runaway.

This unexpected event was studied on sol 48 along with stereoscopic imaging of the arm and scoop and multispectral imaging of a rock called Midgard with an unusual pitted or porous texture. Midgard was behind the lander and was also visible in images taken under the lander with the RAC. More night imaging was undertaken as well. Sol 49 was taken up with work to extend Snow White towards the lander, three scoops wide and 5 cm deep, with the large amount of debris dumped at Croquet Ground. This would provide enough room for testing the sampling mechanism and procedures, and obtaining samples of the hard icy soil. The arm work took longer than expected, temporarily putting the RA in a safe mode, from which it was restored quickly. Additional images were taken to examine the rasp on the scoop and the TECP, and the arm temperatures were monitored and its actuators were tested.

On sol 50 (MY 29, sol 214, or 15 July 2008) a load plate test was performed to prepare for use of the rasp. The rasp, added late in the arm design process to ensure that even the hardest icy material could be sampled, was a small drill mounted on the back of the RA scoop. It protruded through a "load plate" whose rough surface would hold the scoop steady on the surface when pressed down by the arm. The load plate test, repeated several times during the mission, ensured that the arm would be held steady enough to rasp safely and effectively. The rasp was tested at the far end of Snow White for the first time on this sol and the area was imaged before and after rasping (Figure 147). The rasp dug two holes but did not reach pure ice as seen in Dodo–Goldilocks. Some hard soil was delivered to the scoop and imaged with RAC, which showed it to change over a few hours, apparent confirmation that it contained some ice. Images of the right (east) solar panel were taken to monitor dust accumulation, and arm temperatures continued to be monitored on this sol and the next.

On sol 51 Snow White was scraped again to reach deeper and to clean the surface for sampling, but the scoop hit a hard buried object or layer and stopped. The scraped area was left obscured by the arm but a documentation image of it has been assembled from several partial views for Figure 147. Sol 52 was used for further imaging, including observations at midnight with the Sun very low, and on sol 53 the TEGA oven 0 door was opened successfully, fully open this time unlike the previous two attempts. Then the rasp was tested again in Snow White near the holes made on sol 50. First the trench floor was cleaned with two scrapes in each of four columns, and then the rasp dug four small holes to try to reach hard ice-rich soil, but this test sample might not have been clean enough for a reliable analysis. These activities were documented with images including an RAC image of the scoop divot.

Another load plate test was performed on sol 54 and imaged multiple times to make a short animation, and the arm positioning was tested. The TECP was inserted near Vestri and the cameras imaged the surface for a "midnight Sun" image. Since the Sun and the ground needed different exposures and lighting, composite views needed images taken at different times.

Sol 55 was taken up with a 24 hour coordinated atmospheric observation campaign with MRO's MCS instrument, and the TECP data collection continued. On sol 56 the TECP was retracted and a final test of the crucial ice sample collection procedure was performed. Two scrapes were made in each of four columns to clean the trench surface, and the rasp dug two test holes in Snow White, which were partially erased by scraping later in the sol to collect up any ice fragments made by the rasp. The rasped and scraped material was observed to look for changes due to subliming ice. On sol 57 trench preparation continued with 80 scrapes in Snow White, 20 in each of four columns (Figure 147), and the fresh surface was monitored for changes. The goal was to remove as much ice-free soil as possible to maximize ice content in the next sample. TEGA oven 0 and the arm delivery position for that oven were imaged, as well as the WCL 0 arm delivery pose.

Ten more scrapes in four columns were made in Snow White on sol 58 (Figure 148), and the surface was observed frequently to monitor any sublimation. In Dodo–Goldilocks, as relatively pure ice sublimed, the surface receded to form a pit, but in Snow White the hard subsurface layer was dark, not bright, and as the Sun warmed the surface it reverted to a soil-like texture but did not recede. This suggested that the ice occupied spaces between grains in the soil rather than forming a discrete layer. The arm positions in the trench and at TEGA 0 were observed to be sure everything was ready for the next sampling attempt. Then on sol 59 further preparations were made for sampling, to ensure the rasp sample chamber was empty, and imaging tasks included change monitoring in both Dodo–Goldilocks and Snow White.

The TEGA sample was finally attempted on sol 60 (MY 29, sol 224, or 25 July 2008). First the bottom of the Snow White trench was cleared of any loose dry material by scraping twice in each of four columns. Then a 16-hole (four by four) pattern of rasp holes was made in the scraped area to produce a pile of presumably ice-rich debris called Glass Slipper. This was scooped up in the blind (without previous imaging, so it could be collected quickly before all the ice was lost from it), but it stuck in the scoop and could not be delivered to TEGA. Sol 61 began with night imaging of the lidar laser beam, about 02:50 local time, and much of the rest of the sol was taken up with diagnosis of the scoop problem, including a sample sprinkle test on MECA and RAC imaging of both sides of the TEGA, leading to another sampling attempt on sol 62. This sample called Shoes of Fortune was obtained in exactly the same way by scraping and rasping in Snow White followed by a "blind grab," but it also stuck in the scoop and was not delivered.

The next sol was used for atmospheric observations, including the use of the RAC camera to search for dust devils. Then on sol 64 a sample called Wicked Witch was scraped from Snow White, using two scrapes in each of four columns, picked up without prior imaging and successfully delivered to TEGA oven 0. This was not freshly rasped ice like the previous samples but a sublimation lag deposit, material that was presumably cemented by ice when first exposed but which would have lost ice by sublimation over the previous days. In fact, it still contained enough ice for a positive identification by TEGA. Also on sol 64 the arm tested its positioning for delivery to the OM, and late in the evening, less than an hour before midnight, the camera imaged the lidar laser beam again.

Now that Snow White had been sampled, attention could move to new targets. Sol 65 and some of sol 66 were devoted to meteorology. Also on sol 66 a sample called Rosy Red-Sol 66 was collected from the Rosy Red 2 site north of Snow White and delivered to WCL 0. This process was supposed to be repeated on sol 67 for the OM, but the arm failed to touch the surface at Rosy Red 2. Nevertheless, when it automatically moved to the delivery position to drop its sample, a small amount of the previous sample was successfully sprinkled onto the OM for high-resolution imaging, though this was not done until sol 122. This residue from Rosy Red-Sol 66 was named Mother Goose. The arm then started

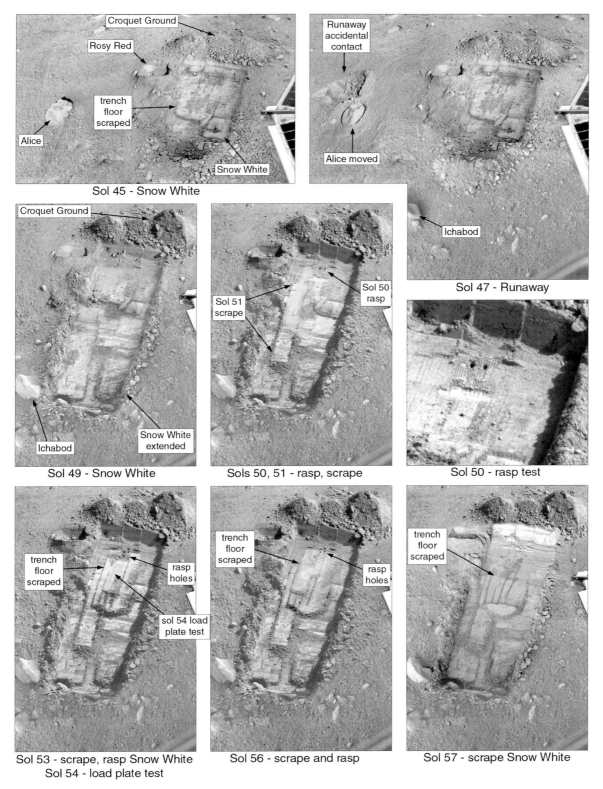

Figure 147. Phoenix surface activities at Wonderland, sols 45–57.

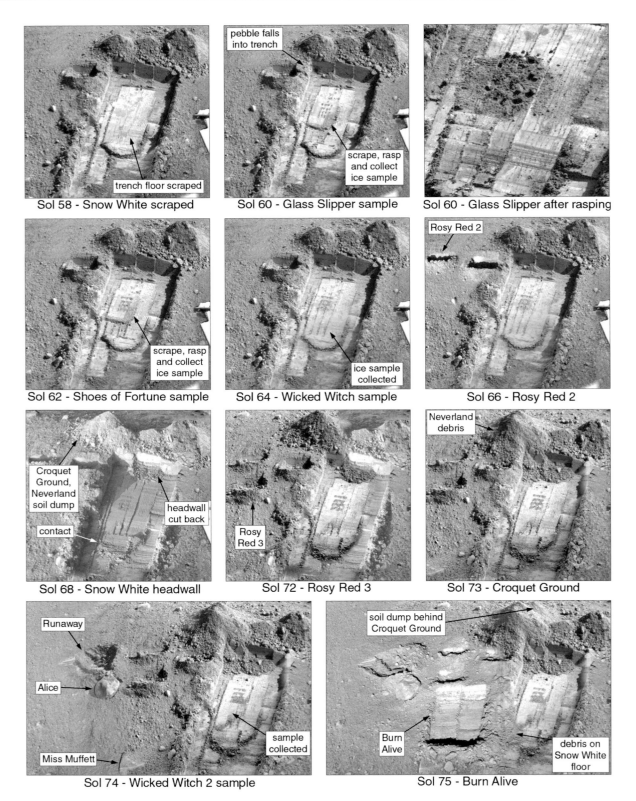

Figure 148. Phoenix surface activities at Wonderland, sols 58–75.

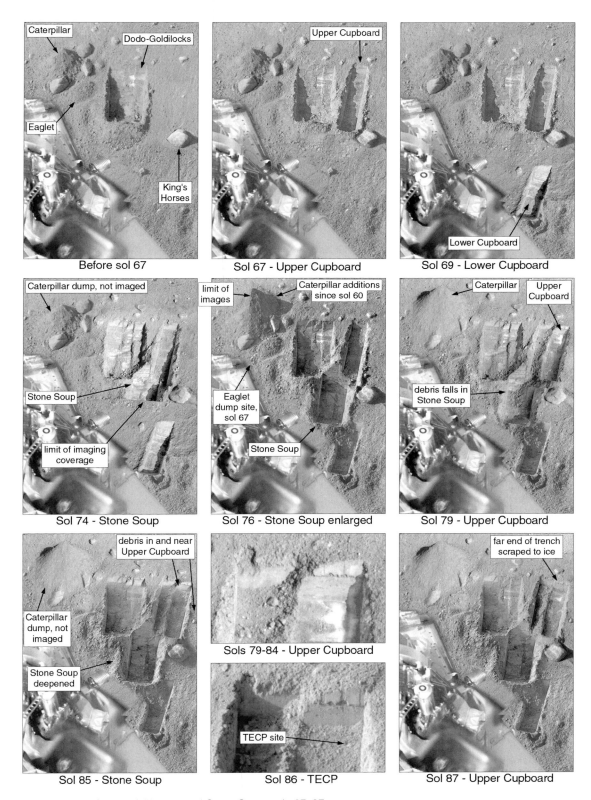

Figure 149. Phoenix surface activities around Stone Soup, sols 67–87.

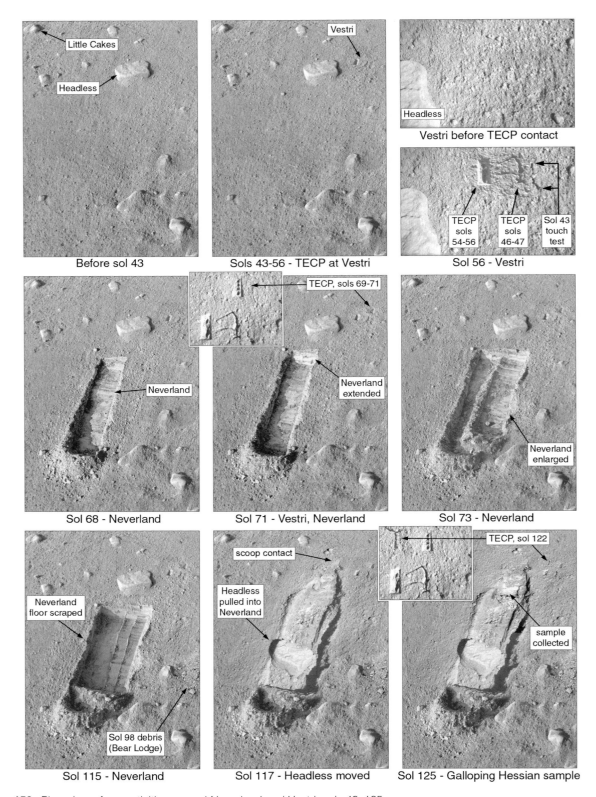

Figure 150. Phoenix surface activities around Neverland and Vestri, sols 43–125

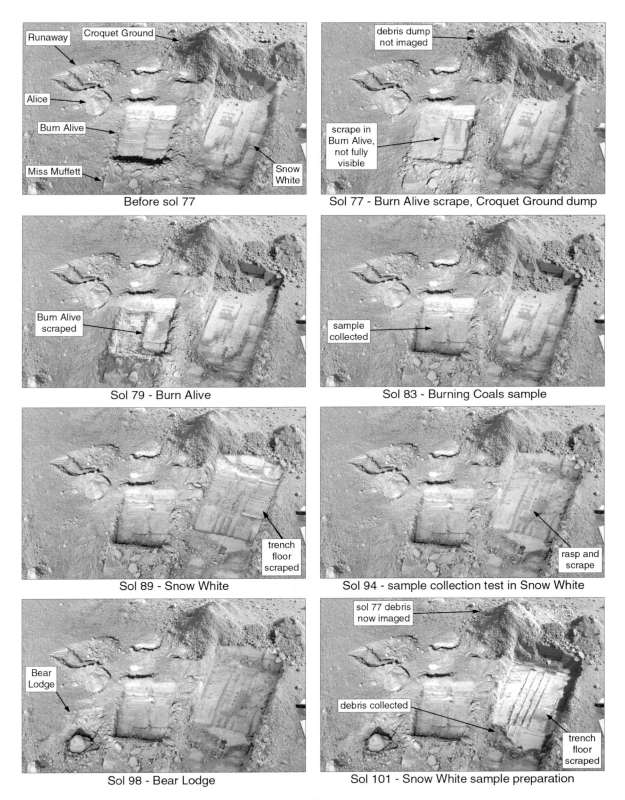

Figure 151. Phoenix surface activities at Wonderland, sols 77–101.

digging a new trench called Upper Cupboard to see how far the ice layer in Dodo–Goldilocks extended (Figure 149). The debris from this dig was dumped on Eaglet according to Arvidson *et al.* (2009), but images taken on sol 76 show only very small changes there and much more at Caterpillar. Ice was found in Upper Cupboard as expected. At the same time the camera took images of another new site to document it before the next trench, Neverland, was begun.

The first dig at Neverland was on sol 68, with debris dumped at Croquet Ground (Figure 150), and that was followed by renewed scraping at the Snow White headwall, the steep outer end of that trench. The headwall was cut back 3 cm, and the scoop also struck the floor of Snow White half-way down the trench. Pictures of the end of the trench nearest to the lander taken on sol 84 showed that the middle of the wall had changed since it was extended on sol 49, apparently struck by the scoop. The new depression was in line with the sol 68 arm stroke and may have occurred unintentionally on that sol. The images of Snow White also showed the Rosy Red 2 sampling location.

Meanwhile, back on Earth, news media were spreading excited stories about important findings from the Baby Bear sample analysis, with apparent implications for the potential for life on Mars. On the following day, sol 69, the arm dug at Lower Cupboard, extending the previous trenching in this region to create a long transect across a polygon interior and trough in order to examine the varying depth to ice across these features. Lower Cupboard was never touched again. Photometric observations of the surface were made at different times of day, coordinated with lidar and MRO. The arm was moved to test positioning for future sample deliveries to TEGA oven 5 and WCL 2, and later the TECP was returned to Vestri again and inserted near the previous TECP sites, but slightly further from the lander. Also on this sol the AFM finally received its first sample from Sorceress, delivered to OM on sol 38.

The AFM was part of MECA, as was the OM. AFM received its samples either from the scoop or as dustfall, dust dropping from the sky. The sample was rotated under the OM to identify particles, and targets would be chosen for AFM ultra-high-resolution imaging. The system performed 59 science observations and 26 calibration tests, and, of the science observations, 28 showed

particles. OM received eight samples (Goetz *et al.*, 2010), listed in Table 39, which were imaged by the microscope on a later sol as indicated here.

On sol 70 (MY 29, sol 234, or 5 August 2008) more photometry studies were undertaken overnight as the seasons changed and the Sun dropped lower. TECP began collecting data and the cameras began looking for signs of overnight frost deposition. Frost was seen in the shadows of rocks on sol 70. The rumors of "life on Mars" were now resolved with the announcement of the discovery of perchlorates in the soil, compounds rich in oxygen and chlorine. These chemicals were unexpected and interesting, but carry no implications one way or the other for the presence of life.

TECP was extracted from the ground on sol 71, and the cameras took more images to look for frost, and to document the Cupboard and Neverland trenches and the headwall of Snow White. Neverland was deepened and extended slightly and its debris was dumped on Eaglet. RAC took images of Snow Queen and Holy Cow while its scoop was full of soil, and also made flat-field images for later data calibration. Sol 72 was taken up with the collection of a surface sample called Rosy Red-Sol 72 from a small new trench called Rosy Red 3, RAC imaging of it in the scoop divot, and its delivery to TEGA oven 5, the oven whose doors had opened partially on sol 25. The Rosy Red 3 sampling area was then documented in images (Figure 148). The Rosy Red samples were all near-surface materials, so they had to be taken from slightly different places.

The next day, sol 73, saw the beginning of a multispectral panorama of the landing site using all color filters, the Happily Ever After panorama (or Happy Pan), which was about 37 percent finished at the end of the mission. Phoenix imaged the Snow White dump pile (Croquet Ground) as well as Snow Queen and Holy Cow, whose icy surfaces were slowly changing (Figure 142). RAC also took images of one of the footpad struts on which small lumps nicknamed barnacles were visible, and seemed to be growing as if drops of brine were attracting water from the atmosphere. Neverland trench was made wider on its southeast side and its debris was dumped on Croquet Ground, some of it falling into Rosy Red. Arvidson *et al.* (2009) identified the dump site as Eaglet, but no change was observed there. TEGA was also imaged again because the Rosy

Red-Sol 72 sample, which had been dropped onto oven 5 the sol before, had not yet entered the instrument.

Vibration of the TEGA screen eventually managed to get some Rosy Red-Sol 72 material into oven 5 on the next day, sol 74. Also on that sol a new excavation began at Stone Soup, a trench between Dodo–Goldilocks and the Cupboard trenches that joined them all to complete the long ice depth transect referred to above. Stone Soup was offset slightly to the left of the Cupboard trenches to avoid the prominent King's Horses rock. The debris was added to Caterpillar. Stone Soup eventually reached a depth of 18 cm without exposing any ice, though some clods of hardened soil were excavated near the bottom of this trench. It could not be made deeper without the arm striking the lander. A sublimation lag sample called Wicked Witch 2 was collected from previous scraping in Snow White for the OM, and the soil dump pile just beyond Snow White was observed again.

Sol 75 began with early-morning frost observations. Then the Rosy Red-Sol 72 debris pile on TEGA oven 5 was imaged to monitor any changes. The clumpy or sticky soil that had caused so many delivery problems might change as water was lost from it. More atmospheric observations were made, and dust devils or similar vortices were observed more frequently and with greater intensity after this date than before. Then part of the Wicked Witch 2 sample was delivered to the OM and the remainder of it was dumped on Croquet Ground. That OM sample would be imaged on sol 122. Next the arm was used to dig at Burn Alive, another new trench just north of Snow White and west of the Rosy Red area, and its debris was also dropped on Croquet Ground (Figure 148). The Burn Alive tailings, pulled towards the lander by the scoop, partly covered the small rock Miss Muffett. Trenches were being dug in many places to map variations in the depth to ice across the surface.

On sol 76 more images of the Happy Pan were taken. The arm was moved to the positions needed for sample delivery to TEGA oven 7 and RAC imaging of ovens 5 and 7 as tests for later activities, and more images were taken of the Rosy Red-Sol 72 pile on TEGA, the Burn Alive trench and the Dodo–Goldilocks–Cupboard–Stone Soup complex of trenches. Some digging was done in Stone Soup, with the excavated soil dumped on Caterpillar. Sol 77 saw additional frost monitoring and more excavation at Burn Alive, with debris dumped at Croquet

Ground, and MARCI saw a cloud of ice crystals pass over Phoenix from the north.

The Happy Pan was continued on sol 78, as well as more frost monitoring. TEGA oven 7 was opened on sol 79 to be ready for a future sample. On the same day the arm was used to scrape the right side of the Burn Alive trench, dumping the debris on Croquet Ground, and to widen Upper Cupboard by one scoop width. Some Upper Cupboard debris fell into Stone Soup, and the rest was dumped onto Caterpillar. Activities were divided among the trenches so that, for any given trench, the digging and planning sols could alternate without losing valuable mission time. Dust devils passed nearby and a drop in humidity was observed.

The next day, sol 80 (MY 29, sol 244, or 15 August 2008), was taken up with imaging for the Happy Pan and for frost monitoring. The dump site Caterpillar was imaged on the same sol. On sol 81 a sample called Burning Coals was supposed to be taken from the Burn Alive trench, but a fault caused the arm to enter a safe mode without collecting a sample. Atmospheric monitoring continued, followed just after midnight by further imaging of the lidar laser beam. The previous night, imaging of the laser on sols 61 and 64 showed a clear atmosphere, but this time more scattering of the laser light was observed, indicating a fog of ice crystals in the cold night air. Later on sol 82 images were taken for the Happy Pan and to monitor frost around the landing site and the trenches at Dodo–Goldilocks and Burn Alive.

The Burning Coals subsurface sample was collected from a depth of about 3 cm in Burn Alive on sol 83 as imaging continued, including atmospheric monitoring, photometric imaging, frost monitoring, stereoscopic imaging of the workspace to create a detailed topographic map of the surface and trenches, pictures of the TEGA magnets, and repeated imaging of a target called Jumping Cow to look for changes associated with frost. This was the only day that the regular frost monitoring images actually showed frost unambiguously, though other observations also revealed frost unintentionally. A RAC scoop shadow test was performed by holding the sample half in Sun and half in shadow to see how temperature affected its stickiness. RAC also observed TEGA oven 7 and the sample on the divot, and the arm tested its sample delivery position. Burning Coals was held in the scoop on sol 84, while further imaging occupied much of the day.

On sol 85 the Burning Coals sample was delivered to TEGA oven 7, and later the arm dug in Stone Soup, clearing out the debris dropped in it on sol 79 and dumping it on Caterpillar. Next, it performed a wrist joint calibration test and imaging continued for the Happy Pan. The TECP was placed in the ground in the Upper Cupboard trench on sol 86, and retracted later in the sol. It was supposed to be placed in the ice exposed on sol 79, and still present as late as sol 84 (Figure 149). Images from sol 85 showed that debris from Stone Soup had been dropped in and around Upper Cupboard as it was being moved to Caterpillar, and this apparently prevented the TECP from contacting ice.

On sol 87 the arm scraped the bottom of Upper Cupboard trench (Figure 149), revealing ice again, and the scoop divot was examined with RAC. Images were also made of Caterpillar and as part of the continuing program of frost monitoring, and the camera also made a large but low-resolution infrared panorama. On sol 88 RAC re-imaged the divot, and then the arm dug in Stone Soup and dumped its debris on Caterpillar again. It also tested delivery of a sample to WCL 3 and RAC imaging of that instrument. The camera took images of Caterpillar and added to the Happy Pan. The arm scraped the outer right side of the Snow White trench on sol 89, 10 times in four columns, and tested RAC imaging of and sample delivery to TEGA oven 0. Images were taken of Stone Soup and Holy Cow on the same sol.

Sol 90 (MY 29, sol 254, or 25 August 2008) was the last day of the nominal mission, and the first time during the mission that the Sun actually dropped below the horizon. The cameras imaged sunrise, looked again for evidence of frost, observed Holy Cow and Cupboard, made a color panorama of the workspace, and continued atmospheric monitoring. Meanwhile, the arm tested imaging and sample delivery to TEGA ovens 0 and 1 and WCL 3. Then it collected a subsurface sample called Golden Goose from the Stone Soup trench. Activities up to sol 90 are mapped in Figure 152.

This sample was held on sol 91 while the camera observed the Sun moving behind the raised arm, taking frames for a short movie of the occultation, something that had also been done by Viking to study atmospheric scattering in the sky near the Sun. Images were also taken for the Happy Pan and for frost monitoring, and of the Stone Soup trench. The Golden Goose sample was

dumped onto Caterpillar on sol 92 so that multispectral images of it could be taken in sunlight, since its source area at the bottom of Stone Soup was always in shadow. More frost monitoring and Happy Pan images were taken, as well as pictures of Snow White for a set of five load plate tests.

After a sunrise movie, sol 93 was devoted to digging at Stone Soup. The debris was dumped on Caterpillar and the cameras again imaged the dumped soil and added to the Happy Pan. On sol 94 the arm tested an ice sample delivery again by scraping the surface in Snow White using two scrapes in each of four columns, followed by making 16 rasp holes to excavate fresh ice. This was scooped up quickly, before the ice could sublime away, for delivery of an ice sample to TEGA oven 0. Despite the fact that this had to be done without locating the sample by imaging, to save time, the test worked very well. Tests suggested that a valve in TEGA intended to regulate the flow of gas through the instrument was leaking, so a plan to deal with the issue had to be developed. Phoenix also imaged Stone Soup and the sky, producing movies of cloud motions. The wind telltale, which was viewed almost every day, moved dramatically on this sol, revealing very windy conditions. Meanwhile MARCI saw several weather fronts with icy clouds pass over Phoenix between sols 93 and 96.

Another subsurface sample called Golden Goose 2 was taken from about 15 cm depth in Stone Soup on sol 95, and the SSI documented the dig, searched for frost and extended the big panorama. One of the sensor tips in the AFM had become worn or broken, but others were ready to take its place. On sol 95 OM imaged the tip break tool, which would remove the bad tip on sol 98. Golden Goose 2 was delivered to WCL 3 on the next sol but did not enter it. TEGA oven 1 was opened, images were taken of Stone Soup and the RAC was used late at night to look at Holy Cow, under the lander. That overnight imaging continued into sol 97, in addition to continued frost monitoring and using the RAC to make a stereo topographic map of the Sindr area (Figure 155). The camera added to the Happy Pan, monitored frost and imaged the Neverland area.

On sol 98 the arm dug a small trench called Bear Lodge just in front of the rock Alice in Neverland, uncovering a buried rock in the process (Figure 151), and dumped the debris on Croquet Ground. The intention

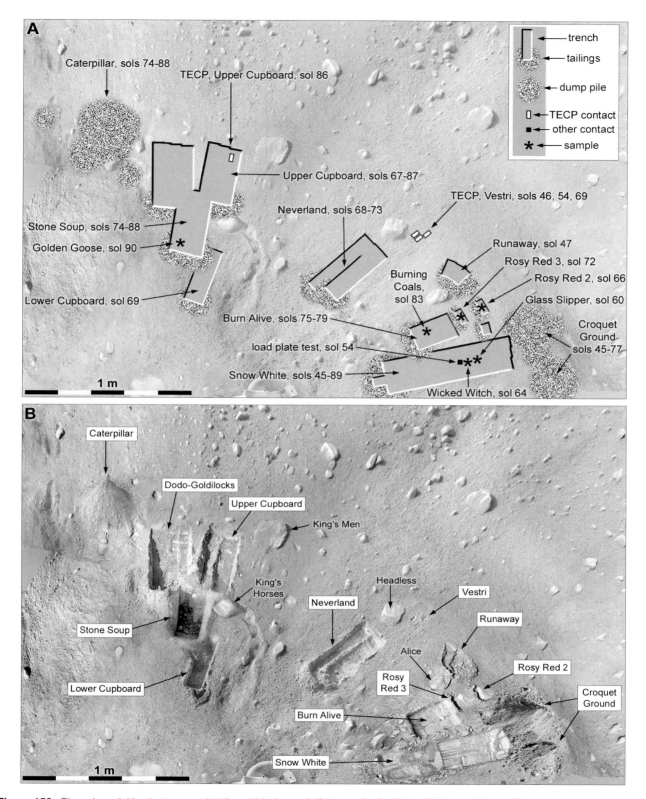

Figure 152. Phoenix activities between sols 45 and 90, the end of the nominal mission. This map follows Figure 146 and is continued in Figure 158.

Table 41. *Phoenix Activities During the Extended Mission*

Sol	Activities
91	Frost search, hold Golden Goose, image Sun occulted by RA, extend Happy Pan, OM and AFM images
92	Frost search, dump sample, extend Happy Pan, image Snow White, TEGA 7 medium heat, load plate test
93	Sunrise movie, extend Happy Pan, RA extend Stone Soup, TEGA 7 medium heat, Caterpillar stereo map
94	RA rasp and scrape Snow White, test delivery to TEGA 0, image Stone Soup, high winds, cloud movies
95	Frost search, RA Golden Goose 2 sample, extend Happy Pan, TEGA pump, OM tip break tool on AFM
96	Open, test TEGA 1, RA Golden Goose 2 to WCL 3, image Stone Soup, night atmosphere and frost data
97	Night images of Holy Cow, frost search, prepare AFM tips, RAC map of Sindr for TECP, TEGA checks
98	Frost search, extend Happy Pan, RA dig Bear's Lodge, deploy TECP at Gandalf, AFM tip break
99	Frost search, retract TECP, RA dig Golden Key sample and deliver to OM, calibrate OM and AFM
100	RAC frost search, map Neverland and Snow White, test delivery to TEGA 1, TEGA pump, AFM data
101	Image sunrise, frost search, extend Happy Pan, collect Golden Goose 3, TEGA pump, OM and AFM data
102	Frost search, prepare WCL for sample, RA deliver Golden Goose 3 to WCL 3, TEGA valve check
103	Image Snow White, clouds and empty scoop, RA place TECP in soil at Sindr, AFM scans
104	TEGA tests, extend Happy Pan, image dust devils, image Snow White, TECP data, communication test
105	Retract TECP, image Dodo–Goldilocks and Snow White, collect Sorceress 2 sample
106	Frost search, TEGA tests and pump, RA place TECP in shadow, TECP data, extend Happy Pan
107	TEGA pump, RA deliver Sorceress 2 to WCL 2, WCL thaw and analysis, dust devil search
108	Imaging of landing site continues, runout sol due to uplink sequencing problem
109	Frost and dust devil searches, RA test sampling from OFB, TEGA checkout, OM and AFM Golden Key
110	TEGA pump, RA dig new Golden Goose sample, deliver to OM, OM images, cloud movies
111	Frost and dust devil searches, TEGA test, AFM test, AFM Mother Goose, TECP at Rosy Red 2, clouds
112	Retract TECP, TECP vertical profile, dust devil search, AFM images of Mother Goose
113	RA Sam McGee sample from Snow White, delivery to TEGA 1 fails, lidar movie, dust devil search
114	Extend Upper Cupboard, image Snow White, extend Happy Pan, cloud movie, AFM Mother Goose
115	OM, AFM images of Mother Goose, extend Happy Pan, scrape Neverland trench, monitor Snow White
116	RA wall failure test at Dodo–Goldilocks, prepare TEGA, scrape Snow White, dust devil search
117	RA pulls Headless rock into Neverland, OM and AFM images, dust devil search
118	Frost search, scrape Snow White, TEGA 0 high heat (practice for OFB), OM and AFM images
119	Frost and dust devil searches, TECP in Upper Cupboard, clouds, image area under Headless, TEGA test
120	RA repeat Sam McGee collection and delivery to TEGA 1 (fails), photometry images, open TEGA 6
121	Image TEGA 6 open door, TEGA prepare for sample, OM images, AFM scans, monitor trenches
122	RA rasp sample from OFB, deliver to TEGA 2 (fails), TECP at Vestri, OM images, dust devil search
123	RAC image TEGA 2 to diagnose fault, change monitoring, dust devil movie, OM images, TECP soil data
124	Atmospheric imaging (runout sol)
125	Retract TECP, dig Galloping Hessian sample, deliver to OM (fails), OM images, RAC dust devil search
126	RA scrape Snow White for Wicked Witch 2, deliver to TEGA 1 (fails), dust devil search, AFM scan
127	Prepare TEGA, RA dig La Mancha trench, dust devil movie, WCL Sorceress sample
128	RA dig new Galloping Hessian sample, deliver to OM, OM images, dig Pet Donkey, cloud movies
129	RA extend Pet Donkey trench, image trenches and Bee Tree debris pile, TEGA 1 high heat (empty cell)
130	TEGA 1 close, RA dig Rosy Red-Sol 130 from Rosy Red N and vibrate sample, TEGA 7 high heat
131	RA deliver Rosy Red-Sol 130 to TEGA 6 (fails), cloud and dust devil movies, survey trenches, WCL tests
132	RA extend La Mancha trench, OM and AFM images, remote sensing, including cloud movies
133	TEGA 1 high heat, RA scrape Pet Donkey trench and remote sensing, including RAC dust devil search
134	TEGA 1 high heat, RA scrape La Mancha trench, WCL Sorceress sample
135	Operations reduced to a minimum as a dust storm passes overhead, SSI optical depth only
136	RA dig Rosy Red-Sol 136 sample, and remote sensing
137	OM and AFM images, remote sensing, night data, dust devils seen about 1–2 km from lander
138	RA deliver Rosy Red-Sol 136 to WCL (fails), remote sensing, TEGA 6 receives Rosy Red-Sol 130

Table 41. (*cont.*)

Sol	Activities
139	Remote sensing, including RAC stereo map of King's Men rock
140	RA dig in Stone Soup, image trenches, TEGA 6 low heat, atmospheric remote sensing
141	Extend Happy Pan, TEGA and SSI atmospheric data, dust devil search, OM and AFM images
142	Image King's Men, atmospheric remote sensing, image lander deck for missing sample from WCL 0
143	RA scrape Snow White to collect ice, OM, AFM images, atmospheric data, TEGA 3 door not open
144	OM data, image Dodo–Goldilocks, Upper Cupboard, Snow White trenches
145	RA dig trenches around King's Men (Ice Man, La Mancha), OM images, TEGA 6 high heat
146	MARDI test (did not turn on), RA dig Upper Cupboard, Stone Soup, OM and AFM images
147	TEGA 6 high heat, RA touch King's Men, WCL sample push, WCL analysis
148	RA scrape in La Mancha, OM and AFM images and atmospheric remote sensing
149	RA scrape Dodo–Goldilocks, park arm with TECP at Alviss, imaging, OM tip break tool aligned
150	RA temperature monitored, image clouds, sky, frost, wind telltale, night lidar, dust storm to north
151	TEGA 3 (blank) high heat, image clouds, sky, frost, wind telltale, last images of mission
152	Shut down arm heaters, low-power safe mode stops science work
153	Phoenix fails to respond to communication attempt, batteries discharged
154	Communications restored, batteries attempt to recharge each day until sol 157, no science activities

Notes: AFM, Atomic Force Microscope; MCS, Mars Climate Sounder (on MRO); OFB, Organic-Free Blank; OM, optical microscope; RA, Robotic Arm; RAC, Robotic Arm Camera; SSI, Surface Stereo Imager; TECP, Thermal and Electrical Conductivity Probe; TEGA, Thermal and Evolved Gas Analyser; WCL, Wet Chemistry Lab.

here was to drag Alice into the new trench with the arm to test procedures for a subsequent move of Headless into Neverland, but the planned test was abandoned to save time. The cameras made another cloud movie and examined WCL 3 with its problematic debris pile. Nothing had yet entered the cell. TECP was inserted into the soil at Gandalf and collected data there, but not for long enough to be useful. Then on sol 99 TECP was retracted and the arm was used to scrape a sublimation lag sample called Golden Key from the Dodo–Goldilocks trench. It was delivered to the OM, and the remainder of the sample, if any, was supposed to be dumped near the western footpad close to Rabbit Hole (Figure 143). It may have fallen on the lander body or beyond the footpad but was not observed. RAC imaged the sample on the divot, and frost monitoring also continued on this sol. The TEGA valve problem was still being diagnosed

The arm was used on sol 100 (MY 29, sol 264, or 5 September 2008) to test a delivery to TEGA oven 1 and the cameras imaged Snow White as part of a continuing program of surface monitoring. Many images of Snow White taken after this date showed frost forming in the trench, but it was rare on undisturbed surfaces. Sunrise

was imaged at the start of sol 101, and more images were taken to look for frost, to add to the Happy Pan and to document the surface in Stone Soup and Snow White. The arm scraped 15 times in four columns in the outer part of Snow White, creating a pile of loose material that was left there until sol 105. It then dug a sample called Golden Goose 2 from Stone Soup, a repeat, as the previous sample could not be analyzed, and moved it over WCL 3 for delivery to the instrument on the next day. The scoop was imaged over WCL on sol 102 and then dumped its sample. Images were also taken to monitor frost and the trenches Snow White, Stone Soup and Dodo–Goldilocks.

On the next sol the empty scoop was imaged to see if any sample remained in it. Pictures were also taken of Snow White and to make a movie sequence of clouds passing overhead. The TECP was placed in the soil at Sindr, and OM imaged its sol 99 sample. The Happy Pan was continued on sol 104, and among its images the first visible dust devils of the mission were seen passing in the distance southwest of the lander. Other images were taken of Snow White, and the TECP remained in the soil, but, as at Gandalf, no useful data were obtained here at Sindr.

The TECP was retracted on sol 105 and any dirt on it was dropped on Caterpillar. Then the arm obtained a sample called Sorceress 2 for WCL from the scrapings made in Snow White on sol 101, after the delivery and RAC imaging positions had been tested. Images were taken of the Dodo–Goldilocks and Snow White trenches. Imaging took up the next sol as well, continuing the frost monitoring and adding to the Happy Pan. TECP was placed in a shadow for temperature monitoring. More images to search for dust devils were taken on sol 107, and the Sorceress 2 sample was delivered to WCL 2.

Sol 108 was a runout sol used for pre-planned remote sensing, as the sol's instructions had not been transmitted properly. Then on sol 109 the cameras documented a test of arm positioning for delivery to TEGA oven 2 and the collection of a sample from the Organic-Free Blank (OFB), a glass ceramic plate that was certified free of organic contaminants so it could provide a control for the other sample analyses. This was just a test of the OFB sampling procedure, which involved placing the load plate on the blank without collecting a sample. Images were also taken for frost monitoring, to look for changes in Snow White, to examine the TECP probe needles, and for the first deliberate dust devil search of the mission after the accidental imaging on sol 104. This time the search, directed to the southwest, was successful.

Another subsurface sample called Golden Goose was taken from Stone Soup (Figure 153) and delivered to OM on sol 110 (MY 29, sol 274, or 16 September 2008), and the camera took a stereoscopic movie of the wind telltale in motion. Other images were made of cloud motions and the trenches Stone Soup and Snow White, and to search for dust devils, but on this sol none were seen. Dust devil search images were made with the RAC on sol 111 while the SSI cameras took a full 360° panorama of clouds. Other images were taken of the OFB and TECP, Stone Soup and Dodo–Goldilocks, and for the Happy Pan and frost monitoring campaign. Also on sol 111 the RA placed the TECP needles in the soil on the outer wall of Rosy Red 2. Coordinated observations using SSI and MRO were made on sol 112, both measuring surface albedo, as well as images for the Happy Pan and for another unsuccessful dust devil search. The TECP was removed from the soil, any debris in the scoop was dumped on Croquet Ground, and then TECP made an atmospheric measurement. Also on sol 112, OM imaged its sol 110 sample.

On sol 113 the arm tried to collect an ice sample called Sam McGee from Snow White and deliver it to TEGA oven 1. The sampling method was the same one used earlier, with two scrapes in each of four columns to remove any dry lag material, 16 RASP holes and a blind grab of the sample. Once again, nothing was delivered to TEGA. Meanwhile the cameras monitored Snow White and late that night once again imaged the lidar laser beam, again showing evidence of an ice crystal fog (Moores *et al.*, 2011).

A HiRISE image taken on this sol (MY 29, sol 277, or 19 September 2008) showed that winds had lifted and overturned one edge of the parachute some time in the three weeks since the previous orbital view had been obtained. Images taken in 2012 showed that the whole parachute had moved about 30 m to the northeast (Figure 137A), the most dramatic move of any parachute observed on Mars. RAC observed Snow Queen, under the lander, and the footpad strut, which seemed to be covered with droplets of water. This was a surprising and controversial interpretation, implying that very concentrated brines could remain liquid at Martian temperatures.

Happy Pan imaging continued on sol 114, as well as a cloud movie sequence and images of Snow White, while the arm extended the Upper Cupboard trench and dumped its debris on Caterpillar. The Caterpillar debris pile had now grown large enough to cover several nearby rocks and was spilling down into the Rabbit Hole depression (Figure 153). A dust devil search on this sol succeeded again after two unsuccessful attempts.

Happy Pan and Snow White imaging continued on sol 115, and the arm scraped the surface at Neverland to prepare for the sol 117 move of Headless rock. On the next sol the two main Phoenix cameras, RAC and SSI, imaged each other. The arm groomed the western end of the Snow White trench, making two digging passes in three columns and dumping the debris on Croquet Ground. Then it was pressed down on the ground just outside the west wall of Dodo–Goldilocks in an attempt to cause the steep wall to collapse, and images were taken before the contact and afterwards (Figure 153). Soil properties would be obtained from the shape of the collapsed wall. Images were also taken of Snow White

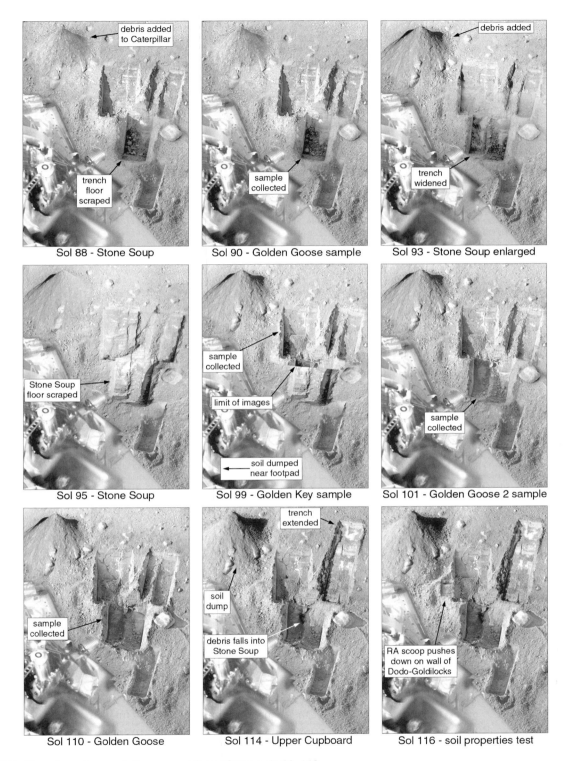

Figure 153. Phoenix surface activities around Stone Soup, sols 88–116.

Figure 154. Phoenix surface activities around Stone Soup, sols 119–149.

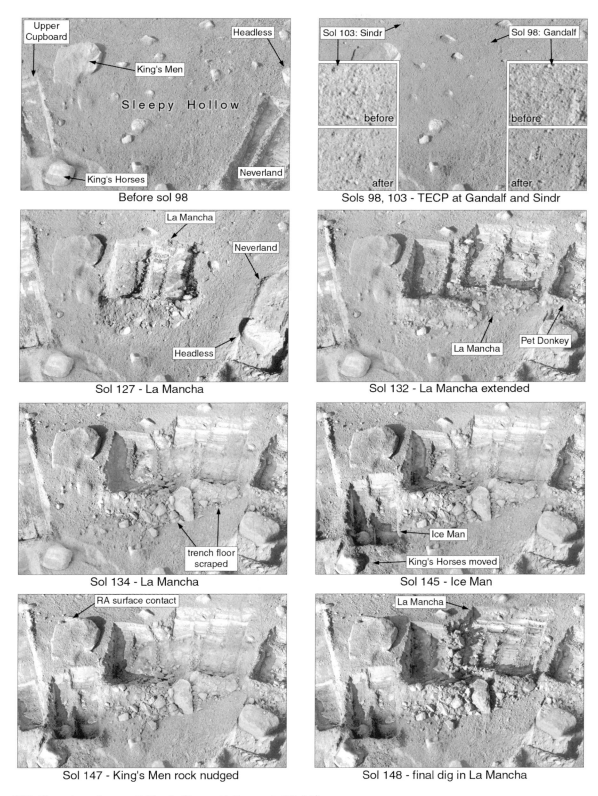

Figure 155. Phoenix surface activities in Sleepy Hollow, sols 98–148.

and the pile of debris dumped on WCL 3 to see if any changes had occurred.

A successful search for dust devils was undertaken on sol 117, and then Headless was pulled by the arm into Neverland trench so that the depth to the ice layer underneath its original position could be observed (Figure 150). The ice level might be higher or lower under a rock, depending on whether it insulated the subsurface or conducted heat downwards. The ice depth was later found to be lower under Headless. As the arm moved Headless, it also made contact with the soil just beyond the rock, leaving a mark seen in Figure 150.

On sol 118 TEGA oven 2 was opened, and 15 RA scrapes in four columns were made in the bottom half of Snow White as well as five load plate tests. The debris from the scraping was dropped on Croquet Ground and images of the trench were taken, as well as pictures of Upper Cupboard and continued frost monitoring. The arm was used to test imaging and delivery positions for TEGA ovens 2, 3 and 6. TECP was used again in the ice found in Upper Cupboard on sol 119, and later retracted, but without obtaining useful data. More frost monitoring images were taken as well as images of the soil under the original location of Headless and of the dump pile Caterpillar. A cloud movie was also obtained on sol 119.

Sol 120 was MY 29, sol 284, or 26 September 2008. The arm was commanded to collect another Sam McGee sample from Snow White and deliver it to TEGA oven 1, but it was again unsuccessful. The method was identical to the earlier ice sampling attempts, and again nothing was delivered. Many images were taken for photometric purposes and the arm tested the TEGA oven 6 imaging position. Then on sol 121 TEGA oven 6 was opened, and images of Upper Cupboard, Dodo–Goldilocks and Snow White were taken to document the trenches. Next, on sol 122, a sample was rasped from the OFB and moved to TEGA oven 2, but the delivery failed. TECP was inserted at Vestri again, and the area previously under Headless was imaged to prepare for later sampling. Another dust devil search was made on that sol, but nothing was seen, and OM imaged two old samples, delivered to it on sols 67 and 75. Sol 123 also included an unsuccessful dust devil search as well as images of OFB and atmospheric studies, looking for ice clouds directly overhead. Sol 124 was another day of atmospheric data collection, a runout sol over the terrestrial weekend.

TECP was retracted from Vestri on sol 125 and then a sample called Galloping Hessian was collected from the place originally covered by Headless rock (Figure 150). It was supposed to be delivered to the OM but the procedure failed. Dust devil searches were made by both SSI and RAC on the same sol but nothing was seen. Then on sol 126 another sublimation lag sample, Wicked Witch 2, was scraped from Snow White for TEGA oven 1, but again delivery was not successful. The sampling consisted of 15 scrapes in four columns in the bottom half of the trench with the sample gathered from the scrapings. The camera made a movie showing clouds moving across the sky. On the next sol a new trench called La Mancha was initiated in Sleepy Hollow, its debris was dropped on Caterpillar, and another successful dust devil movie was made. La Mancha was dug in very cloddy soil, and eventually it reached the icy layer under the surface, but on this sol the arm stalled during the dig and was left in the ground overnight. The next sol was jeopardized by this but its activities were salvaged after great effort by the mission team.

A second Galloping Hessian sample was collected on sol 128 (Figure 157) and delivered to the OM, successfully this time. Later in the same sol a new trench called Pet Donkey was started where Headless had been, to trace the depth of the ice layer. The Pet Donkey debris was dropped behind Runaway trench to create a new dump pile called Bee Tree. Two cloud movies were made and images were taken of La Mancha and Pet Donkey. MARCI saw ice clouds associated with a front passing over Phoenix on sols 128 to 130. The excavation of Pet Donkey continued on sol 129, with its debris dumped on Bee Tree, and images were taken of Bee Tree, Pet Donkey and Snow White. The material from Sleepy Hollow now being dumped on Bee Tree was very blocky, seeming to consist of hard platy clumps rather than the finer-grained or unconsolidated material seen on Croquet Ground and Caterpillar (Figure 157).

Sol 130 was MY 29, sol 294, or 6 October 2008, and the seasons were changing on Mars as they were on Earth, moving towards winter in the northern hemisphere of each world. The lifetime of Phoenix was finite and the mission could end at any time as temperatures dropped and the days shortened. The empty TEGA oven 1 was analyzed without a sample to serve as a control. A surface sample called Rosy Red-Sol 130 was collected

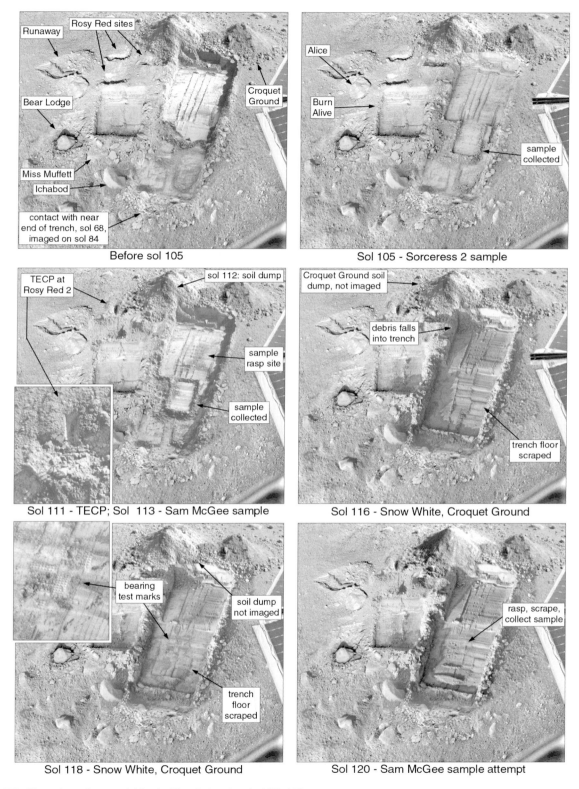

Figure 156. Phoenix surface activities in Wonderland, sols 105–120.

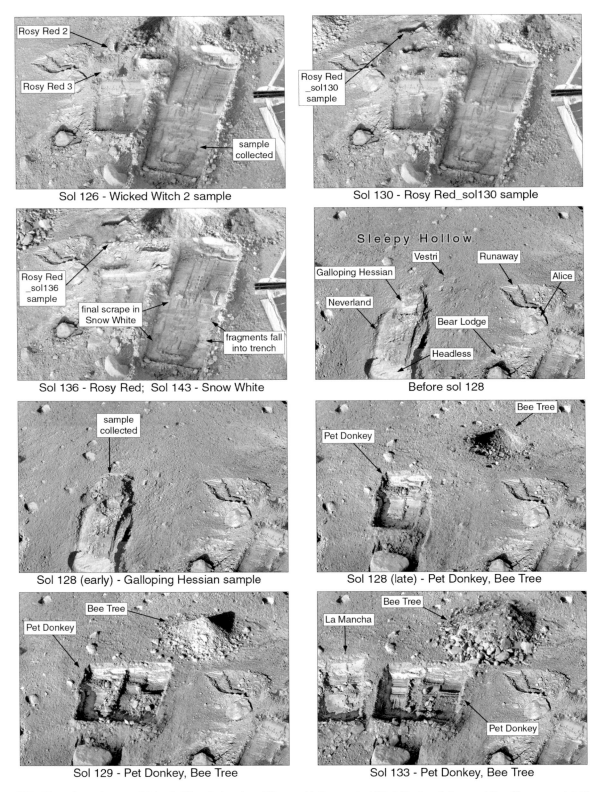

Sol 126 - Wicked Witch 2 sample

Sol 130 - Rosy Red_sol130 sample

Sol 136 - Rosy Red; Sol 143 - Snow White

Before sol 128

Sol 128 (early) - Galloping Hessian sample

Sol 128 (late) - Pet Donkey, Bee Tree

Sol 129 - Pet Donkey, Bee Tree

Sol 133 - Pet Donkey, Bee Tree

Figure 157. Phoenix surface activities in Wonderland and Sleepy Hollow, sols 126–143. A soil dump at Bee Tree on sol 148 was not imaged.

from the Rosy Red area near Snow White, adjacent to the other Rosy Red samples but slightly further from the lander. This pit was called Rosy Red N, where N might be taken to signify New. The sample was "delumped" by vibrating it, and on sol 131 it was delivered to TEGA oven 6, but not successfully. A reduced-resolution panorama was made of the upper part of the work space and other images were made of the WCL 3 debris pile, the Upper Cupboard, Dodo–Goldilocks and La Mancha areas, as well as a dust devil movie and a cloud movie. MARCI saw more ice and dust clouds associated with a front passing over Phoenix on sols 131 and 132.

Cloud movies were also obtained on sol 132, and the La Mancha trench was extended towards Headless rock with debris dropped on Caterpillar, and then imaged. OM imaged the sample it received on sol 128. Then on sol 133 the arm scraped in Pet Donkey again and dropped its debris on Bee Tree (Figure 157), while the camera took images of that operation and RAC made another dust devil search. On sol 134 the arm dug in the La Mancha trench and dumped the debris on Caterpillar, before being positioned over WCL3 to check the arm position for a future attempt to push some of the sample into the instrument. Images of the Bee Tree and Caterpillar dump piles and Snow White trench were made, as well as a successful dust devil search sequence.

A dust storm blew across the area on sol 135 (MY 29, sol 299), an event also seen by the orbiters. Operations were reduced to a minimum to conserve power, but resumed on sol 136 as the arm collected a surface sample called Rosy Red-Sol 136 at Rosy Red 3 for TEGA oven 6 (Figure 157). The arm was left positioned over WCL after a successful delivery to TEGA. The analysis of that sample was not completed before the mission came to an end. On sol 137 the RAC took an image of the sample in the scoop early in the morning, and a successful SSI dust devil search was conducted later. Another good dust devil movie sequence was obtained on sol 138, the last of the mission, in addition to pictures of Upper Cupboard and Dodo–Goldilocks. Images were also taken to document activities, including the attempted but unsuccessful delivery of Rosy Red-Sol 136 to WCL and the now-empty RAC scoop, and an RAC mosaic of the TEGA instrument. On sol 139 RAC imaged the outer side of King's Men in stereo to prepare for an attempt to move it, and more images

were added to the Happy Pan to maximize coverage before the approaching end of the mission.

Sol 140 was MY 29, sol 304, or 16 October 2008. Pictures were taken of TECP and of the Snow White and Stone Soup trenches after some digging at Stone Soup. The debris was dumped on Caterpillar. MARCI saw more frontal ice clouds on sols 140, 146 and 148. On sol 141, images were taken of Snow White and to add to the Happy Pan, and another dust devil movie was made, this time showing numerous events. Movies were made repeatedly during the mission but did not always capture any moving dust plumes. RAC pictures of Holy Cow and Snow Queen were taken on sol 142, the last images taken underneath the lander. More pictures were taken to see if the failed delivery to WCL 0 on sol 137 was due to a poor arm position, to form a mosaic of the spacecraft's deck, and to observe the rock King's Men and the door of TEGA oven 3, which was still closed. That oven was analyzed as a control near the end of the mission. Power was becoming very limited now as the hours of daylight grew shorter and more energy was needed to operate small heaters on the spacecraft. On sol 143 some pictures were taken to look for dust devils and to view the RA scoop after a last grooming activity in Snow White, preparing for ice sampling, which did not happen because TEGA oven 3 did not open. The grooming consisted of 15 scrapes in four columns in the bottom half of the trench. A panoramic set of images was taken to document the mission's trenching activities on sol 144.

On sol 145 a trench called Ice Man was dug downhill from and south of King's Men (Figure 155) with the intention that King's Men would be pulled into it as Headless had been pulled into Neverland earlier. A cloud movie sequence was also obtained, and the TECP was placed near the soil to measure the near-surface air temperature. On the next sol Phoenix imaged the TECP and the Stone Soup area, and the RA dug in Upper Cupboard and Stone Soup with debris dropped on Caterpillar. MARDI, the descent camera, which had not been used during landing, was tested now. It would see an area near Holy Cow from a new perspective, and the small microphone it carried might also record some sounds. Unfortunately, it failed to power up properly and could not be operated. Then on sol 147 the arm moved King's Men a few millimeters by touching it on the outer edge, demonstrating that it was not fixed in place, but the mission ended before the rock could be pulled into the Ice Man trench. Then the arm was pressed down on the sample pile on WCL 3 in a last, unsuccessful, attempt to push soil into the oven.

The penultimate dig of the mission, a scraping of the trench floor, took place at La Mancha on sol 148 (Figure 155), and a cloud movie sequence was made. There might be no more sampling, but as long as imaging was possible the trench floors could be monitored for changes. The La Mancha debris was dropped on Bee Tree. On the next day, sol 149, the arm scraped the Dodo–Goldilocks trench floor and into Stone Soup, dropping its debris on Caterpillar. It was then parked with the TECP probes in the soil at Alviss, though no data were obtained, and SSI imaged the Stone Soup and Dodo–Goldilocks trenches. The arm was not moved again and its heaters were turned off to save power for other spacecraft functions. TEGA operations also ended at this time. Activities up to this time are mapped in Figure 158.

The end of the mission was now rapidly approaching. If power had permitted, RA might have been used again to scrape La Mancha before parking TECP in a new location, but this was not possible. On sol 150 (MY 29, sol 314, or 26 October 2008) atmospheric data were collected by the lander, while in orbit the MARCI instrument on MRO observed a small dust storm on the edge of the polar ice cap north of Phoenix. This quickly developed into a regional dust storm, a threat to the lander. Temperatures dropped to the lowest ever encountered on Mars, $-96\,°C$ at night and $-45\,°C$ during the day, as the spreading dust storm obscured the Sun at the landing site on sol 151. A small amount of atmospheric data were still obtained, but on sol 152 (MY 29, sol 316) the spacecraft began shutting down survival heaters to save power. This was not enough to help, however, and a low-power safe mode effectively ended operations.

On sol 153 the battery power was so low that there were no communications with Earth, but on the next morning the rising Sun supplied enough power to awaken the lander, and it tried to charge its batteries and communicate with passing orbiters. As the Sun sank late in the sol, the low power shut the lander off again. This was called "Lazarus mode." For several sols after that the same sequence unfolded, with some hope that

limited operations might be resumed, including meteorology and possibly a few images. The last brief communication was on sol 157, 2 November 2008 (MY 29, sol 321). After several days of attempted communication, the mission was declared over on 10 November. The last attempts by the fleet of orbiters to listen for transmissions was on 29 November, just before conjunction.

Over 25 000 images were returned to Earth by Phoenix's cameras, the last on sol 151. The Canadian Space Agency's weather station provided a continuous record of meteorological conditions throughout the mission, the most complete set of weather data collected on Mars at that time, including temperature, pressure, wind and humidity. It also observed frost, hazes, clouds, dust devils and snow falling from the clouds. Phoenix also made simultaneous, coordinated, ground and orbital observations of surface albedo with MRO to help compare observations from different instruments.

The first dust devil observation on sol 104 was a serendipitous sighting in a set of panorama images (Ellehoj *et al.*, 2010). After this, several dust devil search campaigns were undertaken by the SSI and the RAC. Typical SSI searches involved taking up to 50 images of the horizon, usually in the afternoon, and dust devils were seen on seven of them (sols 109, 114, 117, 127, 134, 137 and 138). They were not seen on sols 110, 112, 122, 123 and 125. Some 76 sightings, including 37 individual dust devils, were counted, all seen towards the southwest and most of them apparently moving from west to east. The prevailing winds suggested by orbital data are generally from the northwest to the southeast in this area. In addition, 11 other events may have been clouds of dust picked up by strong wind gusts rather than actual dust devils, which involve convection and rotation.

As the next spring arrived at the Phoenix landing site, Mars Odyssey listened for a signal from the lander as it passed over the site several times in mid-January 2010. If its systems recovered and the solar panels generated enough power, Phoenix might have transmitted briefly in an automated "Lazarus mode." Nothing was heard at that time. Then in February a different approach was tried, with Odyssey transmitting to Phoenix. Recovery was not expected and indeed nothing was heard, but the effort was deemed worth making. Efforts ceased in April 2010, and MRO HiRISE images suggested that at least the western solar panel might have been separated from the lander body under the weight of carbon dioxide ice. The parachute was almost hidden by dust deposited over the winter, unlike the parachutes of previous missions at lower latitudes, which remained visible for a decade or more, though later wind gusts moved it and shook dust off it (Figure 137A). The Viking parachutes were difficult to identify after 30 years of dust accumulation, though Mars Pathfinder's was still clearly visible 15 years after landing when the first HiRISE images of them were taken. Polar winter processes evidently deposited dust in larger quantities than elsewhere, suggesting that if the Mars Polar Lander parachute had deployed in 1999 it might not have been visible in the first HiRISE images taken more than six Earth years later.

18 February 2009: Dawn Mars flyby

Dawn was a Discovery mission launched on 27 September 2007, to rendezvous with and orbit the two large main-belt asteroids 1 Ceres and 4 Vesta. It entered orbit around Vesta on 16 July 2011, and departed on 5 September 2012, scheduled to arrive at Ceres in February 2015. On 18 February 2009 (MY 29, sol 425) at 00:28 UT the spacecraft made a gravity-assisted flyby of Mars, primarily intended to increase its orbital inclination to match that of Vesta. The flyby was used to calibrate some instruments, including a camera provided by the German space institute DLR (Rayman and Mase, 2010).

Dawn approached Mars from the night side and flew over the northern hemisphere at a closest-approach distance of 542 km. As it passed over Tharsis it was supposed to take images of a swath several hundred kilometers wide between the Tharsis Montes and Olympus Mons. An hour later Mars Express would image the same area. Since Mars has been so well studied, the images would serve to calibrate the performance of Dawn's camera. Photographs of Mars would also be taken from greater distances for a week after the flyby, and spectrometers on both spacecraft would also acquire data for calibration purposes. This plan was cut short when the spacecraft entered a safe mode caused by a software error on the spacecraft, but several images were taken early in the sequence. Dawn's GRAND (Gamma Ray and Neutron Detector) instrument also took data

during approach, closest approach and departure, which could be compared with Mars Odyssey data to calibrate the instrument. A Visible and Infrared Spectrometer dataset was lost during the safe-mode event.

The strip of images (Figure 159) shows the northwest edge of Tempe Terra at a resolution of about 55 m/pixel in the infrared part of the spectrum. The northeastern part of the mosaic was near the terminator, and much of the apparent detail is caused by clouds, but a few surface features are visible, including a long fracture near 52° N, 287° E and several nearby craters.

2009: Cerberus

Bruce Banerdt of JPL presented details of Cerberus, a proposal for a New Frontiers (mid-sized planetary mission program) Mars geophysical mission, at the 40th Lunar and Planetary Science Conference (Banerdt *et al.*, 2009). It would have used three duplicates of the Phoenix lander to form a network. The science goals were to understand the interior structure, composition and history of Mars using seismometry, electromagnetic sounding and rotational dynamics data, and atmospheric dynamics. The three Cerberus landers might have functioned as nodes in a larger network if other missions (the Humboldt station then planned for ExoMars, but possibly also MarsNEXT and MetNet) also flew in the appropriate period. The Cerberus landers would launch together early in 2016, land in the equatorial zone with a spacing of 3000–4000 km and operate for a full Mars year.

A landing site selection study was performed for Cerberus by Matt Golombek, Eldar Noe and Veronica Hanus of JPL, drawing on extensive experience with previous missions, including Pathfinder, MER and MSL. This study (Golombek *et al.*, 2009), kindly provided by Bruce Banerdt, identified numerous sites which are illustrated in Figure 160. The main constraints on site location were elevation and latitude, in common with many other site selection studies. Site elevations were to be no higher than -2.4 km in the MOLA topographic map, with -2.8 km preferred as a safer upper limit. Analysis of the entry system performance suggested that this might be relaxed to -1.0 km or even 0 km. The latitude range was from 10° S to 20° N. Landing ellipses

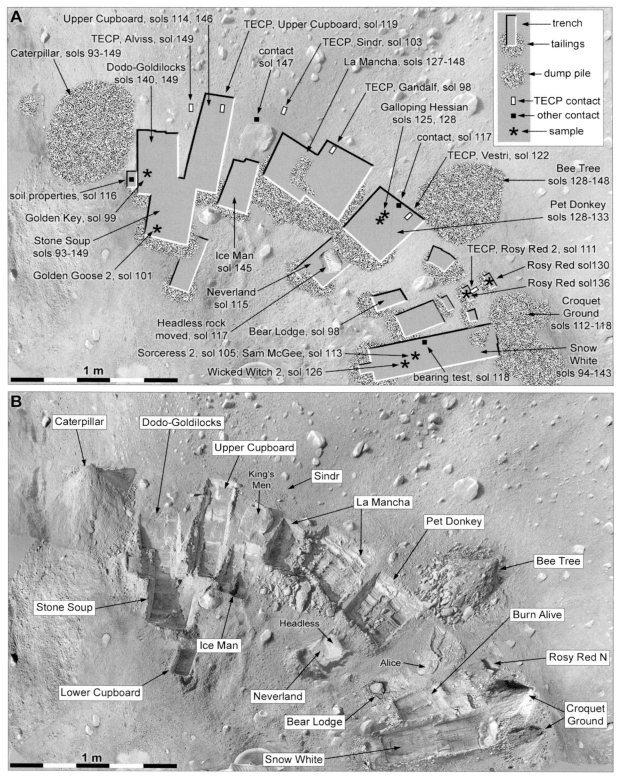

Figure 158. Phoenix activities between sol 91 and the end of the mission. This map follows Figure 152.

were 150 by 30 km, oriented roughly east–west, and they should combine to form a triangular array with sides on the order of 1000–3000 km long. The triangles need not be equilateral but their side lengths should not differ by a factor greater than two. Ellipses should experience relatively benign winds and have reasonable slopes and rock size distributions based on comparisons with other landing sites, which by now gave a good understanding of what was desirable.

The study identified 36 ellipses in the Isidis–Utopia–Elysium region and 43 in the Chryse–Arabia–Meridiani region. Because of differences in ellipse size, orientation and allowed elevation, these are not the same as MER ellipses (Figures 1 and 2), though many occur in similar locations. All these ellipses are identified in Figure 160. The candidate ellipses formed clusters in many areas, and triangular arrays were designed to link the clusters that best matched the spatial requirements of the network, as shown in Figure 160 and listed in Table 42. In the Utopia region, ellipses in western Isidis or north of Gale crater could not be fitted into networks without making arrays that were too elongated. The Chryse region offered more combinations of suitable clusters. Several feasible arrays might be constructed by selecting different combinations of ellipses in those clusters, as specific ellipses were favored or rejected according to safety considerations.

The Utopia array sketched by Golombek *et al.* (2009) had side lengths of about 985, 985 and 1790 km, with nodes in eastern Isidis Planitia, southern Utopia Planitia (overlapping the old Viking A-2 site at Tritonis Lacus) and Elysium Planitia. At Chryse, two triangular arrays were identified, sharing a common node in Arabia Terra. The smaller triangle had sides of 1775, 1560 and 1530 km, and the larger triangle had sides of 3000, 2050 and 2515 km. All these side lengths are approximate because alternative nearby ellipses might be substituted at any node. The Chryse NW ellipses are south of the Viking 1 landing site, and the Meridiani node is west of Opportunity's landing site. The larger Chryse triangle was preferred over the smaller one, but the comparison with the Utopia triangle was more uncertain. All nodes looked satisfactory, but, while the Utopia network had the advantage of lower elevations, the Chryse network was closer to the potentially active Tharsis region.

2010: Obama space policy

The 2004 Vision for Space Exploration proposed by President George W. Bush was directed specifically at the Moon, with Mars as a distant goal. In 2009 a Presidential Commission chaired by Norman Augustine considered problems with the implementation of the Vision and some possible alternatives (Augustine *et al.*, 2009). In 2010 President Barack Obama outlined a reformulated human spaceflight program, changing the hardware development plans and removing the Moon from the initial mission plan. Human missions to near-Earth asteroids would begin in 2025, and a first human Mars mission might begin in about 2035. It would stay in orbit, as a large Mars lander would not have been developed at that time, and would probably involve landings on Phobos and Deimos. The new program also involved technology demonstration missions to the Moon, asteroids and Mars, including demonstrations of aerocapture and precision landing systems, which would be necessary for human exploration. In 2010 a NASA request for ideas relating to these Mars missions suggested that one might fly in 2016 or 2018, but funding was not made available for them.

8 November 2011: Phobos Sample Return Mission

This mission, originally called in Russian Fobos-Grunt (Phobos-Soil), was directed at Phobos but would also make some Mars observations. The spacecraft was at first intended to launch in October 2009, but technical and funding issues delayed it until 2011. It carried China's first Mars probe, Yinghuo-1, and was originally intended to carry one or two MetNet Mars landers developed by the Finnish Meteorological Institute. The 2011 opportunity was dynamically less satisfactory than 2009 and the extra mass of MetNet had to be omitted. Another passenger on the Phobos Sample Return Mission was an experiment from the US-based Planetary Society called Living Interplanetary Flight Experiment (LIFE), which sent 10 types of micro-organisms and a natural soil colony of microbes on the three-year round trip. The results were intended to help determine if meteorite-riding organisms could have spread life throughout the Solar System.

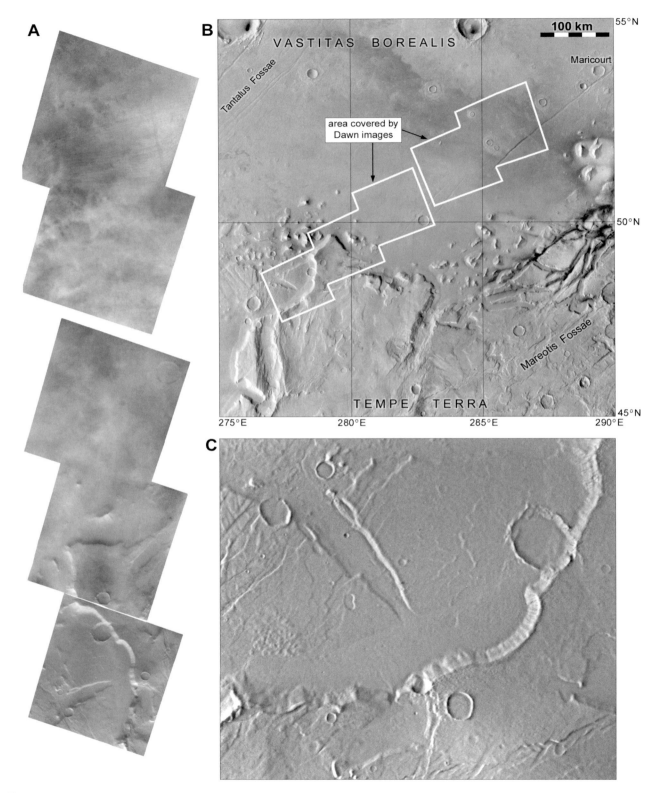

Figure 159. Dawn images taken during the flyby on 18 February 2009. **A:** Dawn images. **B:** Location and orientation of the image strip. **C:** Enlargement of the southwestern image. The area is about 80 km wide. Image courtesy NASA/JPL/MPS/DLR/IDA and the Dawn Flight Team, processing by P. Stooke.

Table 42. *Cerberus Network Nodes From Golombek* et al. *(2009)*

42a. Utopia network

Name	Location	Elevation (km)
Isidis Planitia	9.9° N, 92.97° E	−3.7
Utopia Planitia	18.4° N, 108.6° E	−3.4
Elysium Planitia	13.0° N, 124.1° E	−3.0

42b. Chryse network

Name	Location	Elevation (km)
Arabia Terra	19.2° N, 2.6° E	−2.0
Chryse NW	18.5° N, 312.0° E	−3.3
Chryse South	9.7° S, 343.6° E	−1.2
Meridiani Planum	2.1° S, 354.1° E	−1.4
Ares plateau	14.4° N, 333.2° E	−2.2

The mission was launched from Baikonur at 20:16 UT on 8 November and placed in a 207 by 347 km parking orbit inclined 51.4° to the equator. Some 2.5 hours later its upper stage was to raise the apogee to 4160 km, and a second burn at perigee would send the probe to Mars. The first rocket burn did not occur, but the solar panels did deploy to generate power and several thruster burns were evident in telemetry. Communication was difficult, but ESA facilities, including a ground station in Perth, Australia, helped establish contact. Despite this assistance, the spacecraft could not be commanded and it never left Earth orbit. The spacecraft re-entered the atmosphere on 15 January 2012 over the southeastern Pacific Ocean, Argentina or Brazil.

When recovery still seemed possible, several alternative missions were considered, including landing on Phobos without sample return or flight past a near-Earth asteroid. Activities in Mars orbit would have included observations of the planet's atmosphere and surface, the dynamics of dust storms, and radiation, dust and plasma in Mars orbit. Intended Phobos operations are described in the section on "Phobos and Deimos."

8 November 2011: Yinghuo-1

Yinghuo-1, the first Chinese mission to Mars, was to be carried to Mars on the Russian Phobos Sample Return Mission (Fobos-Grunt) spacecraft, which was lost after launch on 8 November 2011. Yinghuo-1 was a 110 kg cuboid, 0.75 by 0.75 by 0.60 m, with two solar panels. It was intended to observe the planet's surface and satellites, measure its magnetic field and examine atmospheric water loss. Fobos-Grunt and Yinghuo-1 would arrive at Mars together in 2012 and enter a very elliptical equatorial orbit, about 800 by 80 000 km, inclined no more than 5° to the equator of Mars, with a 72 hour period. After three orbits the Chinese probe would separate from Fobos-Grunt and remain in that orbit while the Russian spacecraft dropped closer to Mars and Phobos. No orbit change capability would be provided on Yinghuo-1, and a lifetime of roughly one Earth year was anticipated. Yinghuo-1 might have been the first mission since Viking to be able to make close observations of Deimos. Similar plasma instruments were mounted on both Yinghuo-1 and Fobos-Grunt to enable coordinated observations in Mars orbit. After launch Yinghuo-1 remained joined to Fobos-Grunt until they re-entered the Earth's atmosphere on 15 January 2012.

MSL landing site selection

The goal of the Mars Science Laboratory (MSL) rover mission was to assess the present and past habitability of the chosen landing site. The site selection process was described by Grant *et al.* (2010) and Golombek *et al.* (2012), and openly documented on a NASA Ames Research Center website, and this description summarizes those sources. Engineering considerations initially limited landing sites to the broad zone between 60° N and 60° S, and to elevations below +2 km in the Mars Global Surveyor MOLA elevation data. Sites should be understood geologically and geochemically, with evidence for past or present habitable conditions, should allow preservation of biosignatures (fossils or chemical remnants of life or its precursors), and water or ice should not be present in the surface materials today to prevent possible contamination. The planetary protection rules for MSL were stricter than for earlier missions, based on the new concept of a "special region" within which water availability might at times be capable of supporting the most desiccation-tolerant terrestrial organisms. Any site with evidence of recent water flow,

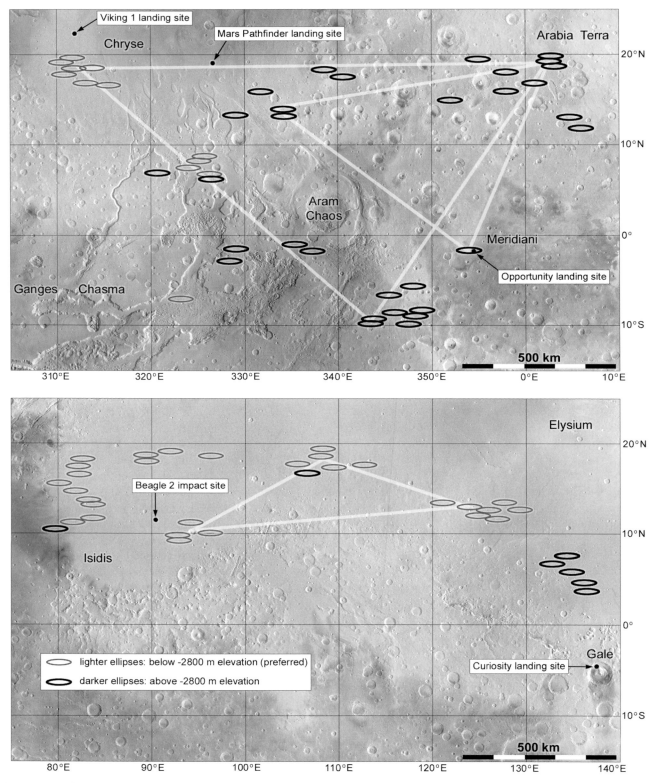

Figure 160. Cerberus landing ellipse candidates and potential arrays from Golombek *et al.* (2009). Array nodes might be located in any ellipse in the cluster surrounding that vertex of the array, pending a more detailed assessment of the ellipses.

including gullies, or with near-surface ice that could be melted by the rover's Radioisotope Thermoelectric Generator (RTG) in the event of an uncontrolled impact, was a "special region" and could not be targeted. This limitation was controversial when first introduced as it precluded numerous very attractive sites.

Proposals for MSL landing sites were sought from the planetary science community in 2006, when launch was anticipated in 2009. A series of open site selection workshops was designed to attract the broadest possible community involvement, and the process to be followed was outlined at the first of these meetings. Of the sites submitted at the first workshop, roughly 40 of the most suitable candidates would be ranked at that meeting. In 2007 a shortlist of about 20 sites would be chosen, reduced to between 5 and 10 sites at a third workshop in 2008. By the end of 2008 a specific landing zone spanning 10° in latitude and 15° in longitude would be selected, possibly containing several ellipses. Finally, in 2009 a fourth meeting would select the final site. This process was extended after a launch delay and the landing zone idea was dropped, but otherwise it was followed fairly closely.

The phyllosilicates or clay minerals mentioned for some sites in Table 43 had been identified from orbit by Mars Express and interpreted as evidence for a wet period very early in Martian history. Bibring *et al.* (2006) described a concept of Martian geological history in which a very early period (Phyllosian) was warmer and wet, producing clay deposits throughout the crust. Later the climate turned colder, with large amounts of acidic surface and ground water depositing salts in layered sediments as seen by Opportunity. This was called the Theiikian (sulfur-rich) period. Finally, Mars became the very cold dry desert it has been ever since, in the Siderikian period. The first period would be the one most conducive to life, so its deposits were high-priority sites for MSL. These time periods do not correspond exactly to the widely used Noachian, Hesperian and Amazonian periods derived from geological mapping based on geomorphology.

Grant *et al.* (2010) presented a list of all sites in a summary of the selection process, ordered by longitude and numbered (Table 43, with some modifications to match workshop presentations). The site numbering from this table is used throughout the site selection discussion, including site designations in Figures 161 to 169 and

Tables 43 to 45. The Mawrth sites in Table 43 are from Grant *et al.* (2010), with additional notes from conference presentations.

The First Landing Site Workshop was held in Pasadena, California, from 31 May to 2 June 2006. Table 43 identifies the sites discussed at that workshop (W1 in the "proposed" column) and Table 44 shows the post-workshop ranking of these sites. Some sites were rejected immediately for being special regions or for exceeding the latitude limits. Each ellipse now became a target for future imaging by the operating fleet of orbiters (MGS, Odyssey, MRO and Mars Express). MGS failed late in 2006, ending its contributions to the MSL site selection process. Figure 161 shows all candidate MSL sites, and Figures 162–165 illustrate each of the first workshop sites in more detail, with ellipses taken from the workshop presentations.

The sites considered at the first workshop included over 90 distinct landing ellipses, and three more sites (A, B and C) were added later but not ranked. At the end of the workshop the 33 preferred sites were ranked in the order shown in Table 44, from top to bottom, and grouped into highest, medium and lower priority. The location listed in Table 44 is taken from a summary table for the workshop, and the refined site is taken from a list of MRO CRISM targets released on the instrument website in the summer of 2007. The CRISM team stated that sites 13, 27, 28 and 31 were no longer being targeted for further imaging. Complete consistency between different versions of these lists is not to be expected.

Modified sites were suggested by several presenters at the 38th Lunar and Planetary Science Conference (LPSC) in March 2007. The engineering constraints had been narrowed by this time to latitudes between 45° N and 45° S, with elevations below +1 km. Latitudes north of 30° N were less desirable because of communication constraints during landing. Four safe sites were identified by Gwinner *et al.* (2007) in Candor Chasma from Mars Express High Resolution Stereo Camera (HRSC) topographic maps, and the same authors also gave coordinates for two ellipses at the Mawrth site, at 22° N, 344° E and 25° N, 340° E. The first is close to the location shown in the ranking table from the first workshop, and the second is close to the site illustrated roughly by J.-P. Bibring at the 2006 Workshop. Paris *et al.* (2007) modified the SW Arabia site (now referred

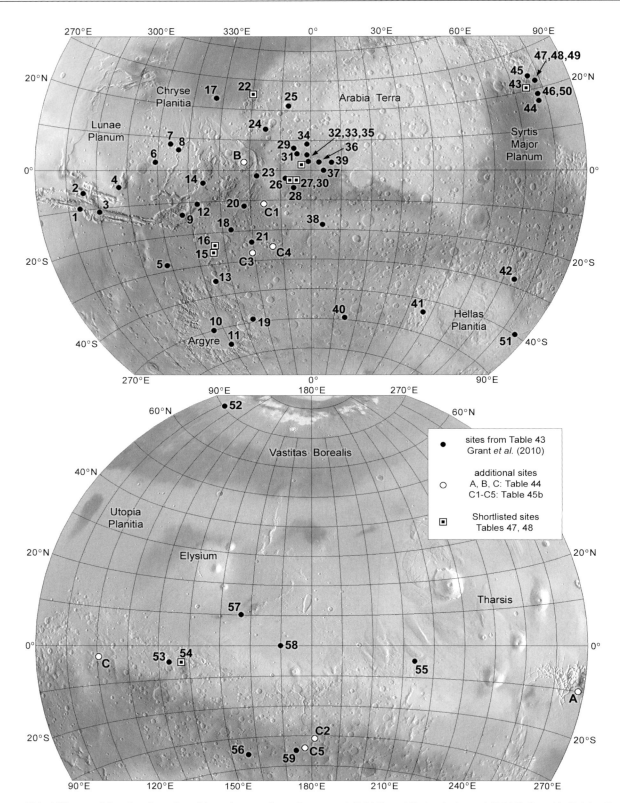

Figure 161. MSL candidate landing sites. Most sites are from Grant *et al.* (2010) and Golombek *et al.* (2012), listed in Table 43, with additions from other tables as indicated. For scale, 10° of latitude or along the equator spans 600 km.

Table 43. *Candidate MSL Sites, Modified From Table 4 in Grant* et al. *(2010)*

Site	Name	Location	Geology	Proposed
1	Melas Chasma	9.8° S, 283.6° E	Paleolake, sulfates	W1
2	Western Candor	5.9° S, 284.1° E	Sulfates, layered deposits	W1
		5.6° S, 285.2° E		W2
3	Eastern Melas	11.6° S, 290.5° E	Layered deposits	W1
4	Juventae Chasma	4.5° S, 297.5° E	Layered sulfates	W1
		4.8° S, 296.8° E		W2
5	Ritchey crater	28.3° S, 308.9° E	Clays, alluvial/fluvial deposits	W2
6	Xanthe Terra	2.3° N, 309.0° E	Delta deposit	Dec. 2009
7	Northern Xanthe	8.0° N, 312.7° E	Hypanis Vallis highlands, valley walls, delta	W1
		6.9° N, 312.8° E	Another site at 8° N, 318° E (floor of Shalbatana)	W1
		11.4° N, 314.7° E	in the presentation	W1
8	Shalbatana Vallis	7.0° N, 317.0° E	Phyllosilicates	W2
9	Eos Chasma Alluvial	13.4° S, 317.5° E	Alluvial fan (13.6° S also considered)	Mar. 2007
10	Argyre	40–60° S, 300–330° E	Old basin bedrock from mountainous rim	W1
11	Argyre	56.3° S, 318.0° E	Glacial/lacustrine features (alternative: 54.4° S,	W1
		55.2° S, 322.4° E	322.2° E) (56.8° S, 317.7° E on debris apron	W1
			mentioned in abstract)	
12	Eos Chasma	10.7° S, 322.0° E	Quartz or silica, aqueous landforms	W1
13	Hale crater	35.7° S, 323.4° E	Gullies (several sites, floor too rough, and a	W1
			"special region" unsuitable for landing)	
14	Valles Marineris	3.8° S, 324.6° E	Floor/walls	Dec. 2009
15	Holden crater	26.3° S, 325.1° E		W1
		26.4° S, 325.0° E	Layered fluvial deposits, lake sediments, alluvial	W1
		26.3° S, 325.4° E	fans	W1
16	Eberswalde crater	24.0° S, 325.6° E	Layered deposits, fan delta, channels	W1
		23.8° S, 327.0° E	(also 24.0° S, 326.3° E)	W1
		23.9° S, 326.7° E		W1
		23.0° S, 327.0° E		W1
17	Tiu Valles	22.9° N, 327.8° E	Fluvial deposits, lake sediments	W2
18	Ladon basin	18.8° S, 332.5° E	Chlorides with phyllosilicates nearby	Dec. 2009
19	Wirtz crater	49.0° S, 334.0° E	Gullies (two sites, "special region" unsuitable for	W1
			landing)	
20	Margaritifer basin	11.7° S, 337.3° E	Fluvial deposits	W1
		12.8° S, 338.1° E		W1
21	Samara Valles	23.6° S, 339.8° E	Valley networks, lake basin, sediments	Mar. 2007
22	Mawrth Vallis			
	site 0	24.5° N, 338.9° E	Noachian layered phyllosilicates	W1
	site 1	24.7° N, 340.1° E	(two sites illustrated at first workshop: 24.9° N,	W1
	site 2	24.0° N, 341.0° E	340.7° E; 26.7° N, 341.3° E)	W1, W3
	site 3	23.2° N, 342.2° E		W1
	site 4	24.9° N, 339.4° E		W3
23	Iani Chaos	1.6° S, 341.8° E	Hematite and sulfate, layered sediments	W1
		2.6° S, 342.2° E		W1
		2.1° S, 342.3° E		W1
24	Margaritifer Terra	13.1° N, 345.3° E	Chlorides (also called Chloride Site 10)	W2
25	Becquerel crater	21.5° N, 351.4° E	Layered deposits (also 21.8° N, 351° E)	W1
		21.3° N, 352.5° E		W1

Table 43. (*cont.*)

Site	Name	Location	Geology	Proposed
26	Chloride west of Miyamoto	3.2° S, 351.6° E	Chlorides (also called Chloride Site 17)	June 2008
27	Miyamoto crater, SW	1.8° S, 352.4° E	Layered deposits, hematite	W1
	Meridiani (also called West	3.5° S, 352.3° E	Chlorides and phyllosilicates, channel deposits	W2
	Meridiani or Runcorn)	3.4° S, 352.6° E	Clays, sulfates, close to hematite-rich material	W2
28	East Margaritifer Terra	5.6° S, 353.8° E	Chlorides, phyllosilicates	Dec. 2009
29	Meridiani Planum bench	8.3° N, 354.0° E	Hematite and sulfate, layered sediments	W1
		7.9° N, 354.0° E		W1
		8.4° N, 354.5° E		W1
30	South Meridiani Planum	3.3° S, 354.4° E	Sulfate plains with phyllosilicate in uplands	June 2008
		3.1° S, 354.6° E		W3
31	Vernal crater (SW Arabia Terra)	6.0° N, 355.4° E	Layered deposits (fluvio-lacustrine?), methane, spring deposits	W1
32	Northern Sinus Meridiani	1.6° N, 357.5° E	Layered deposits, ridges, hematite	W2
33	Northern Sinus Meridiani crater lake	5.5° N, 358.1° E	Layered deposits	W1
34	West Arabia Terra	8.9° N, 358.8° E	Layered deposits	W1
35	Northern Sinus Meridiani	2.6° N, 358.9° E	Layered deposits	W1
36	Northern Sinus Meridiani	1.9° N, 0.4° E	Layered deposits	W1
		3.1° N, 3.3° E		W1
		2.4° N, 3.5° E		W1
37	East Meridiani	0.0° N, 3.7° E	Layers, sulfates and hydrated materials, phyllosilicates nearby	W1
38	Chloride Site 15	18.4° S, 4.5° E*	Chlorides (*longitude should be 2.8° E)	W2
39	Northern Sinus Meridiani	2.4° N, 6.7° E	Layered deposits	W1
40	Southern mid-latitude (SML) craters	49.0° S, 14.0° E	Recent climate-related deposits (flow features, gullies, patterned ground, dissected mantles)	W1
41	Hellas	44.0° S, 46.0° E	Ancient basin bedrock	W1
42	Terby crater	27.4° S, 73.4° E	Hydrated layered deposits (lacustrine?), fluvial and ice-related morphology.	W1
		27.6° S, 74.0° E		W1
		28.0° S, 74.1° E	Safer landing site	W1
43	Nili Fossae trough	21.0° N, 74.5° E	Noachian phyllosilicates, bedrock, clay-rich ejecta, Hesperian volcanics (related sites at Nilosyrtis, 23° N, 76° E and 29.16° N, 72.97° E, ranked 10th at workshop)	W1
44	Northeast Syrtis Major	17.1° N, 75.4° E	Hesperian volcanics, Noachian layered deposits,	W1
		16.1° N, 76.7° E	lava possibly sampled by Mars meteorite	W1
		16.4° N, 77.4° E	(Nakhlite) source crater	W1
		16.3° N, 78.0° E	Diverse mafic volcanics, Noachian layered clays	W1
		16.2° N, 76.6° E	Diverse aqueous alteration minerals on Noachian–	W2
		17.8° N, 77.1° E	Hesperian boundary	Dec. 2009
45	Nilosyrtis	23.0° N, 76.0° E	Phyllosilicates (= site 43b in Figure 164)	W1
46	Nili Fossae crater (Jezero)	18.4° N, 77.7° E	Fan, valleys, layered deposits, inverted channels	W1
47	East Nili Fossae	21.8° N, 78.6° E	Phyllosilicates, mafic rocks	W2
48	Nili Fossae carbonate	21.7° N, 78.8° E	Phyllosilicates, carbonates	Dec. 2009
49	Nili Fossae carbonate plains	21.9° N, 78.9° E	Layered phyllosilicates under sulfates	June 2008
50	Western Isidis	14.2° N, 79.5° E	Escarpment, volatile sink	W1
		18.0° N, 79.6° E		W2

Table 43. (*cont.*)

Site	Name	Location	Geology	Proposed
51	Dao Vallis	38.9° S, 81.2° E	Valley terminus, layered deposits	W1
		39.5° S, 82.7° E		W1
		41.2° S, 84.4° E		W1
		40.7° S, 85.6° E		W1
		41.7° S, 85.8° E		W1
		43.3° S, 86.8° E		W1
52	Vastitas Borealis	70.5° N, 103.0° E	Salt, ice/impact tectonics	Dec. 2009
53	Aeolis Region	5.1° S, 132.9° E	Lobate fan delta	Mar. 2007
54	Gale crater	4.7° S, 137.0° E	Layered deposits, exhumed channels (4.4° S,	W1
		5.7° S, 137.6° E	137.4° E also proposed)	W1
55	Northwest Slope Valleys	4.9° S, 213.7° E	Flood, fluvial morphology (other ellipses at	W1
			4.8° S, 214.1° E and 4.7° S, 212.6° E)	
56	South Terra Cimmeria	35.2° S, 156.1° E	Gullies ("special region," unsuitable for landing,	W1
		35.8° S, 155.6° E	northern crater probably too small for ellipse)	W1
57	Athabasca Vallis	10.1° N, 156.9° E	Dunes, streamlined forms, fissures (second ellipse	W1
			at 9.9° N, 157.3° E	
58	Elysium (Avernus Colles)	1.4° N, 168.7° E	Iron-rich materials at valley terminus	W1
		3.1° S, 170.6° E		W2
		3.1° S, 170.7° E		W1
		0.2° N, 172.5° E		W1
59	Ariadnes Colles	35.0° S, 174.2° E	Phyllosilicates, possible sulfates	W2

Notes: Site 55 is incorrectly located at 146.5° E in all other sources (it was at 146.5° W), and as sites were numbered by longitude it is out of sequence. Site 56 coordinates are from the original workshop presentation. The column at right indicates when the site was proposed. W1 is the First Landing Site Workshop, W2 is the second, and so on. More details are found in Tables 44 to 48.

to as Vernal crater), placing it in the crater at 6°N, 355.5°E. In addition, Litvak *et al.* (2007) reviewed Mars Odyssey GRS/HEND data for the preferred MSL sites and suggested that locations with more evidence of water in surface materials should be considered. Their new sites were numbers 9, 21 and 53 in Tables 43 and 45b. The Samara Valles site (21) is at the Mars 6 site of 1974 (Figure 37 in Stooke, 2012). All these new sites are shown in Figures 162 and 163. Most of these details were presented on conference posters rather than in the published abstracts.

The second MSL Landing Site Workshop was held in Pasadena on 23 to 25 October 2007. The elimination process was accelerated to leave only about five primary sites and some backup candidates, in order to reduce future imaging requirements. Engineering considerations were tightened again, with the zone north of 30°N becoming unfavorable due to communication limitations during entry, descent and landing (EDL). The landing

had to occur within view of MRO to support data relay during EDL, mandating different arrival dates for different latitude zones: 30°N to 15°N, 15°N to 10°S, 10°S to 30°S and 30°S to 40°S. This requirement was dropped in 2008.

The MER site selection procedure had been compromised late in the process when wind modeling ruled out some sites, necessitating the designation of geologically bland low-wind ellipses in Elysium (Figure 3E). To avoid this situation for MSL, sites that met higher safety standards but still afforded high-value science ("safe havens") would serve as backups or conservative selections. Sites below −2 km were preferred. Safe havens required larger landing ellipses (16 km radius), so they had to contain high-value science targets inside the ellipse. The smaller (initially 10 km radius) MSL ellipses could be adjacent to a target ("go-to" sites) if necessary. In fact, the MSL landing ellipses had now evolved to 25 by 20 km to accommodate uncertainties along the

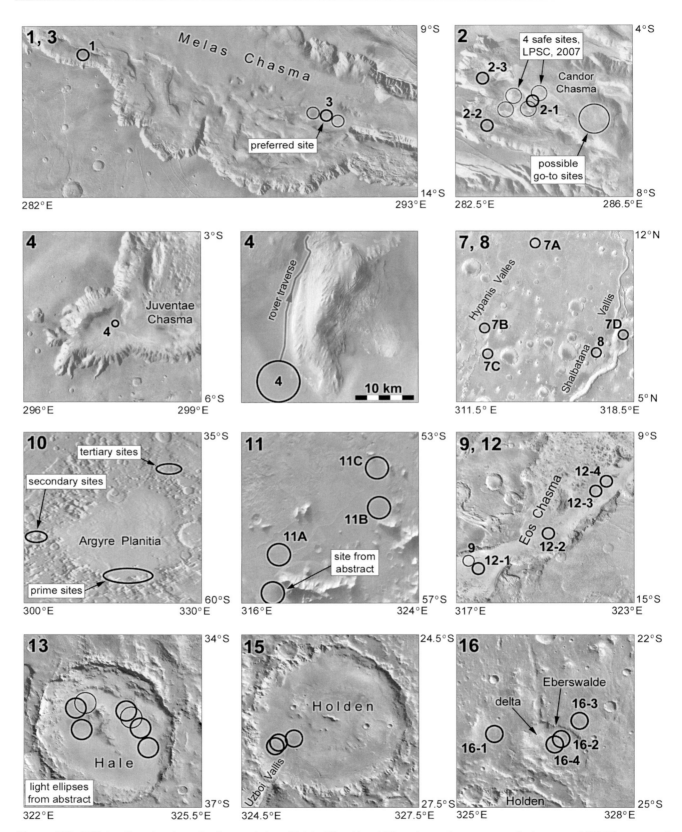

Figure 162. MSL landing sites from the first workshop (Table 43), with additions from other sources. Gwinner *et al.* (2007) proposed safer sites at site 2. Site 4 includes a possible rover traverse. Site 7D was illustrated in a workshop presentation but not listed elsewhere. Site 9 (Eos Chasma) is from Litvak *et al.* (2007). Site 10 (Argyre) includes site 11. For scale in Figures 162–165, 1° of latitude spans 60 km.

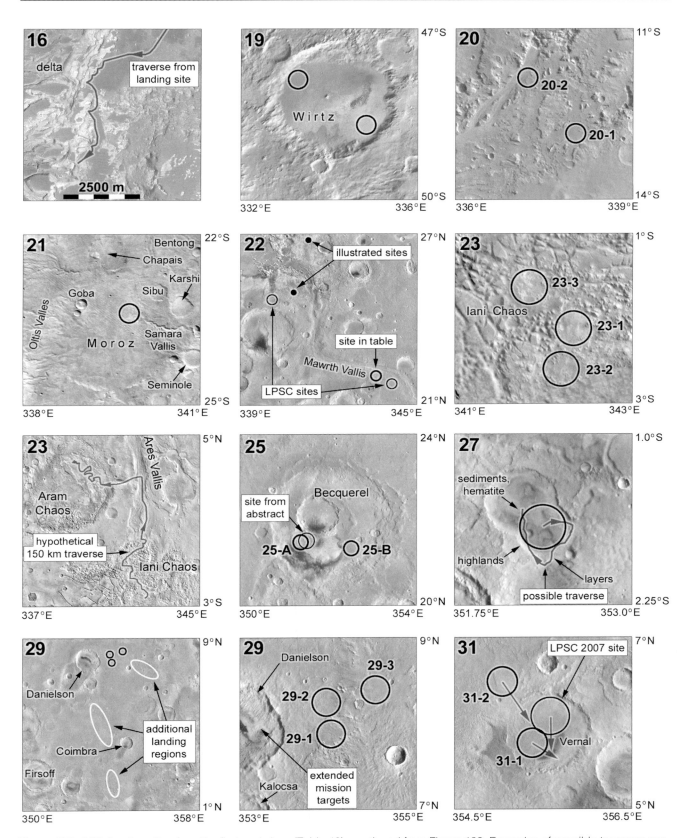

Figure 163. MSL landing sites from the first workshop (Table 43), continued from Figure 162. Examples of possible traverses are shown at sites 16, 23, 27 and 31. At site 22 two locations were illustrated in a presentation, a different location was included in a summary table, and Gwinner *et al.* (2007) showed two different locations at the next LPSC. The site 31 LPSC site is from Paris *et al.* (2007).

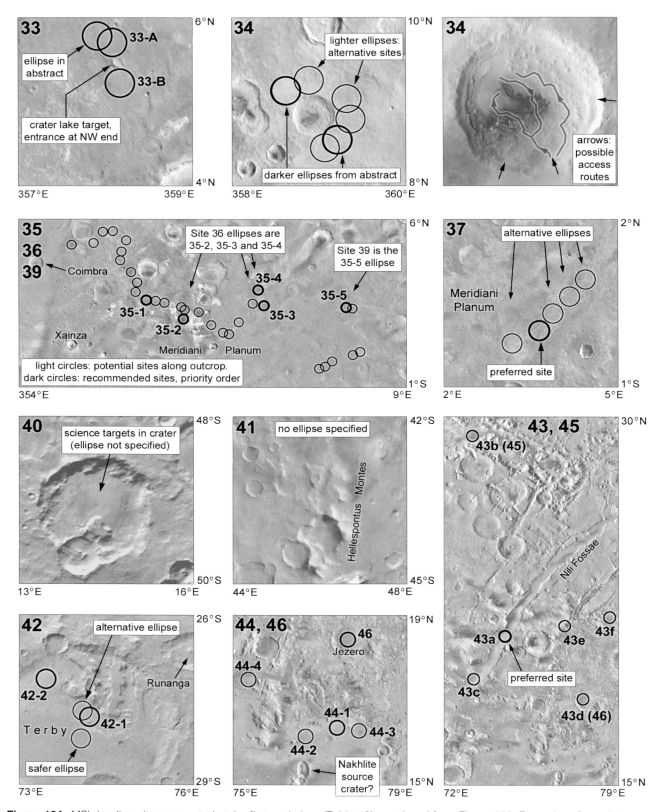

Figure 164. MSL landing sites presented at the first workshop (Table 43), continued from Figure 163. Examples of possible traverses are shown at site 34. Some ellipses shown at site 35 in the workshop were later assigned to sites 36 and 39. Site 43b was also listed as site 45.

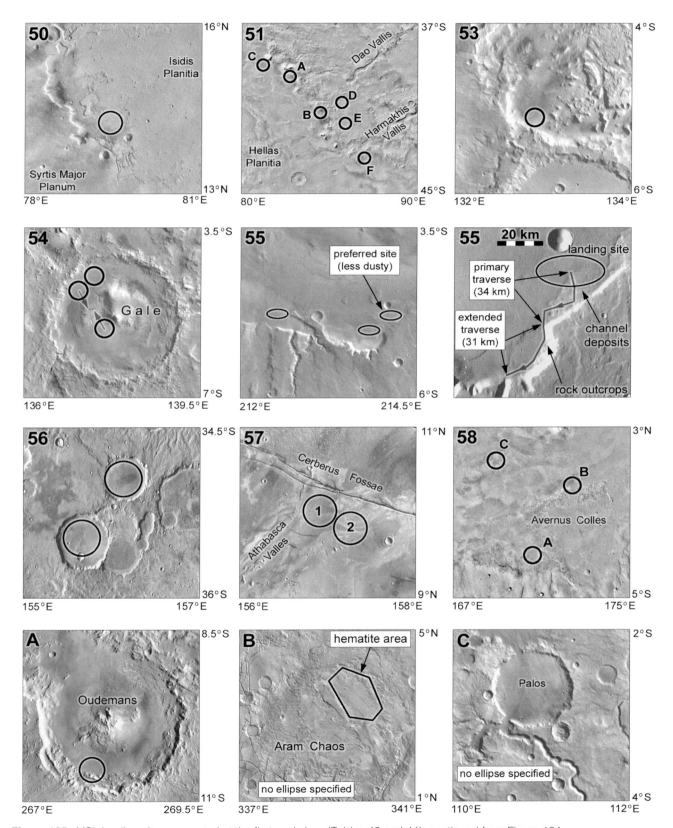

Figure 165. MSL landing sites presented at the first workshop (Tables 43 and 44), continued from Figure 164.

approach track produced by winds. A safe haven site would be needed in each of the three more northerly zones described above.

By the time of the second workshop, 9 sites had been deleted from the 33 ranked in Table 44, leaving 24 still in the running. The deleted sites were Athabasca, Margaritifer Basin, East Melas, Becquerel, Hale crater gullies, Argyre, Northwest Slope Valleys, Meridiani Bench and southern highlands (also referred to as SML, Southern Mid-Latitudes). Further study of the voluminous new datasets, especially the compositional data, resulted in an additional 27 site suggestions. All 51 of these sites are listed in Table 45 and illustrated in Figures 166–170. Comparison of Figures 162–165 and 166–170 reveals changes in many of the sites between the two workshops. These site lists were compiled from two presentations at the second workshop (Grant, 2007; Golombek *et al.*, 2007).

During the workshop, participants promoted their own suggestions and voted for the most promising sites. Matt Golombek identified sites 31, 32, 36 and 37 as possible safe havens, and sites 7, 22-0, 22-3, 27, 42 and 54-south as "über-safe" havens. Sites were rejected for excessive roughness or slope in high-resolution digital elevation models made from HiRISE stereoscopic images, for being poorly understood geologically, or for being unlikely to preserve evidence of habitability. The most southerly sites were downgraded. Finally, a shortlist of the most promising sites emerged (Table 46). Five equatorial or northern sites were preferred, but two interesting southerly sites were retained despite misgivings about operations at those latitudes in winter. Four sites were placed in "purgatory," requiring further study. Safe haven sites in Meridiani were identified in the north and purgatory groups. They were not far from the Opportunity site but offered much more extensive views of stratigraphy and more varied materials.

A revised shortlist of six sites (Figure 170 and Table 47) was circulated on 14 December 2007, and published by Golombek *et al.* (2008). The list included prime and safe haven ellipses, the latter required only if the expected landing precision worsened during further analysis. As the safe haven ellipses were larger, they were more difficult to place and some sites had no safe haven alternative. The prime ellipses were 20 by 25 km, and the safe haven ellipses were 32 by 35 km, elongated

along the entry azimuth. For most sites only one prime ellipse was chosen, but Mawrth still had four possible locations. Site 4 at Mawrth was moved from the position mapped in Figure 167 to a smoother location further from the clay exposures. The site that had been called SW Meridiani was renamed Runcorn (commemorating the geophysicist S. K. Runcorn when it was thought the crater would be named after him) and later became Miyamoto to reflect a newly adopted official name. Jezero was dropped when it was found to be too rocky, and Eberswalde replaced Terby for its greater science potential and to reflect increased concerns about operating at Terby's latitude. North Meridiani was added to the list to guarantee a safe site with high-quality science. These six sites now became the focus of an imaging campaign by MRO.

A third site selection workshop was held from 15 to 17 September 2008. Leading up to it, the site selection group reviewed the sites in Table 47 and considered four more suggested by new data or reappraisals of older data. These four new sites were all similar to locations considered in the second MSL workshop. Gale crater was brought back to add one very low-elevation site to the list, and because new compositional data suggested an interesting variety of surface materials. A new location south of the MER-B Opportunity landing site was substituted for North Meridiani as it offered more geological variety. It was named South Meridiani. The other proposed sites were a deposit of chloride salts west of Miyamoto and a carbonate exposure near Nili Fossae (Figure 171), but they were not accepted. Safe havens were no longer needed as the engineers were now confident they could land in all candidate primary sites. The new shortlist of seven sites and the two rejected candidates are listed in Table 48, and the third workshop considered only these sites.

Presentations at the third workshop dealt with the geological potential of each of the seven sites (counting Mawrth as one site), and some suggested possible traverse routes. The Mawrth region had included four possible ellipses (Figure 170). Sites 1 and 4 were "go-to" sites, located in smooth areas near clay deposits (phyllosilicates) and requiring a drive out of the ellipse to reach the material of most interest. The site 4 seen in Figure 170 was not the same as that shown in Figure 167, which was on the clay deposit. This new site was

Table 44. *Ranked Sites From First MSL Landing Site Workshop*

44a. Highest-priority sites

Site	Location	Refined location	Name	Comments (no. of ellipses)
43	22° N, 75° E	20.93° N, 74.35° E	Nili Fossae trough	Phyllosilicates (2)
15	26.4° S, 325.3° E	26.4° S, 325.3° E	Holden crater fan	Layered materials (3)
42	28° S, 73° E	27.744° S, 74.114° E	Terby crater	Layered materials (3)
22	22.3° N, 343.5° E	22.3° N, 343.5° E	Mawrth Vallis	Phyllosilicates (0)
16	24.0° S, 326.3° E	24.0° S, 326.3° E	Eberswalde crater	Delta (4)
54	4.6° S, 137.2° E	4.6° S, 137.2° E	Gale crater	Layered deposits (2)
2	5.75° S, 284.2° E	5.80° S, 284.17° E	West Candor – East	Sulfate deposits (3)
35, 36, 39	2.7° N, 358.8° E	2.34° N, 6.69° E	Northern Meridiani	Sedimentary layers (5)
4	5° S, 297° E	5° S, 297° E	Juventae Chasma	Layered sulfates (1)
43, 45	23° N, 76° E	29.16° N, 72.97° E	Nilosyrtis	Phyllosilicates (4)
1	9.8° S, 283.6° E	9.81° S, 283.62° E	Melas Chasma	Paleolake (1)

44b. Medium-priority sites

Site	Location	Refined location	Name	Comments (no. of ellipses)
37	0.0° N, 3.7° E	0.01° N, 3.66° E	East Meridiani	Sedimentary layers (5)
57	10° N, 157° E	9.93° N, 156.77° E	Athabasca Vallis	Cerberus Rupes deposits (2)
23	2° S, 342° E	2° S, ~342° E	Iani Chaos	Hematite, sulphate (3)
46	18.4° N, 77.68° E	18.4° N, 77.68° E	Nili Fossae crater	Valley networks, layers (1)
12	11° S, 320° E	10.7° S, ~321.91° E	Eos Chasma	Chert (4)
33	5.6° N, 358° E	5.6° N, 358°E	Meridiani crater lake	Crater lake sediments (2)
44	10° N, 70° E	~10° N, ~70° E	NE Syrtis Major	Volcanics (5)
20	12.77° S, 338.1° E	12.85° S, 338.0° E	Margaritifer basin	Fluvial deposits (2)
3	11.62° S, 290.45° E	11.72° S, 290.72° E	E. Melas Chasma	Layered deposits (3)
51	40° S, 85° E	39.5° S, 82.7° E	Hellas/Dao Vallis	A major valley (6)
7	11° N, 314° E	11.4° N, 314.70° E	Xanthe/Hypanis Vallis	Delta (4)
25	21.8° N, 351° E	21.3° N, 352.52° E	Becquerel crater	Layered deposits (2)

44c. Lower-priority sites

Site	Location	Refined location	Name	Comments (no. of ellipses)
31	2–12° N, 348–355° E	6.01° N, 355.60° E	SW Arabia Terra	Sediments, methane (2)
13	35.7° S, 323.4° E	35.7° S, 323.4° E	Gullies/Hale crater	Gullies – floor too rough (5)
34	8.9° N, 358.8° E	8.9° N, 358.8° E	W. Arabia	Sedimentary rocks (2)
10, 11	56.8° S, 317.7° E	56.8° S, 317.7° E	Argyre	Glacial features (3)
55	0° N, 215° E	c. 0° N, 215° E	Northwest Slope Valleys	Flood features (3)
27	1.8° S, 352.4° E	1.8° S, 352.4° E	West Meridiani	Sediments, hematite (1)
58	1.0° S, 169.5° E	3.1° S, 170.7° E	Elysium/Avernus Colles	High iron abundance (3)
29	7.5° N, 354° E	7.5° N, 354° E	Meridiani Bench	Layered sediments (3)
40	49° S, 14° E	49.03° S, 14.494°E	Southern highlands	Recent climate deposits (0)
50	5–15° N, 80–95° E	18° N, 80–79.60° E	Isidis basin scarp	Volatile sink (1)

44d. Sites added at the 38th LPSC, March 2007

Site	Location	Refined location	Name	Comments (no. of ellipses)
21	24° S, 339.5° E	23.35° S, 339.75° E	Samara Valles	Fluvial, lake sediments (1)
9	13.6° S, 317.4° E	13.60° S, 317.5° E	Eos Alluvial	Layered sediments (1)
53	5.05° S, 132.85° E	5.05° S, 132.85° E	Aeolis fan delta	Fan below crater wall (1)

Table 44. (*cont.*)

44e. Sites also considered but not in final ranking

Site	Location	Refined location	Name	Comments (no. of ellipses)
A	10.0° S, 267.9° E		Oudemans crater	Layers, elevation too high (1)
B	2.5° N, 338° E		Aram Chaos	Hematite in chaos area (0)
C	2.7° S, 110.8° E		Palos crater	Layered materials (0)
19	48.6° S, 334° E		Wirtz crater	Gullies – special region (2)
56	35.5° S, 156° E		Small gullied craters	Gullies – special region (2)
41	44.0° S, 46.0° E		Hellas	Basin rim bedrock (0)

Table 45. Sites Considered at the Second MSL Workshop, October 2007

45a. Sites retained from the First MSL workshop

Site	Name	Location	Description
1	Melas Chasma	9.81° S, 283.62° E	Paleolake
2	West Candor Chasma	5.80° S, 284.17° E	Sulphate deposits
4	Juventae Chasma	4.45° S, 298.09° E*	Layered sulfates
7	Xanthe – Hypanis Vallis	11.4° N, 314.65° E	Layered deposits
12	Eos Chasma	10.7° S, 322.05° E	Chert
15	Holden crater fan	26.32° S, 325.30° E	Layered materials
16	Eberswalde crater	23.85° S, 326.75° E	Delta
22	Mawrth Vallis	24.65° N, 340.1° E†	Phyllosilicates (22–4 in Figure 167)
23	Iani Chaos	2.06° S, 342.41° E	Hematite, sulfate
27a	West Meridiani	1.7° S, 352.39° E‡	Sediments, hematite
31	SW Arabia Terra (Vernal)	6.01° N, 355.60° E	Sedimentary rocks, methane
33	Meridiani crater lake	5.72° N, 358.03° E	Crater lake sediments
34	West Arabia crater	8.45° N, 359.09° E	Sedimentary rocks
37	East Meridiani	0.01° N, 3.66° E	Sedimentary layers
39	North Meridiani	2.37° N, 6.69° E	Sedimentary layers
42	Terby crater	27.744° S, 74.114° E	Layered material
43	Nili Fossae trough	20.93° N, 74.35° E	Phyllosilicates
44	NE Syrtis Major	16.21° N, 76.63° E	Volcanics
45	Nilosyrtis	29.16° N, 72.97° E	Phyllosilicates
46	Nili Fossae crater (Jezero)	18.44° N, 77.58° E	Valley networks, delta sediments
50	Isidis basin escarpment	18.00° N, 79.60° E	Volatile sink
51	Hellas – Dao Vallis	39.5° S, 82.7° E	Valley terminus, layered deposits
54	Gale crater	4.50° S, 137.35° E	Interior layered deposits
58	Elysium – Avernus Colles	3.05° S, 170.60° E	High iron abundance

Notes: * Error in source: 297.09° E would be a better match to the Workshop 1 presentation.

† Not the same as site 22 in the first workshop.

‡ Found unsuitable, so location changed in workshop presentation to site 27b, Table 45b.

Table 45. (*cont.*)

45b. Sites added for consideration at the Second Workshop

Site	Name	Location	Description
2	West Candor Chasma alternative	5.75° S, 285.19° E	Sulfates
4	Juventae Chasma alternative	4.88° S, 297.01° E	Layered sulfates
5	Ritchey crater	28.28° S, 308.53° E	Clays, fan deposit
8	Shalbatana Vallis	7.0° N, 317.0° E	Phyllosilicates
9	Eos Alluvial	13.6° S, 317.5° E	Alluvial fan
17	Tiu Valles	22.9° N, 327.75° E	Chemolithotrophic habitat
21	Samara Vallis	23.55° S, 339.75° E	Valley network channel
22	Mawrth Vallis B1	24.5° N, 338.9° E	Layered clays
22	Mawrth Vallis B2	23.95° N, 341.2° E	Layered clays
22	Mawrth Vallis B3	23.2° N, 342.5° E	Layered clays
24	Chloride Site 10	13.1° S, 345.3° E	Chloride salts in Margaritifer Terra
26	Chloride Site 17	3.2° S, 351.6° E	Chloride salts west of Miyamoto
27b	Southwest Meridiani	3.01° S, 352.1° E	Clays, sulfates
27c	South Meridiani Clays	3.35° S, 352.24° E	Clays, sulfates
27d	Meridiani B3*	3.19° S, 352.20° E	Clays, sulfates
28	Meridiani B4	5.0° S, 354.52° E	Clays, sulfates
32	N. Meridiani alternative	1.5° N, 357.4° E	Sedimentary layers
35	Meridiani B1	3.84° N, 359.04° E	Clays, sulfates
36	Meridiani B2	1.60° N, 3.55° E	Clays, sulfates
38	Chloride Site 15	18.4° S, 4.5° E†	Chloride salts
42	Terby crater alternative	27.4° S, 73.5° E	Clays, possible paleolacustrine
43	Nili Fossae trough alternative	21.73° N, 74.73° E	Clays, mafics
47	Nili Fossae East	21.8° N, 78.6° E	Clays, mafics
53	Aeolis fan delta	5.05° S, 132.85° E	Fan, delta
54	Gale crater alternative	5.66° S, 137.53° E	Interior layered deposits
59	Ariadnes Colles	35.03° S, 174.17° E	Clay-bearing outcrops
C1	Chloride Site 1	11.4° S, 343.4° E	Chloride salts
C2	Chloride Site 2	31.5° S, 180.8° E	Chloride salts
C3	Chloride Site 3	27.9° S, 339.1° E	Chloride salts
C4	Chloride Site 4	25.4° S, 346.6° E	Chloride salts
C5	Chloride Site 5	34.36° S, 177.76° E	Chloride salts

Notes:

* Separate proposal, within overlap area of 27b and 27c.

† Error in Grant *et al.* (2010); longitude should be 2.8° E, from workshop presentation.

smoother for landing but would require a traverse to the southwest to reach the main geological targets. Sites 2 and 3 contained clays inside the ellipses and might need very little driving to reach them. The proponents of Mawrth preferred sites 2 and 4, and eventually chose site 2 as the one that would be voted on at the end of the meeting.

Gale contained a large mound superimposed on its central peak, and the layered sediments of most interest were at the base of the mound. The traverse in Gale (Figure 171) was similar to one illustrated in the previous year (Figure 169), but extended mission options were now added (Edgett *et al.*, 2008). These included crossing the large sedimentary mound to visit a deep canyon on its western side, or driving past that canyon to the southern edge of the mound. At Eberswalde a possible traverse around the ellipse which visited a variety of interesting materials (Figure 172A) was illustrated by Schieber and Malin (2008). Finally, perhaps in an extended mission,

Table 46. *Sites Shortlisted After the Second MSL Workshop*

Category	Sites (numbers in Table 43)
North	Nili Fossae trough (43), Mawrth Vallis (22), Jezero crater (46)
North Safe Haven	Southwest Meridiani Terra (27)
South	Holden crater (15), Terby crater (42)
Purgatory	Eberswalde (16), NE Syrtis (44), Chloride Salts (C1–C5)
Purgatory Safe Haven	East Meridiani (37)

the rover would approach the base of the main delta and climb up onto it. Three possible routes up the sloping delta margin (Figure 172B) were illustrated by Lewis and Aharonson (2008). They offered good stratigraphic possibilities but might not be easy to drive. The Holden site was not changed from the previous meeting, but the possible traverse illustrated at the second workshop was slightly altered (Figure 172C) and extended mission possibilities including Uzboi Vallis flow materials and Holden ejecta were added (Irwin, 2008).

At the end of the workshop, participants voted on the sites (Figure 173), considering geological diversity, degree to which the geological context of samples would be understood, habitability and the feasibility of preservation of any biological materials or markers. The results gave most scientific support to Eberswalde, Holden and Gale, the three crater interior sites. Mawrth and Nili Fossae were less favored, and South Meridiani and Miyamoto were least satisfactory. These results were forwarded to the project, which drew up its formal shortlist of sites in November 2008. Only four sites could be retained to focus data collection and analysis efforts. In addition to the science ranking given above, a safety ranking identified Holden, Gale and Mawrth as the safest sites, Eberswalde less satisfactory, and Nili Fossae marginal. The less scientifically appealing sites were not included in this ranking. Cold temperatures at Holden and Eberswalde were still potentially troubling. The new shortlist consisted of Holden, Gale, Mawrth site 2 and Eberswalde.

Data collection continued in 2009, and the launch of MSL was postponed to 2011 to accommodate hardware and funding concerns. This delay allowed an opportunity to consider an extra site, in case ongoing remote sensing or new interpretations produced a very promising candidate. Suggestions for new sites were solicited in August 2009, resulting in seven proposals (Table 49 and Figure 174). Five of them, the first five in Table 49, met the mission and safety criteria and were studied further (Golombek *et al.*, 2010a), and two of those, Margaritifer Terra and Northeast Syrtis, were selected in January 2010 for further data collection. The Margaritifer ellipse as first suggested was a go-to site with the main science target east of the ellipse. An ellipse centered on the science target, a shallow basin containing chloride and phyllosilicate deposits, was substituted as indicated in Table 49 and Figure 174.

Four locations at Northeast Syrtis were identified by Grant and Golombek (2010) and are shown as white ellipses in Figure 174. The slightly different black ellipses in that map are from Golombek *et al.* (2010a). Ellipse 2 was the favored landing site, and the rover would drive west-southwest from it to the edge of a lava flow emanating from Syrtis Major. This traverse would extend across a major stratigraphic sequence, beginning in ancient clay-rich terrain and ending in an area characterized by sulfates. This site was immediately southwest of the HEM-SAG site at Jezero crater (Figure 129).

The Xanthe site (Popa *et al.*, 2010) was in a crater adjacent to a multilayered delta deposit. The rover would examine silica deposits in the landing ellipse and then drive northwards across the delta into the channel that had fed the delta to examine clays in the channel wall.

Golombek *et al.* (2010a) summarized progress to that time at a MEPAG meeting on 17–18 March 2010 in Monrovia, California. The four main sites on the shortlist all had very attractive characteristics, but none were without problems, so the authors compared the main risks and rewards of each site. The Eberswalde site combined concerns with slopes, rock abundance and its latitude with a well-understood depositional environment whose clay deposits directly addressed MSL science objectives. At Gale there were potential problems with rocks and slopes at the edge of the landing ellipse, and the science goals required a traverse outside the ellipse. Although promising sulfates and clay layers were present, they were in a poorly understood depositional setting of uncertain age. Mawrth had some concerns about slopes and an uncertain geological setting, but

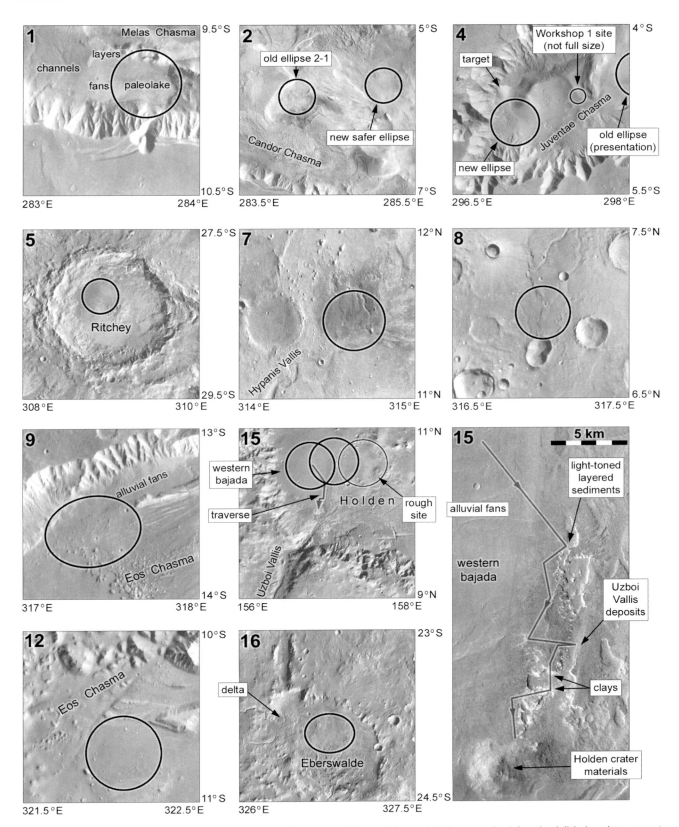

Figure 166. MSL landing sites presented at the second workshop (Table 45). An old ellipse at site 4 (partly visible here) appears to be an error in the original presentation. Images are THEMIS infrared mosaics with inverted shading, except the site 15 traverse background, which is a THEMIS visible image mosaic. For scale in Figures 166–170, 1° of latitude spans 60 km.

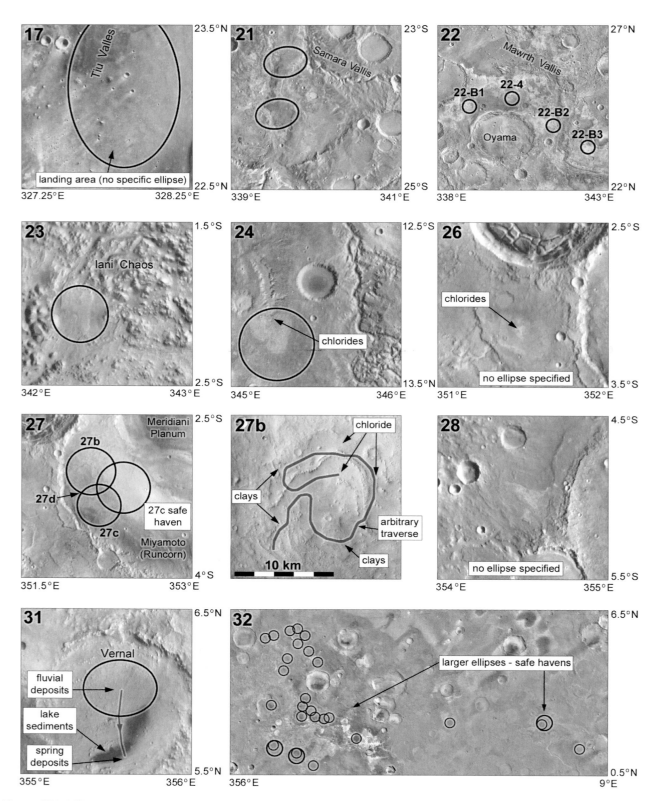

Figure 167. MSL landing sites presented at the second workshop (Table 45). Images are THEMIS infrared mosaics with inverted shading, except the sites 27b, THEMIS visible image V18723001, and 32, Viking MDIM mosaic.

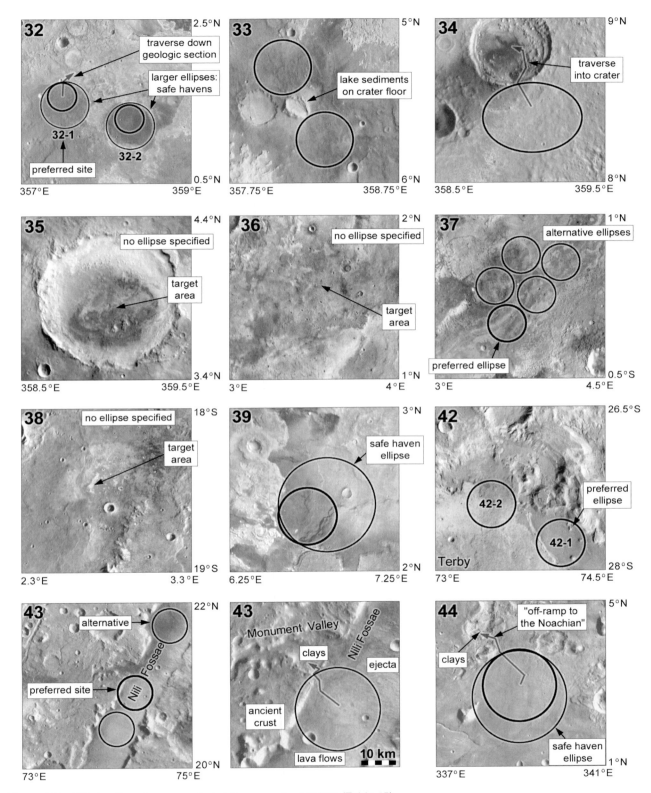

Figure 168. MSL landing sites presented at the second workshop (Table 45).

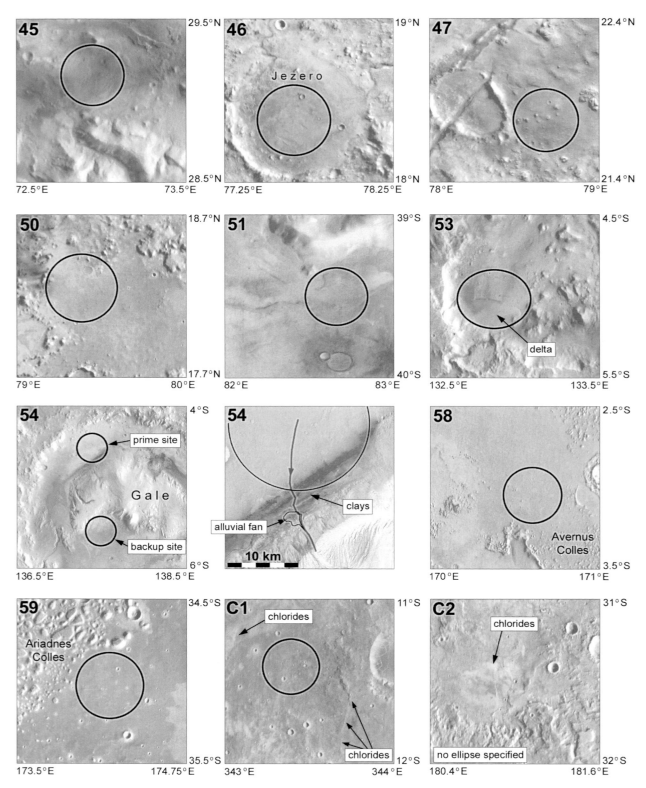

Figure 169. MSL landing sites presented at the second workshop (Table 45).

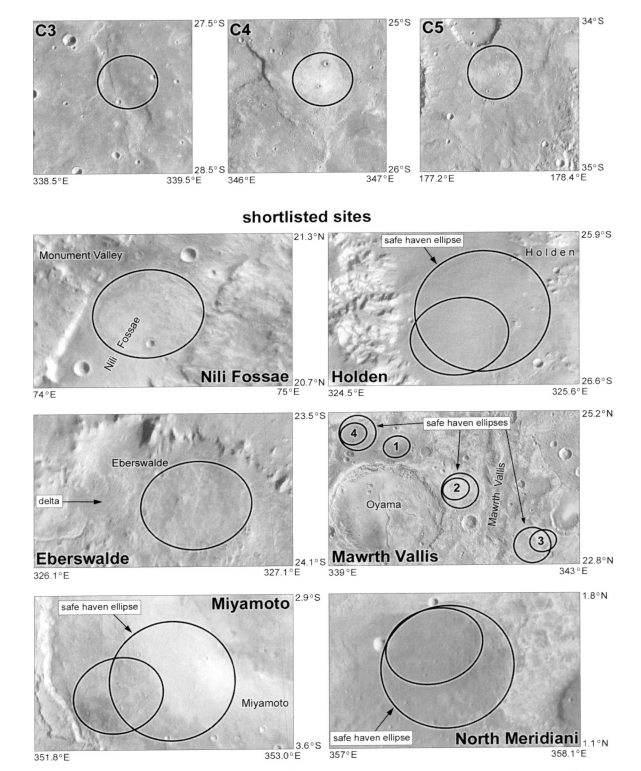

shortlisted sites

Figure 170. MSL landing sites presented at the second workshop (top, Table 45), and the post-workshop shortlist (bottom, Table 47). For scale, 1° of latitude spans 60 km.

Table 47. *Revised Shortlist of MSL Sites, December 2007 and May 2008*

Name (number)	Type of ellipse		Location	Elevation (m)	Geology
Nili Fossae trough (43)	Primary		21.01° N, 74.45° E (*21.00° N, 74.45° E)	−608	Noachian phyllosilicates
Holden crater fan (15)	Primary		26.38° S, 325.08° E (*26.37° S, 325.10° E)	−1940	Fluvial layers and phyllosilicates
	Safe haven		26.25° S, 325.21° E	−2137	
Mawrth Vallis (22)	Site 1 Primary		24.65° N, 340.1° E (*24.65° N, 340.09° E)	−3093	Noachian layered phyllosilicates
	Site 2	Primary	23.99° N, 341.04° E (*24.01° N, 341.03° E)	−2246	
		Safe Haven	23.95° N, 341.11° E (*24.01° N, 341.11° E)	−2254	
	Site 3	Primary	23.21° N, 342.43° E (*23.19° N, 342.41° E)	−2187	
		Safe Haven	23.12° N, 342.20° E	−2268	
	Site 4	Primary	24.85° N, 339.42° E (*24.86° N, 339.42° E)	−3359	
		Safe Haven	24.88° N, 339.78° E	−3355	
Eberswalde (16)	Primary		23.86° S, 326.73° E	−1450	Delta
SW Meridiani (Miyamoto) (27)	Primary		3.51° S, 352.26° E (*3.34° S, 352.26° E)	−1807	Phyllosilicates, sulfates?
	Safe haven		3.09° S, 352.59° E (*3.28° S, 352.44° E)	−1958	
N. Meridiani (32)	Primary		1.58° N, 357.48° E (*1.57° N, 357.49° E)	−1289	Layered sulfates
	Safe haven		1.48° N, 357.55° E	−1301	

Note: * Revised coordinates announced on 27 May 2008. Site 22-4 in Table 45 and Figure 167 is Mawrth site 1 here. Site 4 in this table is a new location.

Table 48. *Revised MSL Site Shortlist, July 2008*

Name	Location	Elevation (m)	Geology
Nili Fossae trough (43)	21.00° N, 74.45° E	−608	Noachian phyllosilicates
Holden crater fan (15)	26.37° S, 325.10° E	−1940	Fluvial layers, phyllosilicates
Mawrth Vallis (22)			
Site 1	24.65° N, 340.09° E	−3093	Noachian layered phyllosilicates
Site 2	24.01° N, 341.03° E	−2246	
Site 3	23.19° N, 342.41° E	−2187	
Site 4	24.86° N, 339.42° E	−3359	
Eberswalde (16)	23.86° S, 326.73° E	−1450	Delta, phyllosilicates
Miyamoto (27)	3.34° S, 352.26° E	−1807	Phyllosilicates?
S. Meridiani (30)	3.05° S, 354.61° E	−1589	Sulfates, phyllosilicates
Gale crater (54)	4.49° S, 137.42° E	−4451	Layered sulfates, phyllosilicates

Other sites considered but rejected

Name	Location	Notes
Chloride site west of Miyamoto (26)	3.1° S, 351.6° E	Insufficient geological diversity
Nili Fossae carbonate (48)	21.9° N, 78.9° E	Good, but Gale preferred

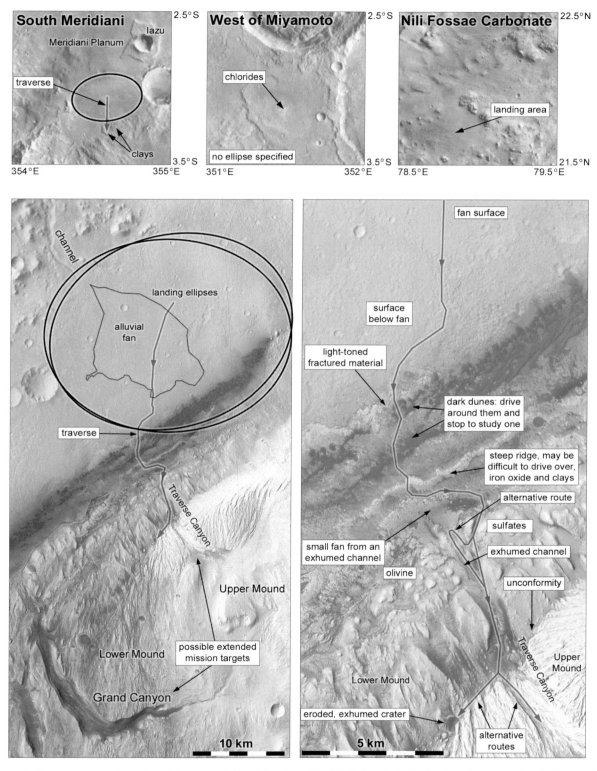

Figure 171. Three new sites introduced for the Third MSL Landing Site Workshop in 2008 (top) and proposed activities at Gale crater (bottom). For scale at the new sites, 1° of latitude spans 60 km.

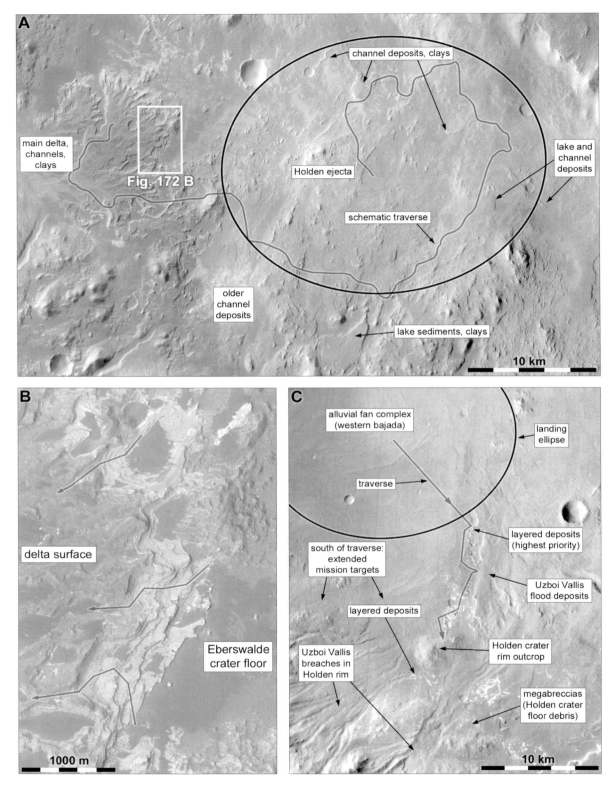

Figure 172. Third MSL workshop sites at Eberswalde and Holden with suggested traverses. **A:** Eberswalde. **B:** Three possible routes onto the Eberswalde delta surface. **C:** Holden. **A** and **C** base images are THEMIS visible mosaics; **B** base image is HiRISE image PSP_004000_1560.

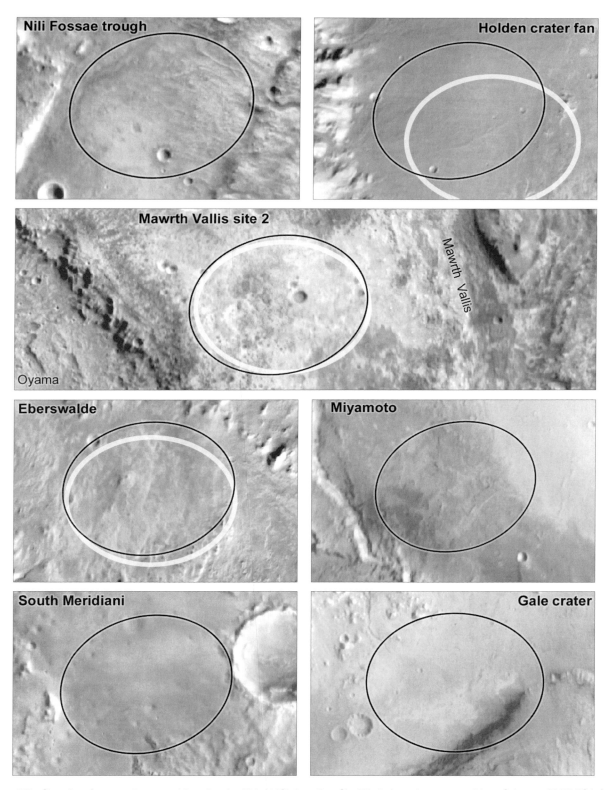

Figure 173. Shortlist of seven sites considered at the Third MSL Landing Site Workshop. Images are Mars Odyssey THEMIS infrared mosaics with reversed shading. For scale, ellipses are 20 by 25 km across. White ellipses show small changes to three of the four final ellipses in 2011 (Table 50). The Gale ellipse did not change.

Table 49. *New MSL Sites Suggested in 2009*

Site	Name	Location	Notes
48	Nili carbonates	21.7° N, 78.8° E	Good, but aeolian drifts or dunes may limit mobility
44	Northeast Syrtis	16.7° N, 76.9° E	Diverse minerals, Noachian–Hesperian boundary
6	Xanthe Terra	2.3° N, 309° E	Fan delta deposit in Chamichel crater
28	Margaritifer Terra	5.59° S, 353.52° E	Chlorides and phyllosilicates, site moved to 5.644° S, 353.82° E and renamed East Margaritifer
18	Ladon basin	18.8° S, 332.5° E	Chlorides
52	Vastitas Borealis	70.5° N, 103° E	Targets ice in Louth crater. Planetary protection hazard and extreme latitude work against it
14	Valles Marineris	3.8° S, 324.6° E	Canyon floor, wind hazards, long traverse, few minerals

offered ancient clay science targets within its ellipse. Holden had no safety concerns except its high southern latitude, and its geologic context was well understood, but its layered clay science targets required a traverse outside its ellipse.

The site selection schedule now involved data collection for the two newly suggested sites at East Margaritifer and Northeast Syrtis and a decision in May 2010 regarding the addition of one of them to the short-list. The decision was to remain with the four original sites, as NE Syrtis was too rugged and East Margaritifer not sufficiently interesting. Over the next year the sites and specific science targets would be discussed at several meetings before final selection by NASA Headquarters in April 2011.

The Fourth MSL Landing Site Workshop was held in Monrovia, California, on 27 to 29 September 2010. The four sites were considered in more detail, a process aided by the immense amounts of information now available from HiRISE and CRISM. Engineering analyses now suggested that all sites were acceptable. Although exact landing points were unpredictable, a possible traverse from a typical landing site at Mawrth (Figure 175) was illustrated by Loizeau *et al.* (2010). They also identified a large number of targets within the ellipse or nearby to its east, and potential longer-range goals for an extended mission, including crater ejecta and fluvially redeposited clays in Oyama crater to the west and Mawrth Vallis to the east.

A second possible traverse at Gale crater (Figure 175C) was identified by Anderson and Bell (2010), and Edgett (2010) mapped another version of one of them (Figure 175D). An earlier traverse by Anderson is also

shown in Figure 169. In such target-rich sites, there were many possible routes. At Eberswalde, Lewis and Aharonson (2010) described the putative lake into which the delta had grown, including an eastern basin connected by a now inverted channel with the main lake basin. A spillway from the main lake basin into an eastern basin in Eberswalde, indicated by the channel, may have acted to stabilize water levels in the lake and allow the delta time to grow. Rice *et al.* (2010b) described several potential routes from the central part of the Eberswalde ellipse to the delta, with different targets along the routes. The most direct route was 13.8 km long, but a southern route 17.1 km long or a northern route 21.6 km long each offered more geologic targets.

The four ellipses described above were formally accepted by the MSL Project as the final shortlist in June 2010 and the latest coordinates for them (Table 50; white ellipses in Figure 173) were given by Golombek *et al.* (2011a). The Holden ellipse had been moved significantly, placing its center nearer the areas of greatest interest. The Gale ellipse was essentially unchanged.

As summarized by Golombek *et al.* (2011a):

Eberswalde crater contains a delta with phyllosilicates, a potentially habitable environment that is particularly favorable to the preservation of organic materials. Holden crater contains finely layered phyllosilicates suggesting deposition in quiet fluvial or lacustrine setting with a well understood context. Mawrth Vallis exposes an ancient preserved layered stratigraphic section providing an opportunity to characterize early wetter conditions in the Noachian. Gale crater offers access to diverse stratigraphy, including interbedded sulphates and phyllosilicates in a 5 km high mound that reflects deposition during changing environmental conditions.

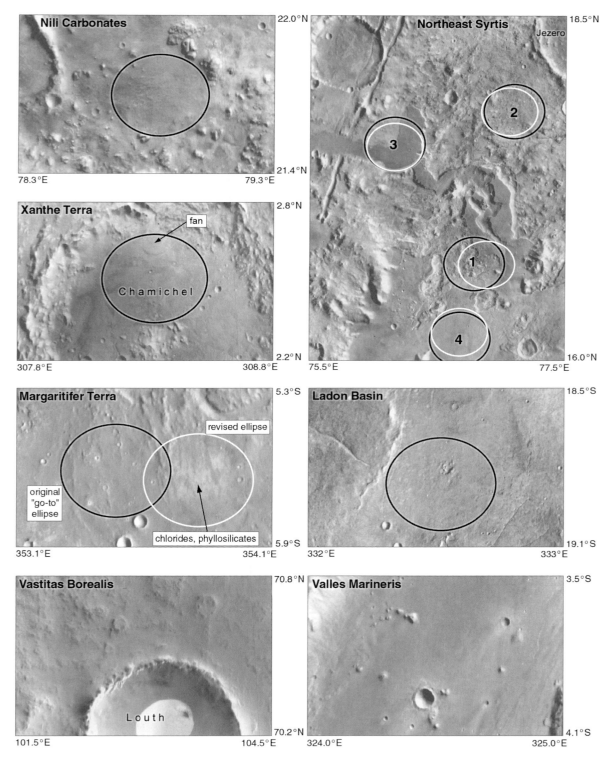

Figure 174. New sites suggested for MSL in 2009. Black ellipses are from Golombek *et al.* (2010a), white ellipses from Grant and Golombek (2010). No ellipses were identified for the bottom two sites. For scale, 1° of latitude spans 60 km and most images are 0.6° (36 km) high.

Table 50. *Final Four MSL Sites From Golombek* et al.
(2011a)

Site	Location	Elevation (m)
Eberswalde	23.8953° S, 326.7426° E	−1435
Gale	4.4868° S, 137.4239° E	−4444
Holden	26.4007° S, 325.1615° E	−2177
Mawrth	23.9883° N, 341.0399° E	−2245

At the Fifth MSL Landing Site Workshop in May 2011, supporters of each site presented their arguments and identified possible traverses or science targets. McKeown and Rice (2011) illustrated a very generalized traverse at Eberswalde, commencing near the middle of the ellipse and moving past the delta and back into the ellipse, where it would encounter additional targets in an extended mission. At Mawrth, Mangold *et al.* (2011) identified a traverse leading eastwards across the Oyama ejecta towards and into Mawrth Vallis, and several science targets along that route (Figure 176A). Mawrth may have been covered with ejecta from Oyama crater just to its west, or severely disrupted by that impact (Sumner, 2011), an observation that now made it seem less compelling since the geological context of any sample would be uncertain. Holden traverses were the same as shown before, but another variation on Gale's traverse was shown (Anderson *et al.*, 2011) and extended mission targets were identified (Figure 176C). One significant difference between the various Gale traverses illustrated in Figures 175 and 176 is their path around or over a fan deposit emanating from a canyon (Traverse Canyon in Figure 171, also called Entrance Canyon) in the upper mound.

In addition to these workshop presentations, Golombek *et al.* (2011b, 2012) described thorough studies of traversability in the ellipses to be sure the objectives could be met. Many possible routes, not illustrated here, were explored using Geographic Information Systems (GIS) and the best available HiRISE-derived digital topographic maps. This level of landing site analysis had never been possible before. Analyses included estimating for each site the fraction of the ellipse consisting of inescapable traps, places such as a crater floor or mesa summit which could accommodate a safe landing but from which the rover could not escape.

After the meeting the shortlist was reduced to two sites, Gale and Eberswalde, and eventually Gale was chosen by NASA in June 2011. It had significant geological diversity, its geology was reasonably well understood, and it was on the opposite side of the planet from Opportunity, so there would be no communication conflict. That was important because it now seemed very likely that Opportunity would still be functioning at Endeavour crater at the beginning of Curiosity's mission. The Gale site was similar to an ellipse considered for Mars Surveyor 2001 (Figure 181 in Stooke, 2012), and not far from a MER candidate ellipse (Figure 177). The large sedimentary mound inside Gale was informally named Mt. Sharp by the science team, and at about the same time it was given the formal name Aeolis Mons by the IAU's Working Group for Planetary System Nomenclature (WGPSN). Both names are used here. The WGPSN gave the name Robert Sharp to an eroded crater at 4.2° S, 133.4° E, only 240 km west of Gale. These names commemorate Robert Sharp (24 June 1911–25 May 2004), the Caltech geologist who had worked on the Mariner 4, 6, 7 and 9 missions to Mars.

26 November 2011: Mars Science Laboratory (Curiosity)

The Mars Science Laboratory (MSL) was a large rover originally planned for launch in 2009, but delayed until 2011 by hardware and software development difficulties. The name Curiosity was chosen for MSL in May 2009 after a competition among US schoolchildren, won by Kansas student Clara Ma. MSL was launched on 26 November 2011 with the goal of examining the habitability of Mars at the Gale crater landing site shown in Figure 177. The rover landed on 6 August 2012, and after a checkout period it made its first drive on 22 August. During the primary mission lasting one Mars year, the rover was driven 7.9 km, and collected and analyzed many rock and soil samples, including four using its full range of analytical instruments.

The 750 kg rover was 2.8 m long with a maximum width at the wheels of 2.7 m. The rover body consisted of a warm electronics box (WEB) holding the electronic systems, with a Radioisotope Thermoelectric Generator (RTG) mounted at the rear. This produced 125 W at the

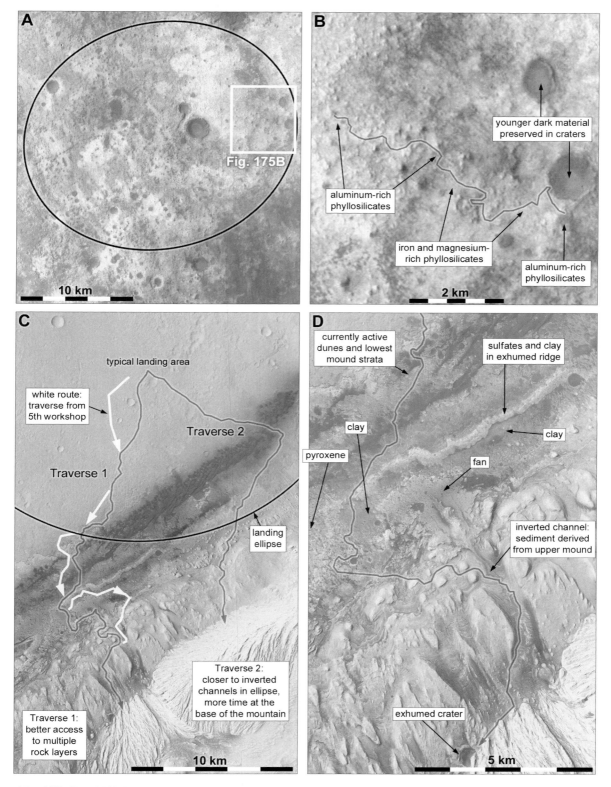

Figure 175. MSL Fourth Workshop traverse plans. **A:** Mawrth ellipse. **B:** Possible traverse, from Loizeau *et al.* (2010). **C:** Gale traverse options, from Anderson and Bell (2010) and Anderson *et al.* (2011), from the Fifth Workshop. **D:** Another variation on the Gale traverse, from Edgett (2010).

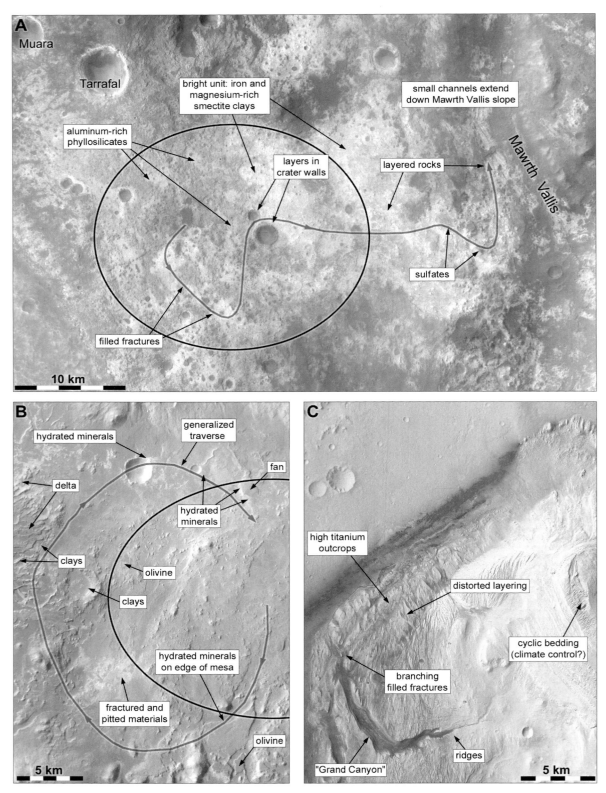

Figure 176. Plans from the Fifth MSL Landing Site Workshop. **A:** Mawrth traverse (Mangold *et al.*, 2011). **B:** Eberswalde traverses (McKeown and Rice, 2011). **C:** Gale extended mission targets (Anderson *et al.*, 2011).

time of landing, not to power the rover directly but to charge batteries from which Curiosity took its power as needed. The advantage over solar panels was that the recharging was not dependent on slope or season and continued day and night. The six rover wheels were mounted on a rocker-bogie suspension and, as on MER, the two front and two rear wheels had individual steering motors. Four pairs of hazard-avoidance cameras (Hazcams) were mounted on the body, facing forwards and backwards. A descent imager (MARDI) was mounted under the front of the rover, facing downwards to take images during descent and on the surface. The Dynamic Albedo of Neutrons (DAN) experiment, a Russian instrument at the rear of the rover body, measured the water content of the surface by observing the speeds of neutrons emitted by DAN and reflected off the surface or released from the ground by cosmic ray strikes. Fast-moving neutrons had been scattered by heavy atoms such as iron, but slow-moving neutrons had scattered off small atoms, especially hydrogen.

A robotic arm with a fist-like turret containing sample collection devices and scientific instruments was mounted at the front of the rover. It carried the Mars Hand Lens Imager (MAHLI), a close-up color camera more capable than the Microscopic Imager on MER, an Alpha Particle X-ray Spectrometer (APXS), a scoop used to collect soil samples for analysis, a rock drill, and CHIMRA (Collection and Handling for In-Situ Martian Rock Analysis), the sample processing and delivery system, which took material from the scoop or drill and deposited it in the analytical instruments.

A Rover Equipment Deck on top of the WEB supported a mast (Remote Sensing Mast, RSM) carrying four Navigation Cameras (Navcams), two high-resolution stereoscopic Mast Cameras (Mastcams), and the Chemistry and Camera (ChemCam) instrument. ChemCam was a combination of a high-resolution camera system (Remote Micro-Imager, RMI) and a Laser-Induced Breakdown Spectrometer (LIBS). It directed a powerful laser at a target up to 7 m from the rover, and obtained spectra of the short-lived plasma derived from the target to analyze its composition. It could also record spectra passively from distant targets. The camera obtained high-resolution images of the target, usually before and after the laser shot. Typically one observation would consist of many shots made in quick succession, which gradually created

a small pit in the target while sampling at different depths. ChemCam also frequently made multiple observations in a line or raster pattern (Figure 189) to explore variations across a target.

Inside the rover body were two sophisticated analytical instruments, the Sample Analysis at Mars (SAM) instrument suite, which analyzed carbon compounds, and the Chemistry & Mineralogy (CheMin) X-ray diffraction/X-ray fluorescence instrument, which identified minerals. In addition, a Radiation Assessment Detector (RAD) was mounted on top of the rover deck together with inlet ports for the two internal instruments. A meteorological instrument system, the Rover Environmental Monitoring Station (REMS), had components on the mast and the deck. An MSL Entry Descent and Landing Instrument (MEDLI) gathered temperature and pressure data during atmospheric entry. Its sensors and power system were mounted in the heatshield but its data were transferred to the rover for later transmission.

Curiosity's mission was to determine whether the Martian surface was in the past or is now a habitable environment for life. The landing site in Gale crater had been chosen as its geological record might have preserved evidence of former environmental conditions, allowing past habitability to be assessed. The goal of assessing current habitability depended largely on radiation measurements and might have been addressed anywhere on Mars. The mission was planned to operate on Mars for at least a full Mars year. Its science objectives were to study organic carbon compounds and any chemical building blocks of life, to identify possible effects of biological processes, to measure the chemical, isotopic and mineral composition of the surface materials, to elucidate the processes that formed and modified rocks and soils, to assess long-term atmospheric evolution, to explore the availability and cycles governing water and carbon dioxide, and to measure the surface radiation environment.

MSL launched from Cape Canaveral on 26 November 2011 at 15:02 UT and was placed in a circular Earth parking orbit for a short time before its upper stage set it on a Mars-bound trajectory. A computer reset three days into the flight required some reprogramming later but had no long-term effects. During its eight-month cruise to Mars, MSL was controlled by a solar-powered cruise stage, with trajectory corrections on 11 January, 26

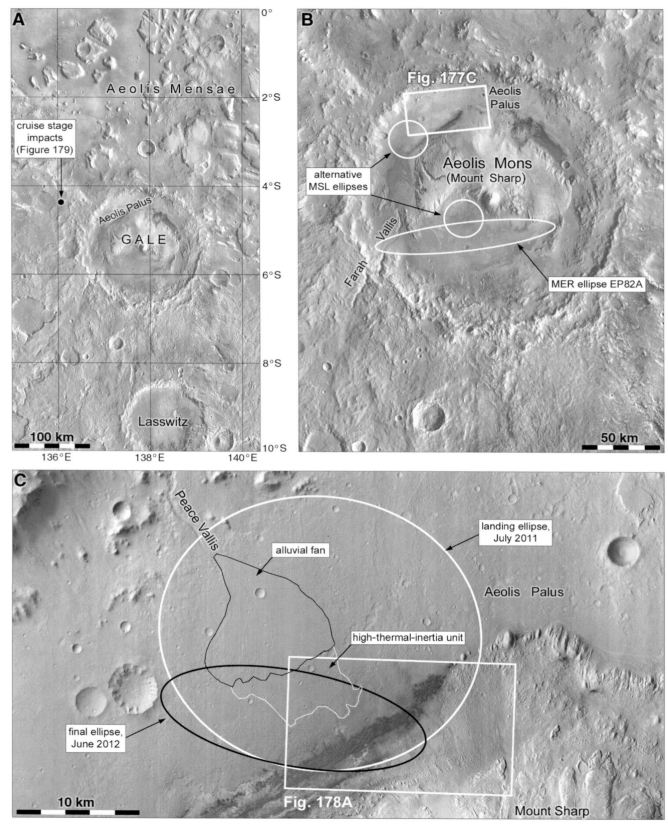

Figure 177. Mars Science Laboratory (Curiosity) landing site in Gale crater. **A:** Gale crater and vicinity. **B:** Gale crater showing ellipses considered earlier for MER and MSL. **C:** The Curiosity landing ellipse. The alluvial fan and high-thermal-inertia unit are described in the text. **A** and **B** are Mars Odyssey THEMIS infrared mosaics with reversed shading. **C** is a Mars Reconnaissance Orbiter CTX mosaic.

March, 26 June and 28 July. The first correction put MSL on a path to enter the atmosphere, allowing the upper stage launched with it to miss the planet and avoid contaminating it. The third correction moved the landing point closer to Aeolis Mons (Mt. Sharp) to make use of a smaller landing ellipse (Figure 177C).

MSL's Radiation Assessment Detector (RAD) began measuring radiation on 6 December 2011, and concluded on 14 July 2012. It measured five large solar outbursts (solar energetic particle or SEP events) and a lower background level of particle radiation. The data would help plan shielding requirements for future human crews. Data collection stopped on 13 July and resumed after landing. Instruments were tested during cruise, including MAHLI, which took a picture using its own light source on 20 April, and the twin computer systems were reconfigured beginning on 17 July. The rover batteries were fully charged from the cruise stage's solar arrays on 24 July.

Figure 177 shows the landing site in Gale crater. The ellipse (Figure 177C) was reduced in size from 20 by 25 km to 7 by 20 km in June 2012 and moved closer to Mt. Sharp, as confidence in the accuracy of the trajectory grew. Figure 117C also shows the outline of an alluvial fan fed by Peace Vallis, a channel on the north wall of Gale, and a high-thermal-inertia area interpreted to be exposed bedrock at the outer end of the fan. Curiosity would land near the edge of the exposed bedrock and examine it at Yellowknife Bay.

MRO's MARCI instrument observed dust clouds south of Gale before the landing, but they did not develop into a problem. Mars Odyssey had been moved in its orbit to provide direct data transmission during landing, which took place just after earthset at the landing site. Curiosity approached Mars on 6 August 2012 (using Universal Time) or 5 August in California (MY 31, sol 319) at 5.9 km/s and discarded its cruise stage 10 min before entering the atmosphere. The entry phase began over 4° S, 126.7° E at an altitude of 125 km with the spacecraft enclosed in its aeroshell and heatshield. Two tungsten weights were ejected to offset the vehicle's center of mass and enable it to steer during entry, and they and the cruise stage struck the surface west of Gale crater (Figure 179). The spacecraft first sensed the atmosphere at an altitude of 125 km and was slowed by friction, protected by its heatshield, until it reached a velocity of

0.58 km/s at an altitude of 10 km. A series of S-curve maneuvers steered by thrusters helped to slow its descent. Six more weights (entry ballast masses) were ejected to restore balance, and struck the surface 12 km east of the landing site (Figure 181).

The heatshield was discarded as the parachute deployed, slowing the vehicle to 100 m/s at an altitude of 1.8 km, at which point Curiosity's descent stage ("skycrane") dropped out of the backshell and ignited its engines. It flew sideways to avoid the falling parachute and descended to a height of 19 m, at which point the rover was lowered down a 7.5 m long set of cables. The vehicle now descended slowly to the surface, touching down at 05:32 UTC (01:32 EDT) on 6 August, or 15:00 local time at the landing site. The descent stage cut its cables, tilted and accelerated again, flying northwest and crashing 20 s later about 700 m from the rover. A cloud of dust raised by the skycrane impact was seen in the first rear Hazcam image taken after landing. MRO obtained a HiRISE image showing the spacecraft descending on its parachute (Figure 178C) with the falling heatshield below (not illustrated). The descent imager (MARDI) on the bottom of the rover took 1500 images from heatshield ejection until the landing, including one frame that captured the heatshield impact 1600 m east of the rover landing site.

A detailed description of Curiosity's activities follows, illustrated in Figures 185 to 203 and summarized in Tables 51 to 54. The entire rover route is shown at a standard scale in maps plotted on an orthorectified HiRISE image base with a 100 m square grid for scale (Figures 185 and 194 and succeeding maps, each one identified in the caption to the previous map). Sites of more complex operations are depicted on larger-scale maps based on HiRISE images or reprojected rover panoramas (*e.g.* Figures 187, 188 and 190). A key to map coverage is shown in Figure 184. The tables and descriptions of Curiosity's activities are derived from documents provided by NASA's Planetary Data System, especially the Analyst's Notebook documents, augmented by mission press releases, brief status reports by Ken Herkenhoff and colleagues at USGS, and by descriptions in Vasavada *et al.* (2014), Arvidson *et al.* (2014b) and Erickson and Grotzinger (2014).

The landing day was called sol 0, as on Viking and Phoenix. The Pathfinder and MER missions landed on sol 1. Several images were transmitted from the Hazcams immediately after landing to give a first look at the landing site, including one showing the dust raised by the descent stage impact. These first images were taken through dust covers designed to keep the camera optics clean. The next few sols were spent on vehicle and instrument testing and verification before any science activities could start. The first Navcam and MAHLI test images were taken on sol 1, the MARDI descent images were transmitted in low-resolution format and the HGA was deployed. An oblique HiRISE image revealed the landing hardware components on sol 1, and a simultaneous CTX image showed the entry ballast mass impact points at the foot of Mt. Sharp 12 km east of the landing site. The RTG was warmer than expected, possibly because dust was blown onto it during landing.

On sol 2 the camera mast was deployed and a full Navcam panorama and deck panorama were taken. The site was free of large rocks and would be relatively easy to drive over. On sol 3 the Mastcam took a full color panorama of the site, transmitted first at low resolution and only later at full resolution as time permitted. REMS was experiencing some problems, but everything else seemed to be in good working order. Navcam images showed that the descent rocket exhaust had stripped dust off the surface in several shallow pits, later called scours, around the rover. An initial thought was that they might be the cleanest surface anywhere for APXS to analyze, but fears of contamination prevented APXS use in the area affected by the exhaust plumes.

Instrument and systems checkout continued through sol 4, and was then paused for a software update. The original flight software was designed for cruise, landing and initial operations, but now much of that was no longer needed and could be replaced with driving and instrument operations software. The new instructions were uploaded on sol 3, installed on sol 4 and used first on sol 5, before Curiosity reverted to the old software for additional testing. Everything looked good, so the backup computer system was updated with a file upload on sol 7 and testing on sol 8, and finally normal work resumed on sol 9. A better HiRISE image of the site was obtained on sol 5 and a second for a stereoscopic pair on sol 12.

Curiosity's RAD radiation detector was intended for use on the surface of Mars, but it was also operated for most of the cruise to assess radiation levels for future human missions. Its data suggested that a crew member would accumulate roughly the permissible lifetime exposure during a typical expedition to Mars and back (Zeitlin *et al.*, 2013). After landing, the radiation level was half that in space. A Navcam portrait of the rover deck showed many pebbles, not just dust, thrown onto the rover during landing. This was not a problem for the sample analysis instrument inlets, but the REMS sensors were unprotected and the wind sensor was not operating. It might have been damaged by a flying pebble. This was the only problem experienced during landing.

Now future activities could be planned. The science team had divided the landing ellipse area into sections for preliminary geological mapping (Figure 183), and the section containing the landing site was named the Yellowknife Quadrangle (a quadrangle is a standard map sheet) after a region with many ancient rocks near that Canadian city. A crater near the landing site was officially named Yellowknife after the landing (Figure 178B). Place names here were taken from terrestrial geological sites and rock units associated with each quadrangle name. The main science target was about 8 km southwest of the landing site, but the team had to decide whether closer targets were worth stopping at. An interesting geological target was visible in HiRISE images about 500 m east of the rover (Figure 178B), a "triple junction" where three distinctly different types of material were found together. This area, called Glenelg, became the first major science target after an initial characterization of the landing site. The name, a palindrome, reflected the fact that the route would turn back along its path because Glenelg and the main science targets at Mt. Sharp lay in opposite directions from the rover. If a deposit of fine material was found along the way, it might be scooped up to test the sample delivery system.

ChemCam was heated to decontaminate it on sol 9 and the team looked for good targets for it within 3 m of the rover. The first ChemCam target was a small angular rock called Coronation, chosen more for target practice than for its science value and referred to initially as N165. The Curiosity team initially used numbers rather than names for many features, assigning names mainly for those chosen for further study. Coronation was imaged by the RMI on sol 12 and analyzed by LIBS

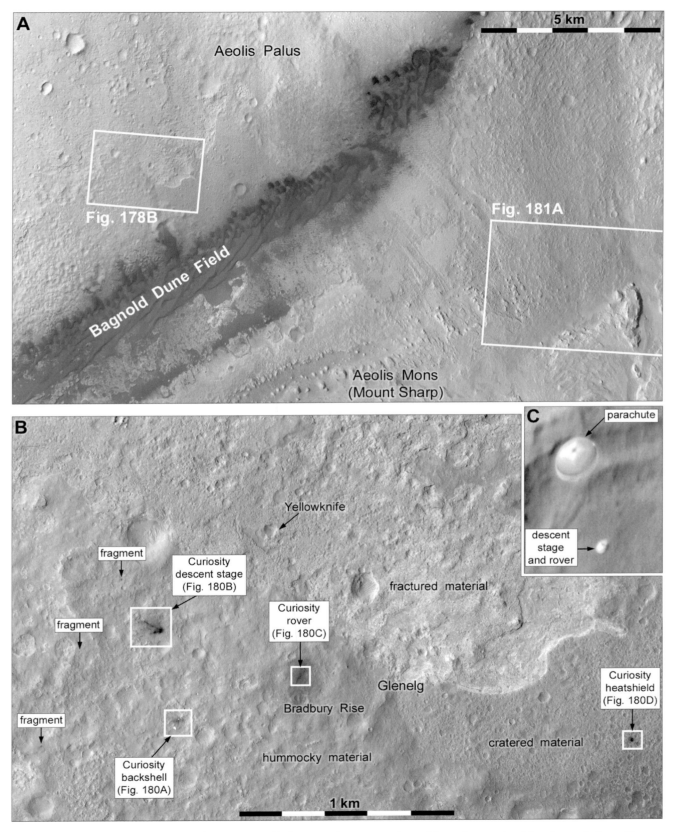

Figure 178. Curiosity landing site. **A:** Context map. **B:** Spacecraft components and three fragments of the descent stage. Three geological units, cratered, fractured and hummocky materials, meet at Glenelg. Yellowknife crater lends its name to the landing region. **C:** Parachute seen during the descent. Images are a CTX mosaic (**A**) and HiRISE images ESP_028401_1755 (**B**) and ESP_028256_9022 (**C**).

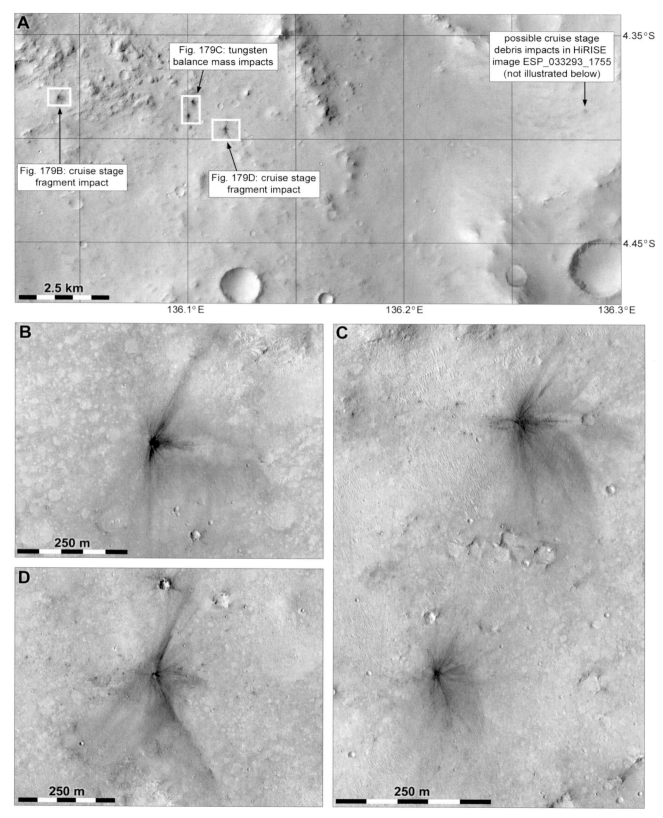

Figure 179. Impacts of the MSL cruise stage and tungsten balance masses ejected before atmospheric entry. **A:** Mosaic of CTX and HiRISE images showing locations of the impacts. **B:** Part of HiRISE image ESP_029601_1755 showing a cruise stage fragment impact site. **C** and **D:** Parts of HiRISE image ESP_029245_1775 showing two scars produced by balance mass impacts (**C**) and the impact site of another cruise stage fragment (**D**).

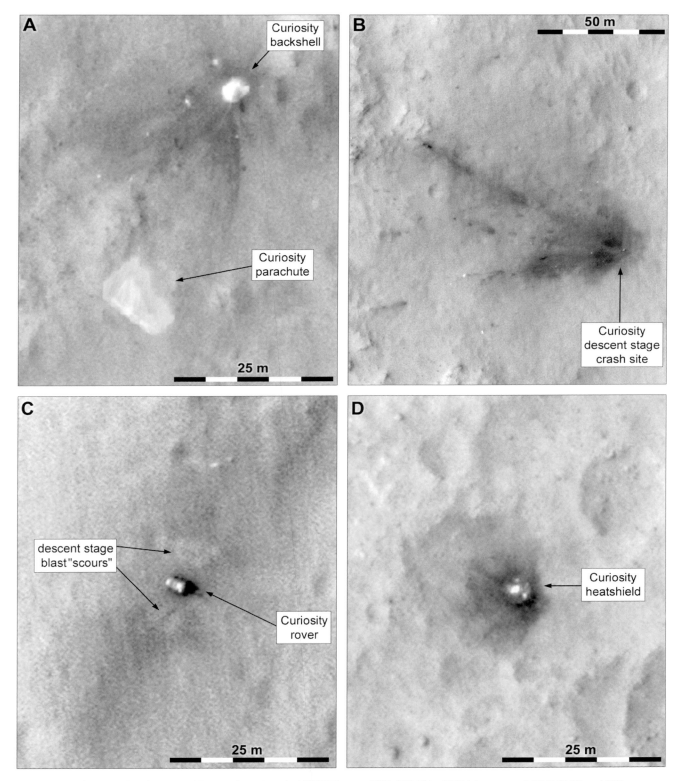

Figure 180. Curiosity hardware and impact locations in HiRISE image ESP_028401_1755 taken on sol 12 (MY 31, sol 331, or 17 August 2012). The descent stage crash site (image **B**) is 125 m wide, the other three are 50 m wide.

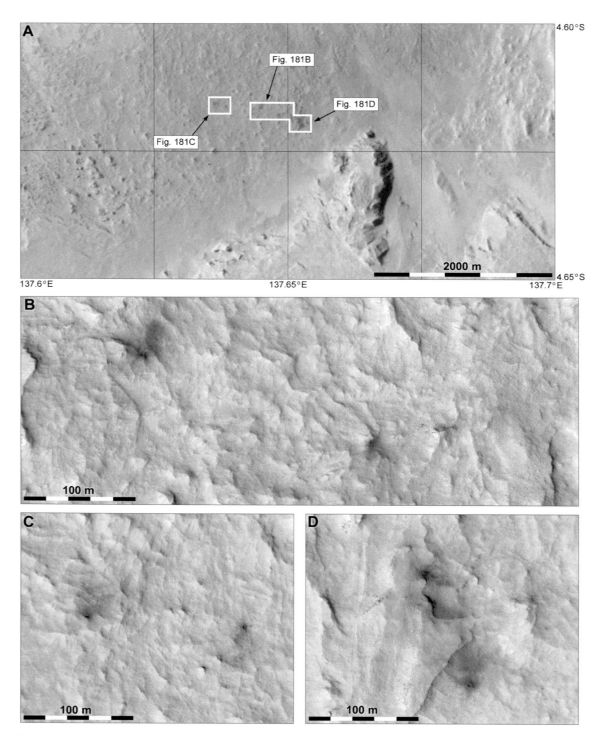

Figure 181. Impacts of the six tungsten balance masses ejected during descent. **A:** Part of CTX image D03_028269_1752_XI_04S222W showing the impact area. **B–D:** Parts of HiRISE image ESP_030524_1755 showing more details of the impact sites. Each small image contains two impact sites, identified by their dark ejecta.

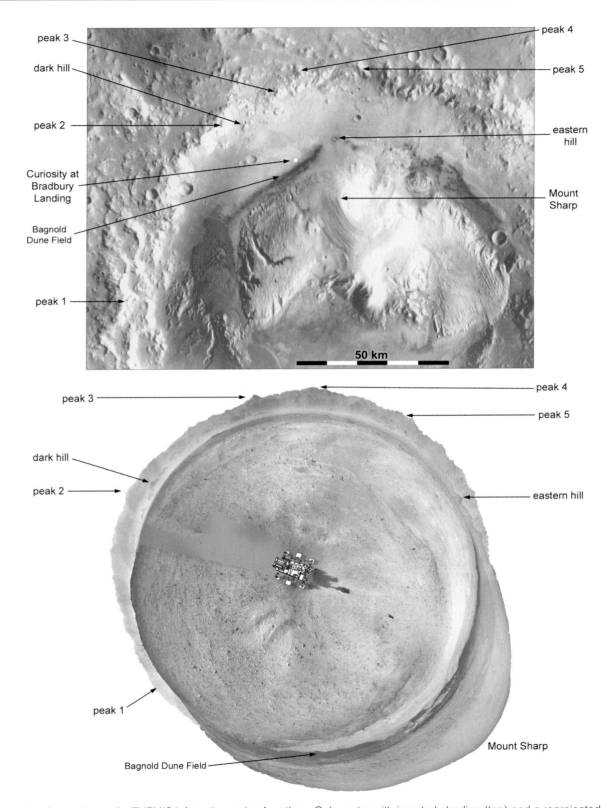

Figure 182. Comparison of a THEMIS infrared mosaic of northern Gale crater with inverted shading (top) and a reprojected panorama from Bradbury Landing (bottom).

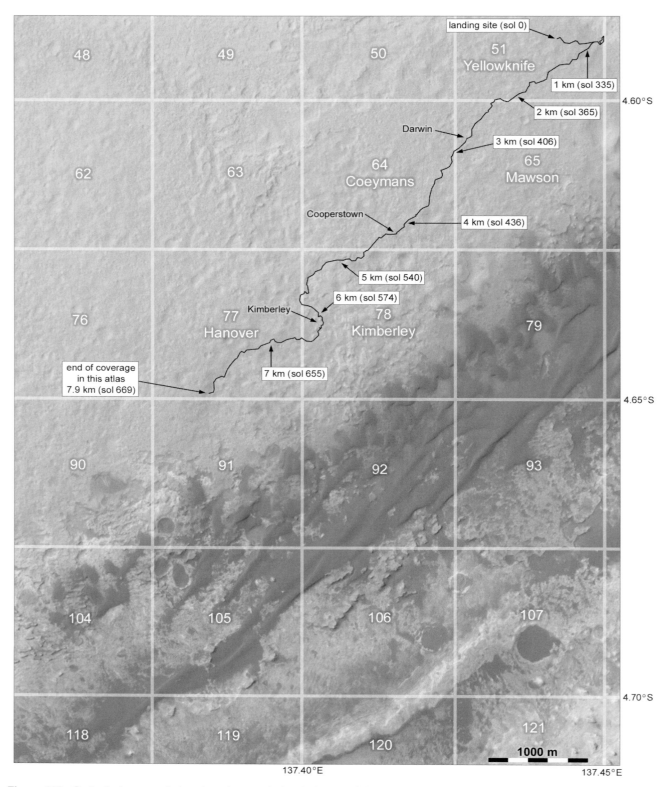

Figure 183. Curiosity traverse during the primary mission (sols 0–669). The white grid identifies the quadrangles (map areas) designated for geological mapping and mission planning.

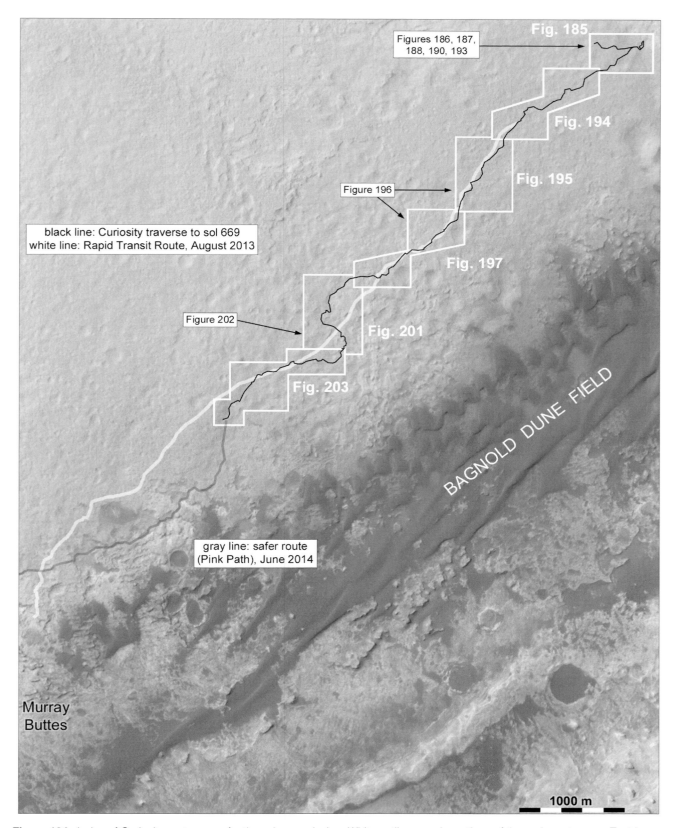

Figure 184. Index of Curiosity route maps for the primary mission. White outlines mark sections of the main route map. Text boxes identify larger-scale maps with additional details. The base is a mosaic of CTX images. The Rapid Transit Route was the planned path to reach Murray Buttes; the safer route was modified to reduce wheel damage.

on sol 13, the first analysis of the mission. Meanwhile the high-resolution Mastcams were taking images of Mt. Sharp.

The instrument arm was first extended on sol 14, raised high and then returned to its stowed position, and Hazcam images were taken to see if the rover moved during the test. The instrument inlet covers on the deck, the Dust Removal Tool (DRT) and the CHIMRA sample handling system were also checked now. The rover had landed facing slightly south of east, with two distinct scour marks dug by the descent stage rocket exhaust on each side of the rover (Figure 187). One in particular exposed bedrock resembling conglomerates seen later at Link and Hottah. This pit was named Goulburn Scour and became another early ChemCam target on sol 14. The scour names came from northern Canadian geology but were selected to suggest burning or dragons to reflect their formation.

On sol 15 the wheels were moved to test steering. All four corner wheels could turn and they were now wiggled individually and left facing straight ahead. One sol later the rover drove for the first time, moving 4 m to the east, then 2 m backwards to the northeast. This drive was dedicated to Jake Matijevic, a rover engineer, who had died just a few days earlier on 20 August 2012, and the team now intended to drive about 100 m to get out of the contaminated area and then look for a fine-grained wind drift large enough to sample safely.

The landing site was now named Bradbury Landing (or Bradbury Station) after author Ray Bradbury (*b.* 22 August 1920, *d.* 5 June 2012), a long-time supporter of space exploration, who had been born 92 years earlier on the date of the rover's first drive. Another candidate to be commemorated here had been NASA's first female astronaut, Sally Ride (*b.* 26 May 1951, *d.* 23 July 2012). On sol 18 the SAM instrument performed its first atmospheric analysis, taking so much power that nothing else could be done on that sol. ChemCam fired its laser at a patch of disturbed soil called Beechey in the rover's tracks on sol 19, making five small pits in a short straight line or transect to look for grain-by-grain variability in the surface material.

A recording of a message from NASA Administrator Charles Bolden was transmitted from Curiosity to Earth on sol 21, the first human voice transmission from Mars, and the rover was driven a few meters to place its DAN instrument over Goulburn Scour. Then it made a turn on sol 22, used ChemCam on Goulburn and began the drive to Glenelg. The new plan was to locate a fine-grained rock such as a basalt outside the contaminated area, and use it to cross-calibrate ChemCam and APXS and test MAHLI. ChemCam experienced a commanding error on sol 22, which prevented its use for a few sols, but the high-resolution Mastcam cameras were tested. On sol 24 Curiosity was driven over a small rock to see if the onboard hazard-avoidance routine would recognize it as a threat, but it did not. After the 21 m drive the data downlink via MRO was partly blocked by Mt. Sharp, limiting data for planning the next sol's drive, so sol 25 was used mostly for imaging, including a sequence that detected high, thin clouds. RAD, REMS and DAN data were collected as on most sols during the long traverses to come.

The sol 26 and 29 drives were each 30 m long, passing a high-albedo rock outcrop called Link on sol 26 during a detour to the southeast to avoid rough terrain. Longer drives were permitted as the team became more familiar with their new rover. Visual odometry (visodom) was tested on sol 26, the drill was checked and CheMin ran an empty cell analysis as rover checkout continued, and ChemCam was now ready for use again. RAD data from sol 27 were examined to see if a solar event on the previous sol was detectable, but nothing was seen. ChemCam analyzed a rough-surfaced outcrop called Link, and SAM performed an atmospheric analysis overnight and into sol 28, when another cloud movie was also made.

Now the main rover operating systems and driving procedures had been tested, but more systems would be tested here and three major commissioning tests remained: arm placement on a target, sample analysis and rock drilling. The arm had been moved on sol 14 but the mission scientists did not want to test it on the ground while Curiosity was still close to the potentially contaminated landing area. On sol 29 the arm was put through several motion tests but it was still too close to the landing site to touch the ground. Visodom had been checked on sol 26 and was now approved for future use. Arm tests continued on sol 30 and planners continued to look out for a basalt-type rock, scoopable soil and layered rock outcrops. By sol 31 the turret, the "hand" on the arm containing the drill, the soil scoop,

APXS and MAHLI were supposed to be checked, but a thermal problem with the arm delayed that test for a sol. CHIMRA, the sample handling system, was also tested on sol 32.

The MAHLI microscopic imager was checked on sol 33 by taking images with and without its dust cover, the first time it was used without the cover. Similarly, on that sol the Mastcam made its first "tau" measurement of atmospheric opacity using a method borrowed from MER, imaging the Sun to see how much of its light was blocked by suspended dust. Mastcam also made images of the spectacular walls of Gale crater and Peace Vallis (Figure 177C). Then on sol 34 MAHLI made a dramatic panoramic mosaic of the underside of the rover and its wheels, and APXS was used for the first time on its calibration target. Sol 35 was slow because of very limited communications, but on sol 36 the arm was tested in positions needed for delivering samples to the instrument inlets on the rover deck.

This series of tests ended on sol 37 with the arm moving to the observation tray, a flat white disk on which samples could be dropped for very high-resolution imaging, and with another test of CHIMRA. A Phobos grazing transit of the Sun was observed on sol 37 and Mastcam imaged targets, including Gibraltar, a large faceted boulder west of the rover. On sol 38 Curiosity drove over a small outcrop, pausing for a DAN measurement over it, then drove on and stopped beside another outcrop called Hottah, which like Link was interpreted as a conglomerate, evidence for the vigorous flow of water down from the crater walls (Williams *et al.*, 2013). The conglomerates were presumably associated with the alluvial fan on the crater floor, fed by water from the crater walls flowing down Peace Vallis. Hottah was examined with Mastcam and ChemCam on sol 39, and a cloud movie was taken with the Navcam, taking advantage of its wider field of view.

Curiosity drove towards Glenelg every day now, and on sol 40 it began taking DAN measurements of hydrogen in the surface materials at intervals of about 10 m along the traverse. DAN had been used only sporadically before that. The meteorology instruments detected numerous brief pressure fluctuations during this season in Gale, suggestive of the passage of dust devils, though imaging from orbit showed no dust devil tracks in Gale. In contrast, Spirit's landing site in Gusev was crossed by many tracks (Figure 4C) and the Opportunity and Phoenix sites by a smaller number. Curiosity's Navcams detected one very faint dust devil against the crater's distant rim on sol 41, the only one seen in many searches during the entire primary mission. Because most of these pressure fluctuation events seemed not to raise dust, they were usually termed "convective vortices" rather than "dust devils." A partly filled crater called Tinney Cove was passed now, and both satellites were seen transiting the Sun on sol 42 as Curiosity drove on towards Musk Ox, a promising rock target for instrument calibration.

The rock was approached on sol 43 and renamed Jake Matijevic after the JPL rover engineer, who was also commemorated at Opportunity's site (Figure 102). Curiosity moved closer to the rock on sol 45 for multispectral imaging, and on the next sol it used MAHLI, APXS and ChemCam for the mission's first *in situ* observations (ISO) of a rock target, in order to compare the two very different measures of composition. This was the first surface contact with the arm. ChemCam observed 14 locations, and APXS two places over two sols, finding the target to be an alkaline volcanic rock of a type not encountered before on Mars (Stolper *et al.*, 2013). Also on sol 45 Mastcam obtained an image of Phobos in the late afternoon sky to see if it could be used like the Sun to estimate atmospheric dust levels. Its irregular crescent shape was clearly visible (Figure 209A).

The second APXS measurement was made on sol 47 as well as more MAHLI images, and ChemCam was used again after a short drive backwards as Curiosity departed on sol 48. The rover drove away facing backwards as it took several images of Jake Matijevic, testing its ability to estimate distances during the departure and driving backwards because the large RTG on the back of the rover partially blocked the rear view. Activities on sols 49 and 50 illustrated the science strategy used by all rovers during long traverses. After the sol 49 drive the region within reach of ChemCam's laser was imaged, and on sol 50 ChemCam was used on targets chosen from those images before the next drive began. APXS and MAHLI activities were planned in a similar way, as they were used more frequently later in the mission. On sol 51 the arm was manipulated to test sample delivery procedures, and images of the region ahead were taken to look for drifts suitable for sampling with the scoop. Several patches of fine material were found in these

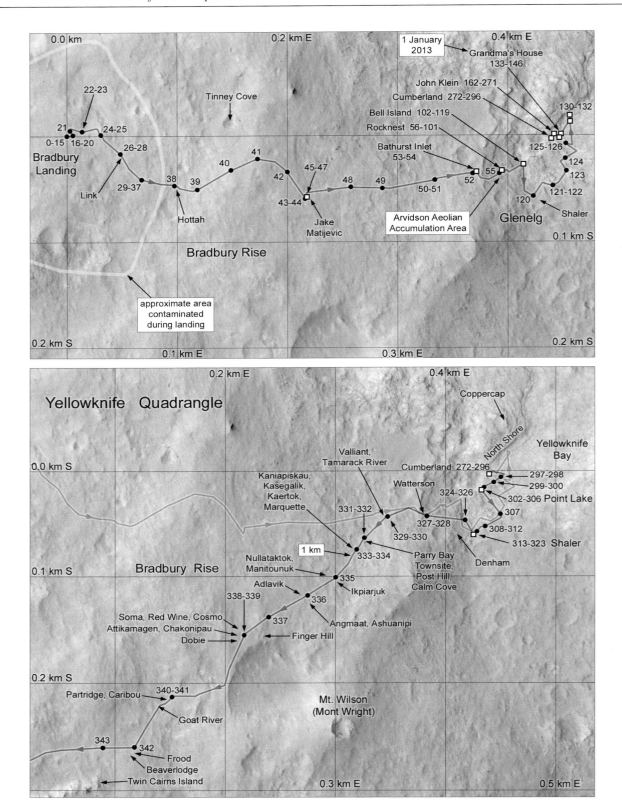

Figure 185. Curiosity route map, section 1. This map is followed by Figure 194 and additional details are shown in Figures 186, 187, 188, 190 and 193. In all rover route maps, black dots are end-of-drive locations, white squares with dark borders are stops for *in situ* observations (ISO). The grid lines are 100 m apart.

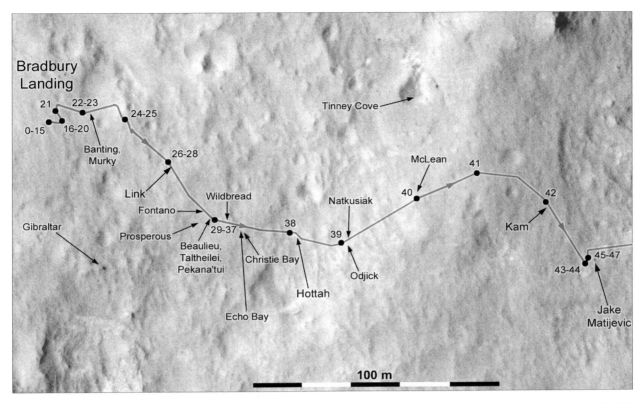

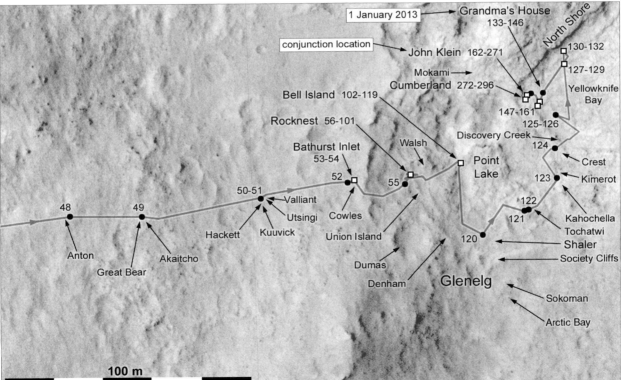

Figure 186. Curiosity activities between Bradbury Landing and Glenelg, sols 0–296. White squares are *in situ* science stops.

Table 51. *Curiosity Activities, Landing Site to Shaler, Sols 0–120*

Sol	Activities
0–2	Land, Hazcam images (0), test images and deploy HGA (1), deploy mast, Navcam panorama (2)
3–7	Mastcam panorama (3), instrument checkout (4), flight software upload (5), install (6) and test (7)
8–12	Install (8) and test (9) backup software, instrument checkout (10–11), ChemCam images N165 (12)
13–14	ChemCam on N165, now called Coronation (13) and Goulburn (14), unstow and test arm (14)
15–16	Test steering – wheel wiggle (15), first drive and drill check (16), imaging
17–18	B-side computer checks (17, 18), SAM analysis of atmosphere (18), imaging
19–20	Test Mastcam (19, 20), ChemCam Beechey (19), APXS atmosphere (20), imaging
21–24	Drive, use DAN on Goulburn (21), drive east (22, 24), test Mastcam (23) and mobility system (24)
25–26	ChemCam tests (25), drive east, visodom test, drill test, CheMin empty cell analysis (26)
27–30	SAM atmosphere (27, 28), drive east, APXS software update (29), test arm and drill (30)
31–33	Anomalies prevent science (31), test turret, vibrate CHIMRA (32), MAHLI cover check (33)
34–35	Test arm, MAHLI wheels (34), retract arm, test Mastcam video, ISO APXS calibration target (35)
36–37	Arm tests and ChemCam recovery (36), arm and CHIMRA tests, Phobos transit video (37)
38–43	Drive east (38–43), SAM heater tests (40, 42), dust devil seen (41), Deimos transit video (42)
44–47	MAHLI rover plaque, flag (44), approach Jake Matijevic (45), ISO Jake Matijevic (46–47), imaging
48–51	Drive to Glenelg (48–50), test sample handling procedures, SAM atmosphere (51), imaging
52–54	Drive to Glenelg, ISO atmosphere (52), approach Bathurst (53), ISO Bathurst Inlet and Cowles (54)
55–58	Drive to Rocknest (55), approach Rocknest drift (56), scuff drift (57), ISO scuff (58), imaging
59–61	Move to Rocknest drift (59), MAHLI range-finding (60), collect Rocknest sample 1 (61), imaging
62–66	Assess rover fragment (62, 63), clean CHIMRA (64, 65), ISO fragment (65), collect sample 2 (66)
67–69	Dump sample 2, MAHLI rover fragment (67), collect Rocknest sample 3, ISO rover fragment (69)
70–72	CHIMRA sample 3 processing, delivery to tray (70) and to CheMin (71), CheMin analysis (71, 72)
73–74	ISO tray, clean CHIMRA, SAM atmosphere (73), collect sample 4, MAHLI CheMin funnel (74)
75–77	CHIMRA sieve sample, empty CheMin (75), ISO tray (76), sample 4 to CheMin and tray (77)
78–80	Sample to tray, CheMin sample 4 (78, 80), prepare CHIMRA, SAM atmosphere (77, 79), imaging
81	Clean CHIMRA, CheMin sample 4, MAHLI REMS UV sensor, SAM atmosphere, imaging
82–84	CheMin Rocknest sample 4 (82), prepare SAM (83), MAHLI self-portrait (84), imaging
85–86	MAHLI self-portrait 2, SAM blank analysis (85), ISO Et-Then, CheMin analysis 2E (86), imaging
87–88	Computer maintenance, CheMin analysis (87), ISO Portage and LaBine, SAM blank analysis (88)
89–90	ISO Portage, empty CheMin (89), ISO Et-Then, clean and test ChemCam, test SAM delivery (90)
91–92	Imaging, ISO tray, calibrate ChemCam, prepare SAM (91), anomaly prevents science (92)
93–94	Collect Rocknest sample 5, MAHLI tray, SAM inlet and dump site, recover ChemCam, SAM sample 1 delivery and analysis (93), CheMin sample 5 delivery and analysis (94)
95–96	ISO sample 5 on tray and prepare SAM (95), MAHLI SAM inlet and deliver sample 2 to SAM (96)
97–99	CheMin sample 5 (97, 98), prepare SAM (97), SAM sample 3 delivery (98) and analysis (99)
100–101	Move to Rocknest 3, calibrate ChemCam, CheMin sample 5, drill tests (101), imaging
102–103	ISO Rocknest 3 and drive to Point Lake (102), clean ChemCam, test drill (103), imaging
104–110	CheMin Rocknest sample 5 (104), SAM atmosphere (105), imaging (105–110)
111–113	Move to Bell Island, CheMin sample 5 (111), check ChemCam, prepare SAM (112), test drill (113)
114–118	Deliver sample 4 to SAM (114, 116), SAM analysis and ISO Bell Island Target 9 (117), imaging
119–120	CheMin Rocknest 5 (119), calibrate ChemCam (119–120), drive to Shaler and recover SAM (120)

Notes: APXS, Alpha Particle X-ray Spectrometer; CHIMRA, Collection and Handling for In-Situ Martian Rock Analysis (sample handling and distribution system); CheMin, Chemistry and Mineralogy instrument; DRT, Dust Removal Tool; ISO, *in situ* observations (APXS and MAHLI); MAHLI, Mars Hand Lens Imager; REMS, Rover Environmental Monitoring Station; SAM, Sample Analysis at Mars instrument.

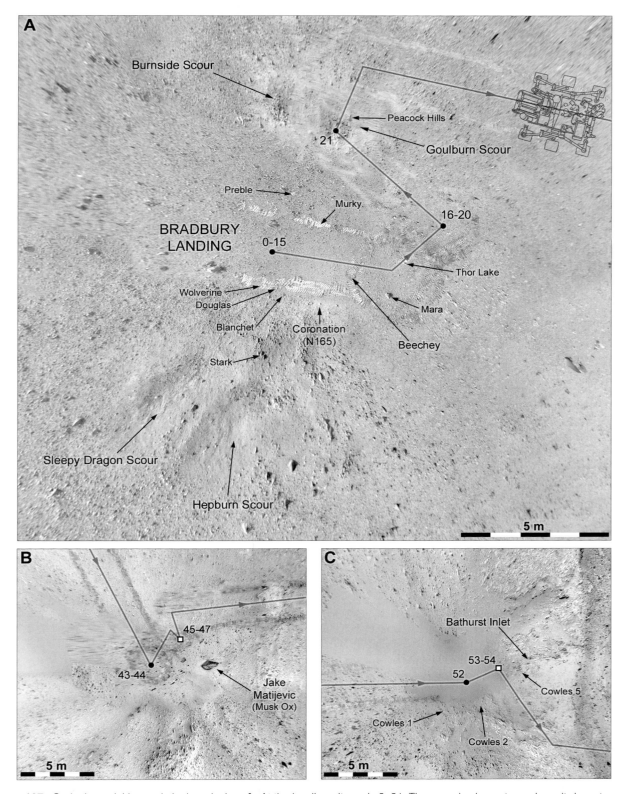

Figure 187. Curiosity activities early in the mission. **A:** At the landing site, sols 0–21. The rover is shown to scale as it departs on sol 22. **B:** Near Jake Matijevic, sols 43–47. **C:** Near Bathurst Inlet, sols 52–54.

images where the terrain sloped down into Glenelg, prompting the name "Arvidson Aeolian Accumulation Area" for the potential sampling site. Ray Arvidson was involved with Curiosity as well as Opportunity at this time.

SAM made an atmospheric analysis late on sol 52, and on the next sol, after a dust devil search, Curiosity paused at Bathurst Inlet, a rocky ledge overlooking Glenelg. APXS and ChemCam obtained data on its composition on sol 54, and MAHLI imaged Bathurst and a nearby location called Cowles 5. This was the start of spring ($L_s = 180$) and the start of the anticipated dust devil season, but nothing was observed after the faint detection on sol 41. DAN was turned on and off to check neutron counts at different times of day. The rover drove down the rough sloping terrain on sol 55 (MY 31, sol 374) and reached its first sampling target, a small dust drift in the wind shadow of a rocky mound now called Rocknest (Figure 188A). Drifts were less common here than at Spirit's landing site and this was the first suitable example along the route.

The final approach to the drift was on sol 56. The sandy material would be run through the sampling and analysis systems twice to remove any lingering terrestrial contamination before a third sample would be analyzed. Was the drift material too coarse for CHIMRA? To see what lay beneath the surface and assess its suitability for sampling, Curiosity scuffed the drift on sol 57 by rotating the right front wheel on it while the others were held steady. Meanwhile ChemCam observed the rock Rocknest 3 and another dust devil movie was made without result. On sol 58 soil targets in the scuff were examined with MAHLI and APXS while afternoon wind measurements were made to ensure good conditions for sample delivery.

The rover repositioned itself for sampling on sol 59 and took images for a long-baseline stereo panorama of Glenelg. Sol 60 was used for a MAHLI range-finding test, checking focus at different target distances to improve knowledge of the instrument locations over four potential scoop targets, before the first sample was dug out of the drift on sol 61. The scoop was being vibrated to clean itself when a small bright object was noticed in images of the ground near the scoop site, causing a delay while it was identified and checked for safety. The analytical instruments (CheMin, SAM) could be contaminated if rover materials were inadvertently placed in them, so the Foreign Object Debris (FOD) was observed with MAHLI on sols 62 and 63 and determined to be a piece of plastic from the rover. On sol 64 the sample was moved into CHIMRA and vibrated to clean it of any terrestrial contamination, a process completed on sol 65.

A second sample was scooped up on sol 66, and small bright fragments (schmutz) were observed in the new "bite" (scoop trench). They were examined carefully to ensure they were not contaminants before sampling continued, but they seemed to be mineral grains. The second scoop had been dumped on sol 67 before the schmutz were deemed safe on sol 68. Finally, part of a third scoop collected on sol 69 was placed in CHIMRA and vibrated to scour it clean again. Planning around this time assumed that Curiosity would leave Rocknest in about 10 more sols and drive down into the bright fractured material at the lowest level of Glenelg, using one of two routes rover planners had identified.

Operations were briefly interrupted here when MRO entered a safe mode, underscoring the importance of the orbital relays, but on sol 70 part of the sol 69 sample was passed to the CheMin instrument for its first analysis and part was dropped on the observation tray and imaged. The CheMin analysis started on sol 71. Meanwhile the ChemCam laser was being used on many sols, including 810 shots on sol 71 alone, a record at the time. Those shots included a raster pattern across light and dark features on a rock called Zephyr, including a small structure named Stonehenge, which resembled a miniature set of arches. ChemCam was also used on sol 72 to examine Schmutz2, a bright object in the second scoop trench. A large Mastcam panorama was compiled here over sols 71 to 74.

Sol 73 was taken up with CHIMRA cleaning, a SAM atmospheric analysis and MAHLI imaging of the observation tray sample. On sol 74 ChemCam took early-morning data from Crestaurum, a patch of fine sand near Zephyr, to see if frost had formed overnight. A comparison of the composition early in the morning with data taken later in the day during the next sol might reveal the presence and composition of any frost. A fourth sample was collected on sol 74 and run through CHIMRA to clean it again on the next sol, while CheMin was emptied and the empty cell analyzed. Cloud and dust devil search images were also obtained on sol 75. Atmospheric pressure data from REMS hinted at dust devils or dust-free vortices passing by in the late morning or early afternoon on many sols, which might interfere with sample delivery and would be taken into account during planning. The fourth sample was dropped onto the tray and imaged on sols 76 and 77, and part was delivered to CheMin on sol 77, while planners considered a possible move to examine dark rocks around Walsh (Figure 188A).

SAM made more atmospheric analyses over sols 78 to 81, failing to detect signs of the methane hinted at by orbiter and Earth-based observations (Webster *et al.*, 2013), but measuring isotope ratios which suggested that much of the planet's early atmosphere had escaped into space. The CheMin analysis continued and more samples were dropped on the observation tray and imaged to see how they were disturbed by arm movements. CHIMRA was cleaned on sols 79 and 81, with sol 80 mainly devoted to recharging batteries, and the nearby rocks Burwash and Et-Then were examined on sol 82 to target MAHLI imaging on later sols. SAM was prepared for its first sample on sol 83 and ran blank analyses on sols 85 and 88. Complex arm sequences were executed on sols 84 and 85 to take a stereoscopic portrait of Curiosity with MAHLI. Apart from its public relations value, the portrait would be useful later when compared with future images to inspect the rover for changes and to diagnose problems. Other activities slowed for a few sols as Mars Odyssey changed its computer system, reducing communication sessions.

MAHLI and APXS were used on the small rock Et-Then on sol 86, and again three sols later in the scuff, this time on a target called Portage, leaving an APXS front plate impression in the soil (Figure 189). On sol 90 the arm tested a SAM sample delivery and placed APXS on Et-Then, and on the next sol APXS was used on the empty observation tray. Cloud and dust devil movies were made repeatedly here and over much of the mission, though dust devils remained elusive.

A ChemCam fault suspended science activities on sol 92 but they were restored by the next sol, when the fifth and final scoop sample was collected and passed to SAM, the first soil sample for SAM. Another part of the sample went to CheMin on sol 94 and a Phobos transit was imaged. The scoop sites are illustrated in Figure 188B. SAM operations were power-intensive and slowed other activities, but more of the sample was dropped on the observation tray for APXS analysis on sols 95 and 96, allowing comparisons between instruments. SAM received another sample from the same scoop on sol 96 for overnight analysis. CheMin did another analysis on sols 97 and 98, and SAM received another part of the sample on sol 98 for analysis on sol 99. The remaining material in the scoop was kept for future analysis as the rover finally moved on from Rocknest on sol 100, making a short drive which obliterated most of the scoop marks and placed the arm within reach of Rocknest 3, a rock just west of the scuff mark. A regional dust storm in Noachis and Hellas raised dust levels and reduced visibility in Gale crater around sol 100 (MY 31, sol 419, or 18 November 2012).

The drill, the main system still to be used, was checked on sol 101 as the CheMin analysis continued. Then on sol 102 a quick touch-and-go operation placed the APXS on Rocknest 3 for comparison with the ChemCam data from sol 57, after which the rover drove 35 m to Point Lake, at the edge of the fractured rocky surface with high thermal inertia at Glenelg. The drill was checked again on sol 103, and over the next week (the US Thanksgiving holiday) extensive imaging was performed to examine Yellowknife Bay, the name given to the lowest level of the fractured material at Glenelg. These images would be used to plan the next few months of exploration. This survey ended on sol 110 with Mastcam imaging of Shaler, a finely layered and cross-bedded outcrop south of Point Lake which might be explored later.

Curiosity moved slightly on sol 111 to position the rock Bell Island in reach of the arm for study by APXS and MAHLI, and ChemCam analyzed its current sample again. Then SAM was prepared for a new sample on sol 112, the drill was checked again on sol 113, and ChemCam was calibrated. CHIMRA tried to deliver another sample to SAM on sol 114 but failed, eventually succeeding with it on sol 116, and APXS and MAHLI examined Bell Island on sol 117. A SAM analysis on sol 117 was partly successful but continued on sol 118, delaying the drive down the hill, and CheMin did another analysis on sol 119.

Planners now suggested that the first drill would be in Yellowknife Bay and a second might occur at Shaler.

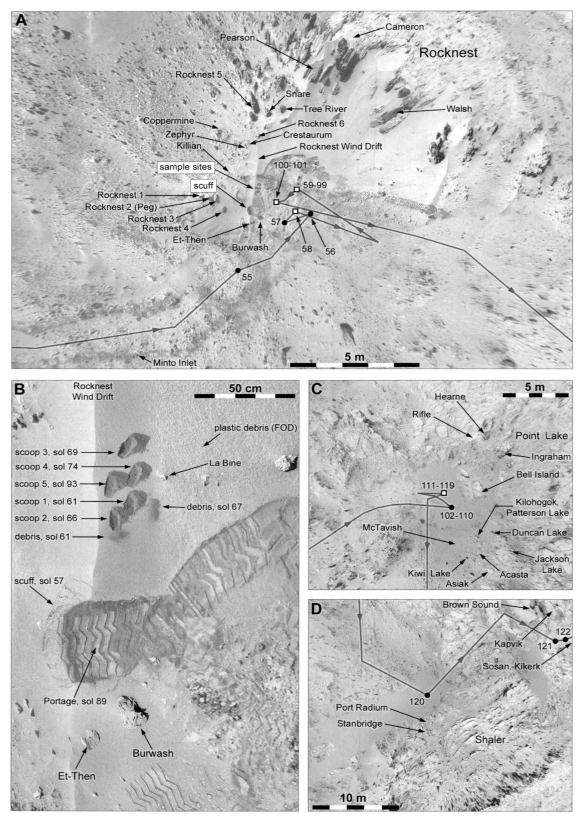

Figure 188. Curiosity activities. **A:** Rocknest, sols 55–101. **B:** Rocknest sample sites, sols 57–93. **C:** Bell Island, sols 102–119. **D:** Shaler, sols 120–122.

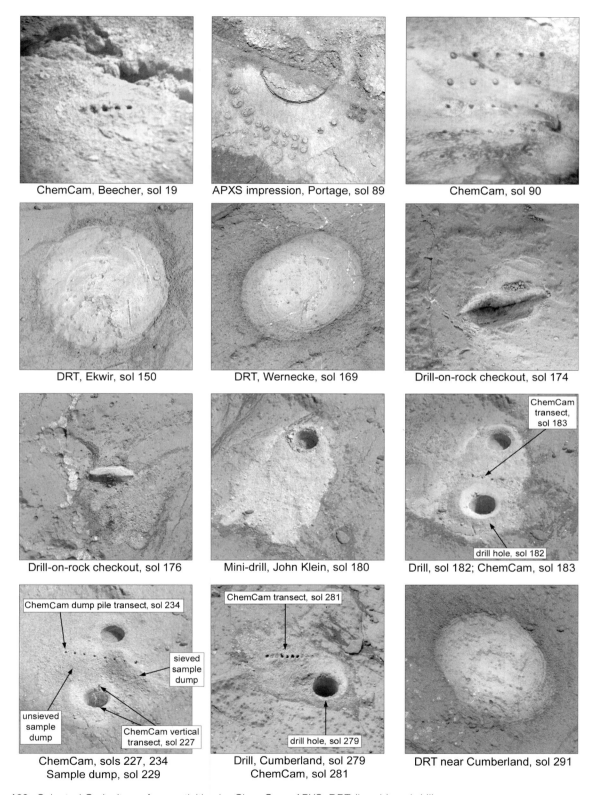

Figure 189. Selected Curiosity surface activities by ChemCam, APXS, DRT (brush) and drill.

Curiosity was driven south to Shaler on sol 120 and northeast to bypass the blocky Point Lake outcrop on sol 121, imaging a rocky hollow called Tochatwi, which resembled a weathered concretion, on sol 122 before descending a series of rocky terraces to the rocky pavement of Yellowknife Bay on sol 125 (MY 31, sol 444). A rock group called Brown Sound, informally referred to as "the seals" for its shape when first seen from Rocknest, was passed on sol 123. CheMin analyzed part of Rocknest sample 5 on sol 121, and ChemCam investigated white calcium sulfate veins in a rock called Crest on the terrace above Yellowknife Bay on sol 125 before the final descent later that sol. The descent came earlier than expected, as planners decided they did not need a toe-dip sequence of drives in and out of the depression to test the path, as Opportunity had done at Endurance and Victoria.

The first stop in Yellowknife Bay, a rock called Sheepbed that contained many veins and small nodules indicative of past episodes of wetting and drying, was observed by ChemCam and Mastcam on sol 126 before Curiosity was driven across the "bay" to its "north shore" on sol 127. The remaining Rocknest sample was dumped on sol 128 and CHIMRA was cleaned before APXS and MAHLI were used on targets called Costello and Flaherty on sol 129. These were also on the Sheepbed layer, the lowest level exposed in Yellowknife Bay. On sol 130 Curiosity approached the rock ledge at Gillespie Lake on the northwest edge of the bay to examine the prominent rock layer above Sheepbed, probing it with APXS and MAHLI on sol 132.

Imaging here included a search for a place to park over the terrestrial Christmas and New Year holiday period. The chosen location was named Grandma's House, just north of Sheepbed, and the time there was spent collecting images for a large panorama while the team chose a target for the first rock drilling operation to complete the long commissioning process for the complex rover. The candidate drill sites were Bathurst, Shaler and here at Yellowknife Bay. The cameras also searched for dust devils and clouds at Grandma's House, while ChemCam analyzed a vein called Rapitan and other targets. The panorama was completed on sol 137, and the rover mainly collected cloud and dust devil images and DAN hydrogen data over the holiday period, one exception being a high-resolution mosaic of Porcupine Plateau, a hill east of Yellowknife Bay (Figure 191).

On sol 147 the rover moved south to the step-like edge of the bay at a region called Selwyn near the future drilling site and a raised line of dark rocks called Snake River. This appeared to be a sedimentary (clastic) dyke, formed as debris collected in an open fracture and eventually became cemented to form a new rock. After a cloud movie on sol 147 and a dust devil search on sol 148, the last CheMin analysis of Rocknest material was performed here. The sample was dumped late on sol 151 and the empty cell analyzed again. Meanwhile the Dust Removal Tool (DRT, equivalent to the RAT brush on Spirit and Opportunity) was used for the first time on sol 150 on a flat rock surface called Ekwir, part of the Sheepbed layer. APXS and MAHLI had observed Ekwir before the brushing on sol 149, as well as the nearby Snake River, and ChemCam targeted a rock called Bonnet Plume. APXS and MAHLI observed Ekwir after brushing as well, looking at the influence of dust on the APXS data.

The location for the first drill activities, just north of Snake River, was chosen now and named John Klein after Curiosity's former project manager, who had died on 27 May 2011, amid new concerns about drilling operations. Some types of rock with substantial water content can liquefy when subjected to frictional heat during drilling, a process called deliquescence, and the salt-rich rocks of Yellowknife Bay might have absorbed enough moisture for this to happen. The drilling would be planned in small stages to examine this possibility, which could seriously compromise the drill. If John Klein was not suitable, Curiosity would return to Shaler instead.

The stratigraphy was becoming understood by now. The fine-grained Sheepbed rocks at the lowest level of Yellowknife Bay were deposited in water and had experienced cycles of wetting and drying to produce small concretions and calcium sulfate veins. Above them was a thicker layer of coarser-grained rock, the Gillespie Lake layer, forming a prominent step around the bay. Over that were the layers at Shaler, which were cross-bedded but too coarse-grained to have been formed by wind, and were therefore probably deposited by flowing water. Above them were the conglomerate layers seen at Link and Hottah.

Curiosity moved forwards on sols 151 and 152 to drive over Snake River, breaking parts of it to expose fresh surfaces for analysis. The contact between units was imaged and ChemCam examined it at Selwyn, while APXS and MAHLI were used at Ungava and Persillon on and below the ledge, respectively. ChemCam also examined John Klein from here on sol 155 and a SAM atmospheric analysis was done overnight. MAHLI imaged a target called Mavor near Persillon on sol 156, but an arm fault stopped APXS making an analysis.

The arm operations were repeated on sol 158, this time successfully, and other nearby targets were examined as well, including Western River, which might have displayed cross-bedding. After further arm and Chem-Cam observations on sol 159, Curiosity rolled backwards to look at Bonnet Plume again. ChemCam data and multispectral images of Mavor and other targets, including freshly broken rocks in Snake River, were acquired on sol 160, and APXS and MAHLI were used on Bonnet Plume and other targets on sol 161. Freshly broken surfaces, including Tintina in Snake River, were so white that some images were overexposed and had to be retaken on sol 162 before Curiosity drove to John Klein.

The new target area had many small protruding veins (Figure 190C), and the rover was driven over them on sol 163 to estimate their strength and to expose fresh material in broken faces for analysis. Curiosity was supposed to back off the veins but the drive stalled, so the reverse drive was made on sol 164 with an additional wheel wiggle to break the veins more effectively. The crushed veins were studied with multispectral imaging and ChemCam on sol 165, and APXS was placed at four spots around Sayunei, an area of undisturbed veins. MAHLI also observed Sayunei, including at night, late on sol 165, illuminating the surface with both white and ultraviolet (UV) light to test for fluorescence, an effect diagnostic of certain minerals. Night observations also included MAHLI imaging of stars to test its usefulness for astronomical observations.

On sol 166 Curiosity finished ISO work at Sayunei, used ChemCam on John Klein and then drove to the drill area to select science targets. Plans to use APXS and MAHLI on sol 167 were delayed when the uplink of instructions failed, but they worked on the next sol at potential drill and DRT sites, and ChemCam's RMI camera examined the drill bit. The DRT brushed a site called Wernecke on sol 169 for ISO work followed on sol 172 by a ChemCam analysis. The step-by-step drill

preparations began with tests of the pressure applied by the arm, and how that might change overnight through diurnal temperature changes. Those "pre-load" check-outs were performed late on sol 170 at four locations (Figure 190C), each one leaving marks on the rock where the drill's two flanking posts touched the surface. Bonnet Plume was studied with ChemCam from a distance on sol 171 to compare results with the closer observations made earlier. Meanwhile the sample system was being prepared as CheMin did an empty cell analysis on sol 172 and CHIMRA was cleaned and SAM prepared on sol 173. Also on sol 172, ChemCam's RMI made very high-resolution images of the drill bit.

The percussive drill was first used on sol 174 to check its hammering motion without rotation, a process called drill-on-rock. It made a dent in the last pre-load position at Thundercloud, but a software error cut the test short, so it was repeated nearby on sol 176 (Figure 190C). ChemCam fired its laser at Wernecke on sol 175 to try to hit a small dark nodule. A new MAHLI self-portrait and a wheel inspection mosaic were made on sol 177 and then the first rotary drill test, termed mini-drill, was attempted on sol 178. It stalled before contacting the surface and was repeated on sol 180. This was not deep enough to collect samples for analysis, but it confirmed that the drill worked properly.

Two sols later the first full drill hole, 6.4 cm deep and 1.4 cm wide, was made just a few centimeters from the mini-drill site and a sample of rock powder was collected. The drilling moved rocks and debris, making many small changes near the site, which were documented in images now as Mastcam and ChemCam analyzed the drill tailings. The new sample was moved from the drill to CHIMRA and the scoop on sol 184, but a problem with the sieve on a duplicate of the sampling system on Earth caused a week-long delay. Throughout this period large panoramas were compiled, cloud and dust devil searches continued, and ChemCam was used repeatedly on nearby targets, including Cumberland, a rock nearby that became the next drill site, and several small bowl-shaped hollows, some containing white material.

The sample in the scoop was sieved on sol 194, and CheMin and SAM eventually received samples on sols 195 and 196, respectively, to begin their analyses. The samples contained clays and other materials suggestive of a wet and habitable environment, but organic compounds were present in very small quantities, if at all. A software update was scheduled soon and the files were uplinked on sol 198, but planners decided to postpone the update until after conjunction. On sol 199 a second sample was passed to SAM, for analysis on sol 200 (MY 31, sol 519, or 27 February 2013), but this was interrupted by a flash memory fault, necessitating a switch to the backup computer system on sol 201 while the problem was resolved. The cause may have been a cosmic ray strike.

The rover was gradually brought back to operational status over the next few sols, but was shut down again on sol 207 as a large coronal mass ejection (CME), plasma expelled from the Sun two days earlier, reached Mars. This was a precautionary measure, but Curiosity weathered the event without problems and was soon back in operation, testing its systems under the control of the new computer system, and diagnosing and repairing the other computer. The first new images were taken on sol 215. The backup (B-side) computer was booted up on sol 217 after being patched on sol 214, but it entered a safe mode and had to be checked again. It operated properly after sol 219 and was fully functional and controlling Curiosity on sol 225.

A triple-sized sample of John Klein was delivered to SAM and analyzed on sol 224 in the hope that a more certain detection of organic materials could be made. Another SAM delivery and analysis occurred on sol 227, and ChemCam made a series of laser shots up the visible wall of the drill hole and into the tailings pile. The remaining sample was dumped onto the ground between the two drill holes on sol 229, and ChemCam made a transect across the dump pile (Figure 189) on sol 234. Then the arm was stowed for conjunction and the last images before the break were taken on sol 235. The REMS environmental sensors and RAD and DAN collected data over the four-week break between sols 236 and 260, and Curiosity transmitted a daily beep via Mars Odyssey to indicate all was well.

Curiosity emerged from conjunction on sol 261, and over sols 263–266 the software update was installed and tested. This would improve Curiosity's autonomous navigation abilities for the long drive to Mt. Sharp, which would begin soon. Images on sol 267 showed that the wind had moved dust on the surface during the break

in operations, including the drill tailings. Navcam tests were conducted on sols 268 and 269, and the Mastcam viewed cracks at The Narrows on sol 269. The next sol was taken up with MAHLI images of the drill hole, ISO of raised ridges at McGrath, a MAHLI self-portrait of the rover, a cloud movie and CheMin analysis of part of the drill sample, which continued into sol 272. Sol 271 also included a dust devil search and the conclusion of a Mastcam panorama of Mt. Sharp.

The team had now selected its next drilling target, the nodule-rich rock Cumberland, which was roughly 3 m west of John Klein. The whole process should be much faster now everything had been tested, and the team hoped to leave Yellowknife Bay by about sol 300. On sols 272 and 274 the rover moved towards Cumberland, the first drives since sol 166, while the Navcam searched for clouds over Mt. Sharp on sol 273 and for dust devils on sol 274. ChemCam investigated potential drill sites at Cumberland and Kazan on sol 274 and concretions and bowl-like hollows on sol 275 as MAHLI imaged the wheels to be sure the rover was stable for drilling.

The APXS was placed on Cumberland on sol 276, fracturing one of the hard nodules in the outcrop, MAHLI viewed the future drill site and a pre-load test was conducted to prepare for drilling. A new Cumberland target was examined with ChemCam and ISO on the next sol as CheMin prepared for the new sample. Atmospheric observations dominated sol 278, a SAM analysis and passive ChemCam collection of sky spectra. That passive mode was calibrated by using it on white paint on Curiosity on sol 281. The Cumberland hole was drilled on sol 279 with MAHLI observations before and after drilling, and the sample was passed to the scoop. SAM baked an empty cell to prepare for its sample on sol 280, and on the next sol lines of ChemCam shots were made across the drill tailings on sol 281. It was important to characterize the newly exposed tailings as quickly as possible.

The new drill sample was passed to SAM on sol 281 and to CheMin on sol 282, and MAHLI looked at both sample inlets on the latter sol. Curiosity searched for dust devils again on sol 283, still seeing none, checked the ChemCam alignment and performed ISO of the tailings and drill hole. On the next sol ChemCam passively examined the sky again for composition and aerosol properties, SAM analyzed the atmosphere and a

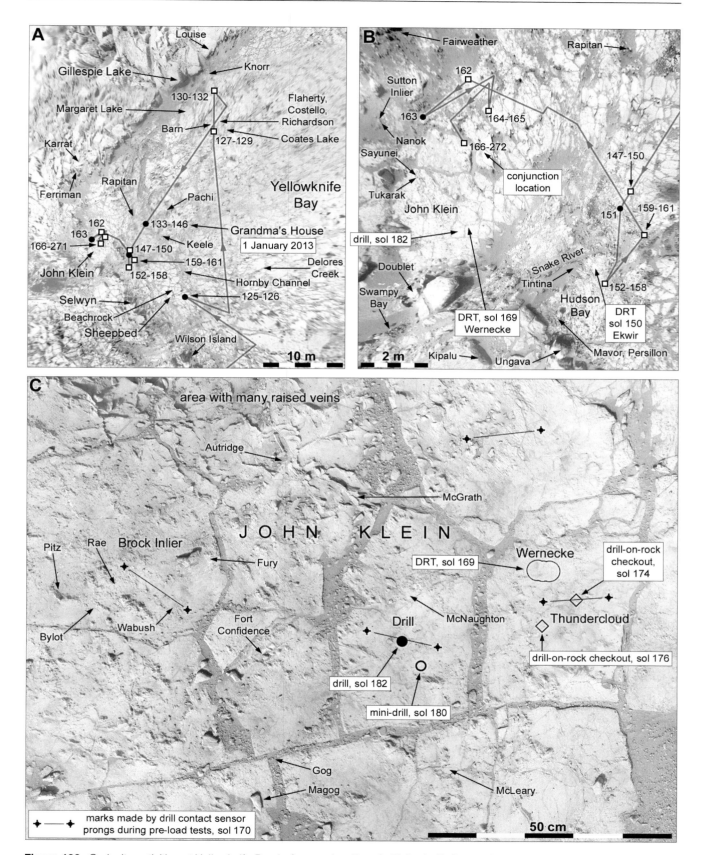

Figure 190. Curiosity activities at Yellowknife Bay before conjunction. **A:** Yellowknife Bay, sols 125–272. **B:** John Klein, sols 147–272. **C:** Drill site, sols 169–182.

Table 52. *Curiosity Activities in Yellowknife Bay Area, Sols 121–328*

Sol	Activities
121–125	CheMin Rocknest 5 sample (121), drive (121–124), enter Yellowknife Bay (125), imaging
126–127	Imaging, ChemCam Sheepbed and clean ChemCam (126), drive to North Shore and SAM test (127)
128–130	Empty and clean CHIMRA (128), ISO Costello, Flaherty (129), drive to ledge (130), imaging
131–133	Imaging, ISO Gillespie Lake (132), drive to Grandma's (or Grandmother's) House (133)
134–140	Clean ChemCam (134, 140), ChemCam Rapitan (135), imaging over holiday break
141–148	Imaging, move to Snake River (147), CheMin last Rocknest sample and clean ChemCam (148)
149–150	ISO Snake River, Ekwir 1, Ekwir 2 (149), DRT and ISO Ekwir 1, ISO Grit (150), imaging
151–152	Move to contact area (151, 152), empty and analyze ChemMin cell (151), imaging
153–156	Imaging, ISO Ungava, Persillon (154), SAM atmosphere (155), clean ChemCam (156)
157–159	Imaging, ISO Mavor, Tindir, Nastapoka (158), move to Bonnet Plume and ISO Twitya (159)
160–161	ISO Tintina, Tindir Lip (160), ISO Bonnet Plume, Hudson Bay, Yukon (161), imaging
162–164	Drive to John Klein, crush veins (162–164), ISO Hay Creek dust free (162), clean ChemCam (164)
165–167	ISO and night UV images of Sayunei (165), move to John Klein, ISO Sayunei (166), imaging
168–169	ISO drill and DRT targets, RMI drill tip (168), DRT and ISO Wernecke, ISO Brock Inlier (169)
170–172	Drill pre-load tests (170, 171), clean ChemCam (171), CheMin empty cell (172), imaging
173–176	ISO Wernecke 3, MAHLI Autridge, prepare CHIMRA and SAM (173), drill checkout (174, 176)
177–178	ISO tray, MAHLI self-portrait, SAM preparation (177), mini-drill attempt fails (178)
179–180	MAHLI range test, ISO calibration target, imaging (179), mini-drill and MAHLI hole (180)
181	ISO Divot 2, MAHLI McLeary, McNaughton and McGrath, SAM test, imaging (181)
182–183	Drill John Klein and MAHLI hole (182), ChemCam tailings (183)
184–193	Transfer sample (184–193), imaging, anomalies prevent science (191)
194–195	Sieve sample (194), deliver John Klein sample to CheMin and analyze, prepare SAM (195)
196–197	Deliver sample to SAM and analyze (196), repeat CheMin analysis (197), imaging
198–199	Software changes, prepare SAM (198), deliver sample to SAM and analyze (199), imaging
200–222	Computer anomaly (200–221), switch to B-side computer (214–217), test instruments (215, 222)
223–224	B-side computer updated, prepare SAM (223), deliver sample to SAM and analyze (224)
225–226	B-side computer use begins (225), CheMin Klein sample, prepare SAM (226), imaging
227–229	Deliver sample to SAM and analysis (227), imaging, CheMin Klein sample, clean CHIMRA (229)
230–231	CheMin Klein sample, ISO drill hole (230), MAHLI Katherine, Dolly and SAM atmosphere (231)
232–233	CheMin John Klein sample (232), MAHLI images REMS sensor (233), imaging
234–261	ChemCam sample pile dumped on sol 229, then stow mast (234), conjunction science (236–261)
262–268	ChemCam calibration (262), imaging, software update (263–267), CheMin John Klein sample (268)
269–271	Imaging, camera tests (268–269), ISO drill hole, McGrath (270), CheMin Klein sample (270–272)
272–276	Move to Cumberland (272, 274), imaging, prepare SAM (275), ISO Cumberland (275–276)
277–280	CheMin empty cell, ISO Cumberland (277, 279), drill Cumberland (279), prepare SAM (280)
281–282	SAM analysis of Cumberland sample (281), CheMin analysis (282), MAHLI inlets (282)
283–286	ISO drill tailings and hole (283), sample drop video (284), prepare SAM (285), SAM analysis (286)
287–289	ISO tailings, CheMin analysis (287), imaging, prepare SAM (288), CheMin sample (289)
290–292	SAM sample (290), DRT and ISO Cumberland, ISO Narrows 3 (291), MAHLI hole at night (292)
293–299	CheMin Cumberland sample (293), imaging, drive to Point Lake (295–299), DAN transect (299)
300–304	CheMin sample (300), approach Point Lake (301, 302), ISO Alligator Head (303) and Balboa (304)
305–314	SAM atmosphere (305), drive to Shaler (307–313), CheMin sample (310), image sunset (312)
315–321	CheMin dark frame (315), drive to Shaler and image Phobos (317), SAM atmosphere analysis (321)
322–323	ISO calibration target and Aillik (322), ISO Shaler targets and Howells, and REMS sensor (323)
324–328	Drive out of Glenelg (324, 327), update SAM software (324–326), ChemCam cleaning (327)

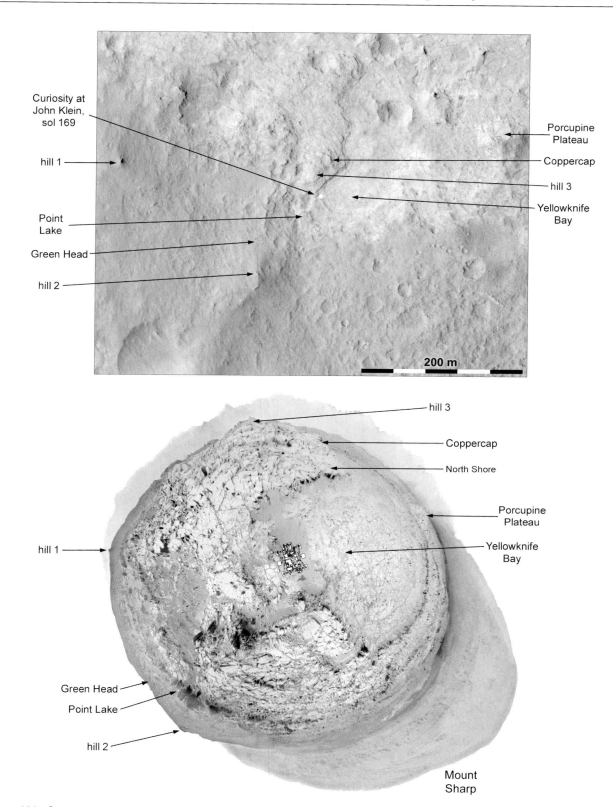

Figure 191. Comparison of HiRISE image ESP_028401_1755 (top) and a reprojected panorama from John Klein (Yellowknife Bay) on sols 168 and 169 (bottom).

Mastcam video was made of the sample delivery process. ChemCam had many uses, including imaging distant objects at resolutions higher than Mastcam could achieve, a capability it tested on sol 285 for use during the long traverse to Mt. Sharp. The targets Ferriman, Karrat and Green Head were imaged for this test. In its LIBS mode it also provided vertical transects of composition on a raised ridge called Lady Nye and the visible inner wall of the drill hole. Mastcam imaged a bright broken rock called Horsethief Creek, completing a busy sol.

Another broken rock called Lumby Lake was imaged on sol 286 as more of the sample was passed to SAM and analyzed, and APXS was placed on the hole tailings. CheMin analyzed its sample on sols 287 and 289, and on the intervening sol ChemCam examined Lady Nye again, Lumby Lake was imaged and a dust devil search was performed. ChemCam observed the tailings during the warmth of the day on sol 289 and then early the next morning to look for signs of water molecules moving in and out of the soil as temperatures varied. It also analyzed a pebble called Miette, and Navcam looked for clouds and observed the area to the south, where DAN would soon be used across a stratigraphic boundary.

SAM analyzed the Cumberland sample on sol 290 as ChemCam examined Lady Nye again and Mastcam viewed a target for the DRT. That target was viewed by MAHLI before and after brushing with the DRT on sol 291, but the brushes were then seen to be bent, and the DRT was not used from now on while the issue was studied. SAM analyzed the atmosphere and MAHLI looked at a target in The Narrows, one of the soil-filled cracks in the outcrop. MAHLI was used during the night on sol 292 on the tailings and The Narrows target, including observations with the UV light source, and the SAM atmospheric analysis continued. The work at Cumberland was coming to an end now, with CheMin analysis of its sample, and ChemCam observations of Cumberland and a raised ridge called Duluth on sol 293. Sol 294 finished the ChemCam work with studies of a concretion in Cumberland and two more raised ridge targets, Sibley and Christopher Island.

Curiosity drove away from Cumberland on sol 295, taking DAN data along the traverse. It was heading for Point Lake, a prominent rock outcrop skirted on its southern side on sol 120 during the approach to Yellowknife Bay (Figures 188C and 193A). The rover stopped after moving 6 m when a wheel stalled, and on the next sol it made a cloud movie and a dust devil search and used ChemCam for passive sky observations while imaging an area called Tindir where it would make its DAN transect across the stratigraphic boundary. On sol 297 it backed up to free the stalled wheel and drove 19 m, beginning the DAN transect, and on the next sol it imaged the transect area again as well as the target Roswell Harbour, and used ChemCam on Mesabi and Cape Strawberry. The DAN transect continued across the geological boundary between the Sheepbed and Gillespie units at Tindir on sol 299, and on sol 300 (MY 31, sol 619, or 10 June 2013) the rover searched for dust devils again and imaged Point Lake for target planning. The north end of the outcrop was preferred over the south, and the goal was to determine if the rock was sedimentary or volcanic.

After a drive on sol 301, the rover made its final approach on sol 302, aiming for a target called Alligator Lake, which was easier to reach than a preferred target nearby. Mastcam imaged veins or nodules at Ekalulia and two targets also analyzed by ChemCam, Knob Lake and Athole Point. Curiosity was tilted 9°, and on the next sol MAHLI, ChemCam and Mastcam viewed many targets while planners checked the rover's stability for arm operations. It seemed to be stable, so APXS was placed on the target Balboa and MAHLI imaged Measles Point and other targets. Over the next two sols SAM analyzed the atmosphere, a dust devil search was carried out, and many targets were scrutinized by the cameras and ChemCam. More methane than usual was detected by SAM here for the first time. There was always a very small amount in the atmosphere, but on sol 306 the amount was elevated by a factor of about 10 compared with the previous measurement on sol 292. The new measure was made during the day and the others at night.

The last target in this area was the cross-bedded outcrop Shaler, seen briefly on sol 120. Curiosity drove towards it on sols 307 and 308, observing the contact between the Point Lake and Gillespie units. A drive on sol 309 was cut short after only 2 m when the rover tilt limit was exceeded while crossing a soil ridge, but ChemCam examined veins or cracks called Rove and Ramah, a harder unit above Rove, and Mastcam viewed Piling, the lowest layer of Shaler. CheMin was still holding its sample and part was analyzed on sol 310 while MARDI imaged the surface under the rover and Navcam made a cloud movie. ChemCam examined Michigamme and Piling on the next sol as Mastcam imaged Society Cliffs, the abrupt ledge topping Shaler (Figure 193D). Of the three terrain units that met at the "triple junction" of Glenelg (Figure 178B), two had been studied in detail but the cratered unit was observed only from here, and Society Cliffs might be considered part of the edge of that unit.

Sol 312 was spent imaging many targets, including the setting Sun and looking for changes at Rocknest, and ChemCam examined several targets and used RMI to view Castle Mountain, a hill on the lower slopes of Mt. Sharp (Figure 200A) as a test of future distant observations. Then Curiosity drove along the foot of Shaler on sol 313 to study it more closely, also making an overnight SAM atmospheric analysis to check the previous methane measurement. This was elevated, but half the sol 306 value. On the next sol the rover viewed the outcrop and made a cloud movie, but images showed that the middle left wheel was not touching the ground, so the rover might not be stable enough to use its arm. ChemCam analyzed several targets in Shaler on sol 315 and Navcam looked for dust devils again, but still finding none. More ChemCam targets were examined on sol 316, and RMI viewed a distant block called Sokoman on the cratered unit (Figure 186).

Sol 317 began with a ChemCam analysis of the soil, and then Curiosity drove to a new study area nearby, stopping with all six wheels firmly on the ground. A cloud movie was made over Mt. Sharp, and a sunset zenith movie showing clouds, part of an ongoing campaign to monitor cloud movements. The next few sols were taken up with detailed characterization of the outcrop in images and with ChemCam, with additional searches for dust devils and clouds and a SAM atmospheric analysis on sol 321. On sol 322 ChemCam examined Husky Creek, a block fallen from Society Cliffs, the ledge above Shaler, and a white vein called Fabricius Cliffs. MAHLI and APXS were calibrated and then used for ISO of Aillik, a fine-grained sandstone.

As this work wrapped up, the planners were looking ahead. Curiosity should reach the main science area at Mt. Sharp as soon as possible, so a direct path called the "Rapid Transit Route" (Figure 184) had been chosen

rather than a path that meandered between many enticing sites on the plains. Nevertheless, some science would be done along the way, and the team had identified five waypoints where a limited time might be spent on analysis. After leaving Shaler the rover would go either to Bradbury Landing, possibly continuing to the parachute and backshell, or directly to Waypoint 1.

Work at Shaler ended on sol 323 with cloud and dust devil movies, ChemCam and MAHLI work at many targets, multispectral images of Fabricius Cliffs to look for signs of hydration, and APXS on a coarse unit called Howells. Then on the next sol the arm was stowed and Curiosity drove north, leaving the Glenelg area by the same route it had entered 200 sols earlier. New software for SAM was uploaded on sols 324 and 326. Sol 325 was the US holiday on 4 July 2013 (MY 31, sol 642), and the rover spent it and the next sol imaging its surroundings.

ChemCam examined nearby rocks on sol 326, and on the next sol RMI viewed the distant target Ameto (Figure 200A) where the mission science team expected to ascend Mt. Sharp just east of an alluvial fan. Then Curiosity drove again, and after the drive it used MAHLI in its stowed position to image the landscape out to the horizon. The image was at an awkward angle, but a series of these images taken after every drive would document the long traverse. MARDI would do the same with images taken under the rover.

Cloud and dust devil imaging followed on sol 328, and on the next sol more imaging, including clouds and Mastcam views of a target called Dumas, and ChemCam observations of several targets. Curiosity drove again later on sol 329, climbing out of Glenelg near its sol 50 location. Memory was nearly full and had to be managed carefully to avoid problems like those encountered by Spirit early in its mission. On sol 331 ChemCam was used on the rocks seen on sol 50, and the old and new tracks were compared to assess the effect of wind on disturbed surfaces. After another dust devil search Curiosity drove southwest, heading through the rolling landscape towards Mt. Sharp, though there had been some interest in visiting the parachute and backshell, or possibly a small fragment of the descent stage which had impacted 600 m west of the parachute (Figure 178).

Sol 332 was a sol for imaging, including a cloud movie, and ChemCam use. Typical sol plans in this period involved imaging and ChemCam observations of

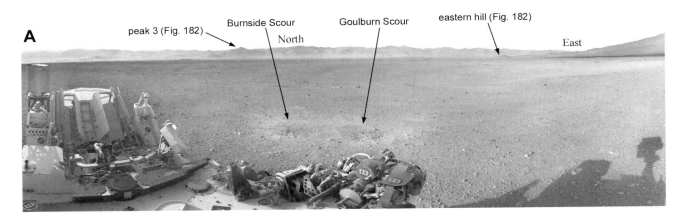

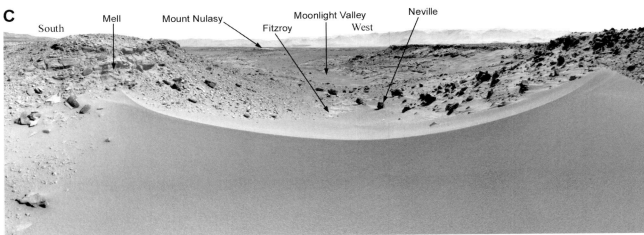

Figure 192 (both pages). Curiosity panoramas. **A:** Landing site, sols 2–13. **B:** Yellowknife Bay, sols 168–176. **C:** Dingo Gap, sol 528. **D:** Kimberley, sol 621.

Figure 192 (continued)

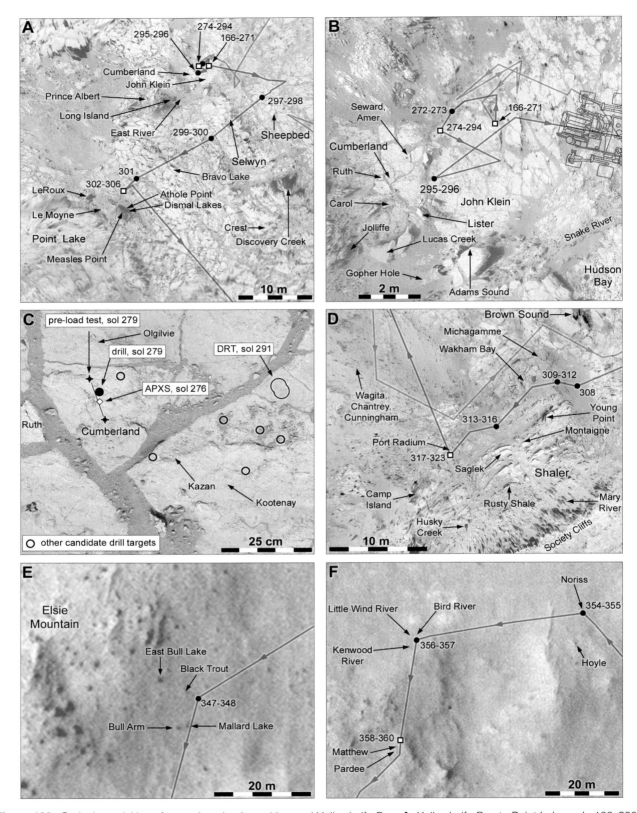

Figure 193. Curiosity activities after conjunction in and beyond Yellowknife Bay. **A:** Yellowknife Bay to Point Lake, sols 166–306. **B:** Cumberland, sols 166–296. The rover is partially shown at the correct scale. **C:** Drill site, sols 276–291. **D:** Shaler, sols 308–323. **E:** Elsie Mountain, sols 347–348. **F:** Matthew, sols 354–360.

interesting targets at the start of the sol, based on the previous day's pictures, followed by a drive with images taken to monitor progress and record features along the route, and ending with a survey of the new location to plan the next day's observations or drive. A drive on sol 333 was cut short by a steering fault, but the next one on sol 335 was successful and carried Curiosity past 1000 m of travel after pre-drive observations of targets, including Marquette and Kaertok. Sol 336 was similar, beginning with observations of Nullataktok, Ikpiarjuk and Mani-tounuk before driving 33 m. That pattern continued for three sols with targets including RMI images of Mont Right and Finger Hill, and Mastcam and ChemCam observations of a dusty hollow called Adlavik and an outcrop called Red Wine.

New operating software with enhanced autonomous driving capabilities had been transmitted to the rover around sol 265 and was first used on sol 340 for a drive of 101 m, a record at that time, aided by a particularly good view ahead after the previous sol. Curiosity approached a rocky ridge called Twin Cairns Island on sol 343, one of many outcrops and rock piles on high points in this area. Curiosity passed too far to the south in this very hilly area to see the landing site or parachute, but the rover reached the same longitude as Bradbury Landing during the sol 345 drive and quickly moved further west (Figure 194). The radiation detector, environmental monitor and DAN instrument were used on almost every sol during the drive. From this area the rims of craters west of the landing ellipse (Figures 177C and 198) could be seen, though the atmosphere was dusty now and the view was much clearer 150 sols later.

Curiosity drove past Twin Cairns Island on sols 344 and 345, taking Mastcam images of it but not stopping. The rocky hilltop unit was much more extensive further to the southwest and could be examined there. Targets called La Reine and Tete Jaune were observed on sol 346, and a dust devil movie was made on sol 348. These observations were reduced in frequency after this, as no dust devils were being seen, and indeed no tracks were obvious at Gale in orbital images. On sol 348 the rover reached a boulder-studded hill called Elsie Mountain (Figure 193E), and early on sol 349 ChemCam and Mastcam observed large rocks called Black Trout, Bull Arm, Mallard Lake and East Bull Lake. Drives around this time were typically about 70 m long.

ChemCam was calibrated on sol 350, and very early on sol 351 (MY 32, sol 1) the Mastcams observed Phobos occulting Deimos in the night sky. Later that morning Mastcam imaged Nulliak, Waterton, Kenamu and an overhanging ledge called Yellorex before moving on again. A full 360° panorama was made with Mastcam on sol 352, and repeated sporadically along the traverse to Mt. Sharp. SAM was prepared for another analysis of the Cumberland sample on sol 352, and performed it on sol 353 as ChemCam observed the targets Hector and Howey and Mastcam looked at Hudson. ChemCam was calibrated again on sol 355 and a Navcam cloud search was made over Mt. Sharp. This sol came one Earth year after landing.

ChemCam and Mastcam examined a conglomerate called Noriss on sol 356 before the rover was driven over a dust ripple, imaging its tracks and more conglomerates called Bird River, Kenwood River and Little Bird River on sol 358. MRO was about to undergo an attitude control system switch, preventing its use as a relay for a few sols over a weekend, but Curiosity was still able to make its first brief touch-and-go observations of the long traverse. Touch-and-go operations began with imaging as the rover arrived at the site, followed the next morning by ISO before the rover drove on later that sol. Previously the rover had spent much longer at any site where the arm was used like this, but these touch-and-go procedures were common with MER and would now be used by Curiosity, increasing the scientific value of the long drive. The touch-and-go activities occurred on sol 360 at a target called Matthew, as well as images of Pardee and the REMS UV sensor, to measure dust accumulation on it.

A partial Phobos solar transit was imaged on sol 363 and ChemCam examined Labyrinth Lake and Buit Lake. A 360° panorama was made with the Mastcam roughly every 300 m along the traverse, including on sol 364, and Curiosity passed the 2 km mark on its sol 365 drive after using ChemCam on Wilkinson and Robin Hood. Then it parked for four sols while another SAM analysis of the Cumberland sample used much of the available power and REMS was reactivated after an outage. The rover also looked for clouds over Mt. Sharp and searched for dust devils. A Phobos transit was imaged on sol 368 with simultaneous Hazcam imaging of the horizon to observe the reduced illumination during

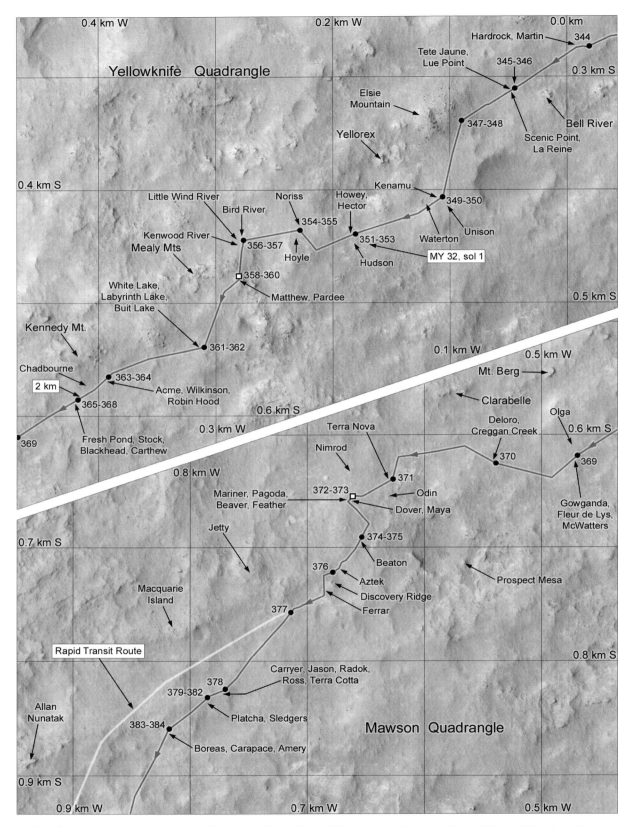

Figure 194. Curiosity route map, section 2. This map follows Figure 185 and is continued in Figure 195. Additional details are shown in Figure 193.

Table 53. *Curiosity Activities, Glenelg to Kimberley, Sols 329–578*

Sol	Activities
329–339	Drive out of Glenelg area, prepare SAM (329), test SAM (331), SAM atmosphere (339), imaging
340–350	Drive towards Mt. Sharp, imaging, SAM atmosphere (341), Phobos transit video (350)
351–352	Coordinated THEMIS–REMS data (351), drive towards Mt. Sharp and imaging, prepare SAM (352)
353–360	SAM Cumberland sample (353), drive southwest, MAHLI REMS UV sensors, ISO Matthew (360)
361–369	Drive, Phobos transit (362, 369), SAM atmosphere (364), sample to SAM (367) and analysis (368)
370–386	Drive southwest, autonav test (372), ISO Dover, Maya (373), SAM delivery (381), analysis (382)
387–392	ISO Spurs, Ruker (387), drive to Panorama Point (388), imaging, drive to Darwin (390, 392)
393–395	Phobos eclipse, prepare SAM, blank analysis (393–4), ISO Bardin Bluffs, Altar Mountain (394–395)
396–408	Move to (396) and ISO (398–400) veins, drive (402–406), prepare ChemCam (407), SAM (407–408)
409–416	Drive (409–413), MAHLI wheels, rover targets (411), SAM sample delivery (414), analysis (415)
417–421	CheMin Cumberland sample, MAHLI REMS sensor (418), prepare SAM (420), SAM blank (421)
422–433	CheMin Cumberland (423, 425), drive, prepare SAM (427), SAM blank (428), CheMin test (432)
434–441	SAM atmosphere (434), drive to Cooperstown (436–440), 100 000th laser shot (439), SAM test (441)
442–446	ISO Pine Plains, Rensselaer (442), software update (444–446), safe mode (446)
447–465	No science (447–452, 457–461), drive (453–455), wheel stall (455), sample to SAM (463, 464)
466–472	Clean ChemCam (466, 468, 471), RMI Hematite Ridge (467), ISO Oswego, MAHLI wheels (472)
473–477	CheMin Cumberland (473, 477), SAM atmosphere (474), MAHLI wheels, REMS sensor (476)
478–487	Software update (478–484), ISO Poughquag (485), dump sample (486), ISO dump pile (486–487)
488–502	MAHLI wheels (488, 490), CheMin Cumberland (488), RAD/REMS science (495–501), imaging
503–509	ISO Nedrow, Morehouse (503), SAM atmosphere (504), ISO Oneida (506), MAHLI wheels (508)
510–515	ISO Lowerre, Larrabee (510), Kodak, Clinton (512), ChemCam Harrison (514), MAHLI wheels (511–515)
516–521	ChemCam and ISO Oscar (516), drive, image Doran, Togo (520), drive, MAHLI wheels (521)
522–526	Test SAM (522), ISO King (523), MAHLI wheels (524), SAM atmosphere (525), ISO Reedy (526)
527–531	Drive, scuff Dingo Gap (528), MAHLI wheels (529), ISO Barker, Argyle, Dampier, Crossland (531)
532–537	MAHLI wheels (532, 537), dune toe-dip (533), imaging, cross dune (535), ISO Fitzroy, Halls (537)
538–540	ChemCam frost search, Collett, Mussell, SAM atmosphere (538), image Moonlight Valley (540)
541–544	Clean ChemCam (541), Image Mt. Nulasy (542), frost search (543), MAHLI wheels, SAM inlet (544)
545–551	MAHLI wheels (545–549), ChemCam water test (545), image Junda, Scrutons (548), ISO JumJum (550)
552–557	Image Kylie, Wilson Cliffs (552), MAHLI wheels (552–554), SAM combustion tests (555–557)
558–560	Image Mt. Amy, ISO Johnny Cake (558), MAHLI wheels (559), ISO Secure (560)
561–564	CheMin tests (561), MAHLI wheels (561–564), RMI dunes (562), ISO Monkey Yard, Crowhurst (564)
565–574	MAHLI wheels (566–569, 572, 574), drive, ISO O-tray (571), SAM methane tests (572)
575–578	Dump CheMin sample, calibrate SAM (575), clean CHIMRA (576), check CHIMRA sieve (578)

the transit, an observation made previously by Viking Lander 1 on its sols 415, 419 and 423 (Stooke, 2012). Another transit was seen on the next sol, and REMS observed the surface temperature to compare with simultaneous measurements by THEMIS on Mars Odyssey.

Now the mission moved out of restricted sols again and could drive nearly every day. ChemCam and Mastcam viewed the conglomerate Gowganda on sol 370 before driving to rock outcrops called Clarabelle and Deloro. They were observed with ChemCam and Mastcam early on sol 371, but the observations were curtailed to allow a record drive of 110 m, made possible by good visibility ahead. On the next sol the rover crossed into the Mawson map quadrangle (Figure 183). Douglas Mawson (1882–1958) was a prominent geologist and Antarctic explorer, and place names here were taken from Antarctic geology, including Terra Nova, a possible duricrust target on sol 372.

Up to this point Curiosity had driven only where the Navcams could see clearly ahead, making stereoscopic images for topographic mapping, but the next sol's drive included the first test of autonav software on the B-side computer. This had not been used yet because of a problem with the analysis of stereoscopic B-side Navcam images. The B Navcams were mounted below the A Navcams and their pointing was found to vary slightly with temperature. After a complex recalibration, the software was finally ready for testing, complicated by a desire to use the arm after the drive. The autonav test was flawless and the procedure could now be used to increase driving distances. Drives would no longer end where the stereo mapping was cut off by a ridge, but could continue under autonav into areas not previously seen by the rover.

The sol 372 drive successfully demonstrated the autonav procedure, so it could now be used to drive beyond areas visible in the surface images. Sol 373 included ISO on the targets Maya and Dover in front of the rover, followed by a long drive on sol 374 and another sol of imaging. The ChemCam's RMI viewed the sky to calibrate its images. Another Mastcam 360° panorama was taken on sol 375, and a long drive on sol 376 used autonav for the first time to drive 10 m over a ridge into unseen territory. This was called Discovery Ridge, and while Curiosity was here the planners on Earth were preparing for work at the first major stop of the long traverse. After imaging Phobos, Deimos and Jupiter early on sol 378, the rover made a long drive, ending with its left rear wheel perched on a rock about 6 cm across, which left the rover too unstable for the arm to be deployed. The plan was to deliver more of the Cumberland sample to SAM for a two-sol evolved gas analysis. The team decided to drive a short distance to a less rocky area on sol 379 and use the arm there.

The sample was delivered on sol 381 and analyzed overnight, using so much power that little other work could be done. Sol 382 began with ChemCam observations of the rock Sledgers and the soil Platcha, followed by a long drive, combining a good view ahead with about 45 m of autonav driving. Long drives were necessary to get to Mt. Sharp in good time, but they limited the observations that could be made along the way. Automated, untargeted, observations were one way to increase science output, so ChemCam now began observing an area on the right side of the rover after most drives. An outcrop of cratered material named MacQuarie Island, similar to that seen at Glenelg, was imaged on sol 385, and a 141 m drive later that sol exceeded Spirit's distance

record of 124 m on its sol 125. This drive, using autonav, was the second longest of Curiosity's primary mission. It ended on a small hill initially called Prairie Dog Hill and later named Panorama Point, with ISO at a rock target called Ruker on sol 387. Some Mastcam astronomy was done here on sol 386, directed at the night sky to calibrate later images of Comet ISON and at an eclipsed and barely visible Phobos occulting the bright star Aldebaran.

A short drive on sol 388 included Mastcam mid-drive and post-drive imaging from the top of Panorama Point for long-baseline stereoscopic viewing of an elliptical depression or basin containing rocky outcrops just south of the viewpoint. A light-toned outcrop called Darwin on its northwestern rim was chosen as the first science stop along the traverse to Mt. Sharp, referred to as Waypoint 1. This and four more waypoints had been chosen from HiRISE images as the long traverse began to explore the various plains units before Mt. Sharp was reached and to place them in geological context. A target called Hart about 50 m north of Darwin was also considered as an ISO site. ChemCam's RMI viewed Darwin from a distance on sol 387 to help plan activities, and long-baseline stereoscopic observations of Darwin and the basin followed. This was said to be the best outcrop area on Bradbury Rise.

Curiosity reached Darwin on sol 392 (MY 32, sol 42) and surveyed the outcrop to plan ISO over the next three sols. Analyses included clasts and matrix in a conglomerate called Bardin Bluffs, a lower layer called Altar Mountain and a small white vein called Eureka. The rover worked late on sol 393 to image Phobos entering eclipse and to view Deimos and Jupiter. Night observations like these were common around this period as the science team prepared to make observations of the approaching Comet C/2012 S1 (ISON). Then on sol 396 the rover drove a short distance to another interesting target, a group of prominent veins, and made photometric observations of the surface to each side of the rover at different times of day to measure the properties and structure of the surface material. ChemCam's RMI and Mastcam made night images of M31 (the Andromeda galaxy) and the star Sirius. MAHLI and ChemCam observed the vein targets Camp Ridge and Beacon Heights on sol 398, followed by ISO of Mt. Bastion, Glossopteris Gully, Shackleton and other targets on sols 399 and 400 (MY 32, sol 50, or 21 September 2013).

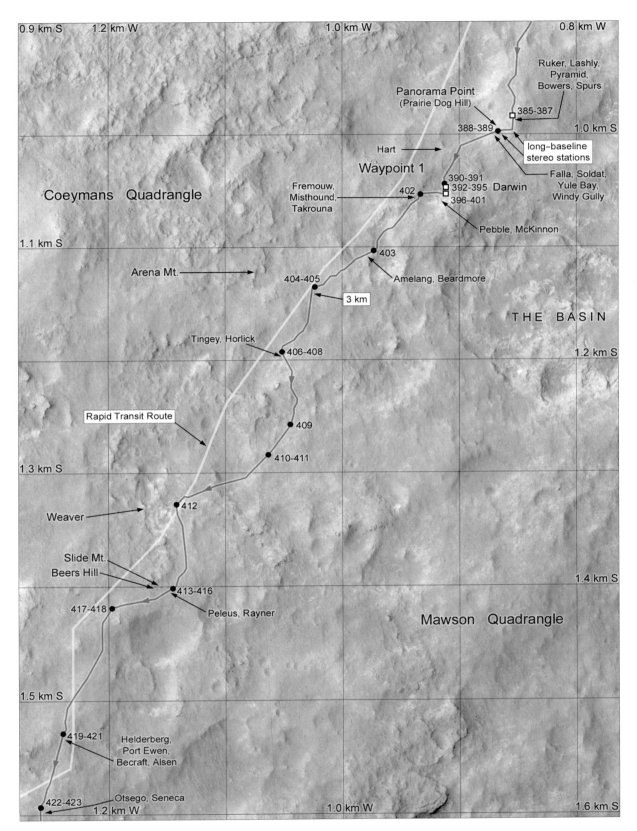

Figure 195. Curiosity route map, section 3. This map follows Figure 194 and is continued in Figure 197. Additional details are shown in Figure 196.

Work here finished with Mastcam images of targets McKinnon, McKelvey and MacKellar, before the rover returned to its Rapid Transit Route to Mt. Sharp.

The 1100 m trek to the next waypoint began on sol 402 and would not include any ISO stops along the way to save time. A drive on sol 404 was cut short as autonav failed to find a safe path past a group of rocks, its autonav-commanded backward drive capability having been temporarily suspended, and the rover stopped in an orientation that prevented the high-gain antenna from receiving instructions for sol 405 from Earth. Every set of commands included "runout" instructions for use if the next sol's commands were not received, so they were carried out on sol 405 and driving resumed on sol 406.

Work was somewhat limited for a few weeks by a political crisis in Washington, which led to a severe restriction of government operations. Activities on the Mars missions continued, as many workers at JPL and elsewhere were contractors not directly employed by NASA or other government agencies, but some team members were prevented from working. Driving continued, but a desire to move quickly limited other observations, as did cooler temperatures during the approaching southern winter season on Mars, which required more power for heating and left less for science.

New software was installed on sol 406. More Mastcam images of the sky were made to prepare for Comet ISON on sol 408, and MAHLI imaged the wheels on sol 411. On sol 412 Curiosity entered the Coeymans map quadrangle (Figures 183 and 195), where names were taken from geological features of the northeastern United States. Because analyses of the Yellowknife Bay samples had not found significant traces of organic compounds, a triple-sized sample of Cumberland was dropped into SAM on sol 414 for analysis on sol 415. A relatively fresh rocky crater here was imaged on sol 417 and a very precariously balanced rock called Slide Mountain was viewed from different directions. It might have been left in its rather unstable state as finer-grained soil was blown out from under it. CheMin was analyzing Cumberland samples over this period, and ChemCam's RMI imaged Dunn Hills, a target to the northwest, which was described as the edge of the Peace Vallis alluvial fan, but was closer to the rover than any visible parts of the fan (Figure 200).

ChemCam and Mastcam observed targets at nearly every stop along the traverse, and on sol 426 Mastcam also viewed dunes in the bottom of a nearby crater, a target called Carlisle Center. Planning for Waypoint 2, also referred to as MR-9 (the ninth location in a list provided by team member Melissa Rice), was in progress now. MR-9 was renamed Cooperstown, and only one or two sols would be spent at the outcrop. The planners wanted to spend as little time here as possible and there had been talk of not stopping at all.

Plans for the terrestrial weekend over sols 428–430 included a blank SAM analysis on sol 428 and Mastcam imaging of a large rock called Mt. Marion. Drives on sols 429 and 431 were interspersed with imaging during restricted sols, and on sol 432 Curiosity tested communications with MRO and Mars Odyssey while CheMin analyzed a sample. If the rover could do both simultaneously, some time would be saved along the route. A SAM night analysis of the atmosphere was done on sol 434 after a daytime RMI observation of West Falls, the canyon on Mt. Sharp just east of the one Curiosity was expected to climb later.

The terrain was rougher here than it had been earlier and drives were shorter, but the rover was driven daily over sols 436 to 440 to get to the second waypoint, the low rocky ledge called Cooperstown, as soon as possible. Before driving on sol 438, Mastcam looked back to image Manorkill, a layered outcrop behind the rover which would not be seen again. By sol 440 the rover was close enough to use ChemCam and Mastcam on the rocks below Cooperstown before moving up to the low scarp for detailed work. Over sols 441–443 ChemCam and Mastcam observed targets, including Pocono, and the arm was used for ISO at Pine Plains and Rensselaer before being stowed on sol 443. Also on sol 443, ChemCam analyzed a tooth-like clast called Deep Kill protruding from the scarp.

A new version of the flight software was uploaded on sol 444 and installed and tested over two sols, but on sol 446 a communication session with MRO was interrupted as Curiosity's computer rebooted. After tests on sol 448 revealed a problem in the newly uploaded software, which would take some time to correct, Curiosity returned to its old software on sol 452 and continued regular operations on sol 453. That sol included imaging Palisades, just above Cooperstown, and a 47 m drive away from Waypoint 2, followed by a longer drive on sol 454. A long scarp was visible about 1.5 km northwest of Curiosity's current location (Figure 200A), and Mastcam imaged a section of it called Schenectady on sol 455, after which the rover drove again. This drive was cut short by a wheel fault, and a planned drive on the next sol was canceled after an electrical short was detected in the rover. The rover engineers collected data to diagnose the problems on sol 458.

Back on Earth, the former JPL Director and prominent Mars researcher Bruce Murray, who had died on 29 August 2013, was commemorated by assigning his name to features at both Curiosity's and Opportunity's landing sites. In Gale crater the newly named feature was a group of hills at the entrance to the base of Mt. Sharp, which now became Murray Buttes (Figure 184).

Curiosity returned to work on sol 462, first transmitting recent data and then transferring more of the Cumberland sample into four SAM cups for later analysis so CHIMRA could be emptied. MAHLI imaged the rover wheels on sol 463. The electrical problem was now considered relatively benign, so Curiosity drove again on sol 465 after ChemCam and Mastcam observations of Greylock, a rock excavated by the impact that formed a 15 m diameter crater nearby, and other targets. Over sols 466–468 ChemCam was heated to remove any contamination, CheMin analyzed the Cumberland sample, and a SAM atmospheric analysis found elevated methane again. RMI imaged the distant target Hematite Ridge (Figure 200), a prominent feature on the approach to Mt. Sharp, on sols 467, 468 and 475, and Mastcam took multispectral images of it on sol 468. Another three-sol plan, taking the rover through the US Thanksgiving holiday, included MAHLI wheel imaging and ChemCam studies of the conglomerate Rock Stream and Skunnemunk, a dark rock, on sol 469, a drive with autonav on the next sol and an imaging sol.

The team had decided to bypass Waypoint 3, and renumbered Waypoints 4 and 5 accordingly, but several issues caused a delay near the original Waypoint 3. The wheels were deteriorating faster than expected, particularly the thin metal surface between the more robust cleats, which now showed many dents and several holes and cracks. Curiosity was crossing a rough surface with many protruding rock fragments, which damaged the wheels, and although no restriction on mobility was anticipated, the wheels would be monitored more carefully in future. MAHLI images of the wheels from beneath the rover and Pancam images from above documented the state of the wheels many times in the next few months, including sets of observations separated by small moves to compile full wheel maps.

Curiosity made another long drive on sol 472 after performing ISO on the small rock Oswego and taking more wheel images. On sol 474 another SAM measurement of elevated methane was recorded. After a drive on sol 474 the rover imaged its new location on the next sol, then viewed its wheels again with MAHLI on sol 476 before a short drive on sol 477 to a nearby site where the remaining Cumberland sample would be dumped and examined. A flat rock called Poughquag was chosen for the sample dump. ChemCam analzsed it on sol 478 and RMI viewed a ripple called Stockbridge to check for changes during the forthcoming software update. More RMI images on sols 486 and 488 showed no changes.

A revised version of the rover flight software was uploaded on sol 479 and tested until sol 484. Science activities resumed on sol 485, beginning with ISO on Poughquag, where the sample would be dumped. ChemCam analyzed the rock first, in the process blowing dust off the rock around the target point (Figure 199). On sol 487 the sample was dumped on Poughquag and the dump pile was imaged by MAHLI (Figure 199). CHIMRA was cleaned after having carried the Cumberland sample for over 200 sols. Arm operations had been restricted while the sample was stored, but now it could be used more often or for more complex operations.

ChemCam and Mastcam observed the Cumberland sample pile on sol 488 (Figure 199), and MAHLI imaged the wheels twice, separated by a 30 cm drive. Meanwhile the mission team decided to limit future drives to no more than 20 m for the time being and to continue to monitor the wheels. Testing on Earth showed that even extremely damaged wheels could still drive successfully, lessening immediate fears about the wheels on Mars. Over sols 489–491 some ChemCam and Mastcam observations were made, followed by a full wheel imaging campaign and searches for clouds and dust devils. Curiosity drove 20 m on sol 494 as the mission prepared for a break.

Sol 500 was MY 32, sol 150, or 1 January 2014. Curiosity made only RAD and REMS environmental measurements between Christmas and New Year (sols 495–501), but the pace picked up on the next sol, with ChemCam and Mastcam observations of targets including Onondaga and RMI views of a damaged area on the right middle wheel. Sol 503 began with images of nearby targets including Ashokan and Chittenango, as well as a dust devil search. A drive on sol 504 was followed by more MAHLI and Mastcam wheel images, and a SAM atmospheric analysis, which could be compared with ChemCam sky spectra taken on sol 505. SAM found elevated methane again. Also on sol 505 RMI examined large filled fractures in bluffs seen across the dune field to the east. Sol 506 began with MAHLI and APXS data from the rock Oneida, and then the rover drove 23 m, imaging its new location and its wheels during and after the drive.

RMI was used again on sol 507, taking advantage of its high resolution to look for changes in the distant dune field to the east. Early the next morning Mastcam images were taken of the stowed arm to see if frost formed on it, and later MAHLI viewed the wheels again before a short drive. An outcrop called Balmville was imaged by RMI and Mastcam on sol 509, and the team looked for a good location to take a full sequence of wheel images. The next ISO targets were the rocks Lowerre and Larrabee on sol 510, and MAHLI again viewed the wheels, and on the next sol Curiosity drove 25 m as the drivers planned a "surge" of stepped-up driving to get out of the rocky area and into easier terrain.

This location was good for wheel observations, so a short drive intended for sol 512 was not needed, and Curiosity could study several targets, performing ISO on Clinton and Kodak, and ChemCam observations of Harrison, a rock containing large feldspar crystals. The next sol was devoted largely to taking a full set of wheel images, and DAN measured subsurface hydrogen as the Navcams looked for clouds. Sol 514 was spent examining the rocks Harrison and Sparkle with ChemCam and Mastcam, but a drive was precluded by a command fault. Another set of MAHLI wheel images were taken on sol 515 before Curiosity drove about 30 m southwards. The rover was now moving into the Kimberley Quadrangle (Figures 183 and 197), with names taken from a geologically ancient region of Australia. A rock called Oscar was targeted on sol 516 for ChemCam, Mastcam and ISO, and after a sol of additional imaging two more

drives occurred on sols 518 and 519, with more wheel imaging each day.

The drive on sol 519 brought Curiosity close to an area of complex topography with several long winding scarps cutting the surface of Aeolis Palus. The rover had to descend from the upper surface to the lower across one of the scarps, using one of several possible descent paths. At about the same time, mission planners decided to deviate from their original Rapid Transit Route (Figure 184) to Mt. Sharp to follow smoother surfaces which would be gentler on the wheels. The planned route extended south from here, but it was rough in places and crossed a steep scarp at Mt. Disaster (Figure 197). A gap in the scarp, initially called The Chute and later Dingo Gap, was seen to the west on sol 519, but HiRISE images showed it contained a large drift. On sol 520 Mastcam and ChemCam investigated two rocks called Doran and Togo, and, after cloud and dust devil searches and more wheel images, the rover drove south again.

Curiosity took five sets of MAHLI wheel images on sol 521, separated by short drives, and then drove 11 m south and imaged the surroundings. No new holes were seen in the wheels. Mastcam and ChemCam observed a conglomerate, Dougalls, and other targets, including Airfield and Carribuddy, on sols 522 and 523, and a rock called King was examined by MAHLI and APXS. The planned route continued south, but now the rover planners decided to turn west towards the gap in the scarp (Figures 196 and 197). Wheel imaging on sol 524 was followed by a drive towards Dingo Gap, as it was now named, with a cloud search and a night SAM atmospheric analysis on sol 525 and ISO on a rock called Reedy on sol 526. The SAM methane reading here was the highest of the primary mission, though still a small trace at nine parts per billion of the atmosphere (Webster *et al.*, 2013). The rover also took MAHLI images of the damaged REMS sensor on sol 526, and then drove 15 m towards Dingo Gap for closer imaging to see if the route was passable. It moved closer again on sols 527 and 528, driving up onto the dune to test its firmness and to view the terrain on the other side, and then backing up to show its tracks in the dune.

Two concerns here were that the dune might be dusty and soft like Opportunity's Purgatory drift, and that the rugged terrain west of Dingo Gap might block direct rover communications with Earth, but the route through

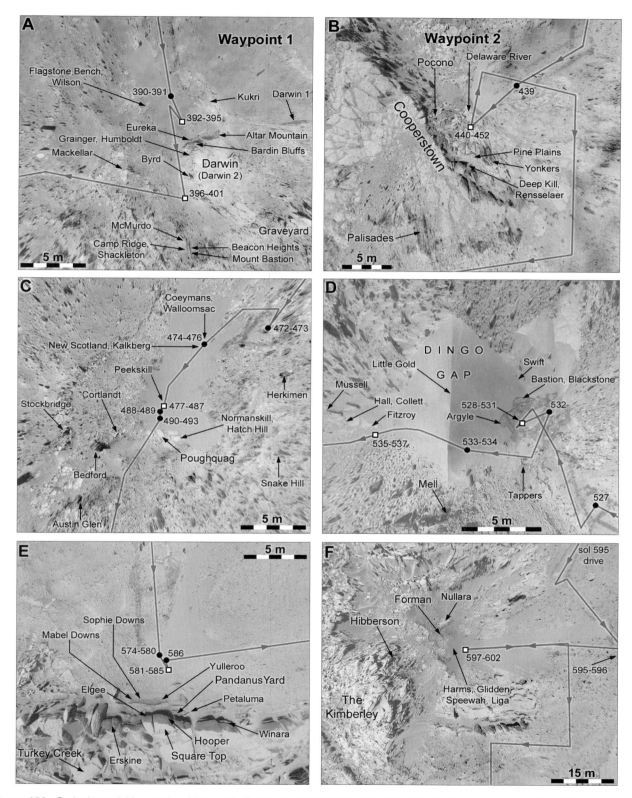

Figure 196. Curiosity activities on the drive to Mt. Sharp. **A:** Waypoint 1, sols 390–401. **B:** Waypoint 2, sols 439–452. **C:** Original Waypoint 3 area, sols 472–493. **D:** Dingo Gap, sols 527–537. **E:** North Kimberley, sols 574–586. **F:** East Kimberley, sols 595–602.

the valley looked smoother than the surroundings. If the dune could not be crossed, the rover might cross the mesa above the gap. Mastcam and ChemCam examined targets in the tracks on the dune on sol 529, and a curious round hollow rock called Tappers, nicknamed "fire ring." Mastcam imaged Earth in the evening sky and made panoramas of the view to the west on sol 530, followed by ISO at Dampier and Argyle, targets in the tracks, and the undisturbed area Barker, on sol 531.

Curiosity rolled up onto the dune at Dingo Gap on sol 533, allowing its left front wheel to toe-dip just over the crest. The rover drove down the other side on sol 535 (Figure 196D), bringing it into Moonlight Valley, where prominent veins cut outcrops on the rocky valley floor. Vein targets called Hall, Collett and Mussell and the bedrock at Fitzroy were analyzed with APXS and Chem-Cam over sols 537 and 538, and ChemCam tried to detect frost at Cascade Bay by comparing sol 537 day-time spectra with measurements at dawn on sol 538. ChemCam also examined Neville, a block of the outcrop forming the top of the mesa that had broken loose and rolled downhill.

The next drive brought Curiosity out into Moonlight Valley and the landscape of rocky hills and valleys beyond. Mastcam imaged the rock layers in the valley walls on sol 540, looked for outcrops on Mt. Nulasy on sol 542, and sought evidence of frost on the surface at sunrise on sol 543. Mastcam and ChemCam examined targets Bindi and Nita on sol 544, and also looked for frost again at Sandy Dam on sol 544 and early on sol 545 before driving resumed. MAHLI wheel images were taken on sols 544 and 546 to document any changes, and on sol 544 MAHLI also imaged a location on the surface from several different angles to examine its photometric properties. Having left the Rapid Transit Route, Curios-ity was now on the Pink Path. Analysis of wheel dynam-ics suggested that the front wheels would be more damaged than the rest by sharp rocks, as the images confirmed, but backward driving would spread the damage more evenly.

A path had now been selected that passed through a rough area at Kylie to reach Waypoint 3, also referred to as KMS-9, but now named Kimberley after its map quadrangle. Possible sites of interest had been chosen by team members and identified by their initials. KMS-9 was the ninth site identified by Katie M. Stack, then a graduate student at the California Institute of Technology working with Curiosity project scientist John Grotzinger. Waypoint 4 was also referred to as FC-9, the ninth site identified by JPL cartographer Fred Calef. It was just a coincidence that these waypoints and MR-9 (Waypoint 2, Darwin) all happened to be the ninth sites in three independent lists.

The surface became smoother as the rover left the valley, making longer drives more feasible, so on sols 547 and 548 two long backward drives carried the rover through Violet Valley, testing the backward driving pro-cedures, which would be needed more in future. The big RTG blocked part of the rearward view for hazard avoid-ance, but that issue was worked around. Curiosity stopped mid-drive on sol 548 to image the Shaler-like layered scarp at Junda and another outcrop called Scru-tons. Junda resembled the lower "striated" unit (named from its appearance in HiRISE images, but better described as "layered" when seen at the surface) seen at Kimberley and Kylie.

More MAHLI wheel images were taken on sol 549, and Curiosity moved a few meters south to examine a conglomerate outcrop named Bungle Bungle. This extension of the Junda outcrop was analyzed with MAHLI and APXS at a target called JumJum early on sol 550, after which the rover took a Mastcam panorama and then moved around a dusty drift to look out over the next valley. RMI viewed the dark dunes from here, and on sol 551 Mastcam viewed the little crater Goat Paddock and Magotty, a conglomerate. On sol 552 Curiosity was driven down into the valley, flanked by Wilson Cliffs on its north side, and imaged the cliffs on the next sol. The targets Angelo and Barbwire, a vein, were also viewed on sol 553, and now the route planners decided not to drive over Kylie, which was too rugged to cross.

In order to understand the broader stratigraphic con-text of Kimberley, where the striated unit was topped by hills capped by a harder rock, this area was now exam-ined carefully. After MAHLI wheel imaging and Chem-Cam observations of Corkwood on sol 554, the striated unit and its contact with Wilson Cliffs was imaged at high resolution over two sols. Another target was War-ton, part of the cap unit at the top of the cliffs. Here SAM undertook energy-intensive analyses of some Cumber-land material remaining in the instrument, which precluded other work until sol 558. On that sol Mastcam and ChemCam viewed a rock called Mt. Amy containing veins, and ChemCam, MAHLI and APXS examined soil at Johnny Cake.

Curiosity's 60 m drive on sol 559 skirted Kylie and approached the next big topographic step, where a small crater on a cliff to the south formed a sloping ramp (Figure 201). Images showed that the path to that ramp was very rough, so, after ISO of a rock called Secure on sol 560, the rover turned west to another possible ramp, reaching it on sol 561. CheMin was prepared for a new analysis on sol 561, and MAHLI imaged the wheels on the next sol, as Mastcam and RMI viewed the dunes at the base of Mt. Sharp. The rover imaged its wheels on sol 563 and then drove carefully to the ramp, taking care to avoid hazards to the wheels. After performing ISO on Monkey Yard, a pitted rock thought to be from the plateau cap rock, Curiosity was driven down the slope on sol 564. It stopped on a surface too tilted for the intended MAHLI wheel imaging, so it drove again on sol 565.

MRO suffered an unexpected computer problem and was placed in safe mode on Curiosity's sol 566, as the rover stood at another gap between rocky areas unable to move because of the loss of its data relay. A planned drive on sol 567 was replaced with remote sensing of the nearby rocks Pillara, Ranford and Yampie, but after the wheel images were taken two long drives on sols 568 and 569 brought the rover within sight of its next waypoint, Kimberley, also called "The Kimberley" after its Australian counterpart. Curiosity had strayed briefly into the next quadrangle here (Figure 183), but returned to the Kimberley Quadrangle on sol 569. That drive included a stop along the way to image Jurgurra, an outcrop of the striated unit at the base of the mesa called Emu Point. The rover recharged its batteries on the next sol.

Kimberley was a triangular outcrop with several distinct rock layers, topped by three prominent rocky hills called Mt. Christine, Mt. Joseph and Mt. Remarkable. Work at Yellowknife Bay had suggested that cosmic rays destroyed organic molecules to depths of 1 m or more over tens of millions of years, but that they might be detected closer to the surface in an area where rock layers were being eroded. Wind-blown sand wore away the edges of exposed rocks, causing scarps or ledges to retreat, so areas close to retreating scarps had only

recently been exposed to cosmic rays (Farley *et al.*, 2013). The Cumberland sample contained more organic material than the John Klein sample, as it had been exposed for a shorter time. Kimberley offered a chance to test this hypothesis.

Rover planners found no interesting features for ISO at the sol 569 location, so APXS analyzed the now-empty observation tray on sol 571 to help interpret data when the next sample was placed on it. The surrounding cliffs were imaged as well, and on sol 573 a SAM methane measurement was performed. This showed only 5 percent of the amount recorded on sol 526 despite being run in a special mode in which carbon dioxide was extracted from the air sample, enriching the concentration of any remaining gases significantly. Driving and wheel imaging continued until sol 574, with ChemCam analyses of Lennard and King Leopold early on sol 574, and later that sol Curiosity reached the northern edge of Kimberley's striated unit, the oldest material exposed there. Instruments were calibrated on sol 575 as planners selected a target for ISO.

A thick coarse-grained layer formed a prominent step in the outcrop, and a target on it named Square Top was examined with ChemCam and Mastcam on sol 576. CHIMRA was supposed to be prepared for future samples by cleaning out any remnants of Cumberland material later on sol 576, but an arm fault stopped it as images were being taken. CHIMRA was eventually cleaned on sol 578, and RMI made images of its sieve to ensure that the strong vibration during cleaning had not damaged it. The mast with its cameras and Chem-Cam had not been stowed as usual because of the sol 576 fault, and on sol 579 the rising Sun passed closer to ChemCam than mission rules dictated, though not close enough to damage it. Later that sol the instrument analyzed Square Top, Yulleroo and Petaluma.

The Mastcam made a large panorama of the dramatic scenery here on sol 580, and on sol 581 the rover used ChemCam on targets Egan, Elgee, Erskine and Elvire before moving 3 m closer to the target area. Curiosity prepared SAM for future use on sol 582 and RMI viewed the dark dunes to the south. Square Top was a target for ISO on sol 583 and ChemCam analyzed Hooper while a Navcam cloud search was made over Mt. Sharp.

MAHLI and APXS made several measurements up the vertical face of the rock layer at Square Top on sol

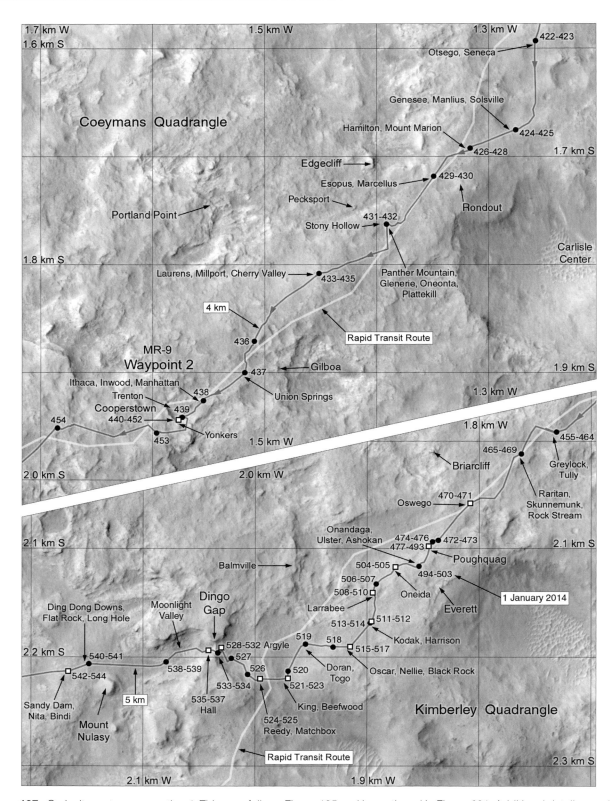

Figure 197. Curiosity route map, section 4. This map follows Figure 195 and is continued in Figure 201. Additional details are shown in Figure 196.

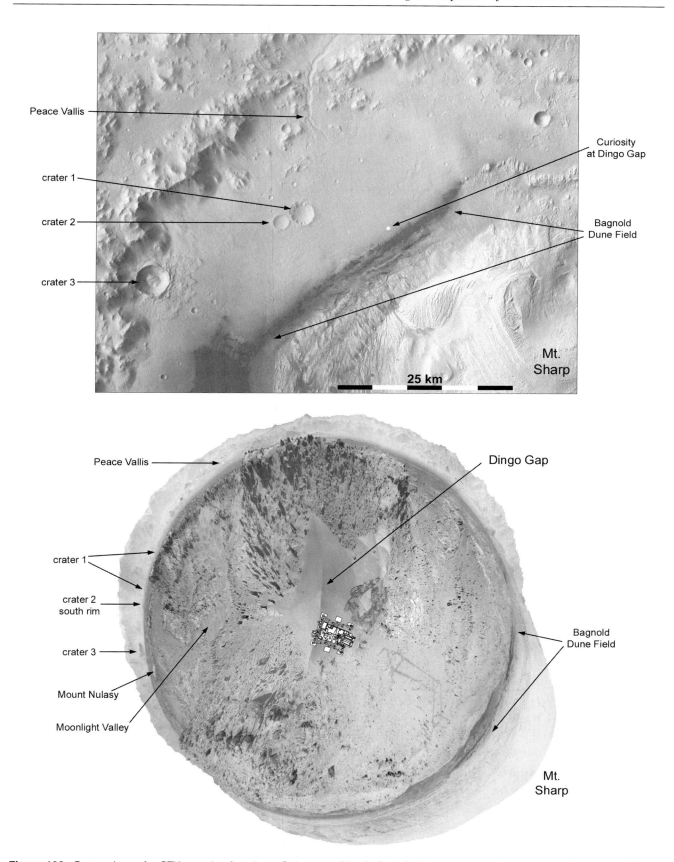

Figure 198. Comparison of a CTX mosaic of northern Gale crater (Aeolis Palus) with a reprojected panorama taken on sol 533 at Dingo Gap.

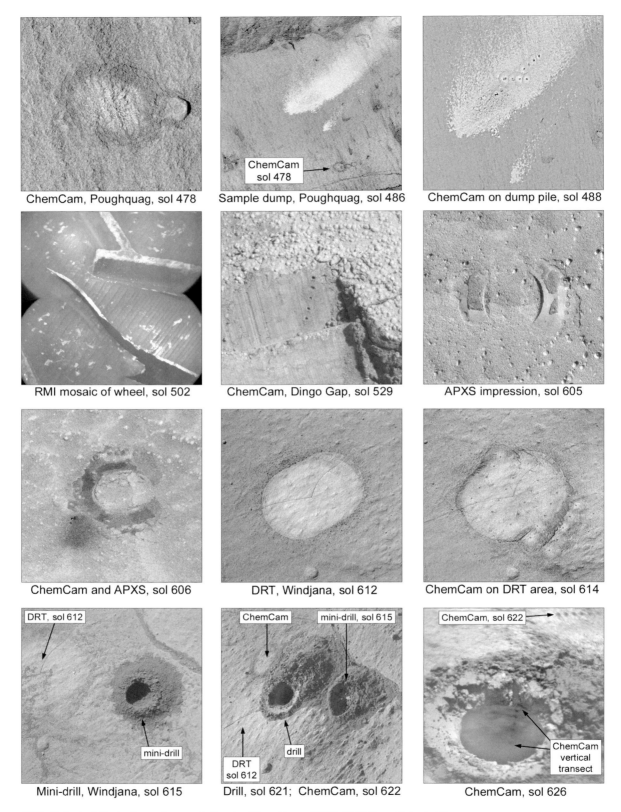

ChemCam, Poughquag, sol 478   Sample dump, Poughquag, sol 486   ChemCam on dump pile, sol 488

ChemCam
sol 478

RMI mosaic of wheel, sol 502   ChemCam, Dingo Gap, sol 529   APXS impression, sol 605

ChemCam and APXS, sol 606   DRT, Windjana, sol 612   ChemCam on DRT area, sol 614

DRT, sol 612

mini-drill

ChemCam

mini-drill, sol 615

drill

DRT
sol 612

ChemCam, sol 622

ChemCam
vertical
transect

Mini-drill, Windjana, sol 615   Drill, sol 621; ChemCam, sol 622   ChemCam, sol 626

Figure 199. Selected Curiosity surface activities by ChemCam, APXS, DRT (brush) and drill.

583, as ChemCam and Mastcam observed a target called Hooper. MAHLI also peered into the shaded area under an overhanging ledge on sol 584, examined small sand slump features named Mabel Downs, Sophie Downs and Sally Downs (named from the geology of the Kimberley region in Australia), and viewed Pandanus Yard, part of the vertical edge of Square Top. ChemCam analyzed Mabel Downs and other targets here, and APXS was placed on Square Top overnight on sol 585. Then Curiosity backed away from the ledge for imaging on sol 586 and CheMin analyzed an empty cell, and on sol 587 a Mastcam mosaic was made to look for changes since the rover arrived at Square Top. Later that sol MAHLI imaged the wheels and the rover drove away to find a drilling location.

Curiosity now drove east and south around the outcrop, pausing to image rocks and a small trough called Brookings Gorge on sol 588 and for ChemCam to examine the soil target Chirup early on sol 589. The Kimberley outcrops were imaged at many locations before reaching the main science target at the foot of Mt. Remarkable on sol 606. That region first became visible on sol 589, and on sol 590 RMI viewed Cone Hill, part of the base of Mt. Remarkable. Another trough called Tickalara was a MAHLI target on sol 591, and other features here were ChemCam targets on sol 592. The troughs here were places where soil had fallen into cracks between buried rock slabs. Standard observations also continued, including a dust devil search on sol 590, a full 360° Mastcam panorama on sol 592 and a cloud search on sol 593.

ChemCam and Mastcam characterized rocks on sol 594, including the layered rock McHale and a conglomerate called McSherrys. On sol 596 the command uplink failed and pre-planned runout observations were made. Then the rover was driven up to the edge of the outcrop on sol 597 for detailed observations over a weekend, but the whole three-day plan was lost and was partially recovered over the next two sols. Mars was at opposition around this time, so from Curiosity's point of view Earth was at inferior conjunction, very close to the Sun in its sky, and communications were sometimes compromised. Sol 600 was MY 32, sol 250, or 14 April 2014. ISO were made on sol 601 at the target Liga, MAHLI was used on Speewah and ChemCam was used on other targets as SAM was prepared for a new sample.

A sandy ripple called Forman was analyzed by ChemCam on sol 602.

The science team met at Caltech on 15–17 April 2014 to plan the activities at Kimberley and their future path to Murray Buttes and Mt. Sharp. One drilling option, site A, was on the rock west of this location (Figure 202A), but it was dropped in favor of site D, later named Windjana. Site A was the backup, less favored because it would require driving over rough rocky outcrops. At site D four possible drill site options were identified (Figure 202B), and on sol 603 Curiosity drove to the closest of them to image it. On sol 605 ChemCam was used on a sandy target called Babrongan, Mastcam viewed the Bickleys outcrop and APXS was used overnight on the soil target Lagrange. After ChemCam analyzed Lagrange on sol 606, another drive brought the rover to the preferred site for a detailed assessment of the drill area. Late that night the rover took Mastcam images of Phobos, Deimos and the asteroids 1 Ceres and 4 Vesta.

The sandstone outcrop at site D, part of the middle layer of strata exposed at The Kimberley, was chosen for drilling to examine the nature of the cement holding the grains together. There were two favored locations, labeled DO1 and DO2 in Figure 202B. ChemCam examined targets at both locations on sol 608 to inform the choice between them. ChemCam examined targets Jarrad and Cow Bore on sol 609, Mastcam took more wheel images, and then the rover approached DO1, now called Windjana. The rover slipped unexpectedly at the end of the drive, but seemed to be stable enough to drill. The wheel surveys showed that the strategy of driving on smoother surfaces worked well, but tests on Earth had shown that even very heavily damaged wheels could still drive well, so the issue was less of a concern now. Images of the sky behind Mt. Remarkable at sunset were taken on sol 610 and ChemCam was calibrated by viewing the sky. The outcrop here was not continuous, as it had been in Yellowknife Bay, and contacts between layers were often masked by dust drifts, making the site less suitable for the proposed analysis at recently exposed rock surfaces.

Activities at Windjana began on sol 611 as the stability of the rover was checked before moving the arm. ChemCam analyzed Stephen, a possible exposed vein, and the targets Kevin's Dam, Thangoo and Blina, and

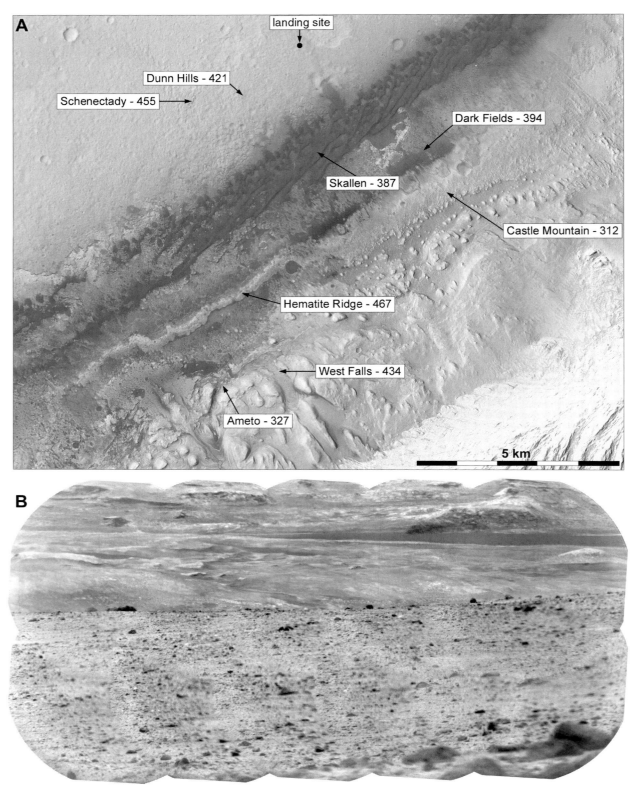

Figure 200. Curiosity long-range observations. **A:** Features on Mt. Sharp and in Aeolis Planum named as targets of long-range Mastcam and ChemCam RMI observations. The number identifies the sol of observation. The base is a CTX mosaic. **B:** ChemCam RMI mosaic of Ameto, sol 327.

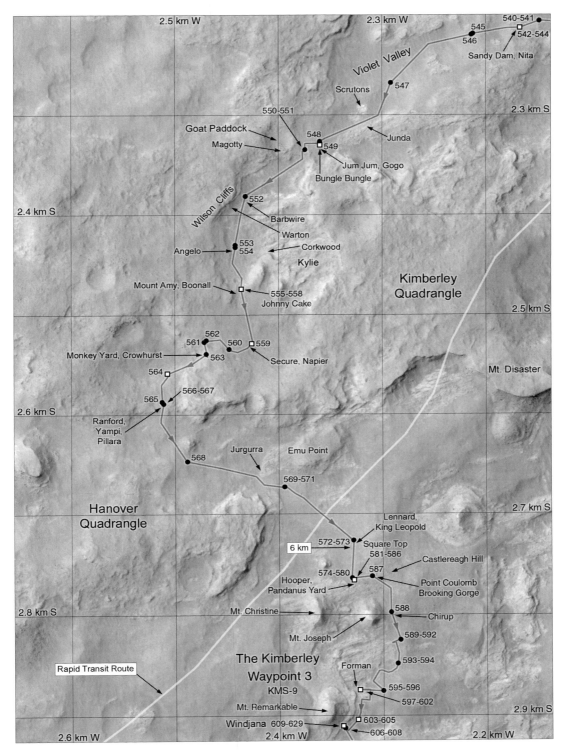

Figure 201. Curiosity route map, section 5. This map follows Figure 197 and is continued in Figure 203. Additional details are shown in Figures 196 and 202.

APXS was cleaned. The rock target was brushed on sol 612, the first use of the DRT since Cumberland, with ISO before and after brushing, and on the next sol MAHLI took another self-portrait of the rover, and the arm was placed on the drill target for a pre-load test, as done previously at John Klein and Cumberland. That evening Phobos was observed setting to measure how much its light was obscured by the atmosphere.

On sol 614 ChemCam made a raster of LIBS observations across the DRT area to measure composition, and on sol 615 a mini-drill test, a shallow hole, was drilled near the DRT to check hardness and stability. The drilling vibration caused slumping in sand slopes around the Windjana area, and the drill tailings were much darker than those in Yellowknife Bay (Figure 199). After the mini-drill test, a MAHLI observation was stopped by a fault with the arm deployed, and that blocked a planned ChemCam analysis of the mini-drill tailings and other work on sol 616. The rover was recovered on sol 617 and MAHLI was tested to be sure it was safe to use. On the next sol ChemCam was used on Beagle, a big rock that had rolled down the side of Mt. Remarkable from its summit area, and Mastcam imaged the tailings. Chem-Cam was used on the tailings on sol 619, and that night Phobos was imaged as it rose.

The drill collected its sample on sol 621 in the brushed area and passed it to CHIMRA on sol 623, from which it was delivered to CheMin later that sol and to SAM on the next sol. CheMin analyzed its sample on sol 625. The tailings were examined by APXS and Chem-Cam on sol 622. ChemCam was used on the rock adjacent to the tailings on sol 622 and made a vertical transect up the side of the drill hole on sol 626, followed by MAHLI night and UV imaging late on sols 628 and 629. Stephen (Figure 202B) was an ISO target on sols 627 and 629, including a raster of four APXS positions on the latter sol, and a ripple called Broome was imaged for change detection studies on sols 624 and 628. Meanwhile the descent imager, MARDI, was given updated software on sol 629 to enable it to measure motion between frames, giving it the same kind of visual odometry capability as already performed by the other cameras. Work here wrapped up with imaging and ChemCam studies of several targets, and MAHLI completed its pre-drilling self-portrait by re-imaging the drill area to show the hole and tailings.

On sol 630 Curiosity was driven south from Windjana to an outcrop which displayed a very prominent change of orientation between two sets of layered rocks, resembling a classic unconformity on Earth, but perhaps related to two episodes of delta or fan-like deposition. This was part of a long survey of outcrops throughout the drive, which eventually established that the plains were built by a series of events forming overlapping alluvial fans, lake deposits and deltas. After Mastcam imaged this area on sol 631, the rover drove south and west across smooth areas between outcrops and depressions, imaging a layered rock called Moogana on sol 632 and using APXS on a dust-free rock called Wift on the next sol.

On sol 635 RMI made one of its series of dune mosaics, monitoring one spot when it was visible to look for changes. Mastcam viewed Wesley Yard and that night Phobos was imaged to measure sky opacity. The next sol included ChemCam analysis of Lamboo and a drive that brought Curiosity out of the Kimberley quadrangle and into Hanover (Figure 183). The drive paused over sols 637–640 for RMI and Mastcam images of Lebanon and Littleton, two large pitted rocks resembling iron meteorites, and Mastcam and MAHLI images of the wheels. No significant new wheel damage had been observed since the rover had moved onto smoother terrain. SAM analyzed the atmosphere overnight on sol 637, and RMI imaged the dune field again on sol 639. Another RMI target was the 5 km diameter crater (crater 3 in Figure 198) at the foot of the western wall of Gale, 30 km west of the rover. This was imaged on sol 641, just before Curiosity drove on from Lebanon.

The rover drove west over the next few sols. The team was eager to leave the landing ellipse before the end of the primary mission, if possible, but paused for Chem-Cam observations of Exeter, a rock with a vein, on sol 646, and to image a double crater called Occam Pond. On sol 647 a rough-surfaced rock called Hitchcock was imaged, and APXS was used on a conglomerate called Furnace Flats early on sol 649. The path here had returned to the original Rapid Transit Route (Figure 184) after the detour to reach Kimberley (Figure 203). Soon after sunrise on sol 650 Curiosity observed a transit of Mercury across the Sun, the first planetary transit ever observed from Mars. Two prominent sunspots were also visible, each about one pixel across in the Mastcam

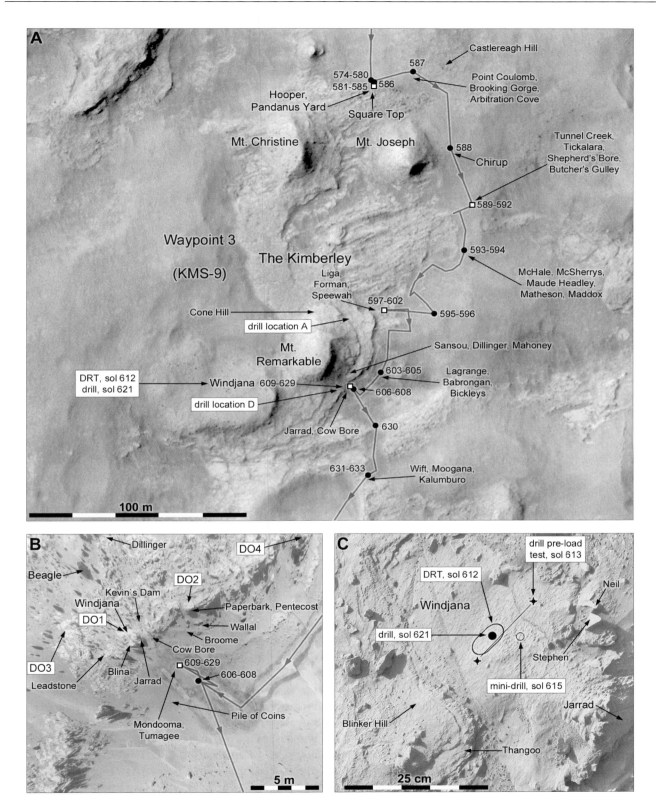

Figure 202. Curiosity activities on the drive to Mt. Sharp. **A:** Kimberley, sols 574–633. **B:** Waypoint 3, sols 606–629. DO1, DO2, etc. are drill option 1, drill option 2, etc. locations. **C:** Windjana, sols 609–629.

Table 54. *Curiosity Activities, Kimberley to Mt. Sharp, Sols 576–669*

Sol	Activities
576–582	Image north Kimberley, clean CHIMRA (576), move to outcrop (581), prepare SAM (582)
583–585	ISO Square Top (583–585), ChemCam Hooper (583), MAHLI Pandanus Yard, other targets (584–585)
586–588	Move back and CheMin empty cell (586), drive, imaging (587–588), MAHLI wheels (587)
589–591	ChemCam Chirup soil, drive (589), prepare SAM and imaging (590), calibrate ISO, ISO Tickalara (591)
592–600	ChemCam outcrop (592, 594), drive (593–597), imaging, runout science (594, 598–600)
601–603	ISO Liga, Speewa, prepare SAM (601), MAHLI ChemCam window, SAM blank (602), drive (603)
604–606	Image possible drill sites (604), ISO Lagrange soil (605), drive, Phobos, Deimos, asteroid images (606)
607–610	Drive, MAHLI wheels, ChemCam Jarrad, Cow Bore (609), imaging, calibrate ChemCam (610)
611–612	Check stability for drilling, prepare APXS (611), DRT and ISO Windjana (612), imaging
613–614	MAHLI self-portrait, drill pre-load test, Phobos imaging (613), ChemCam DRT area (614)
615–621	Mini-drill, ISO Windjana (615), runout (616), MAHLI tests (617–618), imaging, drill Windjana (621)
622–625	ISO drill tailings, prepare SAM (622), CheMin Windjana sample (623, 625), SAM Windjana (624)
626–629	ChemCam hole (626), ISO Stephen (627, 629), ISO hole day (627), night (628–629), move back (629)
630–636	Drive south and west (630–631, 634–636), CheMin sample (630–632), ISO Wift (633)
637–639	SAM atmosphere (637–638), RMI iron meteorites Lebanon and Littleton (637, 640) and dunes (639)
640–644	Sample to CheMin (640), CheMin sample (641), MAHLI wheels (640, 641), drive west (641, 643–644)
645–650	CheMin sample (645), drive (646, 649), ISO Furnace Flats (649), Mercury transit, clean SAM (650)
651–653	Drive with MARDI movie (651), prepare SAM (652), ChemCam Winnipesaukee, SAM sample (653)
654–660	Drive (655–658), CheMin sample (656), imaging, calibrate CheMin, MAHLI REMS and wheels (660)
661–666	Drive (661–665), Phobos, Deimos images (662), SAM test (664), MAHLI CheMin inlet (666)
667–669	MAHLI wheels (667), drive southwest (668–669), stop at Bloods Brook, end primary mission (669)

images. Mercury was much smaller, but darkened its pixel enough to be faintly visible.

Transmission of data was slowed here as the MCS instrument on MRO recovered from a fault. MARDI made a movie sequence of the ground passing beneath it on sol 651 to test its new motion detection capabilities, though the images were overexposed, and ChemCam examined a dark hollow structure called Winnipesaukee in a nearby outcrop on sol 653. Several features like this had also been seen at Yellowknife Bay. Meanwhile CheMin performed another analysis of the Kimberley sample overnight on sol 651, and SAM received a sample on sol 653 for analysis overnight. Several targets, including Meetinghouse and Albee, were examined by ChemCam on sol 654, and Podunk on the next morning, before an 86 m drive on sol 655 brought Curiosity past the 7 km mark on its traverse. The Goldstone DSN station suffered a fault which limited communications with Curiosity, but NASA's Mercury orbiter MESSENGER gave up some of its allotted time to help its Martian cousin.

MAHLI imaged the wheels through a full turn on sol 660, continuing to monitor their condition. Navcam dust devil searches continued, still unsuccessfully, always looking north across the plains, and Curiosity viewed Phobos and Deimos again on sol 662. Apart from imaging and ChemCam targeting of targets like Kittery on sol 656 and Monroe on sol 658, relatively little science was attempted in order to devote the maximum effort to driving. The length of the drive to Mt. Sharp was having an effect on planning for the Mars 2020 rover, which was intended to collect and cache its samples within one Martian year of landing. The landing site would have to provide worthwhile samples close to the landing point.

The rugged terrain required careful route planning, and around sol 660 two possible routes were considered before the southern one was chosen (Figure 203). This smoother route, known as the Pink Path for its color on mission planning maps, is shown in full in Figure 184. In homage to Robert Frost's famous description of two paths diverging before him (Frost, 1916), a narrow valley here was named Robert Frost Pass. Curiosity drove through it on sol 664, imaged the stratigraphy of its sides, and entered Moosilauke Basin, where it skirted a bright

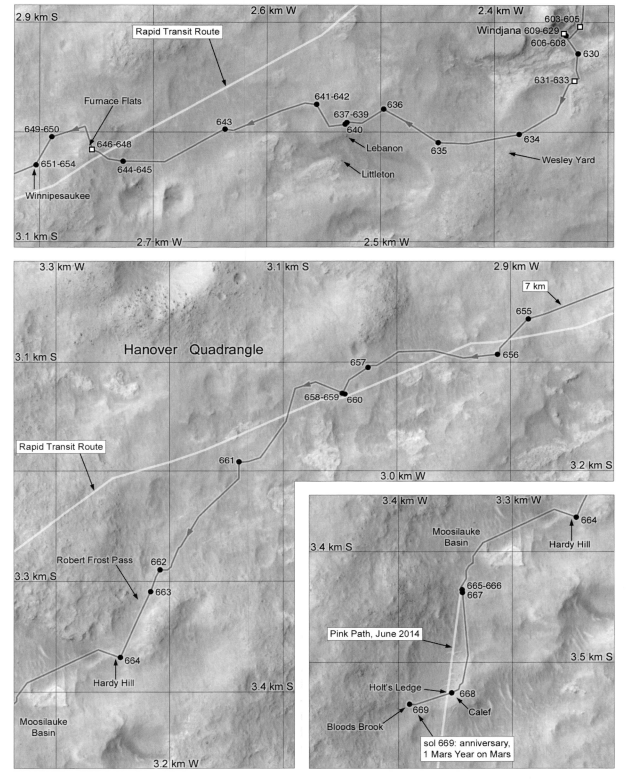

Figure 203. Curiosity route map, section 6. This map follows Figure 201 and concludes route mapping in this atlas. Additional details are shown in Figure 202.

outcrop area and turned south. A drive of 142.5 m on sol 665 was the longest of Curiosity's primary mission.

On sol 667 MAHLI imaged the wheels again, and on sol 669 ChemCam investigated a drift called Holt's Ledge and Calef, a rock named after a geological unit in New Hampshire, but which also shared its name with Fred Calef, one of the mission's cartographers. After those observations, the rover made a short drive on sol 669 (MY 32, sol 320, or 25 June 2014) to a nearby ridge, from which the path ahead could be surveyed. This concluded Curiosity's primary mission and the first Mars year of operations in Gale crater. The rover had been driven 7.9 km and was still about 3 km from Murray Buttes and the effective base of Mt. Sharp.

2. Phobos and Deimos

Spirit and Opportunity

The Mars Exploration Rovers, Spirit and Opportunity, observed Phobos and Deimos frequently during their missions. Images were taken showing the little moons crossing the disk of the Sun, causing partial eclipses at the rover sites. Others were taken showing the satellites in the sky, rising or setting and entering eclipse. These images had scientific value as well as public relations interest, first, by improving knowledge of the satellite orbits and, second, in the rising, setting and eclipse observations, by measuring atmospheric opacity. As light from the moons passed through the atmosphere during rising or setting, it would be dimmed by layers of dust or haze, and during eclipses the moons probed the edge of the shadow of Mars, again probing variations in atmospheric opacity. Figure 204A shows some representative images of Phobos and Deimos from the MER missions. Phobos is resolved well enough to reveal its nonspherical shape and the large crater Stickney in a super-resolution composite of images taken just before entering eclipse on Spirit sol 675. Both satellites were imaged crossing the disk of the Sun on several occasions. Here Phobos is seen transiting on Opportunity sol 47 and Deimos on Spirit sol 420. Although not noted in Stooke (2012), the Viking Landers had also imaged Phobos and Deimos to measure atmospheric opacity at night, though the disks were not resolved (Pollack *et al.*, 1977).

Mars Reconnaissance Orbiter

Mars Reconnaissance Orbiter (MRO) did not approach either satellite closely, but the exceptional resolution of HiRISE allowed for spectacular imaging of the two distant moons, though necessarily covering only the Mars-facing regions. CRISM observed Phobos and Deimos to collect calibration and compositional data.

HiRISE obtained two-color images of Phobos at a resolution of 6 m/pixel with stereo coverage for topographic mapping (Figures 204C and 204D). The two Phobos images are PSP_007769_9010 and PSP_007769_9015, taken 10 min apart on 23 March 2008. The enlarged section of PSP_007769_9015 in Figure 204D shows large rocks and grooves centered at 20° S, 340° E. Deimos was observed in color at 20 m/pixel in images ESP_012065_9000 and ESP_012068_9000 (Figure 204E), taken 5.6 hours apart on 21 February 2009. These HiRISE images are better than any other available coverage in some areas of each satellite. Figure 205A shows the surface coverage of HiRISE images on both satellites, including an area seen only in light reflected off Mars.

The Phobos CRISM data consisted of three observations taken on 23 October 2007, revealing Stickney, Hall and several smaller craters. The resolution was 350 m/pixel. Deimos, viewed on 7 June 2007, at 1200 m/pixel, was seen well enough to show two bright areas along its major ridges around longitude 300° E, but the phase angle and low resolution made craters invisible. Figure 204B shows CRISM observation FRT00002992 of Phobos and FRT00002983 of Deimos. Compositional data tended to support the captured asteroid hypothesis for the origin of the satellites, rather than a Mars ejecta origin. Phobos was partly covered by slightly red material identical to Deimos and possibly derived from it as ejecta (Murchie *et al.*, 2008), and Stickney appeared to have excavated a less red material from the interior of Phobos. Fraeman *et al.* (2012) reported similar conclusions using data from both CRISM and the OMEGA spectrometer on Mars Express (Figure 205 in Stooke, 2012; Figure 209C in this atlas).

2007: Rosetta Mars flyby

The European Space Agency's (ESA's) Rosetta spacecraft flew past Mars (Figure 206) on 25 February 2007, (MY 28, sol 389). On 24 and 25 February it took 70 images of Phobos (Pajola *et al.*, 2012b), comprising

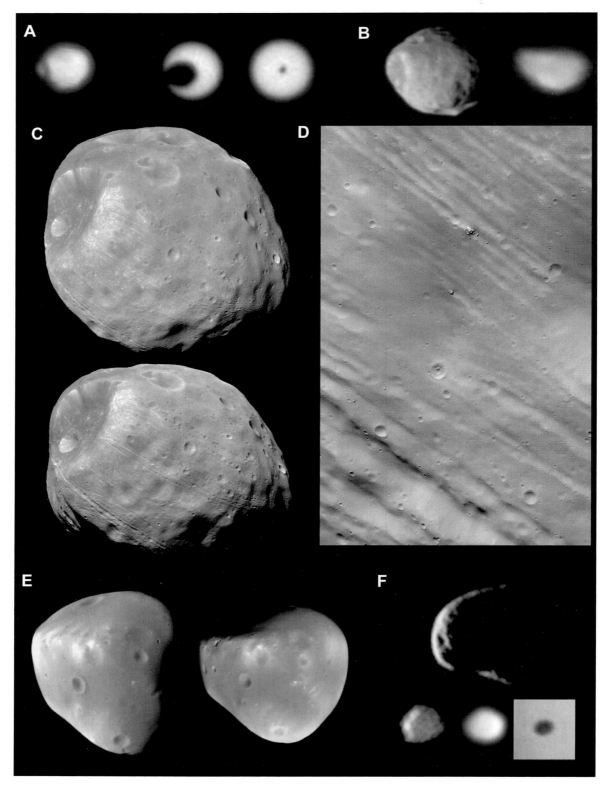

Figure 204. Phobos and Deimos images from MER, MRO and Rosetta. **A:** MER images of Phobos and both satellites transiting the Sun. **B:** MRO CRISM images of Phobos and Deimos. **C:** HiRISE Phobos images. **D:** Detail of HiRISE Phobos image PSP_007769_9015 centered at 20° S, 340° E. **E:** HiRISE Deimos images. **F:** Rosetta images of Phobos, showing a crescent view with a large crater at 15° S, 20° E (top), a crescent illuminated by light reflected off Mars, a full disk view of the anti-Mars region and a view of Phobos over Arabia Terra (bottom left to right).

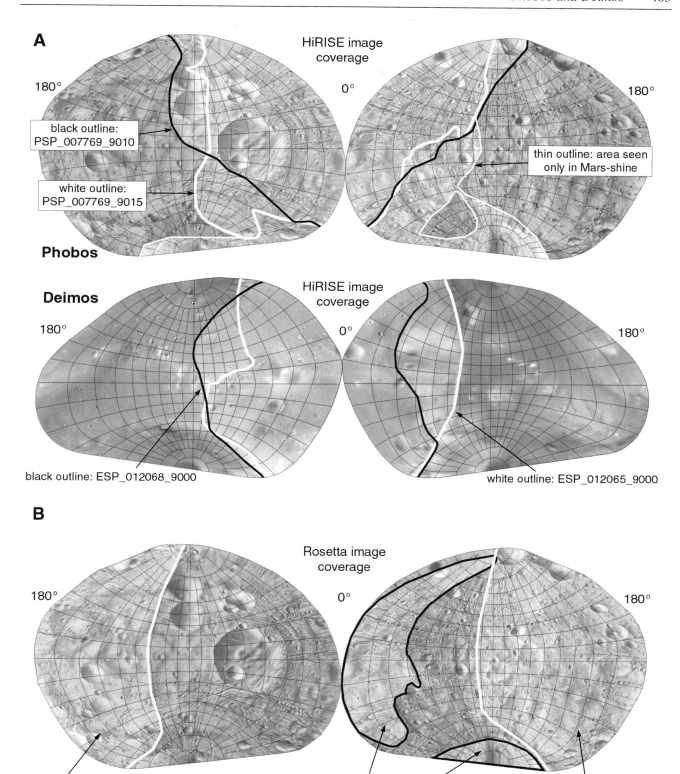

A

HiRISE image coverage

180°   0°   180°

black outline: PSP_007769_9010

white outline: PSP_007769_9015

thin outline: area seen only in Mars-shine

Phobos

Deimos

HiRISE image coverage

180°   0°   180°

black outline: ESP_012068_9000

white outline: ESP_012065_9000

B

Rosetta image coverage

180°   0°   180°

pre-flyby images

post-flyby images

pre-flyby images

Figure 205. MRO and Rosetta image coverage on Phobos and Deimos. **A:** Areas covered by MRO HiRISE images on the two satellites. **B:** Areas covered by Rosetta low-resolution images on Phobos. Imaged areas have lighter shading.

six multispectral sequences with its OSIRIS instrument's Narrow Angle Camera (Figure 204F). The first showed the side of Phobos opposite Mars at very low resolution and a nearly full phase, and the five later sets showed the leading side of Phobos as a narrow crescent at higher resolution. Another sequence followed Phobos as it emerged from behind Mars, only barely resolving the satellite but showing a bright sunlit crescent and a larger area illuminated by sunlight reflected from Mars. Images of Phobos and its shadow crossing the disk of Mars were also obtained. Pajola *et al.* (2012b) described the images and interpreted the multispectral data to suggest the presence of the mineral pyroxene on its surface, and similarities to some dark outer Solar System asteroids. Figure 205B outlines the areas covered by the Rosetta multispectral data.

Gulliver

Gulliver was a Discovery mission proposal in 2006 and 2010, which would collect and return to Earth a regolith sample from Deimos (Britt *et al.*, 2005; Britt, 2010). The name refers to *Gulliver's Travels* (Swift, 1726), the story that suggested Mars might have two small moons long before they were discovered.

After entering an orbit around Mars, 40 km higher than Deimos, the Gulliver spacecraft would be overtaken by the little moon and would transition to an elliptical retrograde 40 by 60 km orbit around Deimos. When the mass of the moon had been determined, Gulliver would circularize its retrograde orbit at 40 km to map the surface at 1 m/pixel resolution and gather shape and gravity data. A set of candidate sampling sites would be chosen from these observations. Then Gulliver would

drop to a 12 km orbit for candidate site mapping at 30 cm/pixel and the sampling sites would be chosen. The spacecraft would touch down but not land, a process called touch-and-go sampling. It would descend to an elevation of 500 m and then drop until contact was made, but hop back off the surface a second later, carrying about 10 kg of regolith scooped up by a rotating brush system (Behar *et al.*, 2003). The process would be documented with continuous imaging.

A project presentation listed five potential sampling regions, and the sub-Mars point was also mapped by P. Stooke at the request of the proposers. These early suggestions (Table 55 and Figure 206) would have been subject to revision during the mission as better images became available. Sites 1, 3 and 4 in particular were in areas of low-resolution and high-Sun Viking imaging. Region 2 is similar to an Aladdin site (Figure 202 in Stooke, 2012).

Phobos Sample Return Mission landing site selection

This Russian sample return mission to Phobos was developed to follow the partial success of Phobos 2 (Sagdeev and Zakharov, 1989). Shingareva and Kuzmin (2000) referred to a potential landing site east of Stickney, in the area of the highest-resolution MOC image (Figure 203A in Stooke, 2012). This location was desirable because of the excellent image coverage and because Mars would be visible in the sky above the landing site. Galimov (2004) illustrated this site at 15° N, 348° E (Figure 207), near a dramatic boulder roughly 25 m across, which cast a long shadow in the MOC image (Figure 208A), sometimes referred to as "the monolith.

Table 55. *Gulliver Mission Sampling Regions on Deimos*

Region	Location	Description
1	38°–42° N, 172°–208° E	Anti-Mars region, poorly imaged, low albedo
2	0°–8° S, 294°–308° E	Ridge area, similar to Aladdin target but mostly low albedo
3	14°–20° N, 52°–60° E	Small bright streaks on gentle ridge
4	36°–42° S, 92°–98° E	Faint bright streak extending into southern depression
5	24°–26° N, 282°–312° E	Smooth low-albedo facet with small craters
6	0° N, 0° E	Sub-Mars point, map requested for planning purposes

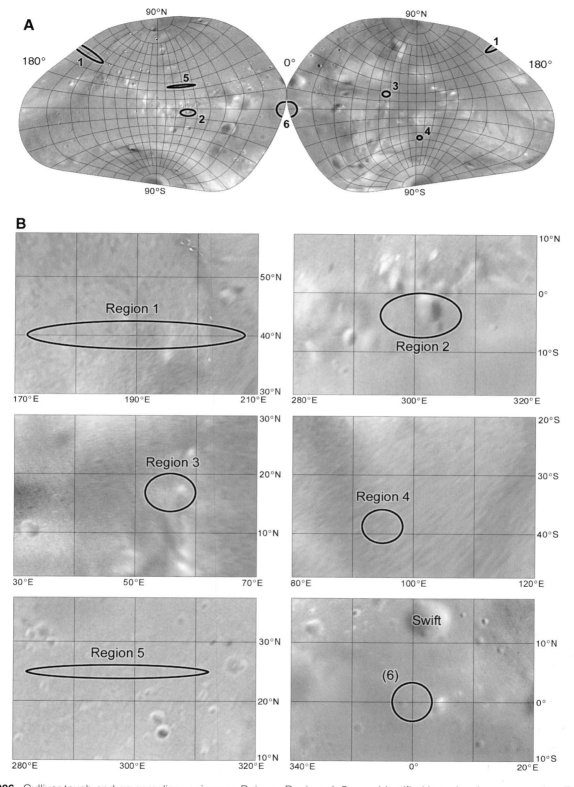

Figure 206. Gulliver touch-and-go sampling regions on Deimos. Regions 1–5 were identified in a planning presentation. Region 6 was also referred to, but the size and shape of the ellipse is conjectural if one was intended at that location. Scale varies on this irregular object, but typically 10° of latitude spans roughly 1000 m on Deimos.

The mission was originally expected to be launched in September 2007, to leave Phobos in August 2009 and to return to Earth with its sample in June 2010. Its launch was delayed by budgetary and other issues until 2011.

Kuzmin and Shingareva (2002) and Kuzmin *et al.* (2003) defined a much larger landing area, better reflecting planning uncertainties, and moved it to an area south of the first site where relatively smooth terrain was available. The general landing zone was described as extending between 10° N and 40° S and between 0° and 50° E, and two landing ellipses (circles 8 km in diameter) relatively devoid of obstacles were selected in this region. The centers of the two circles were at 20° S, 45° E and 13° S, 38° E, only 2.5 km apart (Figures 207 and 208B). Specific landing sites much smaller than these large circles would have been chosen later, but assuming they were at the centers of the circles the spacecraft would approach them from a point at 10° S and 80° E, east-northeast of the landing sites.

Kuzmin and Zabalueva (2003) described a smaller landing zone, extending only from 10° S to 25° S and 30° E to 50° E or roughly the eastern half of the southeastern landing ellipse. That paper also refers to interruptions of lander power during solar eclipses (an eclipse of the Sun by Mars as seen from Phobos, or an eclipse of Phobos as seen from Mars). Not mentioned, but also significant, would be the loss of communication with the lander as Mars blocked radio transmissions. Mars would pass between the lander and the Sun (or Earth) on many orbits, varying seasonally, but if the lander is on the Mars-facing side of Phobos those occultations happen during the day, interfering with operations. If the lander is on the anti-Mars side of Phobos, the occultations happen at night and have minimal impact on the mission. This consideration made landing sites in the anti-Mars longitudes more desirable.

The general area suggested for the anti-Mars region (Basilevsky and Shingareva, 2010) extended from 20° N to 20° S and from 100° E to 150° E. The smoothest region in that area was an 8 km diameter circle centered at 0° N, 127.5° E, and the first specific sites identified were at 5° N, 130° E and 5° S, 125° E, between the Phobos 2 target area and the trailing side targets for Aladdin (Figures 200 and 202 in Stooke, 2012). These coordinates were measured on Viking mosaics based on the coordinate system of Simonelli *et al.* (1993), and the

same locations would be at 5° N, 133° E and 5° S, 128° E in the Mars Express coordinate system of Wählisch *et al.* (2010). Those sites were chosen for their smooth appearance in Viking images, but the illumination angle in those images was quite high. Although high-quality Mars Express images became available in the years preceding the launch date, these sites remained beyond the terminator and could not be reassessed. Eventually, a smooth region nearby was observed in stereoscopic Mars Express images taken on 23 July 2008 (MY 29, sol 221), and as a result the targets were moved into that area, later named Lagado Planitia (Lorenz *et al.*, 2011).

The new landing region extended from 0° N to 30° N and from 120° E to 150° E (Willner *et al.*, 2010), or from 10° N to 30° N and from 120° E to 140° E (Basilevsky and Shingareva, 2010). An 8 km circle centered at 15° N, 130° E became the landing ellipse, and new specific sites (Figure 208C) were identified at 11° N, 134° E and 17° N, 140° E in Mars Express coordinates (Basilevsky *et al.*, 2008), or at 12.5° N, 129° E and 18.5° N, 136° E in Viking coordinates. These were on the edge of one of the areas imaged at highest resolution by Viking (Figure 198C in Stooke, 2012). All landing sites shown in Figures 207 and 208 are plotted relative to surrounding features as illustrated in source documents rather than using coordinates.

8 November 2011: Phobos Sample Return Mission

The Phobos Sample Return Mission, also referred to in Russian as Fobos-Grunt (Phobos-Soil), was intended to collect a sample of regolith from Phobos and return it to Earth. The spacecraft was launched from Baikonur at 20:16 UT on 8 November 2011 on a Zenit booster and successfully placed in its parking orbit, 287 by 202 km high. The upper stage was to fire twice, first to raise the apogee to 4150 km and again at the next perigee to depart from Earth orbit. The upper stage failed to ignite and place the spacecraft on its Mars transfer trajectory after an attitude control problem triggered a safe mode. Despite many efforts to regain control of the spacecraft, including the use of ESA ground stations in South America, Australia and the Canary Islands, it and the upper stage burned up over the South Pacific near Chile on 15 January 2012 after 1097 orbits.

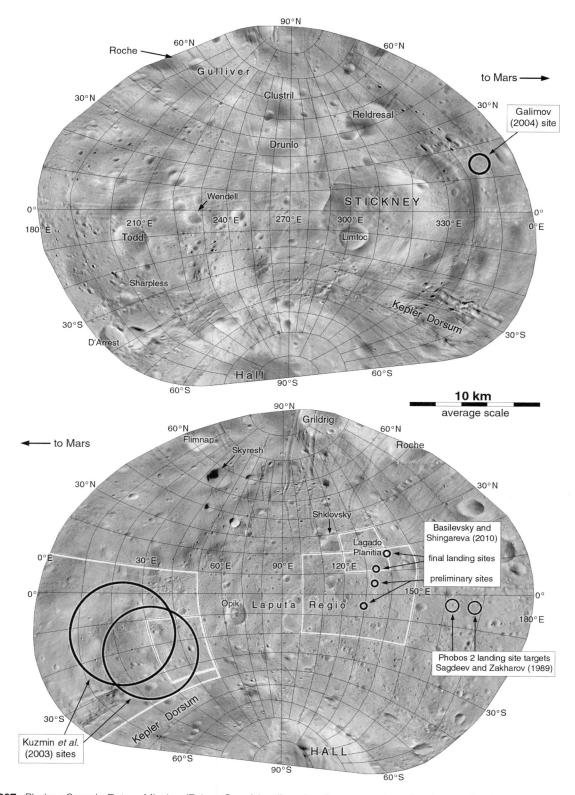

Figure 207. Phobos Sample Return Mission (Fobos-Grunt) landing sites from several studies. Laputa Regio is the region without grooves on the trailing side of Phobos. White outlines delineate the landing regions described in the text. The name Reldresal was spelled incorrectly in Stooke (2012).

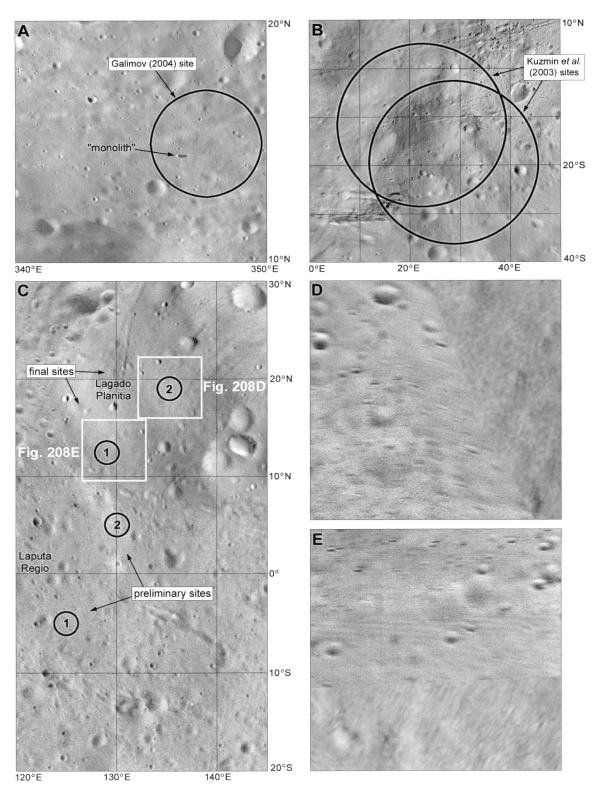

Figure 208. Enlarged views of Phobos Sample Return Mission (Fobos-Grunt) sites. **A:** From Galimov (2004). **B:** From Kuzmin *et al.* (2003). **C:** From Basilevsky and Shingareva (2010). **D** and **E:** The final landing sites, partly covered by Viking images 244A06 and 244A07, the highest-resolution images available. For scale, 10° of latitude spans roughly 2 km.

The journey to Mars was intended to take about 10 months, including up to three trajectory corrections. Several different versions of the following schedule have been published, so this is only a rough guide to the expected activities. Fobos-Grunt would arrive at Mars in September or October 2012 and brake into a near-equatorial initial orbit of roughly 800 by 79 000 km with a period of about 72 hours. The braking rocket module and the Chinese orbiter Yinghuo-1 would be separated and left in that orbit. Some studies of Mars would be made over a period of up to three months while the orbit was adjusted, matching its inclination to that of Phobos and raising periapsis to 6500 km. In December 2012 an engine burn would make the orbit circular with a radius of 9910 km, just over 500 km outside the orbit of Phobos, with an 8.3 hour period. This observation orbit would allow the satellite to pass beneath the spacecraft at intervals, and within a month the orbit would be adjusted again to be roughly synchronous with Phobos. The landing would take place early in 2013.

Immediately after landing, the sampling system would collect a soil sample and put it in the return capsule. If communications were lost, the sampling and return would be carried out automatically. After images of the sampling area were received on Earth for planning, a robotic arm with a claw-like attachment would collect up to 20 scoops of soil with a total mass of about 100 g, and a drill would provide a different sample. The two methods increased the likelihood of successful sampling on the unknown surface. Cameras would record the sampling activities. The solar-powered lander also carried a seismometer.

When sampling was completed, in about April 2013, the return capsule would be ejected by springs, rising to a safe height before igniting its rocket to escape from Phobos. This would protect the lander from rocket exhaust damage, so it could continue observing the landscape and analyzing samples with its own instruments for as much as a year. The return capsule would move to an orbit 300 km closer to Mars than Phobos, with a 7.2 hour period, and wait for several months until orbital geometry was suitable for the return journey. In August 2013 it would be placed in a highly elliptical 72 hour orbit, then adjust its inclination and lower its periapsis again, reversing the maneuvers that it followed to reach Phobos. In September 2013 it would leave Mars and coast for 11 months, correcting its trajectory as needed, before arriving at Earth in mid-August 2014 and landing in Kazakhstan. Only a 7 kg return capsule would return to Earth from the original vehicle mass of 13 500 kg.

8 October 2011: Yinghuo-1

Yinghuo-1 was a small Chinese Mars orbiter, designed to be launched and carried to Mars with Fobos-Grunt. It was destroyed with that spacecraft during atmospheric re-entry on 15 January 2012. Yinghuo-1 was a cuboidal box 75 by 75 by 60 cm in size, with solar panels extending 7.8 m from side to side. Its mass was 110 kg. It would be released into a highly elliptical near-equatorial orbit about 800 by 79 000 km, from which it could observe Mars, its satellites and the planet's atmosphere. It might have been able to obtain unique images of Deimos, which had not been imaged at high resolution since the Viking missions.

26 November 2011: Mars Science Laboratory (Curiosity)

Curiosity, a large rover formally named Mars Science Laboratory, was launched on 26 November 2011 and landed in Gale crater, Mars, on 6 August 2012. The Mast Cameras imaged Phobos and Deimos on several occasions (Figure 209A), at different phases in the night sky and during solar transits (eclipses). On Curiosity's sol 393 Phobos was imaged as it entered the shadow of Mars. These images allowed the orbits of the two moons to be refined. Many details were resolved on Phobos, but Deimos could only be seen as a somewhat irregularly shaped object (Figure 209B). The Remote Microscopic Imager (RMI), part of the ChemCam instrument, would have been able to resolve Phobos and Deimos as well, but this observation was not made during the primary mission.

Other Phobos and Deimos mission proposals

Many proposals have been made for exploration of Phobos and Deimos, and some are summarized here.

Where landing sites are suggested, they are illustrated in Figures 209C and 209D.

Netlander was a proposed Mars surface network mission accompanied by an orbiter, studied by the French space agency CNES (Centre National d'Études Spatiales) with possible contributions by NASA (Figure 118A). It was intended to launch in August 2005 and arrive at Mars in July 2006. A later version of this network mission, called Mars Premier, would have launched in 2007. Mars Premier was intended as a forerunner of a Mars sample return mission, and, apart from its network components, it might have tested automated rendezvous hardware and procedures in Mars orbit. The Russian firm Lavochkin studied a 30 kg Phobos lander to be delivered by Mars Premier (Ball *et al.*, 2009).

A Europlanet Landing Site Workshop (European Space Agency, 2011) discussed site selection procedures for potential lunar and Martian missions. Phobos was used as an example for various future small-body mission targets, including the asteroid Eros. Two areas were described as potentially interesting landing areas (Figure 209D) without further published justification. A broad region east of Stickney would include Stickney ejecta and more typical surface material as well as access to some of the most prominent groove structures. A small region along the sharp western rim of Stickney was also illustrated. The landing would have to occur on a level surface west of the rim or on the crater floor below the rim rather than on the steep slope illustrated in the report. The prominent downslope movement of debris on the crater wall could be studied here, possibly including very fresh material recently exposed by landslides.

Another ESA study (Renton, 2006) examined a possible Deimos sample return mission. The goal was to collect 1 kg of Deimos regolith and return it to Earth. Up to 10 percent of the sample might consist of Mars crater ejecta. The mission might land for sample collection, touch down for a moment and lift off again, probably at several different sites, or use a projectile to generate a debris cloud from which a sample could be collected without landing. The touch-and-go approach was favored for this study. No specific sampling locations were identified, as the surface was said to be homogeneous, though this ignored potential differences between ridges and facets or bright and dark markings. The only specific requirement was that fresh craters should be avoided.

Ball *et al.* (2009), Lee and Martin (2011) and Lee *et al.* (2012) briefly summarized several mission studies, some of which are described below. Apart from the intrinsic value of each mission's data, any of them would also contribute to plans for future human exploration.

The Canadian Space Agency studied a Phobos mission called Phobos Reconnaissance and International Mars Exploration (PRIME), which consisted of an orbiter and a lander. The orbiter would orbit Mars but remain very close to Phobos and study the satellite for up to three months, in particular seeking evidence of hydrogen in its regolith. The landing would be a "rock docking" immediately adjacent to the 25 m diameter rock "monolith" at 15° N, 348° E seen in the best MOC image and also proposed as a landing target for Fobos-Grunt (Figures 208A and 209C) by Galimov (2004). The lander would operate for up to a Mars year, imaging its landing site, analyzing the regolith composition and monitoring Mars.

A mission called ASAPH, the Advance Smallsat (or Spacecraft) Assay of Phobos Hydrogen, studied at the NASA Ames Research Center, was a small orbiter that would examine the composition of Phobos, especially looking for hydrogen. It would also image the surface at high resolution and look for dust in the vicinity of Phobos, and at the end of its mission it might attempt a landing, as the Near Earth Asteroid Rendezvous (NEAR) spacecraft had done on asteroid 433 Eros on 12 February 2001. No landing site was identified.

Hall (Lee *et al.*, 2010) was a mission concept developed by NASA Glenn Research Center, the SETI Institute and the Mars Institute, derived in part from the Canadian PRIME mission study. Solar electric propulsion would help make the mission possible, and the name "Hall" commemorated both Asaph Hall, discoverer of Phobos and Deimos, and Edwin Hall, a physicist whose work contributed to the development of electric propulsion. A large circular solar array to power the electric thrusters was derived from an array developed for NASA's Orion human crew vehicle. This sample return mission would target two locations on Phobos and one on Deimos, and one on Phobos was identified, the large rock "monolith" seen in the highest-resolution MOC image or a similar block, as already suggested for Fobos-Grunt and PRIME. This was in the "blue" material associated with ejecta from the large crater Stickney, and the second sample from Phobos would be collected from a redder area. The Deimos target was not identified, but would be another large block. Samples would be collected from the blocks themselves and from the adjacent regolith.

The Mars Multiple Moon-landing Mission (M4) derived from a Discovery asteroid-landing mission proposal called Amor. The M4 spacecraft would be able to land at two places on Phobos, in the red and blue units near the rim of Stickney, and at one place on Deimos in a brighter (less space-weathered) area on an old crater rim (Lee *et al.*, 2011). Target areas would be close to large blocks, and a schematic diagram of the mission suggested the sites illustrated in Figure 209D. During flight it would obtain multispectral data from both moons for compositional analysis and site selection. At the landing site it would use multispectral imaging for analysis of nearby boulders and regolith at varying resolutions, make microscopic images of grains and crystals in rocks using an instrument on a robotic arm. A combination neutron and gamma ray spectrometer would measure elements in the surface during flight and at each landing site. M4 was viewed as a less expensive alternative to Hall.

Ball *et al.* (2009) described a small orbiter and lander mission called Mars Phobos and Deimos Survey (M-PADS). Two small spacecraft would make use of solar electric propulsion to orbit and characterize both moons and to land on one of them. A comprehensive global dataset would be obtained from both moons, including morphology, composition, magnetic properties and internal structure. At the landing site, possibly not chosen until the mission's global studies had been made, the spacecraft would use a suite of analytical instruments to determine surface composition.

GETEMME was an ESA M-class mission study, variously described as "Gravity Experiment with TimE Metrology on Martian satEllites" (Lainey and Le Poncin-Lafitte, 2010) or "Gravity, Einstein's Theory, and Exploration of the Martian Moons' Environment" (Oberst *et al.*, 2012). The GETEMME spacecraft would examine both Phobos and Deimos in detail with images, altimetry and spectrometry, deploy its landers on the satellites and then move into a 1500 km altitude Mars orbit. The orbiter would then use a laser to measure ranges to the moons, and its own position would be

similarly measured by lasers on Earth, for a full Martian year. The precise tracking of the spacecraft and satellite orbital motions would address issues of fundamental physics, including relativity, and the structure and dynamics of Mars and its satellites. GETEMME could be carried to Mars by a larger mission if it was not flown on its own.

The 2010 version of GETEMME had two landers called Romeo (for Phobos) and Juliet (for Deimos), each carrying a laser reflector and other instruments. In the 2012 mission design, the landers were small passive laser reflectors, two for each moon, dropped from low altitude during a close flyby after the orbital mapping phase for each satellite. After a January 2020 launch, solar electric propulsion would deliver the spacecraft to Mars two years later, and GETEMME would enter a circular low-inclination orbit, from which it would rendezvous with Deimos. When Deimos had been mapped, the two landers would be dropped during a low pass, and the process would be repeated at Phobos. The orbiter would then drop to its circular 1500 km altitude orbit for the laser ranging phase of the mission. The lander descents would be unpowered and the touchdowns relatively slow but uncontrolled. After a few bounces, the landers would open a protective lid, righting themselves in the process, and would obtain images and thermal and magnetic data for a few hours, powered by a battery. After the batteries failed, the landers would function merely as passive laser reflectors. Landing sites would be within $5°$ of the equator and roughly $20°$ east and west of the sub-Mars point (Figure 209D) on both satellites.

Merlin (Rivkin *et al.*, 2011; Murchie *et al.*, 2011) was a proposed Deimos landing mission which would launch in about 2017 to extend the results of Fobos-Grunt to its neighboring moon. During an orbital phase, multispectral and high-resolution cameras would study the moon's geology, surface properties and shape, while radio science probed the interior. After landing, a robotic arm would deploy an alpha particle X-ray spectrometer to measure elemental composition, a miniature Raman spectrometer to study mineralogy, and a color microscopic imager to examine regolith properties. Cameras would provide operational support and characterize geological processes at the landing site. Several potential landing sites, each 500 m across, would be planned in advance, but a final site would be selected using new

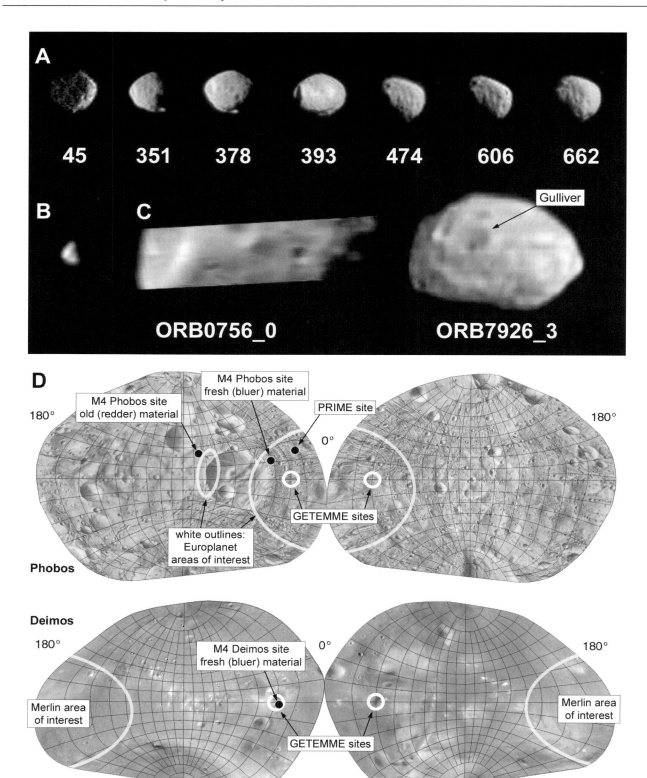

Figure 209. Images and landing targets for Phobos and Deimos. **A:** MSL images of Phobos, identified by sol number. **B:** MSL image of Deimos from sol 351. **C:** Mars Express OMEGA images of Phobos. **D:** Future mission sites on Phobos and Deimos.

data from the remote sensing phase of the mission. Sites illustrated in the mission descriptions were no more than plausible locations, but they suggested a preference for the anti-Mars region of Deimos, as had been the case for Fobos-Grunt (Figure 207). This region of Deimos was only poorly imaged by Viking.

Phobos Surveyor was a proposal to fly several small landers to explore Phobos, or in future to any small body without an atmosphere (Pavone *et al.*, 2013). An orbiter would deploy one or more small landers, each a sphere with many external spikes, sometimes likened to a hedgehog or sea urchin. The propulsion consisted of internal flywheels mounted in a mutually orthogonal configuration, which could be spun in various ways to enable the vehicle to hop or roll in a microgravity environment. The landers were 40 cm in diameter, with a mass of 5 kg. A reference mission at Phobos involved an orbiter at the Mars–Phobos L1 Lagrange point, carrying a stereoscopic camera system and compositional instruments, and several landers each with an X-ray spectrometer and a microscope for regolith studies, among other instruments. The illustrated mission concept included a lander touching down near 5° N, 312° E on the floor of Stickney crater and hopping over the crater floor to examine bright and dark areas (Figure 211). The surface mission would end in Limtoc crater, and other hoppers might then go to other locations.

Cartographic innovations for Phobos and Deimos

Phobos and Deimos were the first distinctly nonspherical objects in the Solar System to be observed closely by spacecraft, and thus have often been the subjects of cartographic innovations designed to cope with nonspherical objects. The first map of Phobos, and the first for any nonspherical object, was a sketch map by JPL engineer Thomas Duxbury in 1974. Noland and Veverka (1977) followed with a very simple sketch of a few features on Deimos, and Blunck (1977) also published an early map of Deimos. Stooke (2012) summarized the history of this aspect of cartography in connection with the Mariner 9, Viking and Mars Express missions.

Figure 210 illustrates some novel maps of Phobos and Deimos. Figure 210A is a map of Phobos on a projection devised by Professor Lev Bugaevsky of the Moscow

State University for Geodesy and Cartography (MIIGAiK), kindly provided by Irina Karachevtseva. Bugaevsky's projection resembles a modified version of a Mercator projection, with poles added as insets. Stooke (2012) describes this projection and one by John Snyder (USGS) as if they did not use planetocentric coordinates, but this was true only of Snyder's work (Irina Karachevtseva, personal communication, 15 January 2013). Bugaevsky's projection used planetocentric coordinates and provided accurate cartographic representations of the surfaces of triaxial ellipsoids. It has been used for maps of Phobos and Deimos, including those in MIIGAiK (1992), as well as for asteroids and other planetary satellites. Figure 210B is a map of Phobos by Maxim Nyrtsov, also of MIIGAiK, on a form of the Kavrayskiy projection modified by varying the body radius across the map.

Figures 210C and 210D are maps of Deimos by Chuck Clark (Clark and Clark, 2013). Clark's approach, called constant-scale natural boundary mapping, allows a surface to be cut along ridges, valleys or other significant features and opened out into a map. The dividing lines are chosen to assist with the visualization of specific phenomena. For instance, a map of bright streaks on Deimos produced by downslope movement of loose material might be divided along ridges to illustrate the relationship between topography and material flow. A map designed to help select landing sites near potential exposures of sub-regolith material might be cut through the facets and valleys to show the ridge network without interruption.

Human Phobos and Deimos exploration

Studies of human flight to Mars, including Phobos and Deimos, were summarized by Jesco von Puttkamer (2011). The first technical study of human exploration of Phobos, by Steinhoff (1962), envisioned a human-tended outpost on Phobos from which an expedition could be flown down to Mars. Singer (1984) described a human mission to Mars which would spend about six months on Deimos. Its crew of eight would split into two teams, one of which would sample Deimos itself and fly a sortie to Phobos to sample that moon. The other team would operate several robotic missions on the surface of

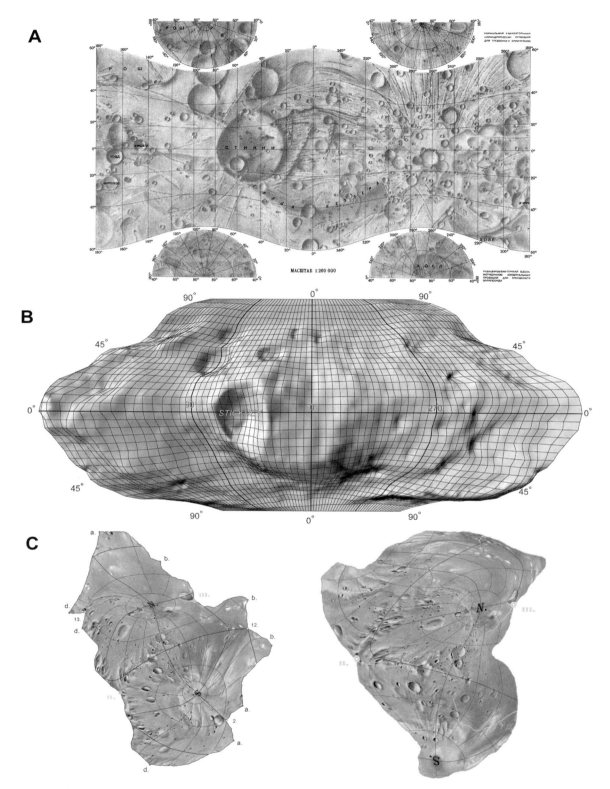

Figure 210. Novel map projections for the Martian satellites. **A:** Phobos, Bugaevsky's projection. **B:** Phobos, Kavrayskiy's projection modified by M. Nyrtsov. **C:** Deimos, two versions of Constant Scale Natural Boundary maps by Chuck Clark.

Mars, including rovers, penetrators and sample return missions, taking advantage of their proximity to drive rovers or operate equipment directly rather than with the long time delays inherent in Earth-based teleoperation (the remote operation of robots). All samples would be delivered to Deimos for return to Earth with the crew. This was called the Ph-D mission, and later versions of it suggested pre-landing supplies and equipment on Deimos to reduce risk.

Another study of this type by Texas Space Services (1986), an entity created by students at the University of Texas for a university project, described a Phobos–Deimos exploration mission involving a robotic precursor to both moons, followed by a human expedition to Phobos. Other components of this scenario were a small space station in Mars orbit and a robotic Deimos lander. The mission description is included here as an example of student work made widely available via the Internet, without which it would probably not have become known outside the university.

The first step was to send a robotic orbiter and lander based on Viking technology to examine the moons. In one scenario, the orbiter would examine Phobos first and then move to Deimos, acquiring global high-resolution images and studying composition, volatiles, regolith properties, gravity and the radiation environment. The images of Phobos would be used for landing site selection, and then the lander would descend to the surface. A separate probe to Deimos was also considered at this stage.

The second step was developed as part of a broader program of Mars exploration, including an orbiting space station and landings on the planet. The satellite scenario involved 11 to 15 extra-vehicular activities (EVAs) at four to eight sites on Phobos over about 22 days (Table 56 and Figure 211). A crew of four would form two teams of two, who would undertake EVAs on alternating days, allowing each crew a rest day between EVAs. After arrival at Mars, the crew would take one day to approach Phobos orbit, and one day to rendezvous with Phobos, land at the first site and check the lander systems. The next two days would be spent on EVAs, one per day by the two teams, and a "safe haven" habitat would be buried with regolith on the third to sixth days after landing to provide a solar flare storm shelter at that first site. On the seventh day, the lander would transit to the second site. After two days of EVAs and a third

contingency EVA if needed, the lander would transit to the third site (11th day) and the same events would be repeated there, and the process would be repeated at the fourth site (reached on the 15th day). Activities at that site would end with a possible contingency EVA on the mission's 18th day since landing, and two more days would be spent on departure, transfer and docking to an orbiting space station.

EVA activities would include sampling and setting up an instrument station similar to an Apollo ALSEP (Apollo Lunar Surface Experiments Package). A hopping mobility unit might be used, though in the low gravity it might need power to fly downwards as well as upwards on its trajectories. Two sets of four sites were selected in case the reconnaissance imaging showed the first set were unsatisfactory, or if the remaining time allowed additional sites to be visited after the primary set. The sites were chosen from Viking images for their usefulness and science value, and ordered so that the route between them was the minimum travel distance, in each set of four sites.

The sites in Figure 211 and Table 56 are based on a map and table in the published report, which do not match each other exactly. Site 8 was in a crater then called Roche but now named Gulliver, because the name Roche is now given to a smaller nearby crater. In early published maps of Phobos based on Mariner 9 images, the name Roche was assigned to three different craters, leading to confusion, which was only resolved when more names were added in 2006.

An additional robotic Deimos lander and rover would also be derived from Viking hardware and operated by astronauts on the Mars orbiting space station, since people would not travel to Deimos in this scenario. The lander, RTG powered like Viking, would carry composition and age dating instruments, and would search for water in the regolith and image the surface. The rover would allow analysis at a distance from the lander's exhaust contamination and would also allow repositioning for optimal communications. Mobility might involve wheels or a hopping vehicle, the latter maybe preferred in this very low-gravity environment. One possible application of this system would be to collect samples and launch them into Mars orbit for retrieval by astronauts. At this date even film return from a camera was considered, as resolution might have been superior to the

Table 56. *Phobos Human Landing Sites From Texas Space Services (1986)*

56a. Primary sitesFigure

Site		Location	Notes
1	Stickney crater (inner crater – ejecta blocks)	10°–20° S, 300°–310° E	Safe haven location, inside Limtoc crater, so burial may be easier. Depth within Stickney offers more shelter from solar radiation. Close to large ejecta blocks
2	Stickney ejecta	0° N, 340°–350° E	Sample ejecta, test mobility with a site close to site 1
3	Major groove concentration	20°–30° N, 340°–0° E	Study grooves, structure, origin, possible outgassing
4	Stickney antipode	0°–10° S, 70°–90° E	Study the area without grooves (Laputa Regio), nearly opposite Stickney

56b. Secondary sites

Site		Location	Notes
5	Southern ridges	40°–60° S, 60°–80° E	Sample ridges (old crater rims) in southern hemisphere
6	Hall crater	60°–70° S, 120°–150° E	Examine possible layers of darker material in the south polar crater's walls
7	Area west of Stickney	20° N–20° S, 240°–270° E	Smooth area in Viking images, then thought to be a younger surface, but an artifact of lower resolution
8	Gulliver crater	50°–70° N, 190°–220° E	Crater near north pole (called Roche in original report)

digital systems then envisaged. No specific landing sites were mentioned, but ejecta blocks like those seen in the highest-resolution Viking images and high-albedo streamers on many slopes on Deimos were considered promising targets.

At the First International Conference on the Exploration of Phobos and Deimos, held at NASA Ames Research Center in November 2007, Stooke (2007a) suggested the use of Phobos, or Deimos if preferred, as a sample cache location. The rationale was that a first human flight to Mars would be operated like Apollo 10 at the Moon to test vehicles, equipment and procedures without the added risk of landing on Mars. In order to give this rehearsal mission an important scientific goal, it would gather samples previously cached on Phobos during a decade or more of robotic exploration. Samples could be collected from several places on Mars and delivered to Phobos, impacting at relatively low velocity protected by braking rockets and airbags. Samples might also be brought to Phobos from Deimos, from one or more of the Mars trojan asteroids, and possibly even from the main asteroid belt, by robotic missions enabled by solar electric thrusters. These sampling missions could be undertaken by several nations or agencies, spreading the cost and risk (Stooke, 2014).

Red Rocks

At the Second International Conference on the Exploration of Phobos and Deimos, held at NASA Ames Research Center in March 2011, Hopkins and Pratt (2011a) presented a Lockheed Martin study of optimum landing sites on Phobos and Deimos for a future human mission. A modified version of the presentation with slightly different site locations was presented later that year at another meeting (Hopkins and Pratt, 2011b). The "Red Rocks" mission to land on a Martian moon would follow the "Plymouth Rock" mission to a near-Earth asteroid in a series of stepping-stone missions leading from Earth orbit to Mars. Planners had often suggested that astronauts on the little moons could operate rovers or sample return landers on Mars more effectively than people on Earth because of the negligible radio signal time delay. This remote operation of robots is referred to as teleoperation.

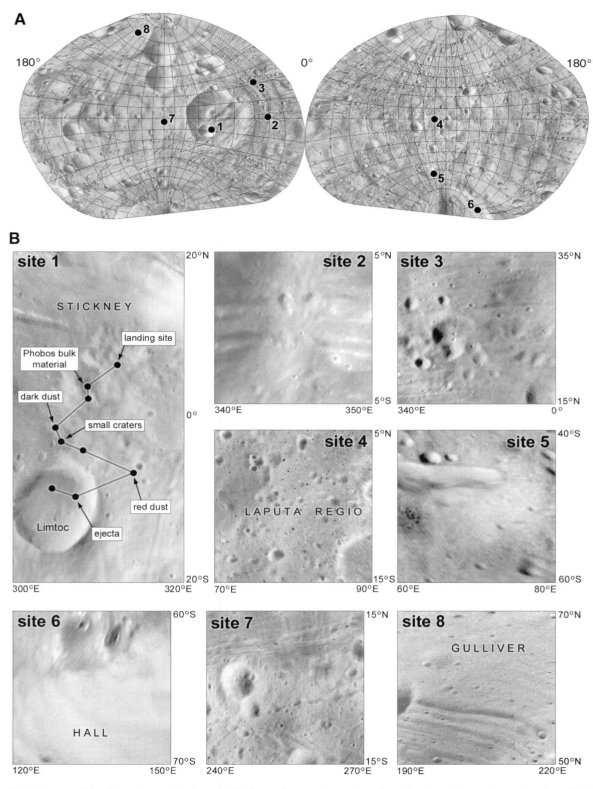

Figure 211. Proposed landing sites on Phobos and Deimos. **A:** Locations of landing sites from Texas Space Services (1986). **B:** Suggested landing sites for the Texas Space Services missions. Site 1 also illustrates the Phobos Surveyor hopping traverse of Pavone *et al.* (2013). For scale, 10° of latitude spans approximately 2 km on Phobos.

Assuming such a mission, this study compared Phobos and Deimos, finding Deimos better in most respects. Using an illumination model that took into account the shapes of the moons and seasonal changes in eclipses and lighting, they identified areas with the longest periods of continuous sunlight within view of Mars. The south poles of both moons fall in depressions, shifting the southern maximum illumination points onto ridges on or near the Mars-facing prime meridians, on Kepler Dorsum in the case of Phobos. The large south polar depression of Deimos would be in darkness during southern hemisphere winter and at least partly shadowed by the 0° longitude hill during summer, making it a useful storage site for cryogenic propellants (Figure 212A).

Teleoperation of landers on Mars requires a direct line of sight. Analyses of the shape models of the satellites identified the area facing Mars, which always provided visibility of the entire disk of Mars, and the anti-Mars area, in which no part of the disk was visible. Because Mars appears so large in the sky of each satellite, especially Phobos (from which the disk of Mars spans 43°), a broad region between those two areas would provide views of part of the disk. These visibility regions are outlined in Figure 212A, along with the areas permanently illuminated at midsummer in each hemisphere.

The best landing sites for teleoperation missions were interpreted as the areas having both full disk visibility of Mars and permanent illumination at midsummer (Figures 212B and 213, and Table 57). Two areas in the northern latitudes of Phobos were identified, and one in the south, as well as one in each hemisphere of Deimos, using the coordinate system of Simonelli *et al.* (1993).

There were several reasons for preferring Deimos. Less total energy was required for the round trip from Earth to Deimos and back, compared with Phobos, largely because Deimos is farther from Mars. Astronauts on Deimos can communicate with landers on Mars at higher latitudes in either hemisphere, compared with Phobos. Individual communication periods last longer from the near-synchronous orbit of Deimos than from Phobos, though gaps between communications passes are longer for Deimos. Communication between Earth and Deimos is interrupted less frequently by Mars occultations. Deimos is eclipsed less often than Phobos. Its eclipse seasons are shorter, and eclipses last for a smaller fraction of an orbit, though each eclipse lasts longer

(84 min, compared with 54 min for Phobos). Finally, the north polar region of Deimos has continuous sunlight for ten months, compared with five months for Phobos. In the south the times are about seven months for Deimos and three for Phobos.

Conversely, Phobos is more geologically varied and interesting than Deimos, and is more likely to have collected Mars crater ejecta. Sample return vehicles launched from Mars can reach Phobos more easily from latitudes below about 40°, but Deimos is easier to reach from higher Martian latitudes.

A solar-powered Deimos lander could operate at the southern landing site with several months of continuous sunlight during the southern summer, then return to orbit for several months during the equinoctial eclipse season, and land again at the northern site for ten months of continual sunlight. Hopkins and Pratt (2011b) described a mission like this which would launch on 17 April 2033, and arrive at Mars on 4 November after a 201 day cruise. After capture into a very elliptical Mars orbit during a close flyby of the planet, the spacecraft would raise its periapsis and change the orbit plane as it reached the first apoapsis. It would arrive at Deimos on 8 November 2033, circularize its orbit and land at the southern landing site for a stay of 114 days. Before the eclipse season commenced, the crew would take off on 2 March 2034 and orbit Mars, remaining in orbit through the northern spring equinox on 11 April. The spacecraft would land at the northern landing site on 20 April 2034, and stay there for 373 days before leaving on 7 May 2035, and after a 199 day cruise the crew would return to Earth on 22 November 2035. The stay at the northern site would exceed the period of continuous sunlight but would ensure the planets were suitably placed for the return journey.

Merrill and Strange (2011) described a human expedition to the Martian satellites which would suit the revised exploration plans adopted by the Obama Administration after the cancelation of NASA's Constellation Program. Supplies and equipment would be prepositioned in a high Mars orbit with a 10 day period before the crew arrived, and a habitat vehicle would be placed in the vicinity of Phobos. The crew would arrive and spend about 40 days in high Mars orbit, retrieving supplies and preparing equipment. Then they would descend to Deimos for about 10 days of exploration

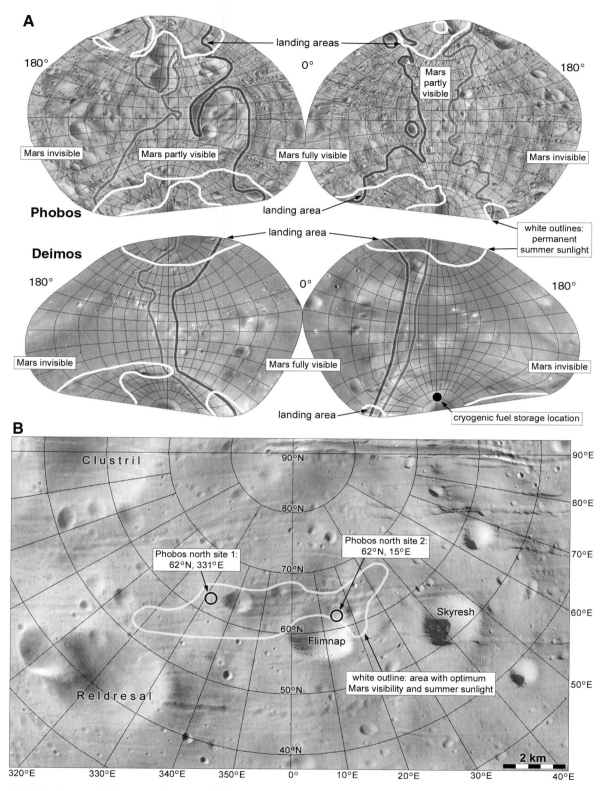

Figure 212. Phobos and Deimos landing sites for teleoperation of Mars surface missions, from Hopkins and Pratt (2011a, 2011b). **A:** Mars visibility and illumination regions. **B:** Phobos northern summer landing sites.

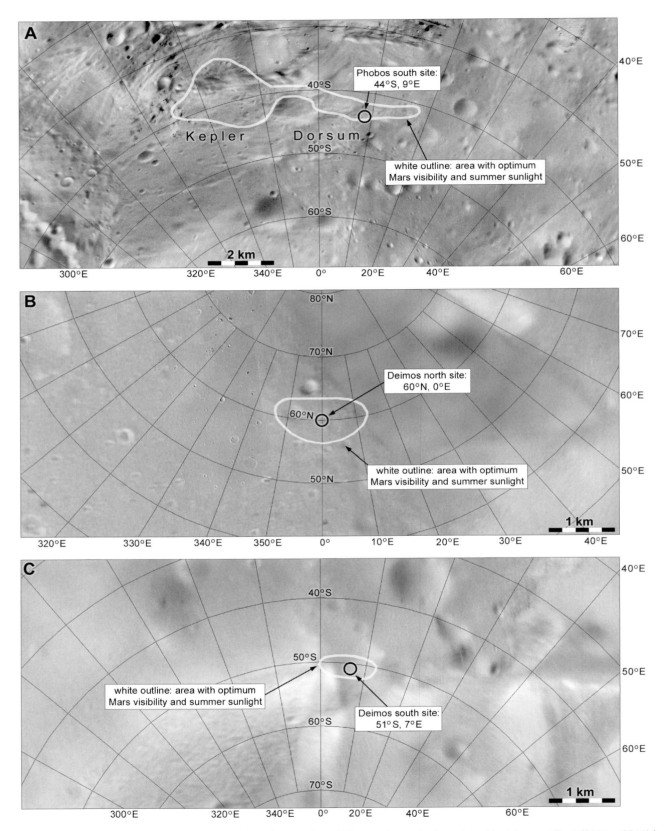

Figure 213. Phobos and Deimos landing sites for teleoperation of Mars surface missions, from Hopkins and Pratt (2011a, 2011b).
A: Phobos southern summer landing site. **B:** Deimos northern summer landing site. **C:** Deimos southern summer landing site.

Table 57. *Landing Sites From Hopkins and Pratt (2011a, 2011b)*

Satellite	Location	Maximum Mars latitude for lander visibility	Lander contact period	Time between communication sessions	Continuous sunlight period
Phobos	62.0° N, 15.0° E 62.3° N, 330.6° E 44.3° S, 9.0° E	65° N–S	4 hours	7 hours	5 months (N) 3 months (S)
Deimos	60.0° N, 0.0° E 51.7° S, 7.6° E	78° N–S	60 hours	72 hours	10 months (N) 7.5 months (S)

before moving on to Phobos. There they would occupy the habitat and make use of additional supplies during a 40 day exploration of Phobos. Finally, the crew would return to the main spacecraft in high Mars orbit and return to Earth. At Earth, the small crew vehicle would perform a direct atmospheric entry, and the main spacecraft would perform a flyby with gravity assist that would allow it to be captured back into a high Earth orbit a year later. After restocking and fueling, it might be used again.

3. Updates to Volume 1

Mariners 3 and 4

Mariners 3 and 4 were intended to make the first NASA flights past Mars. Mariner 3 was launched on 5 November 1964, but failed when its payload shroud could not be ejected after launch, preventing its solar panels from opening to power the spacecraft. Mariner 4 launched on 28 November 1964 and flew past Mars on 15 July 1965 (MY 6, sol 305), transmitting back to Earth the first detailed images of the surface showing topographic features (Figures 9–13 in Stooke, 2012). The following additional information is from NASA (1967a, 1967b) and Olivier de Goursac (personal communication, 27 December 2013).

The television experiment team described several targets that should be covered by the cameras on the two missions. These were the sunlit limb (horizon), areas of high Sun angle for photometric observations, areas near the terminator, which would reveal topography, the edge of the south polar cap and, if possible, the terminator on the polar cap, where the bright surface would reveal subtle topography even near the poorly lit terminator. Specific surface features of interest would be the dark areas Syrtis Major or Sabaeus Sinus, the bright "desert" areas Hellas and Elysium, and seasonally varying features such as Solis Lacus, Trivium Charontis and the dark spots or "oases" in Aethiopis (Figure 214). Syrtis Major was a particularly favored target, but navigation studies showed it was not accessible during the flyby.

The imaging strategy for the double flyby mission resulted in the following plan. Mariner 3 would observe a swath of the planet's surface running roughly northwest to southeast, centered on 190° E and reaching as far south as the south polar cap. Mariner 4 would view a swath extending roughly along the equator between 140° E and 280° E. When Mariner 3 failed, a single swath imaging scheme was designed. It began north of the variable feature Trivium Charontis, which resembled a smaller version of Syrtis Major, at 40° N, 165° E, and extended to the southeast as far as 50° S, 250° E, before turning north again to cross the terminator near 40° S, 270° E. This swath would not reach the south polar cap. During the long cruise to Mars, navigation data showed that the time of the flyby would be 36 min earlier than originally expected. This resulted in a swath about 10° further east than intended, which did not observe Trivium Charontis. These swaths are illustrated in Figure 214.

Mars 3

Mars 3 was launched on 28 May 1971 and landed on Mars on 2 December 1971 (MY 9, sol 568). Transmissions from the surface proved that a landing had occurred, the first successful landing on Mars, but they ceased after only 20 s. The landing site was within about 150 km of 45° S, 202° E, but its location was not known precisely, and it was often assumed that the spacecraft would never be identified in HiRISE images. After HiRISE image PSP_006154_1345 became publicly available in 2008, Russian and other space enthusiasts searched for the lander and its associated hardware, coordinating their efforts through the online forum Novosti-kosmonavtiki.ru.

Vitaly Egorov identified the group of features illustrated in Figures 215 and 216 based on comparisons with the expected distribution and appearance of spacecraft components. A second HiRISE image (ESP_031036_1345) was taken of the area in 2013, and McEwen (2013) expressed guarded optimism that this interpretation might be correct. All other landers have been identified with the help of surface images, which increase the number of features to match with orbital images, or by comparison of images taken before and after the landing. Without those additional sources of information, the identity of these objects is still uncertain. The south side of the parachute candidate feature appears slightly different in the two HiRISE images, as if the fabric is flapping in the wind, as other parachutes are observed to do. Figure 215 locates these objects and Figure 216

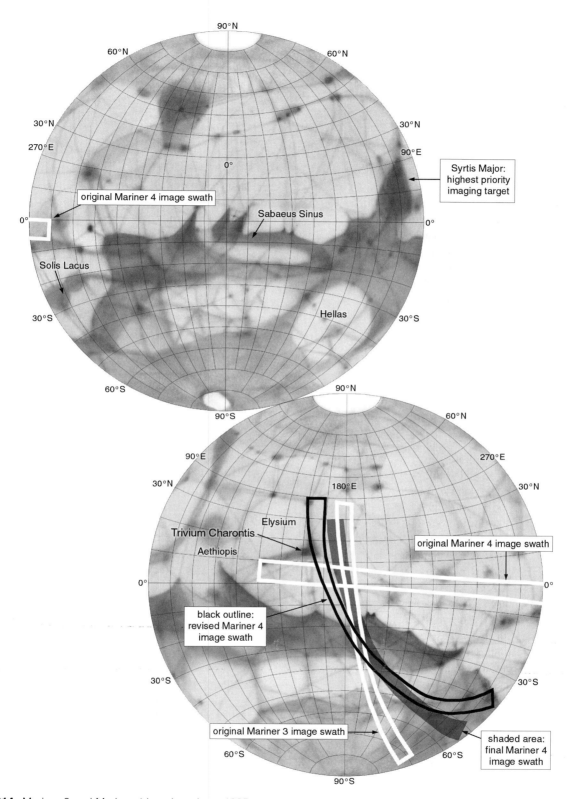

Figure 214. Mariner 3 and Mariner 4 imaging plans, 1965.

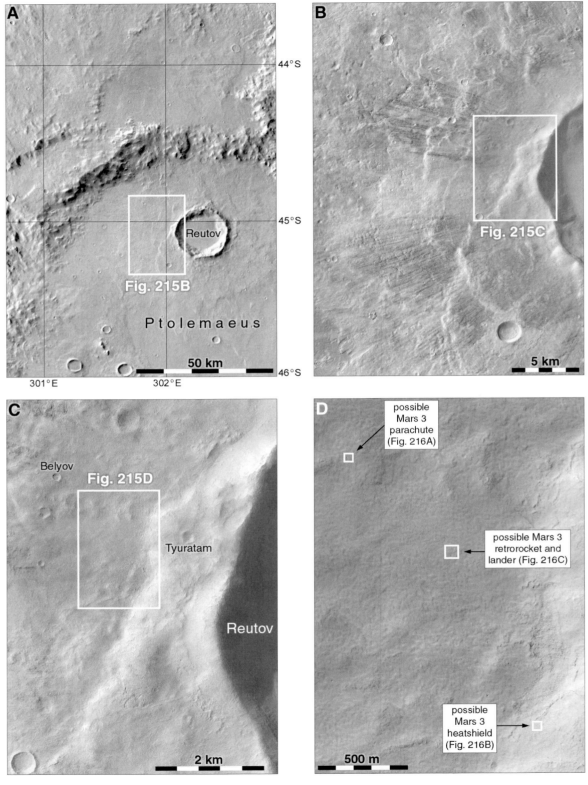

Figure 215. Mars 3 landing area and possible components. **A:** THEMIS infrared mosaic of the landing area, from Figure 23C in Stooke (2012). **B** and **C:** CTX image D04_028834_1347_XN_45S158W of the landing area. **D:** HiRISE image PSP_006154_1345 showing locations of possible components.

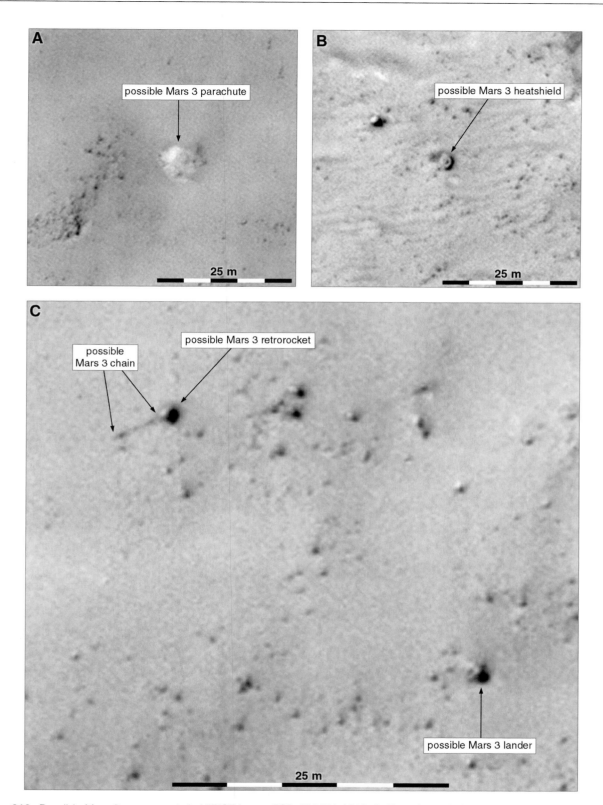

Figure 216. Possible Mars 3 components in HiRISE image PSP_006154_1345. **A:** Parachute. **B:** Heatshield. **C:** Lander and retrorocket with chain.

shows them in more detail. Nearby craters were given names of towns associated with the mission. The spacecraft was built in Reutov, near Moscow, and launched from the Baikonur cosmodrome, near Tyuratam.

Viking landing site selection

Figure 217 is a modified version of Figure 39 in Stooke (2012), showing the first sites suggested for the Viking Landers after the third meeting of the Viking Landing Site Working Group on 2–3 December 1970. Portree (2012) illustrated a contemporary map showing the six sites listed in Table 5b of Stooke (2012) with appropriate orientations for the Viking approach trajectories. The other sites suggested at the meeting have been rotated to match those orientations. Areas thought to be too high for successful operation of the parachute are outlined, using data probably derived from Belton and Hunten (1969).

Mars Odyssey

Mars Odyssey arrived in Mars orbit on 24 October 2001, and continued working throughout the decade covered in this atlas, becoming the longest-operating spacecraft at Mars in December 2010. Its THEMIS instrument mapped the whole surface of Mars (Figure 188 in Stooke, 2012), and it served as a communication relay for the Mars Exploration Rovers, Phoenix and Curiosity.

On 11 July 2012 during a small orbit adjustment, Mars Odyssey's Inertial Measurement Unit (IMU) failed and put the spacecraft in a safe mode. It was recovered a few sols later, and such safe modes were not uncommon during the long mission, most recently on 8 June 2012. A small orbital maneuver on 24 July 2012 moved the orbiter slightly so that it would be able to serve as a communication relay during the landing of Curiosity on 6 August. The relay worked perfectly despite concerns about the recent safe-mode events, allowing direct transmission of data from Curiosity to Earth after Earth had set in the Martian sky as seen from the descending spacecraft.

Mars Odyssey had two flight control computer systems for redundancy, and each was linked to its own set of other spacecraft systems. The recent safe modes and diagnostic studies indicated that the gyroscope system in one of the IMUs was beginning to wear out. Before it failed, the redundant IMU was activated on 5 November 2012, leaving the old system as an emergency backup system if required. This involved a switch to the redundant computer and other related systems, including the rover communication relay, which had not been used during the first 11 years at Mars.

Comet ISON passed within 10 million kilometers of Mars on 2 October 2013, and was imaged by many of the spacecraft at the planet, but only MRO's HiRISE camera obtained useful data. Mars Odyssey did not make observations, as its instruments were unlikely to see the faint comet, and any attempt would further stress its aging systems.

From June 2009 onwards Mars Odyssey had crossed the equator near local noon to provide optimum conditions for THEMIS. On 11 February 2014 the orbit was adjusted to allow it to drift towards morning crossings of the equator. The drift would be stabilized in November 2015, giving THEMIS a chance to observe the surface after dawn and after sunset on opposite sides of its orbit. Mars Odyssey was moved again with a maneuver on 5 August 2014 to ensure the spacecraft would be sheltered by Mars as the tail of Comet C/2013 Siding Spring swept past the planet on 19 October that year. During the flyby, in an attempt to gather thermal and compositional data, THEMIS observed the coma and tail, but they were too faint to detect.

Mars Express

Mars Express arrived in Mars orbit on 25 December 2003, and continued working throughout the decade covered in this atlas. Near-global color imaging and stereoscopic coverage of the surface were obtained by the High Resolution Stereo Camera (HRSC) (Figure 190 in Stooke, 2012), among many other observations, including compositional mapping of the planet. The topographic mapping from stereoscopic images provided much higher resolution than the Mars Orbiter Laser Altimeter (MOLA) data from Mars Global Surveyor, and far greater coverage than was possible from the still higher-resolution HiRISE stereoscopic imaging.

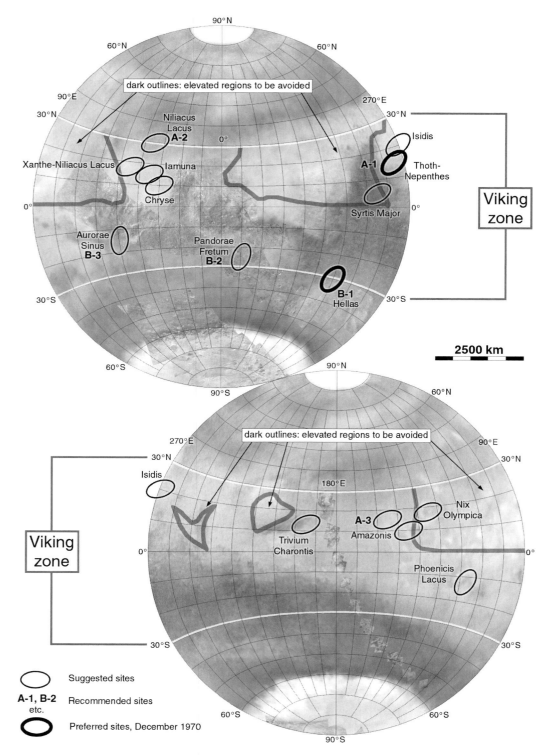

Figure 217. Early proposals for Viking landing sites, 1970. This map updates Figure 39 of Stooke (2012) to incorporate ellipse orientations and elevated areas illustrated by Portree (2012).

Comet ISON observations were made by most of the instruments on Mars Express between 24 September and 3 October 2013, particularly by HRSC on 24 and 30 September and 2 October, but no detections were made. A test consisting of three long-exposure images of Phobos by the Visual Monitoring Camera (VMC) was made on 26 September to prepare that camera for imaging of ISON, but, though Phobos was visible at a range of 8000 km, VMC could not detect the comet.

During the Curiosity rover landing on 6 August 2012, Mars Odyssey transmitted data directly from Curiosity to Earth when Earth passed behind Mars as seen by the rover. In case Mars Odyssey should experience problems during those crucial events, Mars Express also received the rover's transmissions. They could not be transmitted directly to Earth, but were recorded for later use if needed. Mars Express also relayed data from Curiosity occasionally, including ChemCam's RMI images of a rock target at Rocknest taken on 3 October 2012 (Curiosity's sol 57).

A very close Phobos flyby, only 45 km from the satellite, occurred on 29 December 2013. Knowledge of the mass and density might be slightly improved by this, but the very close flyby was also designed to reveal internal variations, if any, in the mass distribution. The previous close flyby had been at a distance of 67 km on 3 March 2010. On 28 April 2014 Mars Express was occulted by Phobos, permitting further refinement of the moon's evolving orbit. OMEGA images of Phobos (Fraeman *et al.*, 2012) are shown in Figure 209C.

The HRSC super-resolution channel (SRC) obtained several images of Comet Siding Spring during the spacecraft's orbit 13 710 on 19 October 2014, and looked for meteors in the atmosphere. The nucleus, only about 500 m across, was not resolved at the range of 137 000 km, but the coma was visible. Among other observations, the dust detector ASPERA measured particles in the comet's tail as it swept past the planet, and the MARSIS radar instrument detected electrons associated with the comet. The spacecraft's orbit was adjusted to minimize dust impacts as the tail passed by, but an earlier plan to turn Mars Express to shelter its body behind the high-gain antenna was not used.

Mars mission data

Table 58. *Mars Impact and Landing Events*

Spacecraft	Event	Arrival date	Mars date[*]	L_s	Location
Zond 2[†]	Impact?	6 August 1965	MY 6, sol 326	154	Unknown
Mars 2	Impact	27 November 1971	MY 9, sol 563	300	45° S, 58° E
Mars 3	Landing	2 December 1971	MY 9, sol 568	303	45° S, 202° E
Mars 6	Impact	12 March 1974	MY 11, sol 40	19	24° S, 340.5° E
Viking Lander 1	Landing	20 July 1976	MY 12, sol 209	97	22.48° N, 312.03° E
Viking Lander 2	Landing	3 September 1976	MY 12, sol 253	118	47.97° N, 134.26° E
Mars Pathfinder	Landing	4 July 1997	MY 23, sol 304	143	19.33° N, 326.45° E
Mars Climate Orbiter	‡	23 September 1999	MY 24, sol 424	211	34° N, 190° E
Mars Polar Lander	Impact	3 December 1999	MY 24, sol 494	256	76.3° S, 165.5° E
Deep Space 2 – Amundsen	Impact	3 December 1999	MY 24, sol 494	256	75.0° S, 164.4° E
Deep Space 2 – Scott	Impact	3 December 1999	MY 24, sol 494	256	75.0° S, 164.4° E
Beagle 2	Impact	25 December 2003	MY 26, sol 599	322	11.6° N, 90.4° E
MER-A lander	Landing	4 January 2004	MY 26, sol 609	327	14.57° S, 175.47° E
Spirit (end of drive)	Drive	25 March 2010	MY 30, sol 146	68	14.60° S, 175.52° E
MER-B lander	Landing	25 January 2004	MY 26, sol 630	339	1.95° S, 354.47° E
Opportunity (sol 3700)	Drive	24 June 2014	MY 32, sol 319	150	2.28° S, 354.63° E
Phoenix	Landing	25 May 2008	MY 29, sol 164	77	68.22° N, 234.25° E
MSL (Curiosity) landing	Landing	5 August 2012	MY 31, sol 319	151	4.59° S, 137.44° E
Curiosity (sol 669)	Drive	25 June 2014	MY 32, sol 320	151	4.65° S, 137.39° E
MSL cruise stage	Impact	5 August 2012	MY 31, sol 319	151	4.41° S, 136.10° E

Notes: [*] Mars dates are based on a calendar proposed by Clancy *et al.* (2000).

[†] Zond 2: possible impact, location unknown, suggested by Murray *et al.* (1967).

‡ Mars Climate Orbiter: approximate location at closest approach, or fragments if any reached the ground.

Table 59. *Mars Flyby and Orbital Events*

Spacecraft	Event	Arrival date	Mars date	L_s
Mars 1	Flyby	19 June 1963	MY 5, sol 237	110°
Mariner 4	Flyby	15 July 1965	MY 6, sol 305	143°
Zond 2	Impact or flyby	6 August 1965	MY 6, sol 326	154°
Mariner 6	Flyby	31 July 1969	MY 8, sol 405	200°
Mariner 7	Flyby	5 August 1969	MY 8, sol 410	203°
Mariner 9	Orbit	14 November 1971	MY 9, sol 550	292°
Mars 2	Orbit	27 November 1971	MY 9, sol 563	300°
Mars 3	Orbit	2 December 1971	MY 9, sol 568	303°
Mars 4	Flyby	10 February 1974	MY 11, sol 10	5°
Mars 5	Orbit	12 February 1974	MY 11, sol 12	6°
Mars 7	Flyby	9 March 1974	MY 11, sol 36	18°
Viking Orbiter 1	Orbit	19 June 1976	MY 12, sol 178	83°

Table 59. (*cont.*)

Spacecraft	Event	Arrival date	Mars date	L_s
Viking Orbiter 2	Orbit	7 August 1976	MY 12, sol 226	105°
Phobos 1	Flyby	25 January 1989	MY 18, sol 647	348°
Phobos 2	Orbit	29 January 1989	MY 18, sol 651	350°
Mars Observer	Flyby	24 August 1993	MY 21, sol 268	125°
Mars Global Surveyor	Orbit	12 September 1997	MY 23, sol 371	179°
Nozomi	Distant flyby	28 August 1999	MY 24, sol 399	196°
Mars Climate Orbiter	Flyby, possible entry	23 September 1999	MY 24, sol 424	211°
Deep Space 1	Distant flyby	10 November 1999	MY 24, sol 471	240°
2001 Mars Odyssey	Orbit	24 October 2001	MY 25, sol 497	258°
Nozomi	Flyby	14 December 2003	MY 26, sol 588	316°
Mars Express	Orbit	25 December 2003	MY 26, sol 599	322°
Mars Reconnaissance Orbiter	Orbit	10 March 2006	MY 28, sol 46	23°
Rosetta	Flyby	25 February 2007	MY 28, sol 389	190°
Dawn	Flyby	18 February 2009	MY 29, sol 425	212°
MAVEN	Orbit	22 September 2014	MY 32, sol 407	200°
Mars Orbiter Mission (Mangalyaan)	Orbit	24 September 2014	MY 32, sol 409	202°

Table 60. *Mars Exploration Chronology by Mars Year*

Mars Year	Events
1	Sol 483: First well-observed global dust storm begins (McKim, 1996)
2	Sol 214: Sputnik 1 in orbit
3	
4	Sol 129: Yuri Gagarin (Vostok 1) in orbit
5	Sol 237: Mars 1 flyby
6	Sol 305: Mariner 4 flyby. Sol 326: Zond 2 flyby or impact
7	
8	Sol 394: Apollo 11 lunar landing. Sol 405: Mariner 6 flyby. Sol 410: Mariner 7 flyby
9	Sol 550: Mariner 9 in orbit. Sol 563: Mars 2 impact, orbit. Sol 568: Mars 3 landing, orbit
10	Sol 156: Mars 2 and 3 end of mission. Sol 220: Mariner 9 end of mission
11	Sol 10: Mars 4 flyby. Sol 12: Mars 5 orbit. Sol 36: Mars 7 flyby. Sol 40: Mars 6 impact
12	Sol 178: Viking 1 in orbit. Sol 209: Viking 1 landing. Sol 226: Viking 2 in orbit. Sol 253: Viking 2 landing
13	Sol 255: Viking Orbiter 2 end of mission
14	Sol 197: Viking Lander 2 end of mission. Sol 311: Viking Orbiter 1 end of mission
15	Sol 448: Viking Lander 1 end of mission
16	
17	
18	Sol 647: Phobos 1 flyby. Sol 651: Phobos 2 in orbit
19	Sol 37: Phobos 2 end of mission
20	
21	Sol 268: Mars Observer flyby
22	
23	Sol 304: Mars Pathfinder landing. Sol 371: Mars Global Surveyor in orbit. Sol 396: Mars Pathfinder end of mission
24	Sol 399: Nozomi distant flyby. Sol 424: Mars Climate Orbiter flyby or atmospheric entry. Sol 470: Deep Space 1 distant flyby. Sol 494: Mars Polar Lander and Deep Space 2 impacts
25	Sol 497: Mars Odyssey in orbit

Table 60. (*cont.*)

Mars Year	Events
26	Sol 588: Nozomi flyby. Sol 599: Mars Express orbit and Beagle 2 impact. Sol 609: MER-A (Spirit) landing. Sol 630: MER-B (Opportunity) landing
27	Sol 94: Opportunity enters Endurance crater. Sol 100: Spirit reaches West Spur, Columbia Hills. Sol 558: Spirit reaches summit of Husband Hill. Sol 592: Opportunity reaches Erebus crater (Olympia)
28	Sol 18: Spirit reaches Home Plate. Sol 46: Mars Reconnaissance Orbiter in orbit. Sol 244: Opportunity arrives at Victoria crater. Sol 277: Mars Global Surveyor end of mission. Sol 389: Rosetta flyby. Sol 583: Opportunity enters Victoria crater
29	Sol 164: Phoenix landing. Sol 257: Opportunity exits Victoria crater. Sol 317: Phoenix end of mission. Sol 425: Dawn flyby
30	Sol 146: MER-A (Spirit) end of mission. Sol 407: Opportunity arrives at Santa Maria crater. Sol 636: Opportunity arrives at Endeavour crater (Cape York)
31	Sol 319: MSL (Curiosity) landing. Sol 444: Curiosity enters Yellowknife Bay. Sol 646: Curiosity departs from Glenelg area
32	Sol 14: Opportunity arrives at Solander Point. Sol 260: Curiosity arrives at Windjana (Kimberley). Sol 367: Opportunity arrives at Cape Tribulation. Sol 407: MAVEN in orbit. Sol 409: MOM (Mangalyaan) in orbit. Sol 433: Comet Siding Spring Mars flyby

Anticipated events

33	Sol 447: Insight landing. Sol 475 (approx): ExoMars Trace Gas Orbiter in orbit, EDM landing
34	Sol 365 (approx): ExoMars 2018 rover landing
36	Sol 35 (approx): NASA Mars 2020 rover landing

Bibliography

Anderson, R. and Bell, J., 2010. The geomorphology of the proposed MSL field site in Gale crater. Presented at the 4th MSL Landing Site Workshop, Monrovia, California, 27–29 September 2010.

Anderson, R., Sumner, D. and Bell, J., 2011. Science targets along a proposed Gale traverse. Presented at the Fifth MSL Landing Site Workshop, Monrovia, California, 16–18 May 2011.

Arvidson, R. E., Deal, K., Seelos, F., *et al.* (the Phoenix Science Team), 2005. Phoenix landing site characterization and certification. Presented at the First Mars Express Science Conference, Noordwijk, The Netherlands, 21–25 February 2005.

Arvidson, R. E., Squyres, S. W., Anderson, R. C., *et al.*, 2006. Overview of the Spirit Mars Exploration Rover mission to Gusev crater: landing site to Backstay Rock in the Columbia Hills. *Journal of Geophysical Research*, v. 111, E02S01, 22 pp., doi:10.1029/2005JE002499.

Arvidson, R. E., Ruff, S. W., Morris, R. V., *et al.*, 2008a. Spirit Mars Rover mission to the Columbia Hills, Gusev crater: mission overview and selected results from the Cumberland Ridge to Home Plate. *Journal of Geophysical Research*, v. 113, E12S33, doi:10.1029/2008JE003183.

Arvidson, R., Adams, D., Bonfiglio, G., *et al.*, 2008b. Mars Exploration Program 2007 Phoenix landing site selection and characteristics. *Journal of Geophysical Research*, v. 113, E00A03, doi:10.1029/2007JE003021.

Arvidson, R. E., Bonitz, R. G., Robinson, M. L., *et al.*, 2009. Results from the Mars Phoenix Lander Robotic Arm experiment. *Journal of Geophysical Research*, v. 114, E00E02, doi:10.1029/2009JE003408.

Arvidson, R. E., Bell III, J. F., Bellutta, P., *et al.*, 2010. Spirit Mars Rover mission: overview and selected results from the northern Home Plate winter haven to the side of Scamander crater. *Journal of Geophysical Research*, v. 115, E00F03, doi:10.1029/2010JE003633.

Arvidson, R. E., Ashley, J. W., Bell III, J. F., *et al.*, 2011. Opportunity Mars Rover mission: overview and selected results from Purgatory Ripple to traverses to Endeavour crater. *Journal of Geophysical Research*, v. 116, E00F15, doi:10.1029/2010JE003746.

Arvidson, R. E., Squyres, S. W., Bell, J. F., *et al.*, 2014a. Ancient aqueous environments at Endeavour crater, Mars.

Science, v. 343, no. 6169, 24 January 2014, doi:10.1126/science.1248097.

Arvidson, R. E., Bellutta, P., Calef, F., *et al.*, 2014b. Terrain physical properties derived from orbital data and the first 360 sols of Mars Science Laboratory Curiosity rover observations in Gale crater. *Journal of Geophysical Research*, v. 119, no. 6, doi:10.1002/2013JE004605.

Ashley, J. W., Golombek, M. P., Christensen, P. R., *et al.*, 2011. Evidence for mechanical and chemical alteration of iron–nickel meteorites on Mars: process insights for Meridiani Planum. *Journal of Geophysical Research*, v. 116, E00F20, doi:10.1029/2010/JE003672.

Augustine, N. R., Austin, W. M., Bajmuk, B. I., *et al.*, 2009. *Seeking a Human Spaceflight Program Worthy of a Great Nation*. Final Report of the Review of U.S. Human Spaceflight Plans Committee. Washington, DC: National Aeronautics and Space Administration.

Ball, A. J., Price, M. E., Walker, R. J., *et al.*, 2009. Mars Phobos and Deimos Survey (M-PADS) – a Martian moons orbiter and Phobos lander. *Advances in Space Research*, v. 43, no. 1, pp. 120–127.

Banerdt, W. B., *et al.* (the Cerberus Science Team), 2009. Cerberus: a Mars geophysical network mission for new frontiers. Presented at the 40th Lunar and Planetary Science Conference, The Woodlands, Texas, 23–27 March 2009. Abstract no. 2485.

Barriot, J.-P., Dehant, V., Folkner, W., *et al.*, 2001. The Netlander ionosphere and geodesy experiment. *Advances in Space Research*, v. 28, no. 8, pp. 1237–1249.

Basilevsky, A. T. and Shingareva, T. V, 2010. The selection and characterization of the Phobos-Soil landing sites. *Solar System Research*, v. 44, no. 1, pp. 38–43.

Basilevsky, A. T., Neukum, G., Michael, G., *et al.*, 2008. New MEX HRSC/SRC images of Phobos and the Fobos-Grunt landing sites. Presented at the 48th Vernadsky–Brown Microsymposium on Comparative Planetology, Russian Academy of Sciences, Moscow, Russia, 20–22 October 2008.

Behar, A., Rivellini, T. and Nicaise, F., 2003. Brush-wheel sampler concept for Gulliver Deimos sample return discovery proposal. Presented to the Canadian Space Agency, Montreal, 5 May 2003. JPL document 20060040069.

Bell, J. F., Lemmon, M. T., Duxbury, T. C., *et al.*, 2005. Solar eclipses of Phobos and Deimos observed from the surface of Mars. *Nature*, v. 436, pp. 55–57.

Belton, M. J. S. and Hunten, D. M., 1969. Spectrographic detection of topographic features on Mars. *Science*, v. 166, pp. 225–227.

Bibring, J.-P., Langevin, Y., Mustard, J. F., *et al.* (the OMEGA Team), 2006. Global mineralogical and aqueous Mars history derived from OMEGA/Mars Express data. *Science*, v. 312, pp. 400–404, doi:10.1126/science.1122659.

Blamont, J. and Jones, J. A., 2002. A new method for landing on Mars. *Acta Astronautica*, v. 51, no. 10, pp. 723–726. PII: S0094-5765(02)00019-X.

Blaney, D. L. and Wilson, G. R., 2000. Scouts: using numbers to explore Mars in situ. Presented at the Concepts and Approaches for Mars Exploration Meeting, Houston, Texas, 18–20 July 2000. Abstract no. 6053.

Blunck, J., 1977. *Mars and its Satellites: a Detailed Commentary on the Nomenclature*, First Edition. Hicksville, New York: Exposition Press.

Bonfiglio, E. P., Adams, D., Craig, L., *et al.*, 2011. Landing-site dispersion analysis and statistical assessment for the Mars Phoenix lander. *Journal of Spacecraft and Rockets*, v. 48, no. 5, pp. 784–797, doi:10.2514/1.48813.

Braun, R. D., Wright, H. S., Croom, M. A., *et al.*, 2004. The Mars airplane: a credible science platform. Proceedings of the IEEE Aerospace Conference, Big Sky, Montana, 6–13 March 2004. Paper IEEE 04-1260.

Braun, R. D., Wright, H. S., Croom, M. A., *et al.*, 2006. Design of the ARES Mars airplane and mission architecture. *Journal of Spacecraft and Rockets*, v. 43, no. 5, pp. 1026–1034, September–October.

Britt, D., 2010. The Gulliver mission: sample return from Deimos. Presented at the European Planetary Science Congress 2010, Rome, Italy, 19–24 September 2010. Abstracts, v. 5, EPSC2010-463.

Britt, D., *et al.* (the Gulliver team), 2005. Sample return from Deimos: the Gulliver mission. Presented at the 68th Annual Meteoritical Society Meeting, Gatlinburg, Tennessee, 12–16 September 2005. Abstract no. 5274.

Cabrol, N. A., Grin, E. A., Carr, M. H., *et al.*, 2003. Exploring Gusev crater with MER A: review of science objectives and testable hypotheses. *Journal of Geophysical Research*, Special Mars Exploration Rover (MER) Mission Issue, no. 8076, doi:10.1029/2002JE002026.

Certini, G. and Ugolini, F. C., 2013. An updated, expanded, universal definition of soil. *Geoderma*, v. 192, pp. 378–379, doi:10.1016/j.geoderma.2012.07.008.

Chojnacki, M., Burr, D. M., Moersch, J. E. and Michaels, T. I., 2011. Orbital observations of contemporary dune activity in Endeavor crater, Meridiani Planum, Mars. *Journal of Geophysical Research*, v. 116, E00F19, doi:10.1029/2010JE003675.

Christensen, P. R., Bandfield, J. L., Clark, R. N., *et al.*, 2000. Detection of crystalline hematite mineralization on Mars by the Thermal Emission Spectrometer: evidence for near-surface water. *Journal of Geophysical Research*, v. 105, no. E4, pp. 9623–9642.

Christensen, P. R., Bandfield, J. L., Hamilton, V. E., *et al.*, 2001. Mars Global Surveyor Thermal Emission Spectrometer experiment: investigation description and surface science results. *Journal of Geophysical Research*, v. 106, pp. 23,823–23,871, doi:10.1029/2000JE001370.

Clancy, R. T., Sandor, B. J., Wolff, M. J., *et al.*, 2000. An intercomparison of ground-based millimeter, MGS TES, and Viking atmospheric temperature measurements: seasonal and interannual variability of temperatures and dust loading in the global Mars atmosphere. *Journal of Geophysical Research*, v. 105, pp. 9553–9571.

Clark, P. E. and Clark, C., 2013. *Constant-Scale Natural Boundary Mapping to Reveal Global and Cosmic Processes*. Springer Briefs in Astronomy. New York: Springer, doi:10.1007/978-1-4614-7762-4.

Cole, S. P., Watters, W. A. and Squyres, S. W., 2012. Structure of Husband Hill and the west spur of the Columbia Hills, Gusev crater. Presented at the 43rd Lunar and Planetary Science Conference, The Woodlands, Texas, 19–23 March 2012. Abstract no. 1134.

Connerney, J. E. P., Acuña, M. H., Wasilewski, P. J., *et al.*, 2012. The global magnetic field of Mars and implications for crustal evolution. *Geophysical Research Letters*, v. 28, no. 21, pp. 4015–4018, doi:10.1029/2001GL013619.

Coradini, A., Grassi, D., Capaccioni, F., *et al.*, 2010. Martian atmosphere as observed by VIRTIS-M on Rosetta spacecraft. *Journal of Geophysical Research*, v. 115, E04004, doi:10.1029/2009JE003345.

Crumpler, L. S., Squyres, S. W., Arvidson, R. E., *et al.*, 2005. Mars Exploration Rover geologic traverse by the Spirit rover in the plains of Gusev crater, Mars. *Geology*, v. 33, pp. 809–812.

Crumpler, L., Arvidson, R., Squyres, S., *et al.* (the Athena Science Team), 2010. Overview of the field geologic context of Mars Exploration Rover Spirit, Home Plate and surroundings. Presented at the 41st Lunar and Planetary Science Conference, The Woodlands, Texas, 1–5 March 2010. Abstract no. 2557.

Crumpler, L. S., Arvidson, R. E., Squyres, S. W., *et al.*, 2011. Field reconnaissance geologic mapping of the Columbia Hills, Mars, based on Mars Exploration Rover Spirit and

MRO HiRISE observations. *Journal of Geophysical Research*, vol. 116, no. E00F24, doi:10.1029/2010JE003749.

Crumpler, L. S., *et al.* (the MER Athena Science Team), 2013a. Field geologic context of Opportunity traverse from Greeley Haven to the base of Matijevic Hill. Presented at the 44th Lunar and Planetary Science Conference, The Woodlands, Texas, 18–22 March 2013. Abstract no. 2292.

Crumpler, L., *et al.* (the MER Science Team), 2013b. Sol 3397 – August 13, 2013. Blog post, Field notes from Mars blog. New Mexico Museum of Natural History & Science, Albuquerque, New Mexico, 13 August 2013.

Crumpler, L., *et al.* (the MER Science Team), 2014. Sol 3623 – April 3, 2014. Blog post, Field notes from Mars blog. New Mexico Museum of Natural History & Science, Albuquerque, New Mexico, 3 April 2014.

Cull, S., Arvidson, R. E., Morris, R. V., *et al.*, 2010. Seasonal ice cycle at the Mars Phoenix landing site: 2. Postlanding CRISM and ground observations. *Journal of Geophysical Research*, v. 115, no. E00E19, doi:10.1029/2009JE003410.

Davy, R., Davis, J. A., Taylor, P. A., *et al.*, 2010. Initial analysis of air temperature and related data from the Phoenix MET station and their use in estimating turbulent heat fluxes. *Journal of Geophysical Research*, v. 115, E00E13, doi:10.1029/2009JE003444.

Dohm, J. M., Baker, V. R., Anderson, R. C., *et al.*, 2000. Martian magmatic-driven hydrothermal sites: potential sources of energy, water and life. Presented at the Concepts and Approaches for Mars Exploration Meeting, Lunar and Planetary Institute, Houston, 18–20 July 2000. Abstract no 6040.

Dohm, J. M., Ferris, J. C., Barlow, N. G., *et al.*, 2004. The Northwestern Slope Valleys (NSVs) region, Mars: a prime candidate site for the future exploration of Mars. *Planetary and Space Science*, v. 52, pp. 189–198.

Domokos, A., Bell III, J. F., Brown, P., *et al.*, 2007. Measurement of the meteoroid flux at Mars. *Icarus*, v. 191, no. 1, pp. 141–150.

Drake, B. G. (ed.), 1998. Reference mission version 3.0, Addendum to the Human Exploration of Mars: the reference mission of the NASA Mars Exploration Study Team. NASA/SP-6107-ADD. June 1998.

Drake, B. G. (ed.), 2009. Human Exploration of Mars, design reference architecture 5.0. NASA Mars Architecture Steering Group report. NASA/SP-2009-566. July 2009.

Drake, B. G., Hoffman, S. J. and Beaty, D. W., 2010. Human Exploration of Mars, design reference architecture 5.0. Presented at the Aerospace Conference, 2010 IEEE, 6–13 March 2010, updated 1 August 2010. IEEEAC Paper no. 1205, version 5.

Edgar, L. A., Grotzinger, J. P., Hayes, A. G., *et al.* 2012. Stratigraphic architecture of bedrock reference section, Victoria crater, Meridiani Planum, Mars. In *Sedimentary Geology of Mars*, Grotzinger, J. P. and Milliken, R. E. (eds), SEPM Special Publication No. 102, pp. 195–209.

Edgett, K. S., 2010. Curiosity's candidate field site in Gale crater, Mars. Presented at the 4th MSL Landing Site Workshop, Monrovia, California, 27–29 September 2010.

Edgett, K. S., Sumner, D. Y., Milliken, R. E. and Kah, L. C., 2008. Gale crater. Presented at the 3rd MSL Site Selection Workshop, Monrovia, California, 15–17 September 2008.

Ellehoj, M. D., Gunnlaugsson, H. P., Taylor, P. A., *et al.*, 2010. Convective vortices and dust devils at the Phoenix Mars mission landing site. *Journal of Geophysical Research*, vol. 115, E00E16, doi:10.1029/2009JE003413.

Ellery, A., Richter, L., Parnell, J. and Baker, A., 2006. A low-cost approach to the exploration of Mars through a robotic technology demonstrator mission. *Acta Astronautica*, v. 59, pp. 742–749.

Erickson, J. and Grotzinger, J. P., 2014. *Mission to Mt. Sharp. Habitability, Preservation of Organics, and Environmental Transitions*. Senior Review Proposal, Sections 1 and 2. Pasadena, CA: California Institute of Technology, April 2014.

Erickson, J. K., Manning, R. and Adler, M., 2004. Mars Exploration Rover: launch, cruise, entry, descent, and landing. Presented at the 55th International Astronautical Congress, Vancouver, BC, 4 October 2004. Paper IAC-04-Q.3.a.03.

European Space Agency, 2011. *Summary Outcome and Recommendations, European Workshop on Landing Sites for Exploration Missions*. Report of a Workshop Held at Leiden, The Netherlands, 17–21 January 2011, released 20 March 2011.

Fairén, A. G., Dohm, J. M., Uceda, E. R., *et al.*, 2005. Prime candidate sites for astrobiological exploration through the hydrogeological history of Mars. *Planetary and Space Science*, v. 53, no. 13, November 2005, pp. 1355–1375.

Farley, K. A., Malespin, C., Mahaffy, P., *et al.* (the MSL Science Team), 2013. In situ radiometric and exposure age dating of the Martian surface. *Science*, v. 343, no. 6169, doi:10.1126/science.1247166.

Farrand, W. H., Bell, J. F., Johnson, J. R., *et al.*, 2014. Observations of rock spectral classes by the Opportunity rover's Pancam on northern Cape York and on Matijevic Hill, Endeavour crater, Mars. *Journal of Geophysical Research*, v. 119, no. 11, pp. 2349–2369, doi:10.1002/2014JE004641.

Fraeman, A. A., Arvidson, R. E., Murchie, S. L., *et al.*, 2011. Testable hypotheses for Opportunity's traverse from Santa Maria to the rim of Endeavour crater. Presented at the 42nd

Lunar and Planetary Science Conference, The Woodlands, Texas, 7–11 March 2011. Abstract no. 2199.

Fraeman, A. A., Arvidson, R. E., Murchie, S. L., *et al.*, 2012. Analysis of disk-resolved OMEGA and CRISM spectral observations of Phobos and Deimos. *Journal of Geophysical Research*, v. 117, E00J15, doi:10.1029/2012JE004137.

Frost, R., 1916. *Mountain Interval*. New York: Henry Holt and Co.

Galimov, E. M., 2004. State of the planetary research in Russia ('Phobos SR' and 'Luna-Glob' projects). In *Proceedings, International Lunar Conference 2003*, International Lunar Exploration Working Group 5, Durst, S. M., Bohannan, C. T., Thomason, C. G. *et al.* (eds). American Astronautical Society Science and Technology Series, v. 108, pp. 23–31. Paper AAS 03-702.

Garvin, J. B., Levine, J. S., Anbar, A. D., *et al.*, 2008. Scientific goals and objectives for the Human Exploration of Mars, 2. Geology and geophysics. Presented at the 39th Lunar and Planetary Science Conference, League City, Texas, 10–14 March 2008. Abstract no. 1343.

Goetz, W., Pike, W. T., Hviid, S. F., *et al.*, 2010. Microscopy analysis of soils at the Phoenix landing site, Mars: classification of soil particles and description of their optical and magnetic properties. *Journal of Geophysical Research*, v. 115, E00E22, doi:10.1029/2009JE003437.

Golombek M. P., 2012. Timescale of small crater modification on Meridiani Planum, Mars. Presented at the 43rd Lunar and Planetary Science Conference, The Woodlands, Texas, 19–23 March 2012. Abstract no. 2267.

Golombek, M. P., Grant, J. A., Parker, T. J., *et al.*, 2003. Selection of the Mars Exploration Rover landing sites. *Journal of Geophysical Research*, v. 108, no. E12, pp. 8072–8120, doi:10.1029/2003JE002074.

Golombek, M. P., Crumpler, L. S., Grant, J. A., *et al.*, 2006a. Geology of the Gusev cratered plains from the Spirit rover transverse. *Journal of Geophysical Research*, v. 111, E02S07, doi:10.1029/2005JE002503.

Golombek, M. P., Grant, J. A., Crumpler, L. S., *et al.*, 2006b. Erosion rates at the Mars Exploration Rover landing sites and long-term climate change on Mars. *Journal of Geophysical Research*, v. 111, E12S10, doi:10.1029/2006JE002754.

Golombek, M., Michalski, J., Noe, E., *et al.*, 2007. General assessment of safety of prospective MSL landing sites. Presentation at the Second MSL Landing Site Workshop, Pasadena, California, 23–25 October 2007.

Golombek, M., Grant, J., Vasavada, A. R., *et al.*, 2008. Down-selection of landing sites for the Mars Science Laboratory. Presented at the 39th Lunar and Planetary Science Conference, League City, Texas, 10–14 March 2008. Abstract no. 2181.

Golombek, M., Noe, E. and Hanus, V., 2009. *Cerberus Mars Network Study*. JPL Report 301-169, 3 February 2009.

Golombek, M., Grant, J., Grotzinger, J., *et al.*, 2010a. Mars landing site selection activities: an update on MSL and future missions. Presented at the 22nd MEPAG Meeting, Monrovia, California, 17–18 March 2010.

Golombek, M., Robinson, K., McEwan, A., *et al.*, 2010b. Constraints on ripple migration at Meridiani Planum from Opportunity and HiRISE observations of fresh craters. *Journal of Geophysical Research*, v. 115, E00F08, doi:10.1029/2010JE003628.

Golombek, M., Grant, J., Vasavada, A. R., *et al.*, 2011a. Final four landing sites for the Mars Science Laboratory. Presented at the 42nd Lunar and Planetary Science Conference, The Woodlands, Texas, 7–11 March 2011. Abstract no. 1520.

Golombek, M., Hoover, R., Sladek, H., *et al.*, 2011b. Update on landing site characterization. Presented at the 5th MSL Landing Site Workshop, Monrovia, California, 16–18 May 2011.

Golombek, M., Grant, J., Kipp, D., *et al.*, 2012. Selection of the Mars Science Laboratory landing site. *Space Science Reviews*, v. 170, pp. 641–737, doi:10.1007/s11214-012-9916-y.

Golombek, M. P., Warner, N. H., Ganti, V., *et al.*, 2014. Small crater modification on Meridiani Planum and implications for erosion rates and climate change on Mars. *Journal of Geophysical Research*, v. 119, no. 12, pp. 2522–2547, doi:10.1002/2014JE004658.

Graf, J. E., Zurek, R. W., Erickson, J. K., *et al.*, 2007. Status of Mars Reconnaissance Orbiter mission. *Acta Astronautica*, v. 61, pp. 44–51.

Grant, J., 2007. Overview of process and goals. Presentation at the Second MSL Landing Site Workshop, Pasadena, California, 23–25 October 2007.

Grant, J. and Golombek, M. P., 2010. Ellipse locations for 2 new sites. MSL Landing Site Memorandum, 5 January 2010. marsoweb.nas.nasa.gov/landingsites/msl/memoranda/sites_jan10/Marg_NE_Syrtis_Ellipse_Locations.ppt.

Grant, J. A., Wilson, S. A., Cohen B. A., *et al.*, 2008. Degradation of Victoria crater, Mars. *Journal of Geophysical Research*, v. 113, E11010, doi:10.1029/2008JE003155.

Grant, J. A, Golombek, M. P., Grotzinger, J. P., *et al.*, 2010. The science process for selecting the landing site for the 2011 Mars Science Laboratory. *Planetary and Space Science*, v. 59, no. 11–12, pp. 1114–1127, doi:10.1016/j.pss.2010.06.016.

Greeley, R. and Thomas, P. E. (eds), 1995. *Mars Landing Site Catalog*. NASA Ref. Publ. 1238, Second Edition. Washington, DC: National Aeronautics and Space Administration.

Greeley, R., Waller, D. A., Cabrol, N. A., *et al.*, 2010. Gusev crater, Mars: observations of three dust devil seasons. *Journal of Geophysical Research*, v. 115, E00F02, doi:10.1029/2010JE003608.

Griebel, H. S., 2006. ARCHIMEDES (Aerial Robot Carrying High-Resolution Imaging, a Magnetometric Experiment and Direct Environmental Sensing Instruments). *Acta Astronautica*, v. 59, pp. 717–725.

Grotzinger, J. P., Arvidson, R. E., Bell, J. F., *et al.*, 2005. Stratigraphy and sedimentology of a dry to wet eolian depositional system, Burns Formation, Meridiani Planum, Mars. *Earth and Planetary Science Letters*, v. 240, pp. 11–72.

Gwinner, K., Hauber, E., Neukum, G., *et al.*, 2007. Contribution of High Resolution Stereo Camera (HRSC) data analysis for landing site selection on Mars. Presented at the 38th Lunar and Planetary Science Conference, League City, Texas, 12–16 March 2007. Abstract no. 1685.

Haberle, R. M., 2003. The science return of the Pascal Mars scout mission. *Proceedings IEEE Aerospace Conference*, March 8–15, 2003, v. 8, pp. 4025–4039.

Haberle, R. M., Catling, D. C., Chassefiere, E., *et al.*, 2000. The Pascal discovery mission: a Mars climate network mission. Presented at Concepts and Approaches for Mars Exploration, Houston, Texas, 18–20 July 2000. Abstract no. 6217.

Haberle, R. M., *et al.* (the Pascal Team), 2003. The Pascal Mars scout mission. Presented at the Third International Conference on Mars Polar Science and Exploration, Lake Louise, Alberta, Canada, 13–17 October 2003. Abstract no. 8075.

Heet, T. L., Arvidson, R. E., Cull, S. C., *et al.*, 2009. Geomorphic and geologic settings of the Phoenix lander mission landing site. *Journal of Geophysical Research*, v. 114, E00E04, doi:10.1029/2009JE003416.

Hopkins, J. B. and Pratt, W. D., 2011a. Comparison of Deimos and Phobos for human exploration and identification of preferred landing sites on Deimos. Presented at the Second International Conference on the Exploration of Phobos and Deimos, NASA Ames Research Center, 14–16 March 2011.

Hopkins, J. and Pratt, W., 2011b. Comparison of Deimos and Phobos as destinations for human exploration, and identification of preferred landing sites. Presented at the AIAA Space 2011 Conference, Long Beach, California, 27–29 September 2011.

Irwin, R., 2008. Notional traverses and science targets in Holden crater. Presented at the 3rd MSL Site Selection Workshop, Monrovia, California, 15–17 September 2008.

Iskander, E., Hartman, E., Ngo, F., *et al.*, 2008. Atromos: Mars companion mission, mid sized polar lander investigation.

Presented at the 6th International Planetary Probe Workshop, Atlanta, Georgia, 23–27 June 2008.

Jakosky, B. M., Westall, F. and Brack, A., 2007. Mars. In *Planets and Life – The Emerging Science of Astrobiology*, Baross, J. and Sullivan, W. (eds), Cambridge and New York: Cambridge University Press, pp. 357–387.

Johnson, J. R., Bell, J. F., Cloutis, E., *et al.*, 2007. Mineralogic constraints on sulfur-rich soils from Pancam spectra at Gusev crater, Mars. *Geophysical Research Letters*, v. 34, L13202, doi:10.1029/2007GL029894.

Johnston, M. D., Graf, J. E., Zurek, R. W., *et al.*, 2007. The Mars Reconnaissance Orbiter mission: from launch to the primary science orbit. Presented at the Aerospace Conference, IEEE 2007, Big Sky, Montana. IEEEAC Paper no. 1001.

Johnston, M. D., Herman, D. E., Zurek, R. W. and Edwards, C. D., 2011. Mars Reconnaissance Orbiter: extended dual-purpose mission. Presented at the Aerospace Conference, IEEE 2011, Big Sky, Montana. IEEEAC Paper no. 1278.

Jones, J. A., Cutts, J. A., Hall, J. L., *et al.*, 2005. Montgolfiere balloon missions for Mars and Titan. Presented at the Third International Planetary Probe Workshop, Anavyssos, Greece, 27 June to 1 July 2005.

Kenney, P. S. and Croom, M., 2003. Simulating the ARES aircraft in the Mars environment. Presented at 2nd AIAA Unmanned Unlimited Conference, San Diego, CA, 15–18 September 2003, doi:10.2514/6.2003-6579.

Kerstein, L., Bischof, B., Renken, H., *et al.*, 2006. Micro-Mars: a low-cost mission to planet Mars with scientific orbiter and lander applications. *Acta Astronautica*, v. 59, pp. 608–616.

Knocke, P. C., Wawrzyniak, G. G., Kennedy, B. M., *et al.*, 2004. Mars Exploration Rovers landing dispersion analysis. Presented at the AIAA/AAS Astrodynamics Specialist Conference, Providence, Rhode Island, 16–19 August 2004.

Kuhl, C. A., 2008. A planetary protection strategy for the Mars Aerial Regional-scale Environmental Survey (ARES) mission concept. NASA/TM-2008-215344, August 2008.

Kuhl, C. A., 2009. Design of a Mars airplane propulsion system for the Aerial Regional-scale Environmental Survey (ARES) mission concept. NASA TM-2009-215700, March 2009.

Kuzmin, R. O. and Shingareva, T. V., 2002. Phobos-Grunt mission. A morphologic characteristic of the proposed landing site. Presented at the Vernadsky–Brown Microsymposium 36, Moscow, Russia, 14–16 October 2002.

Kuzmin, R. O. and Zabalueva, E. V., 2003. The temperature regime of the surface layer of the Phobos regolith in the region of the potential Fobos-Grunt space station landing site. *Solar System Research*, v. 37, no. 6, pp. 480–488.

Kuzmin, R. O., Shingareva, T. V. and Zabalueva, E. V., 2003. An engineering model for the Phobos surface. *Solar System Research*, v. 37, no. 4, pp. 266–281.

Lainey, V. and Le Poncin-Lafitte, C., 2010. The GETEMME mission or when fundamental physics meets planetology – the planetology part. *European Planetary Science Congress Abstracts*, Vol. 5, EPSC2010-60, 2010.

Lamb, M. P., Grotzinger, J. P., Southard, J. B. and Tosca, N. J., 2012. Were aqueous ripples on Mars formed by flowing brines? In *Sedimentary Geology of Mars*. SEPM Special Publication no. 102. Tulsa, Oklahoma: Society for Sedimentary Geology, pp. 139–150.

Landis, G. A., *et al.* (the MER Athena Science Team), 2007. Observation of frost at the equator of Mars by the Opportunity rover. Presented at the 38th Lunar and Planetary Science Conference, League City, Texas, 12–16 March 2007. Abstract no. 2423.

Lebleu, D. and Schipper, A.-M., 2003. The Huygens and Netlander missions: a system overview. Presented at International Workshop, Planetary Probe Atmospheric Entry and Descent Trajectory Analysis and Science, Lisbon, Portugal, 6–9 October 2003.

Lee, P. and Martin, G. L., 2011. Human missions to Phobos and Deimos: what science? Presented at the 2nd International Conference on the Exploration of Phobos and Deimos, NASA Ames Research Center, Moffett Field, California, 14–16 March 2011.

Lee, P., Veverka, J., Bellerose, J., *et al.*, 2010. Hall: a Phobos and Deimos sample return mission. Presented at the 41st Lunar and Planetary Science Conference, The Woodlands, Texas, 1–5 March 2010. Abstract no. 1633.

Lee, P., Jones, T., Jaroux, B., *et al.*, 2011. M4: Mars Moons Multiple-landings Mission. Presented at the 2nd International Conference on the Exploration of Phobos and Deimos, NASA Ames Research Center, Moffett Field, California, 14–16 March 2011.

Lee, P., Hoftun, C. and Lorber, K., 2012. Phobos and Deimos: robotic exploration in advance of humans to Mars orbit. Presented at Concepts and Approaches for Mars Exploration, Houston, Texas, 12–14 June 2012. LPI Contribution no. 1679. Abstract no. 4363.

Le Maistre, S., Rivoldini, A., Dehant, V., *et al.*, 2013. New Mars rotational model from Opportunity radio-tracking and implications to the interior structure of Mars. Presented at the European Planetary Science Congress 2013, London, UK, 8–13 September 2013. Abstracts, v. 8, Abstract no. EPSC2013-978.

Leshin, L. A., 2003. Stardust goes to Mars: the SCIM Mars sample return mission. Presented at the Cometary Dust in Astrophysics Conference, Crystal Mountain, Washington, 10–15 August 2003. Abstracts, p. 43.

Levine, J. S., Garvin, J. B. and Head, J. W., 2010a. Martian geology investigations. Planning for the scientific exploration of Mars by humans, Part 2. *Journal of Cosmology*, v. 12, October–November, pp. 3636–3646.

Levine, J. S., Garvin, J. B. and Elphic, R. C., 2010b. Martian geophysics investigations. Planning for the scientific exploration of Mars by humans, Part 3. *Journal of Cosmology*, v. 12, October–November, pp. 3647–3657.

Levine, J. S., Garvin, J. B. and Hipkin, V., 2010c. Martian atmosphere and climate investigations. Planning for the scientific exploration of Mars by humans, Part 4. *Journal of Cosmology*, v. 12, October–November, pp. 3658–3670.

Levine, J. S., Garvin, J. B. and Doran, P. T., 2010d. Martian biological investigations and the search for life. Planning for the scientific exploration of Mars by humans, Part 5. *Journal of Cosmology*, v. 12, October–November, pp. 3671–3684.

Lewis, K. and Aharonson, O., 2008. Geomorphic aspects of Eberswalde delta and potential MSL traverses. Presented at the 3rd MSL Site Selection Workshop, Monrovia, California, 15–17 September 2008.

Lewis, K. W. and Aharonson, O., 2010. Evolution and stratigraphic architecture of the Eberswalde basin. Presented at the 4th MSL Landing Site Workshop, Monrovia, California, 27–29 September 2010.

Lewis, K. W., Aharonson, O., Grotzinger, J. P., *et al.*, 2008. Structure and stratigraphy of Home Plate from the Spirit Mars Exploration Rover. *Journal of Geophysical Research*, v. 113, E12S36, doi:10.1029/2007JE003025.

Litvak, M. L., Kozyrev, A. S., Malakhov, A. V., *et al.*, 2007. Dynamic albedo of neutrons instrument onboard MSL mission: selection of landing site from HEND/Odyssey data. Presented at the 38th Lunar and Planetary Science Conference, League City, Texas, 12–16 March 2007. Abstract no. 1554.

Lockheed Martin, 2004. Lockheed Martin's systems-of-systems lunar architecture point-of-departure concept. CE&R BAA Open Forum, CA-1 Midterm Briefing, 30 November 2004.

Lockheed Martin, 2005. Lockheed Martin's systems-of-systems lunar architecture point-of-departure concept. CE&R BAA Open Forum, CA-1 (Basic Period) Final Briefing, 1 March 2005.

Loizeau, D., Bibring, J.-P., Bishop, J., *et al.*, 2010. One day, one month, one year at Mawrth Vallis. Presented at the 4th MSL Landing Site Workshop, Monrovia, California, 27–29 September 2010.

Lorenz, C., Basilevsky, A., Oberst, J., *et al.*, 2011. Geology of Phobos-Grunt landing sites: a view from the MEX HRSC images. Presented at the Second Moscow Solar System Symposium, Moscow, Russia, 10–14 October 2011.

Malin Space Science Systems, 2004a. Mars Exploration Rover (MER-A) Spirit landing site. MGS MOC Release no. MOC2-594, 3 January 2004.

Malin Space Science Systems, 2004b. Mars Exploration Rover (MER-B) Opportunity landing site. MGS MOC Release no. MOC2-615, 24 January 2004.

Malin Space Science Systems, 2006. New gully deposit in a crater in the Centauri Montes region: evidence that water flowed on Mars during the past 7 years? MGS MOC Release no. MOC2-1619, 6 December 2006.

Malin, M. C., Edgett, K. S., Posiolova, L. V., *et al.*, 2006. Present-day impact cratering rate and contemporary gully activity on Mars. *Science*, v. 314, pp. 1573–1577, doi:10.1126/science.1135156.

Malin, M. C., Kennedy, M. R., Cantor, B. A. and Edgett, K. S., 2008. MRO CTX spots dust devils at Phoenix landing site. Malin Space Science Systems Captioned Image Release, MSSS-30, 6 May 2008.

Mangold, N., Loizeau, D., Bibring, J.-P., *et al.*, 2011. Targets to address MSL goals and major Martian science objectives at Mawrth Vallis. Presented at the Fifth MSL Landing Site Workshop, Monrovia, California, 16–18 May 2011.

Marsal, O., Harri, A.-M., Lognonné, P., *et al.*, 1999. Netlander: the first scientific lander network on the surface of Mars. Presented at 50th International Astronautical Congress, American Institute of Aeronautics and Astronautics. Paper IAF-99-Q.3.03.

Matousek, S., 2001. Mars scouts: an overview. Presented at 15th Annual AIAA/USU Conference on Small Satellites, Logan, Utah, 13–16 August 2001.

McCoy, T. J., Sims, M., Schmidt, M. E., *et al.*, 2008. Structure, stratigraphy, and origin of Husband Hill, Columbia Hills, Gusev crater, Mars. *Journal of Geophysical Research*, v. 113, E06S03, doi:10.1029/2007JE003041.

McEwen, A., 2013. Could this be the Soviet Mars 3 lander? HiRISE Press Release, 11 April 2013, http://www.uahirise.org/ESP_031036_1345.

McKeown, N. and Rice, M., 2011. Mineralogy of select outcrops at Eberswalde crater. Presented at the Fifth MSL Landing Site Workshop, Monrovia, California, 16–18 May 2011.

McKim, R. J., 1996. The dust storms of Mars. *Journal of the British Astronomical Association*, v. 106, no. 4, pp. 185–200.

Mellon, M. T., Arvidson, R. E., Marlow, J. J., *et al.*, 2008. Periglacial landforms at the Phoenix landing site and the northern plains of Mars. *Journal of Geophysical Research*, v. 113, E00A23, doi:10.1029/2007JE003039.

Mellon, M. T., Malin, M. C., Arvidson, R. E., *et al.*, 2009. The periglacial landscape at the Phoenix landing site. *Journal of Geophysical Research*, v. 114, E00E06, doi:10.1029/2009JE003418.

MEPAG HEM-SAG (2008). Planning for the scientific exploration of Mars by humans. Unpublished White Paper, posted March 2008 by the Mars Exploration Program Analysis Group (MEPAG) at http://mepag.jpl.nasa.gov/reports/index.html.

Merrill, R. G. and Strange, N. J., 2011. Crewed Mars' moons exploration utilizing the 2010 human exploration framework. Presented at the Second International Conference on the Exploration of Phobos and Deimos, NASA Ames Research Center, 14–16 March 2011 [11-019].

MIIGAiK, 1992. *Atlas Planet Zemnoi Gruppy i ikh Sputnikov* (in Russian). Moscow: Moscow State University of Geodesy and Cartography (MIIGAiK).

Miller, D. D., Haberle, R. M., Mitchell, S. J., *et al.*, 2003. An affordable mission design for emplacement of a global network on Mars. *Proceedings of the 2003 IEEE Aerospace Conference*, 8–15 March 2003, v. 1, pp. 1–15, doi:10.1109/AERO.2003.1235056.

Moores, J. E., Lemmon, M. T., Smith, P. H., *et al.*, 2010. Atmospheric dynamics at the Phoenix landing site as seen by the Surface Stereo Imager. *Journal of Geophysical Research*, v. 115, E00E08, doi:10.1029/2009JE003409.

Moores, J. E., Komguem, L., Whiteway, J. A., *et al.*, 2011. Observations of near-surface fog at the Phoenix Mars landing site. *Geophysical Research Letters*, v. 38, L04203, doi:10.1029/2010GL046315.

Morris, R. V., Ruff, S. W., Gellert, R., *et al.*, 2010. Identification of carbonate-rich outcrops on Mars by the Spirit rover. *Science*, v. 329, 23 July 2010, no. 5990, pp. 421–424, doi:10.1126/science.1189667.

Murbach, M. S., Papadopoulos, P., White, B. and Tegnerud, E., 2007. Atromos: a Mars companion mission enabled by advanced EDL concepts. Presented at the 5th International Planetary Probe Workshop, Bordeaux, France, 23–29 June 2007.

Murbach, M., Papadopoulos, P., White, B., *et al.*, 2010. Atromos 2016 – a Mars dual-probe surface mission. Presented at the 7th International Planetary Probe Workshop, Barcelona, Spain, 14–18 June 2010.

Murbach, M. S., Colaprete, A., Papadopoulos, P. and Atkinson, D., 2012. Atromos: a SALMON-class Mars companion surface mission. Presented at Concepts and Approaches for Mars Exploration, Houston, Texas, 12–14 June 2012. LPI Contribution no. 1679. Abstract no. 4230.

Murchie, S., Choo, T., Humm, D., *et al.* (the CRISM Team), 2008. MRO/CRISM observations of Phobos and Deimos. Presented at the 39th Lunar and Planetary Science Conference, League City, Texas, 10–14 March 2008. Abstract no. 1434.

Murchie, S. L., Chabot, N. L., Rivkin, A. S., *et al.*, 2011. MERLIN: Mars-Moon Exploration, Reconnaissance and Landed Investigation. Presented at the 9th IAA Low-Cost Planetary Missions Conference, Laurel, Maryland, 21–23 June 2011.

Murray, B. C., Davies, M. E. and Eckman, P. K., 1967. Planetary contamination II: Soviet and U.S. practices and policies. *Science*, v. 155, 24 March 1967, no. 3769, pp. 1505–1511.

NASA, 1967a. *Mariner Mars 1964 Project Report: Spacecraft Performance and Analysis*. NASA Technical Report 32-882, 15 February 1967. Pasadena, California: Jet Propulsion Laboratory.

NASA, 1967b. *Mariner Mars 1964 Project Report: Television Experiment. Part 1. Investigators' Report. Mariner IV Pictures of Mars*. NASA Technical Report 32-884, 15 December 1967. Pasadena, California: Jet Propulsion Laboratory.

NASA, 2004. *The Vision for Space Exploration*. Document NP-2004-01-334-HQ, 55583main_vision_space_exploration2.pdf, February 2004. Washington, DC: National Aeronautics and Space Administration.

Noland, M. and Veverka, J., 1977. The photometric functions of Phobos and Deimos. II. Surface photometry of Deimos. *Icarus*, v. 30, no. 1, pp. 200–211.

Noreen, G., Komarek, T., Diehl, R., *et al.*, 2002. G. Marconi Orbiter and the Mars relay network. Presented at the 20th AIAA International Communication Satellite Systems Conference, Montreal, Canada, 13–15 May 2002.

Oberst, J., Lainey, V., Le Poncin-Lafitte, C., *et al.*, 2012. GETEMME – a mission to explore the Martian satellites and the fundamentals of solar system physics. *Experimental Astronomy*, v. 34, no. 2, pp. 243–271.

O'Neil, W. J. and Cazaux, C., 2000. The Mars sample return project. *Acta Astronautica*, v. 47 (2–9), pp. 53–65.

Pajola, M., Magrin, S., Lazzarin, M., La Forgia, F. and Barbieri, C., 2012a. Rosetta-Mars fly-by, February 25, 2007. *Memorie della Societa Astronomica Italiana Supplement*, v. 20, pp. 105–113.

Pajola, M., Lazzarin, M., Bertini, I., *et al.* (the OSIRIS Team), 2012b. Spectrophotometric investigation of Phobos with the Rosetta OSIRIS-NAC camera and implications for its collisional capture. *Monthly Notices of the Royal Astronomical Society*, v. 427, pp. 3230–3243, doi:10.1111/j.1365-2966.2012.22026.x.

Paris, K. N., Allen, C. C. and Oehler D. Z., 2007. Candidate landing site for the Mars Science Laboratory: Vernal crater, S.W. Arabia Terra. Presented at the 38th Lunar and Planetary Science Conference, League City, Texas, 12–16 March 2007. Abstract no. 1316.

Parker, T. J., Gorsline, D. S., Saunders, R. S., *et al.*, 1993. Coastal geomorphology of the Martian northern plains. *Journal of Geophysical Research*, v. 98, pp. 11061–11078.

Parker, T. J., Golombek, M. P. and Powell, M. W., 2010. Geomorphic/geologic mapping, localization, and traverse planning at the Opportunity landing site, Mars. Presented at the 41st Lunar and Planetary Science Conference, The Woodlands, Texas, 1–5 March 2010. Abstract no. 2638.

Pavone, M., Castillo-Rogez, J. C., Nesnas, I. A. D., *et al.*, 2013. Spacecraft/rover hybrids for the exploration of small solar system bodies. In *Aerospace Conference 2013 IEEE*, pp. 1–11. New York: IEEE.

Pollack, J. B., Colburne, D., Kahn, R., *et al.*, 1977. Properties of aerosols in the Martian atmosphere, as inferred from Viking lander imaging data. *Journal of Geophysical Research*, v. 82, no. 28, pp. 4479–4496.

Popa, C., Esposito, F. and Colangeli, L., 2010. New landing site proposal for Mars Science Laboratory (MSL) in Xanthe Terra. Presented at the 41st Lunar and Planetary Science Conference, The Woodlands, Texas, 1–5 March 2010. Abstract no. 1807.

Portree, D. S. F., 2012. The earliest candidate Viking landing sites (1970). Wired Science Blog: Beyond Apollo, posted 22 November 2012. http://www.wired.com/wiredscience/2012/11/early-candidate-viking-landing-sites.

Prince, J. L., Desai, P. N., Queen, E. M. and Grover, M. R., 2008. Entry, descent, and landing operations analysis for the Mars Phoenix lander. Presented at the AIAA/AAS Astrodynamics Specialist Conference, Honolulu, Hawaii, 18–21 August 2008.

Putzig, N. E., Mellon, M. T., Golombek, M. P. and Arvidson, R. E., 2006. Thermophysical properties of the Phoenix Mars landing site study regions. Presented at the 37th Lunar and Planetary Science Conference, League City, Texas, March 2006. Abstract no. 2426.

Rayman, M. D. and Mase, R. A., 2010. The second year of Dawn mission operations: Mars gravity assist and onward to Vesta. *Acta Astronautica*, v. 67, pp. 483–488, doi:10.1016/j.actaastro.2010.03.010.

Renton, D., 2006. Deimos sample return technology reference study, executive summary. Report SCI-A/2006/010/DSR, 28 August 2006. European Space Agency, Science Payload and Advanced Concepts Office, Planetary Exploration Studies Section.

Rice, J. W., Greeley, R., Li, R., *et al.* (the Athena Science Team), 2010a. Geomorphology of the Columbia Hills complex: landslides, volcanic vent, and other Home Plates. Presented at the 41st Lunar and Planetary Science Conference, The Woodlands, Texas, 1–5 March 2010. Abstract no. 2566.

Rice, M., Bell, J., Gupta, S. and Warner, N., 2010b. Testable hypotheses and candidate science targets within the Eberswalde landing ellipse. Presented at the 4th MSL Landing Site Workshop, Monrovia, California, 27–29 September 2010.

Rivkin, A. S., Chabot, N. L., Murchie, S. L., *et al.*, 2011. MERLIN: Mars-Moon Exploration, Reconnaissance and Landed Investigation. Presented at the Second International Conference on the Exploration of Phobos and Deimos, NASA Ames Research Center, 14–16 March 2011 [11-014].

Ruff, S. W., Farmer, J. D., Calvin, W. M., *et al.*, 2011. Characteristics, distribution, origin, and significance of opaline silica observed by the Spirit rover in Gusev crater, Mars. *Journal of Geophysical Research*, v. 116, E00F23, doi:10.1029/2010JE003767.

Sagdeev, R. Z. and Zakharov, A. V., 1989. Brief history of the Phobos mission. *Nature*, v. 341, pp. 581–585, doi:10.1038/341581a0.

Schieber, J. and Malin, M., 2008. The depositional setting at Eberswalde crater. Presented at the 3rd MSL Site Selection Workshop, Monrovia, California, 15–17 September 2008.

Schröder, C., Rodionov, D. S., McCoy, T. J, *et al.*, 2008. Meteorites on Mars observed with the Mars Exploration Rovers. *Journal of Geophysical Research*, v. 113, no. E6, E06S22, doi:10.1029/2007JE002990.

Schröder, C., Herkenhoff, K. E., Farrand, W. H., *et al.*, 2010. Properties and distribution of paired candidate stony meteorites at Meridiani Planum, Mars. *Journal of Geophysical Research*, v. 115, E00F09, doi:10.1029/2010JE003616.

Schulze-Makuch, D., Dohm, J. M., Fan, C., *et al.*, 2007. Exploration of hydrothermal targets on Mars, *Icarus*, v. 189, no. 2, pp. 308–324, doi:10.1016/j.icarus.2007.02.007.

Selsis, F., Lemmon, M. T., Vaubaillon, J. and Bell III, J. F., 2005. A Martian meteor and its parent comet. *Nature*, v. 435, p. 581.

Shingareva, T. V. and Kuzmin, R. O., 2000. Some new data on morphology of Phobos surface. Presented at the 31st Lunar and Planetary Science Conference, Houston, Texas, 13–17 March 2000. Abstract no. 1665.

Simonelli, D. P., Thomas, P. C., Carcich, B. T. and Veverka, J., 1993. The generation and use of numerical shape models for irregular solar system objects. *Icarus*, v. 103, pp. 49–61.

Singer, S. F., 1984. The Ph-D proposal, a manned mission to Phobos and Deimos. In *The Case for Mars; Proceedings of the Conference*, University of Colorado, Boulder, CO, April 29–May 2, 1981, Boston, P. J. (ed.), pp. 39–65. Paper AAS 81-231.

Sizemore, H. G. and Mellon, M. T., 2006. Multi-scale variability in the ice-table depth at potential Phoenix landing sites. Presented at the 37th Lunar and Planetary Science Conference, League City, Texas, 13–17 March 2006. Abstract no. 2141.

Sizemore, H. G., Mellon, M. T. and Golombek, M. P., 2009. Ice table depth variability near small rocks at the Phoenix landing site, Mars: a pre-landing assessment. *Icarus*, v. 199, no. 2, pp. 303–309, doi:10.1016/j.icarus.2008.10.008.

Spencer, D. A., Adams, D. S., Bonfiglio, E., *et al.*, 2009. Phoenix landing site hazard assessment and selection. *Journal of Spacecraft and Rockets*, v. 46, no. 6, pp. 1196–1201, doi:10.2514/1.43932.

Squyres, S., 2005. *Roving Mars: Spirit, Opportunity, and the Exploration of the Red Planet*. New York: Hyperion.

Squyres, S. W., Arvidson, R. E., Bell III, J. F., *et al.*, 2004a. The Spirit Rover's Athena science investigation at Gusev crater, Mars. *Science*, v. 305, 6 August 2004, no. 5685, pp. 794–799, doi:10.1126/science.3050794.

Squyres, S. W., Arvidson, R. A., Bell III, J. F., *et al.*, 2004b. The Opportunity Rover's Athena science investigation at Meridiani Planum, Mars. *Science*, v. 306, 3 December 2004, no. 5702, pp. 1698–1703, doi:10.1126/science.1106171.

Squyres, S. W., Arvidson, R. A., Bollen, D., *et al.*, 2006a. Overview of the Opportunity Mars Exploration Rover mission to Meridiani Planum: Eagle crater to Purgatory Ripple. *Journal of Geophysical Research*, v. 111, E12S12, doi:10.1029/2006JE002771.

Squyres, S. W., Knoll, A. H., Arvidson, R. E., *et al.*, 2006b. Two years at Meridiani Planum: results from the Opportunity rover. *Science*, v. 313, 8 September 2006, no. 5792, pp. 1403–1407, doi:10.1126/science.1130890.

Squyres, S. W., Aharonson, O., Clark, B. C., *et al.*, 2007. Pyroclastic activity at Home Plate in Gusev crater, Mars. *Science*, v. 316, no. 5825, pp. 738–742, doi:10.1126/science.1139045.

Squyres, S. W., Arvidson, R. E., Ruff, S., *et al.*, 2008. Detection of silica-rich deposits on Mars. *Science*, v. 320, no. 5879, pp. 1063–1067, doi:10.1126/science.1155429.

Squyres, S. W., Knoll, A. H., Arvidson, R. E., *et al.*, 2009. Exploration of Victoria crater by the Mars rover Opportunity. *Science*, v. 324, no. 5930, pp. 1058–1061, doi:10.1126/science.1170355.

Squyres, S. W., Arvidson, R. E., Bell, J. F., *et al.*, 2012. Ancient impact and aqueous processes at Endeavour crater, Mars. *Science*, v. 336, no. 6081, pp. 570–576, doi:10.1126/science.1220476.

Steinhoff, E. A., 1962. A possible approach to scientific exploration of the planet Mars. In *From Peenemünde to Outer Space – Commemorating the Fiftieth Birthday of Wernher von Braun, 23 March 1962*, Contribution 38, pp. 803–836. Huntsville, Alabama: NASA Marshall Space Flight Center.

Stolper, E. M., Baker, M. B., Newcombe, M. E., *et al.* (the MSL Science Team), 2013. The petrochemistry of Jake_M: a Martian mugearite. *Science*, v. 341, no. 6153, doi:10.1126/science.1239463.

Stooke, P. J., 2007a. Mars sample return via Phobos cache and human retrieval. Presented at the First International Conference on the Exploration of Phobos and Deimos, 5–7 November 2007.

Stooke, P. J., 2007b. *The International Atlas of Lunar Exploration*. Cambridge: Cambridge University Press.

Stooke, P. J., 2012. *The International Atlas of Mars Exploration*, vol 1, *The First Five Decades*. Cambridge: Cambridge University Press.

Stooke, P. J., 2013. MER early traverse mapping: MOC vs HiRISE localization. Presented at the 44th Lunar and Planetary Science Conference, The Woodlands, Texas, 18–22 March 2013. Abstract no. 1396.

Stooke, P. J., 2014. Mars sample return via robotic collection, Phobos cache and human retrieval. Presented at the 45th Lunar and Planetary Science Conference, The Woodlands, Texas, 17–21 March 2014. Abstract no. 1043.

Sumner, D., 2011. Distribution of breccias types in the Mawrth landing ellipse. Presented at the Fifth MSL Landing Site Workshop, Monrovia, California, 16–18 May 2011.

Swift, J., 1726. *Travels Into Several Remote Nations of the World, in Four Parts, by Lemuel Gulliver*. London: Benjamin Motte.

Tamppari, L. K., 2005. The Phoenix mission and its current landing site options. Conference poster, American Astronomical Society Annual Division of Planetary Sciences Meeting, Cambridge, England, 4–9 September 2005.

Texas Space Services, 1986. A manned mission to Mars. Preliminary design review 2, 12 May 1986, Report No. NGT-21-002-080. 'Texas Space Services' (Student Group), The University of Texas at Austin.

Vasavada, A. R., Grotzinger, J. P., Arvidson, R. E., *et al.*, 2014. Overview of the Mars Science Laboratory mission: Bradbury Landing to Yellowknife Bay and beyond. *Journal of Geophysical Research*, v. 119, no. 6, doi:10.1002/2014JE004622.

von Puttkamer, J., 2011. Project Humans to Mars. A look at 60 years of mission studies. Presented at the National Space Society Meeting, International Space Development Conference 2011, Huntsville, Alabama, 18–22 May 2011.

Wählisch, M., Willner, K., Oberst, J., *et al.*, 2010. A new topographic image atlas of Phobos. *Earth and Planetary Science Letters*, v. 294, nos. 3–4, pp. 541–546.

Walker, R. J., Ball, A. J., Price, M. E., *et al.*, 2006. Concepts for a low-cost Mars micro mission. *Acta Astronautica*, v. 59, pp. 617–626.

Wang, A., Bell, J. F., Li, R., *et al.*, 2008. Light-toned salty soils and coexisting Si-rich species discovered by the Mars Exploration Rover Spirit in Columbia Hills. *Journal of Geophysical Research*, v. 113, E12S40, doi:10.1029/2008JE003126.

Webster, C. R., Mahaffy, P. R., Atreya, S. K., *et al.*, 2013. Measurements of Mars methane at Gale crater by the SAM tunable laser spectrometer on the Curiosity Rover. Presented at the 44th Lunar and Planetary Science Conference, The Woodlands, Texas, 18–22 March 2013. Abstract no. 1366.

Williams, R. M. E., Grotzinger, J. P., Dietrich, W. E., *et al.* (the MSL Science Team), 2013. Martian fluvial conglomerates at Gale crater. *Science*, v. 340, no. 6136, pp. 1068–1072, doi:10.1126/science.1237317.

Willner, K., Oberst, J., Scholten, F., *et al.*, 2010. Phobos DTM and coordinate refinement for Phobos-Grunt mission support. *European Planetary Science Congress*, Abstracts, Vol. 5, EPSC2010-222, 2010.

Zeitlin, C., Hassler, D. M., Cucinotta, F. A., *et al.*, 2013. Measurements of energetic particle radiation in transit to Mars on the Mars Science Laboratory. *Science*, v. 340, no. 6136, pp. 1080–1084, doi:10.1126/science.1235989.

Zent, A. P., Hecht, M. H., Cobos, D. R., *et al.*, 2010. Initial results from the Thermal and Electrical Conductivity Probe (TECP) on Phoenix. *Journal of Geophysical Research*, v. 115, E00E14, doi:10.1029/2009JE003420.

Zimmerman, W., Anderson, F. S., Carsey, F., *et al.*, 2002. The Mars '07 north polar cap deep penetration Cryo-Scout mission. *Aerospace Conference Proceedings, 2002*, Vol. 1, pp. 305–315. New York: IEEE.

Zurek, R. W. and Smrekar, S. E., 2007. An overview of the Mars Reconnaissance Orbiter (MRO) science mission. *Journal of Geophysical Research*, v. 112, E05S01, doi:10.1029/2006JE002701.

Index